INTERNATIONAL MINING GEOLOGY CONFERENCE 2024

7–8 MAY 2024
PERTH, AUSTRALIA

The Australasian Institute of Mining and Metallurgy
Publication Series No 2/2024

Published by:
The Australasian Institute of Mining and Metallurgy
Ground Floor, 204 Lygon Street, Carlton Victoria 3053, Australia

© The Australasian Institute of Mining and Metallurgy 2024

No part of this publication may be reproduced, stored in a retrieval system or transmitted in any form by any means without permission in writing from the publisher.

All papers published in this volume were peer reviewed before publication.

The AusIMM is not responsible as a body for the facts and opinions advanced in any of its publications.

ISBN 978-1-922395-38-2

ADVISORY COMMITTEE

Daniel Howe
MAusIMM
Conference Advisory Committee Chair

Mark Berry
MAIG

Marcelo Cortes
MAusIMM

Angela Dimond
MAusIMM

Rayleen Hargreaves
MAusIMM(CP)

Jaime Livesey
MAIG

Danielle Robinson
MAusIMM(CP)

Dale Sims
FAusIMM(CP), MAIG

Rene Sterk
FAusIMM(CP), MAIG RPGeo

Thomas Taylor
MAusIMM

AUSIMM

Julie Allen
Head of Events

Fiona Geoghegan
AAusIMM
Senior Manager, Events

Vicki Melhuish
Manager, Events

REVIEWERS

We would like to thank the following people for their contribution towards enhancing the quality of the papers included in this volume:

Heath Arvidson
MAusIMM

Mark Berry
MAIG

Paul Blackney
MAusIMM

Tim Callaghan
MAusIMM

John Carswell
FAusIMM(CP)

Marcelo Cortes
MAusIMM

Angela Dimond
MAusIMM

Diane Djotaroeno
MAusIMM

Christopher Esterhuizen
MAusIMM

Robert Findlay

Jan Graham

Rayleen Hargreaves
MAusIMM(CP)

Sean Helm

Daniel Howe
MAusIMM

Jaime Livesey
MAIG

James Llorca
FAusIMM(CP)

Hayley McLean
MAusIMM(CP)

Mark Paine
FAusIMM

Geoff Phillips

James Pocoe
MAusIMM

Rebecca Prain

Danielle Robinson
MAusIMM(CP)

Bill Shaw
FAusIMM(CP)

Dale Sims
FAusIMM(CP), MAIG

Rene Sterk
FAusIMM(CP), MAIG RPGeo

Steve Sullivan
FAusIMM

Chris Taylor
MAusIMM

Thomas Taylor
MAusIMM

Jill Terry

FOREWORD

On behalf of the Conference Advisory Committee, we are delighted to welcome you to the AusIMM's International Mining Geology Conference 2024.

The International Mining Geology Conference 2024 is an opportunity for industry experts to come together and explore advancements in mining geology and discover solutions to shared challenges, sustainable practices, and the latest techniques to maximise orebody value, drive productivity and improve decision-making.

Keynotes Cam McCuaig, Head of Geoscience Excellence, BHP; Flavia Tata Nardini, Co-Founder and CEO, Fleet Space Technologies; Paul Hodkiewicz, Director, Samarah Solutions; and Professor Peter Betts, Geological Society of Australia and Monash University will deliver thought provoking presentations. These keynotes are complemented by a high-quality technical program and bustling exhibition floor. This conference is a great chance to collaborate, share ideas and strengthen your professional network.

Whether you are a seasoned industry professional, graduate, researcher, or student, we look forward to welcoming you to the AusIMM's International Mining Geology Conference 2024.

Yours faithfully,

Daniel Howe
MAusIMM

International Mining Geology 2024 Conference Advisory Committee Chair

SPONSORS

Major Conference Sponsor

Supporting Partner

Platinum Sponsors

Gold Sponsors

Industry Challenge Sponsor

Young Professionals Day Sponsor

Silver Sponsors

Industry Challenge Supporter

Networking Function Sponsor

RioTinto

Name Badge and Lanyard Sponsor

Coffee Cart Sponsor

entech.

Note Pad and Pen Sponsor

CUBE CONSULTING

CONTENTS

Challenges of reconciliation – lessons learnt

Reconciliation system implementation for corporate governance reporting – a Minas-Rio case study — 3
J Souza, R Hargreaves, C Morley and W Machado

Decision-making – the mine geologist's value

From resource model to grade control – risks and work arounds — 17
B C Afonseca, D Carvalho and A Chiquini

An ever-evolving resource model as new pieces are added to the Avoca Tank puzzle, Tritton Operations — 39
D Beckett, A Dimond, N Cox, O Gonzalez, A Jackson and B Cox

Leveraging geological logging as pathfinders and proxies in machine learning resource and geometallurgical modelling – benefits, limitations, and pre-screening techniques — 47
D M First, D Mogilny, Y H Rajcoomar and S Vedrik

Adaptive mine planning in open pits – the role of near-real-time orebody characterisation in grade control modelling — 67
S O'Brien and S Maguire

Advancing mine geology through optimised sampling strategies – a case study from Porphyry Underground Mine, Western Australia — 71
R Snell

SBRE workflow development for Nickel West – enabling decisions under uncertainty — 85
C Williams, I Minniakhmetov, G Merello and R Finch

Excellence in mine geology – delivering value

Mineable Shape Optimiser and ore wireframing, analysis of tools for scheduling open pit mining at KCGM — 93
J Ball

Overcoming practical challenges to honour geological controls in short-term modelling of the Kamoa-Kakula copper deposit — 103
H Bananga, G Gilchrist and J Chitambala

Remote sensing LiDAR for enhanced underground geology mapping – a Nova Mine case study — 115
G Boyce

Homogeneity and heterogeneity of crushed gold certified reference materials — 131
J Carter and B J Armstrong

Interrogating the space between surveys – an assessment of deep diamond drill hole survey intervals and the impact on resource confidence at Olympic Dam — 143
A Chapman and D Clarke

Enhancing geometallurgical ore domaining through improved orebody knowledge – a case study from the Sotiel Cu-Pb-Zn-Ag Mine, Spain *R Dale, M Biven, A Claflin, J Gonzalez and M Puerto*	155
Dynamic production grade cut-off – integrating financial understanding into daily mine geology *M Darvall, G Sternadt and M Van Ryt*	173
Rapid drill hole planning for underground delineation drilling at the Platreef PGE-Cu-Ni deposit, South Africa *G Gilchrist and J Chitambala*	181
Use of near mine exploration resource estimates in assessing strategic growth at the Super Pit Kalgoorlie *J Ireland*	187
Mine-scale reflection of a multi-stage regional structural history in the orebody geometry of Cutters Ridge Open Pit at Mungari, Western Australia *R Lumaad Paras, R Gordon, R Gulay and J Strang*	201
Ore control based on value – experiences in design and execution *C Morley*	209
Hidden between the data, small but significant high-grade structural lodes *R Reid, D Magautu, M Lucien, L Gagau, M Mondo and E Mudinzwa*	221
Opening the geologists tool box – building a grade control system *C Rowett, P Rolley and D O'Rielly*	231
Cosmic-ray muon tomography – new developments in mining applications at BHP's Leinster and Olympic Dam mines *D Schouten, J Townsend, E Western, M Owers, J Taylor and A Tunnadine*	243
Paving the way for future value – confidence in Olympic Dam's structural models utilising digital workflows *J Sidabutar, J Taylor and M Goldman*	253
Strategies for enhancing ore control and production at Martabe Gold Mine Indonesia *A Triyunita, N Saala, N Khariyah, M I Hidayat and J Hertrijana*	263
Black rock cave flow model reconciliation and optimised draw strategy *M C Vermaak and C Curtis-Morar*	279
Technological solutions to remnant mining ore control at the Martha Underground Mine *A Whaanga, W Vigor-Brown, A Harvison and J Maxted*	289

Futureproofing mine geology – planning ahead

Copper mineralisation at Mt Isa – unseen insights from seismic surveys after a century of geoscientific study *S Bright, G Turner, A Brown and M Megebry*	301
Optimising workflows in post-pandemic operations – impacts and implications for mine geologists *S Edmond, S O'Brien, A Tod, D Yeates and M Ravella*	309

Machine learning-based silicate alteration – optimising spectrographic core imaging, multi-element geochemical, and dynamic rebound hardness data set to improve the objectivity of alteration domaining 319
M H Rahadi, A Wiguna, M Heriyanto and R Taube

Implementation of the Chrysos PhotonAssay™ Method at Fosterville Gold Mine, Victoria 341
S P Hitchman, S Simbolon, D J Symons, J P Hoare and J B Carpenter

Using representative elemental analysis on conveyed flows to complement grade control – a win-win for geologists and processors 347
H Kurth

Directional change in underground drilling – a case study 355
V Mead, B Sharrock and M Ayris

Open to anything

Exploring deep learning in resource estimation 371
N Battalgazy, R Valenta, P Gow, C Spier and G Forbes

Author index 379

Challenges of reconciliation – lessons learnt

Reconciliation system implementation for corporate governance reporting – a Minas-Rio case study

J Souza[1], R Hargreaves[2], C Morley[3] and W Machado[4]

1. Enterprise Solutions Consultant, Datamine, Belo Horizonte, Brazil.
   Email: jady.souza@dataminesoftware.com
2. Principal Consultant, Snowden Optiro, Perth, WA 6000.
   Email: Rayleen.hargreaves@snowdenoptiro.com
3. FAusIMM(CP), Owner – Reconciliation & Ore Control Knowledge Services (R&OCKS), Perth WA 6000. Email: craig@rockservices.com
4. Geology and Mineral Resource Coordinator, Anglo American, Belo Horizonte, Brazil.
   Email: wander.machado@angloamerican.com

ABSTRACT

Anglo American Minas-Rio operation is a globally renowned iron ore venture situated in the regions of Minas Gerais and Rio de Janeiro, Brazil. Covering the complete mining value chain, it starts with open pit extraction and extends through beneficiation at a dedicated plant. Minerals are transported via a pipeline to port facilities for shipment.

Anglo American is actively implementing a strategic initiative to align with internal corporate governance reporting protocols across all mine sites. The Minas-Rio operation is a continuation of this pioneering effort, adopting an off-the-shelf reconciliation system following successful implementations at Kumba's Sishen and Kolomela iron ore operations. This underscores Anglo American's commitment to continuous improvement through reconciliation and consistent governance standards throughout operations.

With the reconciliation system's deployment into the Minas-Rio production environment in September 2022, the site has yielded substantial benefits. These include faster report generation and streamlined material tracking, contributing to improved oversight of mining's impact on monthly reconciliation.

This initiative's impact extends beyond the site, reaching into broader corporate spheres. Standardised terminologies for report generation enable comparisons of reconciliation outcomes across various operations, reaffirming Anglo American's dedication to uniform reporting practices and providing insights into multifaceted performance.

This paper presents a case study on the implementation of a reconciliation system at Minas-Rio. It outlines the integration of multiple data sources into a centralised data management system and highlight the benefits achieved in terms of enhanced reconciliation practices across the mine value chain. These achievements encompass streamlining monthly reconciliation reporting protocols, the application of business rules to elevate the quality of data inputs, and the centralisation of data to compute reconciliation factors.

INTRODUCTION

Anglo American plc stands as one of the largest global mining companies, operating in 15 distinct regions with a total of 59 sites. It serves as a worldwide producer of platinum, diamonds, copper, nickel, iron ore, metallurgical coal and thermal coal, spanning operations across five continents. The company's administrative centre and corporate representation are headquartered in London, England.

In Brazil, Anglo American has been active since 1973, operating through four business units: Niquelândia, Codemin and Barro Alto, contributing to nickel production, and Minas-Rio, the focus of this study, an iron ore operation.

The Anglo American Minas-Rio operation is an iron ore operation located in the states of Minas Gerais and Rio de Janeiro. Minas-Rio consists of an open pit mine and a beneficiation plant in Minas Gerais, specialising in the production of direct reduction and blast furnace pellet feeds. The mining activity involves traditional methods of drilling, blasting, loading and transportation, with the presence of a high-precision Fleet Management System (FMS). The iron ore, extracted from the open pit operation, is processed in the beneficiation plant and then transported through a 529 km pipeline to the port of Açu in the state of Rio de Janeiro.

Anglo American fully owns Minas-Rio through its subsidiary, Iron Ore Brazil (IOB), except for the port facility, where they hold a 50 per cent stake in a joint venture with Ferroport.

The majority of Minas-Rio's open pit production is located in the municipality of Conceição do Mato Dentro, within the state of Minas Gerais in south-eastern Brazil. This site is approximately 160 km north-east of the state capital, Belo Horizonte (Figure 1). A smaller section of the mine, the mineral processing plant and a portion of the tailings are situated in the adjacent municipality of Alvorada de Minas (Figure 2). Minas-Rio activities are overseen through a subsidiary Anglo American Iron Ore Brazil South America.

FIG 1 – Geographical location of Minas-Rio operation, Brazil.

FIG 2 – Minas-Rio operation and infrastructure (image courtesy of Anglo American).

Situated along the south-east border of the São Francisco Craton, the Minas-Rio iron ore project explores iron deposits within the Serra do Sapo and Itapanhoacanga hosted in a proterozoic metasedimentary sequence located in Serra do Espinhaço belt. Both deposits, host two distinct ore types: friable and hard itabirite/hematite.

The mineral rights in the Serra do Sapo region were initially under the ownership of Vale S.A. (Vale) until 2004. Subsequently, between 2004 and 2006, various owners held the mineral rights for brief periods until June 2006, when MMX (Eike Batista) acquired them. Following the acquisition, MMX conducted extensive exploration activities in the region, encompassing geophysical surveys, geological mapping, drilling campaigns, metallurgical and geotechnical test work and hydrological studies.

During the first half of 2007, Anglo American acquired a 49 per cent stake in the Minas-Rio System and by August 2008, it assumed full control of the entire project through its subsidiary, IOB. In 2019, Anglo American completed the transfer of the Serra do Sapo mineral rights. The construction of the slurry pipeline took place from 2008 to 2014 and in 2014, the shipment of 150 kt of ore marked the commencement of operational milestones. In the fiscal year 2021, the operation achieved a production of 22.9 Mt of ore, with sales amounting to 23 Mt at an average cost of $24 per ton (IBRAM Mineracao do Brasil, 2022; Anglo American, 2012).

MINING RECONCILIATION AND THE CHALLENGE OF STANDARDISATION

Mine reconciliation serves as a robust tool for evaluating an operation's performance with targets, plant recovery and the accuracy of resource estimates. It is often used as an essential key performance indicator (KPIs) for mining activities. This process entails the comparison of measured values, including resource models, reserves, and mine plans, with the actual measured values obtained from the plant's final product, mill feed, and recorded tonnage extracted from the mine (Morley, 2003; Morley and Moller, 2005; Schofield, 2001; Fouet *et al*, 2009).

Reconciliation is a multidisciplinary activity that encompasses the entire value chain, from geology and mine planning to mining operations and metallurgy. It involves comparing tonnes, grade and metal content between two distinct points in the mineral flow. Typically, the outcome of this comparison is considered a 'reconciliation factor', that consists of a simple ratio between two like measures but can also be expressed in absolute differences or percentage differences.

Continuous analysis of reconciliation comparisons allows responsible parties to easily identify technical issues in operations, analyse them and take action to resolve these problems, leading to long-term continuous improvement in production.

Despite numerous studies on reconciliation factors (Parker, 2012; Fouet *et al*, 2009; Hargreaves *et al*, 2022; Shaw *et al*, 2013) and the widely discussed principles of reconciliation in various mining codes for decades, there is still no global convention for reporting reconciliation factors (Hargreaves and Booth, 2019). Each operation presents a unique reporting form and factors may be altered or adapted for each operation (Fouet *et al*, 2009).

F and R factors are tools used in the mining industry to ensure that resource and reserve estimates are as accurate and reliable as possible, considering various uncertainties and factors that may affect the estimation process. The significant challenge in standardising nomenclature lies in the adoption of acronyms such as F factors (Resource or Reserve model comparisons) or R factors (Resource comparisons), for example, which may not apply to all types of operations. For instance, mines without a grade control model for example are not capable of publishing a reconciliation factor that utilises this metric.

RECONCILIATION AT ANGLO AMERICAN

As part of a global initiative, Anglo American has been moving to standardise the reporting format of reconciliation comparisons across its global mining operations. The aim was to introduce a unified nomenclature and a consistent calculation methodology for generating standardised reconciliation comparisons, irrespective of the diverse commodities produced by each operation. This initiative sought to ensure uniformity in reports across all business units of Anglo American, enhancing transparency and comparability.

Within the framework of this global initiative, a reconciliation reporting standard was implemented. This standardised template, known as the Opco Report, is generated on a monthly, quarterly and annual basis. It facilitates temporal reconciliation comparisons for tonnes, grade and metal products, expressed as percentage differences. The report incorporates a traffic lights system, facilitating the quick identification of discrepancies that go beyond anticipated thresholds, these variations require in-depth analysis to address and alleviate the identified differences.

The comparisons within this standardised process encompass reserve models (long-term), grade control models (short-term), budget plans and actual measurements. These measurements include final product metrics from the plant, plant feed, and material addition and reclaim from ore stockpiles, tracked through trucks weightometer accounting.

In contrast to employing acronyms, such as the commonly used F and R factors, Anglo American utilises a descriptive nomenclature for these comparisons. This nomenclature, outlined in Table 1, is accompanied by a clear description of the source of data that is being compared for each reconciliation analysis. For example, the comparison between the resource model and the grade control model has the aim of validating assumptions in the resource model, indicating its local accuracy in relation to the grade control model, which contains more short-term information.

TABLE 1

Available comparisons used by Anglo American operations.

	Reconciliation Relationship	Action	Monitor	Maintain	Purpose
1	Resource Model to Grade Control Model	> 20%	10 to 20%	< 10%	Validates the assumptions in the Resource Model and shows how locally accurate it is.
2	Mine Delivered to Plant Received	> 20%	10 to 20%	< 10%	Verifies the material actually sent to the Plant against Plant results prior to any mass balance calculations.
3	Reserve Model to Plant Accounted	> 20%	10 to 20%	< 10%	Shows the effectiveness of the Reserve Model to predict what will be achieved when the resource passes through the Plant.
4	Grade Control Model to Plant Accounted	> 20%	10 to 20%	< 10%	Indicates the effectiveness of the Grade Control Model to predict what will be achieved by the Plant.
5	Budget to Plant Accounted	> 10%	5 to 10%	< 5%	Shows if the quantity and quality of ore, and the quantity of product specified by the budget has been received and recovered as predicted in the budget.
6	Budget to Mine Delivered	> 20%	10 to 20%	< 10%	Reveals whether the mining activities have delivered the quantities and qualities specified by the budget.
7	Budget to Plant achieved recovery	> 5%	2 to 5%	< 2%	Reports whether the plant has achieved planned recovery.
8	Spatial compliance to plan (monthly and cumulative year-to-date)	< 70%	70 to 80%	> 80%	Confirms whether we mined where we said we would mine, when we said we would, reflecting operational stability.
9	Reconciliation Confidence Rating	< 75%	75 to 90	> 90%	Indicates how much uncertainty exists in the reported reconciliation results.

Some examples of initiatives that have been derived from reconciliation analysis at Anglo American include (Morley and Arvidson, 2017):

- Improvement in definition of ore during ore control processes resulting in increased volumes of ore delivered to the processing plant.
- Identification of compliance to plan issues resulting in changes to production activities to enable the operation to return to the plan.
- Identification of incongruent bulk density values being utilised resulting in misreporting of tonnages.
- Removal of errors in model wireframes that were resulting in misclassification of material types.

MINE RECONCILIATION AT MINAS-RIO

Minas-Rio used to perform reconciliation reporting with an Excel spreadsheet. This was generated through a macro and generated the Opco reconciliation report already adopted by the company as part of the Anglo American standardisation and reporting governance project. The Opco report was generated on a monthly, quarterly and annual basis.

The monthly reconciliation Excel spreadsheet was manually populated with data gathered from various Excel sheets and databases. Despite Minas-Rio having a dedicated resource for Opco report generation, this process was taking approximately four days to complete.

The mine reconciliation process for Minas-Rio aimed to compare various elements across multiple data sources, including surveys of long-term models, grade control models, mine production obtained directly from a centralised database fed by their FMS system and a metal-balanced data on plant feed and production values. The process flow is presented in Figure 3.

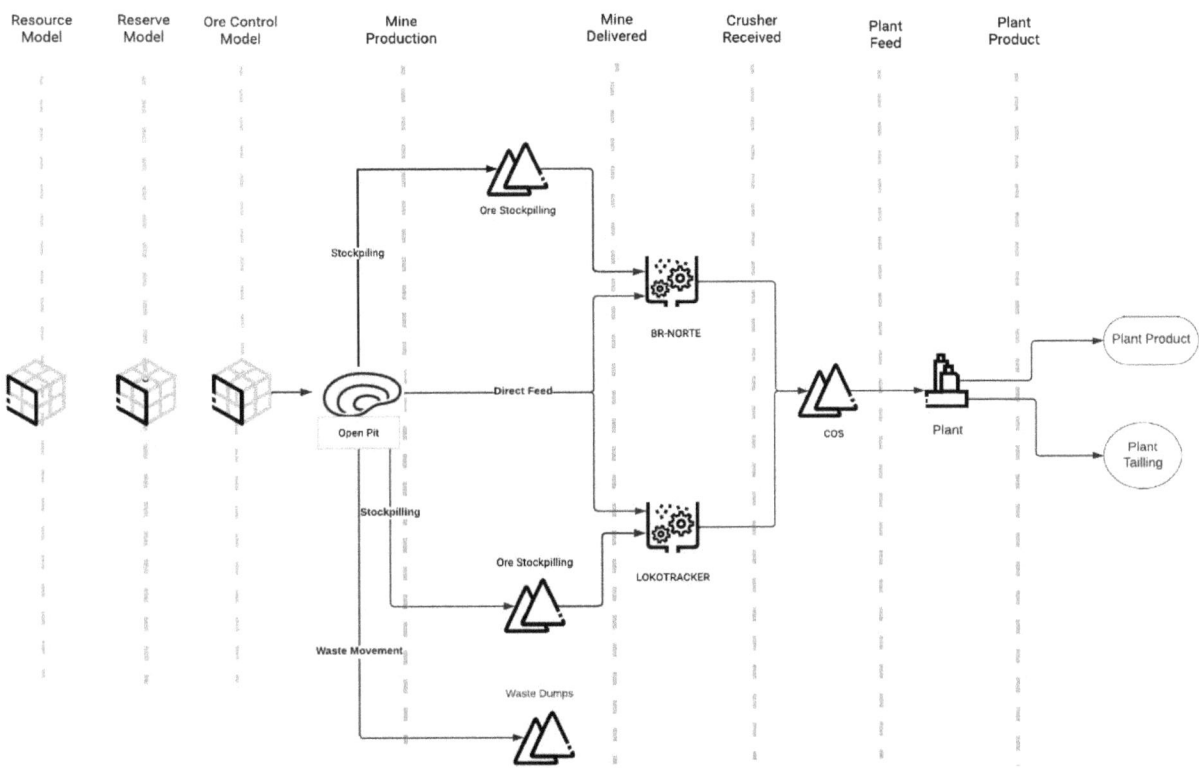

FIG 3 – Minas-Rio material movement flow with standardised metrics.

THE IMPLEMENTATION PROCESS AND OUTCOMES

Minas-Rio implemented reconciliation software for generating the monthly Opco reconciliation report using Reconcilor™ version 9.5, (by Snowden Optiro, Datamine).

The key objectives behind this decision included:

- Automation of report generation using an integrated system, reducing dependency on collecting data from multiple sources and decreasing the time required for monthly reconciliation report generation.
- Facilitate data storage in a centralised database, allowing for auditability and transparency.

The implementation of the reconciliation software for Minas-Rio commenced in May 2022 and concluded in September 2022. A cross-functional team, consisting of experts from various mine data generation departments such as geology, mine planning, mine operations, dispatch, metallurgy and IT, played a crucial role in the successful implementation.

The solution is a web-based system utilising a centralised SQL database (Figure 4). The implementation process involves creating various import routines for different data sources with the goal of integration and automated report generation. For Minas-Rio, these data sources consisted of:

- Geological models and model depletions in csv format generated by Datamine's Studio RM™.

- Mine Plans in csv format generated by Studio OP™ (by Datamine).

- Direct integration with SQL database, specifically for Minas-Rio, a data centraliser named Data Lake utilising Azure Service Principal authentication, that consolidates production data directly from the Modular Fleet Management system and plant data available in Production Accounting™ (by Datamine).

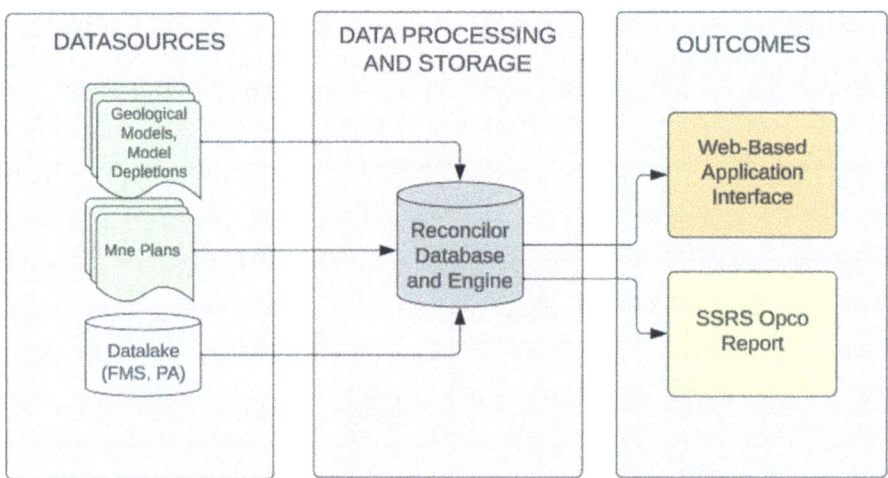

FIG 4 – Different data sources and outcomes of the implementation using Reconcilor™ at Minas-Rio.

The reconciliation database uses SQL Server Reporting Services (SSRS) to generate the Anglo American Opco Report, adhering to the template defined by Anglo American and acceptance limits following company standards with automated generation of monthly, quarterly and annual reports, providing Minas-Rio with a reduction from days to hours in monthly report preparation. The comparisons available in Opco Report are shown below and in Figures 5 and 6:

- Resource Model to Grade Control Model – validates assumptions in the resource model and shows how locally accurate it is.

- Mine Delivered to Plant Received – verifies material sent to the plant against plant results prior to any mass balance calculations. Mine delivered to plant is defined as haulage to the crushers. Plant received is defined as material delivered to the Course Ore Stockpile (COS) after the crushers.

- Mining Model to Plant Received – indicates the effectiveness of the Mining model to predict what will be achieved by the plant.

- Budget to Plant Received – shows if the quantity and quality of ore and the quantity of product specified by the budget has been received and recovered as predicted in the budget.

- Budget to Mine Delivered – Reveals whether the mining activities have delivered the quantities and qualities specified by the budget.

- Budget to Plant Achieved Recovery – Reports whether the plant has achieved planned recovery.

- Spatial Compliance to Plan – Confirms whether mining occurred where the planned mining dictated, reflecting operational stability.

- Reconciliation Confidence – Indicates how much uncertainty exists in the reconciliation results.

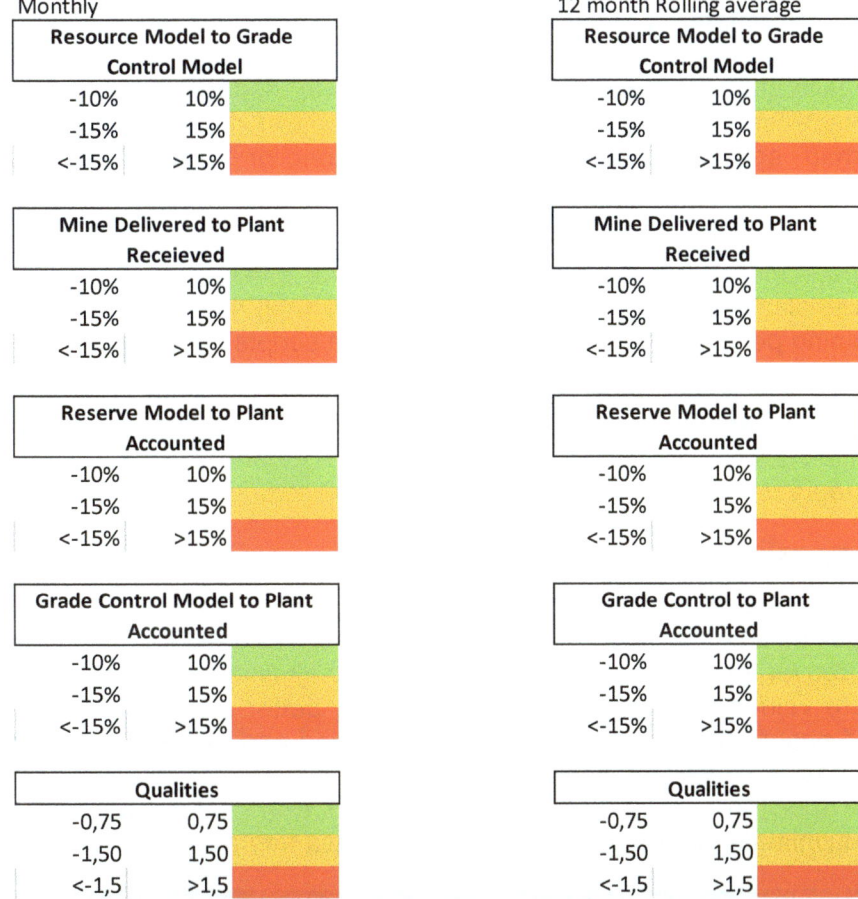

FIG 5 – Percentage acceptance ranges adopted for Anglo American Minas-Rio Opco report, including percentage difference on qualities.

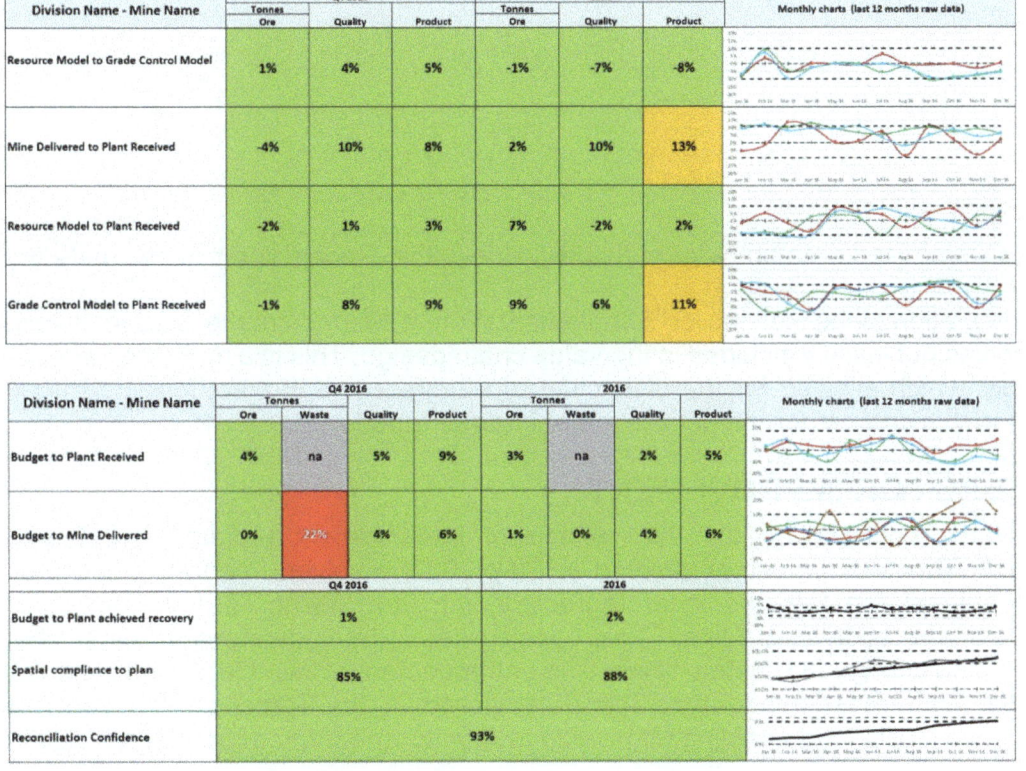

FIG 6 – Example Opco report template used for automatic generation in Reconcilor™ for Anglo American Minas-Rio (Morley and Arvidson, 2017).

CHALLENGES

Throughout the implementation process, challenges were encountered, such as the creation of a Data Lake to act as the centralised data storage and its provision for access to Datamine consultants. To address this challenge, a third-party company was made responsible for creating the Data Lake, maintaining constant communication with representatives of Datamine and Minas-Rio from each department to map specific fields required by Reconcilor™. Effective communication and commitment from all involved areas was crucial for the project's development.

For Reconcilor™, standardising nomenclature across various data sources is a crucial activity, allowing for comparisons of data from different sources at different levels of detail. In Minas-Rio, standardisation was also a challenge for geological models and a solution was found by automating the insertion of mining polygon levels into models through scripts via Studio RM™.

The visualisation of raw data connected in line with Reconcilor™ facilitated easy visualisation, providing transparency and traceability of materials throughout the mine value chain for the geological and mine administration team. This allowed for the quick adoption of action when necessary and a rapid view of improvements implemented throughout the process.

SYSTEM CUSTOMISATION

In consideration of the local geology of the ore-hosting rocks (itabirites) at Minas-Rio, moisture accounting emerges as a crucial factor. Moisture values are assigned for each lithology type present in the deposit. Moreover, given the climatic conditions at the mining site, characterised by significant variability, moisture values undergo annual changes. To better align with Anglo American Minas-Rio's process and ensure the accuracy of tonnes produced and hauled, a customised screen was developed for Minas-Rio, allowing the input of annual moisture values for each lithology type (Figure 7). These moisture values are then applied to the models, model surveys and truck movement data, culminating in the calculation of dry tonnes hauled.

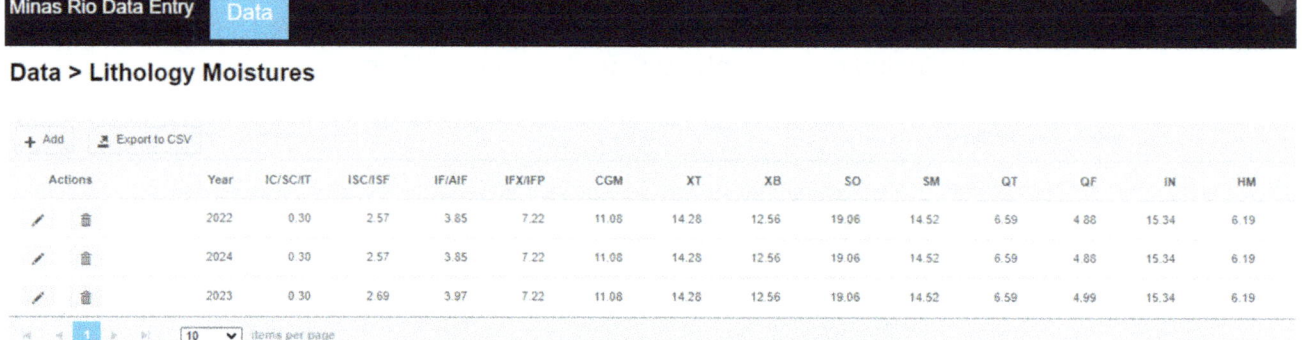

FIG 7 – Customised data entry for moisture value within Reconcilor™ to calculate dry tonnes.

After the completion of the Reconcilor™ implementation, a second crusher was incorporated into the system that was not initially included in the value chain design. This is a mobile crusher of the LOKO-Tracker type, which allowed for the distinction of all direct and indirect feeding material (through Stockpiles) to this crusher for the generation of internal reports.

NEXT STEPS FOR IMPROVEMENT

Minas-Rio have commenced the third phase of improvement projects targeted at improved reconciliation. The first project, completed in May 2023, involved integrating the product tonnes variable calculated through non-linear regression into the Reconcilor™ database and analysis screens. Datamine addressed Minas-Rio's specific needs by implementing minor changes to the core product and configuring custom filters. The enhancement allowed technical consultants to apply custom logic and transformations to the data stored in the Reconcilor™ reporting database, enhancing reportability and enabling Minas-Rio to monitor the trend of product tonnes comparisons over time.

For the second improvement project, Anglo American Minas-Rio collaborated with Datamine to develop a routine for generating depletion models for Resource, Reserve and Ore Control models created in Studio RM™ software. This process included exporting the models to the Reconcilor™ database through a scheduled import. The collaboration with technical teams from Datamine enhanced the monthly model depletion report, providing detailed information at the mine polygon level. This improvement facilitated direct comparisons between dispatch movement loads and survey data for the month, as well as the initial tonnes projected in the mining model design.

During the initial implementation, Minas-Rio had a set of 13 attributes available for physical comparisons. This included dry and wet tonnes, volume, moisture, grade and metal content for iron, aluminium, silica, manganese, calcium, magnesium, titanium and sodium. In the enhancement project, the lists of imported variables in Reconcilor™ were expanded to include 16 new geometallurgical attributes. With the contribution from multi-disciplinary teams, the extra attributes are now provided in the source data for geological and depletion models, truck movements and plant data and are being stored in the database. These variables are now available on the Reconcilor™ analysis screens for internal assessments for the intention of analysing beneficiation plant performance (Figure 8).

FIG 8 – All geometallurgical attributes from the models are available in Reconcilor™ screens and the database for monitoring mining performance, reconciliation analysis and beneficiation plant performance.

As a third and final phase of the reconciliation improvement project, an API is being developed to establish a direct connection between Reconcilor™ and the FMS system utilised by Minas-Rio. This API enables the seamless export of information from mine polygons, complete with geographically referenced positioning for the high-precision GPS units of the FMS system. The exported data includes details on lithology, grade, initial tonnes and the geometallurgical variables. Figure 9 illustrates the logic employed for the online connection between Reconcilor™ and the developed API, facilitating the direct feed of information into the Modular FMS system.

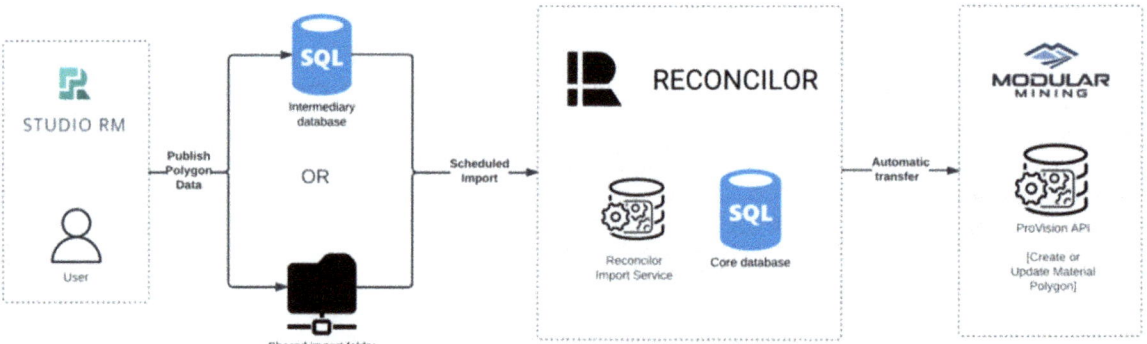

FIG 9 – Schematic diagram of the logic developed by Datamine for Minas-Rio for the integration of Reconcilor™ via Modular FMS system API.

CONCLUSIONS

The successful implementation of Reconcilor™ at Anglo American's Minas-Rio operation marks a transformative step in advancing standardised reconciliation reporting within the mining industry. The transition from labour-intensive, manual Excel-based processes to a sophisticated, web-based system has yielded substantial benefits, positioning Minas-Rio at the forefront of efficient, transparent and governance-driven mining practices.

This case study underscores the pivotal role reconciliation plays in evaluating operational performance throughout the mining value chain. Anglo American's global initiative to standardise reconciliation reporting, exemplified by the Opco Report, has been effectively adopted at Minas-Rio. The descriptive nomenclature and traffic lights system introduced by Anglo American provide a clear, consistent framework for reconciliation comparisons, enhancing transparency and facilitating rapid identification of operational discrepancies.

The challenges encountered during the implementation process, including standardising nomenclature and creating a centralised Data Lake, were met with a collaborative and adaptive approach. The successful customisation of Reconcilor™ to accommodate site-specific conditions, such as moisture accounting, showcases the flexibility of the system to address unique operational nuances.

The multidisciplinary team involved in the implementation, comprising experts from various mine data generation departments, played a pivotal role in ensuring the seamless integration of diverse data sources. The reduction in report generation time from days to hours signifies a significant improvement in operational efficiency, enabling timely decision-making and proactive issue resolution.

Looking ahead, Minas-Rio's commitment to continuous improvement is evident in the ongoing improvement projects. The integration of product tonnes variables, development of routines for generating depletion models and the planned API for direct connection between Reconcilor™ and the FMS system underscore a commitment to staying at the forefront of technological advancements in reconciliation and reporting practices.

In conclusion, the Minas-Rio case study serves as a beacon for the mining industry, demonstrating that embracing innovative reconciliation systems not only streamlines processes but also enhances governance, transparency and comparability across diverse mining operations. As the industry navigates toward increased standardisation, the success at Minas-Rio stands as a testament to the positive impact of technological advancements in fostering a culture of continuous improvement and excellence in corporate governance reporting.

REFERENCES

Anglo American, 2012. Minas Rio Project Iron Ore, p 14 [online]. Available from: <https://minedocs.com/20/Anglo_American_Minas_Rio_Project_Iron_Ore_2012.pdf#page=14> [Accessed: 18 January 2024].

Fouet, T, Hargreaves (née Riske), R, Morley, C, Cook, A, Conti, D and Centofanti, J, 2009. Standardising the reconciliation factors required in governance reporting, in *Proceedings Seventh International Mining Geology Conference 2009*, pp 127–140 (The Australasian Institute of Mining and Metallurgy: Melbourne).

Hargreaves, R and Booth, G W, 2019. Transparency and Standardisation in Metal Reconciliation Reporting, in *Proceedings Mining Geology Conference*, pp 36–42 (The Australasian Institute of Mining and Metallurgy: Melbourne).

Hargreaves, R, Elkington, T, Booth, G W and Shaw, W J, 2022. Mine Reconciliation Standardisation – R Factor Series, in *Proceedings International Mining Geology Conference*, pp 366–374 (The Australasian Institute of Mining and Metallurgy: Melbourne).

IBRAM Mineracao do Brasil, 2022. Anglo American releases 2021 production results [online]. Available from: <https://ibram.org.br/noticia/anglo-american-divulga-resultados-de-producao-de-2021/> [Accessed: 16 January 2024].

Morley, C and Arvidson, H, 2017. Mine value chain reconciliation – demonstrating value through best practice, in *Proceedings Tenth International Mining Geology Conference*, pp 103–116 (The Australasian Institute of Mining and Metallurgy: Melbourne).

Morley, C and Moller, R, 2005. Iron Ore Mine Reconciliation – A Case Study from Sishen Iron Ore Mine, South Africa, in *Proceedings Iron Ore Conference*, pp 311–318 (The Australasian Institute of Mining and Metallurgy: Melbourne).

Morley, C, 2003. Beyond Reconciliation – A Proactive Approach to Using Mining Data, in *Proceedings Fifth Large Open Pit Conference*, pp 185–191 (The Australasian Institute of Mining and Metallurgy: Melbourne).

Parker, H M, 2012. Reconciliation principles for the mining industry, *Mining Technology*, 121(3):160–176.

Schofield, N A, 2001. The myth of mine reconciliation, in *Mineral Resource and Ore Reserve Estimation – The AusIMM Guide to Good Practice* (ed: A C Edwards), pp 601–610 (The Australasian Institute of Mining and Metallurgy: Melbourne).

Shaw, W, Weeks, A, Khosrowshahi, S and Godoy, M, 2013. Reconciliation – Delivering on Promises, presented at 36th APCOM, (Application of Computers and Operations Research in The Mineral Industry).

Decision-making – the mine geologist's value

From resource model to grade control – risks and work arounds

B C Afonseca[1], D Carvalho[2] and A Chiquini[3]

1. MAusIMM(CP), Senior Resource Geologist, Glencore, Brisbane Qld 4000.
   Email: bruno.dedeusafonseca@glencore.au
2. MAusIMM, Principal Geologist, Glencore, Brisbane Qld 4000.
   Email: dhaniel.carvalho@glencore.com.au
3. MAusIMM, Principal Resource Estimation, Rio Tinto, Brisbane Qld 4000.
   Email: ana.chiquini@riotinto.com

ABSTRACT

Throughout the lifetime of a mineral venture, a series of geological models and resource estimates are generated, essential for informing strategic, operational and financial decisions. Each model is tailored to a specific purpose, employing diverse estimation methodologies to achieve distinct objectives. In initial phases of a mineral project drilling is typically conducted with wide spacing, and the primary objective of estimation models is to assess the grade-tonnage relationship on a large scale. At this stage, the focus is not on deciding the final destination of a mining block; this decision will be supported by production models, which will be later estimated with tighter drilling spacing.

When transitioning from resource to grade control models, some critical aspects must be considered to ensure the model effectively represents its intended purpose. The estimation support choice depends on drill hole spacing and the model's specific goal. The volume being estimated is related to the degree of smoothing and estimation error. At the time of mining, grade control estimates commonly rely on a more sizeable database compared to that available during resource modelling. This is known as the 'information effect', which holds implications for ore/waste selection due to the distinct distributions typically associated with estimates derived from varying levels of information. There will always be a level of uncertainty that cannot be resolved at the time of mining because a deposit cannot be sampled exhaustively.

Another critical aspect to consider when transitioning from resource models to grade control ones is the sampling method. As the models evolve to the production stage, alternative sampling methods such as underground channel sampling or open pit blastholes may replace diamond drilling used in exploration and resource definition. If associated with significant sampling bias or different volumes, these methods can impact the inference of spatial continuities and local estimates.

Lastly, the geostatistical framework is another crucial consideration for creating fit-for-purpose models. The selection and definition of techniques and parameters will impact differently the performance of estimates on both local and global scales. This paper discusses these matters from theoretical and practical perspectives, with real-world case studies as illustrative examples. Good practices and alternatives are presented, leveraging traditional built-in mining software tools and self-developed Python scripts.

INTRODUCTION

Estimated grade models are required throughout the entire life-of-mineral ventures, serving specific purposes. Typically, three main types of models are developed to inform decisions as a project progresses to an operation: (i) long-term or resource models, (ii) mid-term models, and (iii) grade control models; although the need for additional estimates at intermediate stages may vary between different projects/operations.

Interim or long-term resource models are first generated in early stages of a mining project, when the available information is limited. The primary objective of these estimates is to appropriately represent the global grade-tonnage relationship. In this context, the specific localisation of the grades is not a primary concern. Instead, the focus is on getting the best assessment of the actual economic tonnes. This information is critical for making well-informed decisions on whether to advance a mining project or not. Differently, medium-term models are developed at operational mines. Mid- and short-term models aim to refine the local estimation of long-term models, given that mining is not planned on the total tonnage but on relatively smaller volumes, such as underground stopes or open pit benches. These models are essential for mine planning, budget allocation and cash flow forecasting for the upcoming operational years. Grade control models represent the final estimates prior to production and are typically generated in consideration of production shapes, such as stope volumes in underground operations or bench heights in open pits. These models drive economic and operational decisions, including determining whether a specific region will be developed or the destination given to the mined material. At this stage, the precise localisation of grades is the main concern. Sending economic material to the waste dump or waste material to the mill will result in financial loss. Grade control models serve as the final opportunity to address any issues that could compromise the effectiveness and selectivity of the mining process.

Given the distinct objectives of these three models, the workflow employed must follow a fit-for-purpose approach. In a simple analogy, we can think of estimating metal grades at unsampled locations as making observations through a telescope. Using the same telescope, an observer can look at a nearby building just across the street or even the moon hundreds of thousands of kilometres away. However, the quality of the observed image in both cases is determined by how the equipment is calibrated. In this context, the ability to adjust focus is what determines the quality of the observations at different distances, whether the target is near or far. We can conceptualise the estimation framework in a similar manner. A toolkit of 'geostatistical telescopes' which includes kriging and its derivatives, is used to make observations of geological properties in tri-dimensional space. The quality also depends on how far we want to look and how the parameters are calibrated. Certain aspects should be considered when designing fit-for-purpose estimates. This paper specifically discusses four aspects that can be managed from the resource evaluation perspective: the support effect, the information effect, the estimation strategy and the sampling method.

The primary consideration is the effect on support. In mining applications, the term support refers to a scale, a dimension. For instance, in a sampling context, the term support refers to the volume of the sampled material. This volume could be as small as a few kilograms in a chip sample or as large as hundreds of kilograms, as is often the case for geometallurgical purposes; in estimation, it could be the dimensions of the grid or block model and in an operational context, it may be thought of as the accumulated production volume. The support effect essentially describes how the variability of a specific property changes when observed at different scales. The averaging of high and low values with increased support leads to a less variable and more symmetric distribution (Isaaks and Srivastava, 1989). In other words, the sample distribution of a certain attribute will have higher variance than the block distribution of the same attribute (Rossi and Deutsch, 2013; Harding and Deutsch, 2019). The support effect becomes evident in various stages and situations in mining. For instance, when reconciling production grades to models, the variability increases as the reconciliation time frame shortens (from a quarter of production to daily or weekly production, for example). Another example is the continuity observed in experimental variograms calculated for the same variable at different supports. As the data support increases, the sill (or variance) decreases and the variogram value increases gradually from smaller to larger distances. The support effect is particularly important in resource estimation because it influences the overall proportion of material below or above the economic cut-off grade. As the support size increases, the variance of the variable distribution decreases (Figure 1).

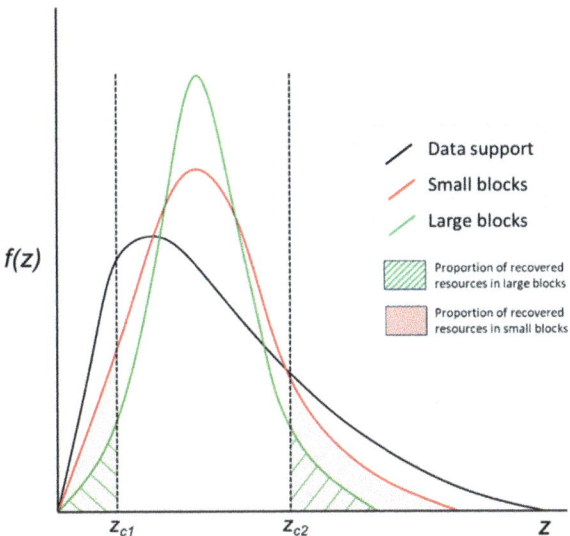

FIG 1 – Grade distributions of data at different support (modified from Harding and Deutsch, 2019).

While there are geostatistical approaches available to address the volume-variance relationship, they share a limitation by not accounting for the information effect. The information effect describes the fact that, at the time of mining, the information used to decide which portion of the deposit is ore and which is waste is based on more information than that available when building the long-term model (Rossi and Deutsch, 2013). Despite the availability of additional data, the real mining block grades are not known during production. Only an estimated value of them, based on production samples, is known, and pay blocks are defined according to this value, not the real grades (Roth and Deraisme, 2001). Therefore, the information effect reflects the potential for misclassification of ore/waste material. Long-term resource models do not account for future information that will be available at the time of mining. Anticipating information to try and minimise ore/waste classification errors is critical. The economic performance of any mining operation will be impacted by misclassification of material (Chiquini, 2018; Chiquini and Deutsch, 2020). Applying empirical approximations to anticipate changes in grade distribution as more information is acquired can be valuable in inferring misclassification rates in future mining operations.

The chosen estimation strategy impacts the estimated grade distribution. Practical considerations in the workflow are discussed to support a better representation of grade continuities in the model. These considerations involve compositing, domaining, the continuity models, the strategy to handle extreme values, the grid size and other aspects. This paper focuses on two aspects that require special attention: the geostatistical framework and the search neighbourhood. While most geostatistical methods exhibit high robustness and suitability across various deposit types and stages of a mining project, there are limitations to be considered when choosing one method over another. For instance, certain methods are designed to reproduce multivariate relationships, while others may better address heterogeneities due to multiple mineralisation events. Certain methods can better handle volume-variance relationships but may fail to make accurate local predictions. Therefore, the model's purpose must align with the chosen method's limitations and strengths. The estimation search definition remains a concern regardless of the geostatistical method selected. The impact of the search neighbourhood on estimates has been extensively discussed in literature, including works by Rivoirard (1987), Vann, Jackson and Bertoli (2003) and Deutsch, Szymanski and Deutsch (2014). The search neighbourhood defines the number of data points and their respective locations and orientations relative to the point being estimated. In other words, it dictates what samples participate in the estimation process. It is important for the estimation plan to align with the intended purpose of the model. For example, the estimation strategy might prioritise either local or global accuracy. In long-term models, the estimates are intended to align with the target histogram and keep smoothing to a minimum. In the context of grade control, the objective is to minimise mean squared error and conditional bias (Deutsch and Deutsch, 2015). Ideally, the conditional bias is eliminated for a final estimate, but this may not always be possible. In practice, obtaining a combination of conditionally unbiasedness and accurate reproduction of the histogram is impossible. Ed Isaaks (2005) called this the 'Kriging Oxymoron'.

Lastly, sample types will be discussed. Commonly, geological databases of mature mining operations contain data from multiple sampling methods. In exploration stages, diamond drilling is generally the most common method of data acquisition. At this stage, the drilling's purpose is to define the dimensions and extents of the mineralised body. As the mining project progresses, additional sampling methods might be considered in operational context. An example is the collection of face channel samples in underground mines. These samples are useful to guide development and inform short-scale variability in grades. However, these types of samples commonly lack rigour in quality and representativeness compared to exploration diamond drilling sampling. This is due to various reasons, such as a rushed mining cycle and adverse conditions (poor lighting, equipment limitations, costs, logistical challenges, safety concerns etc). These factors may contribute to production samples not having the same quality standard, resulting in higher contamination, poor recovery and lack of representativeness. This could affect short-term estimates in underground operations. In open pit operations, it is common to have production sampling performed by RC drilling or blastholes. In this case, achieving representative samples may be more complicated, especially if there is a loss associated with grain segregation and cross-contamination in the RC sampling system (Abzalov, 2016). In the *Sampling Method and Quality* section of this paper, real case studies of bias in different sampling methods in open pit and underground projects are presented.

The upcoming sections will explore each of the four aspects mentioned in this introduction, providing a theoretical background and further illustrating them through real-world case studies.

THE SUPPORT EFFECT

The concept of the effect of support refers to how variability changes depending on different scales. In resource estimation, for both long- and short-term models, the support or volume can have various impacts on estimates results. Such impacts can be related to sampling length, composite sizes, production volumes, block sizes, selective mining unit (SMU), variogram continuity and others. The volume-variance relationship is a key concept used in the determination of the dispersion variance across different volumes by using original data and understanding the variogram model, as described by Parker (1979). Journel and Huijbregts (1978) also expressed the volume–variance relationships through the dispersion variance. This equality is also known as Krige's additivity relationship:

$$D^2(v/V) = \bar{\gamma}(V,V) - \bar{\gamma}(v,v)$$

Where $D^2(v/V)$ is the dispersion variance of grades of small volumes within significantly larger volumes, $\bar{\gamma}(V,V)$ is the mean variogram value for the large volume and $\bar{\gamma}(v,v)$ is the mean variogram value for the smaller volume. As noted by Sinclair and Blackwell (2002), to understand volume-variance relationships, the formula can be rewritten with s representing samples, SMU representing the selective mining unit and D is the deposit:

$$D^2(s/SMU) = \bar{\gamma}(SMU, SMU) - \bar{\gamma}(s,s)$$

Which explains the relationship between samples and SMU, and:

$$D^2(SMU/D) = \bar{\gamma}(D,D) - \bar{\gamma}(SMU, SMU)$$

that explains the relationship between the SMU and the deposit (or estimation domain).

Adding both equations and rearranging:

$$D^2(SMU/D) = D^2(s/D) - D^2(s/SMU)$$

This equation shows that block grade dispersion can be indicated by the variogram and sample grade dispersion. Therefore, with a well-defined variogram model for samples, the dispersion of SMU grades can be estimated using the previous equations. Blocks and samples share the same mean value, leading to the only missing piece being the distribution shape for the relative histogram of SMU grades. Typically, an unbiased histogram of data is used, assuming the distribution of SMU grades maintains a similar general shape. If this shape can be presumed and block grade dispersion is estimated (as discussed previously), the estimated block grade distribution can be ascertained. This method is commonly known as the change of support. Several techniques, such as affine correction and indirect lognormal correction (Journel and Huijbregts, 1978), as well as the Discrete Gaussian Model (DGM) (Chilès and Delfiner, 2009; Machuca-Mory, Babak and Deutsch, 2008), are widely used for calculating this change of support. The DGM is considered a more robust methodology for change of support given it does not rely on the permanence of distribution compared to affine and indirect lognormal corrections (Rossi and Deutsch, 2013). Theoretical background of DGM will be provided in the next section: *The Information Effect*.

Change of support models are vital for evaluating recoverable resources in mining, facilitating the understanding of smoothing levels and assessing mining selectivity. These models are essential for long-term resource recovery planning, considering spatial variability, economic factors and technical constraints. They quantify dilution effects, encompassing geological, operational and internal dilutions due to SMU support. Practitioners use these models to predict SMU-scale grade distributions, guiding parameter calibration. While kriged estimates are generally smooth, restricted searches help adjust smoothing levels. However, a limitation is the focus on internal dilution, without fully incorporating site-specific factors. Change of support models also aid in selectivity analysis, influencing operational strategies and balancing recoverable resource value against selectivity costs.

Three case studies illustrate the impact of the support effect on resource estimates using data from base metal deposits. The examples show how composite size, variogram continuity, block sizes and support influence estimate variance, smoothness, material classification and metal content. Specific details about the deposits, such as names and locations, are withheld for confidentiality purposes, and certain tonnage and grade values are provided for cropped orebody layers and depleted mining areas. Ordinary kriging is employed for all grade estimates.

Composite sizes and variogram continuity

For the first case study, diamond drill hole data from a copper deposit was used. The Cu mineralisation occurs in association with alteration halos within shales. High-grade mineralisation is associated with a silica alteration, followed by a lower grade mineralisation related to a dolomitic alteration.

For the composite test, a block model of 5 × 10 × 5 m dimensions was created and estimated using two different composite sizes: 1.5 m and 25 m. The first size represents the historical sample length, while the second is an exaggerated yet plausible size based on the deposit's dimensions. This significant difference emphasizes the impact of varying support sizes. Both composite tests use the same parameters and variogram models. Figure 2 shows that 1.5 m composites yield estimates that are closer to the input grade distribution, highlighting low and high-grade zones. However, 25 m composites result in oversmoothed estimates, obscuring lower grade zones and lacking data resolution (Figure 3). Despite ordinary kriging's inherent smoothing, the use of a large support leads to oversmoothing.

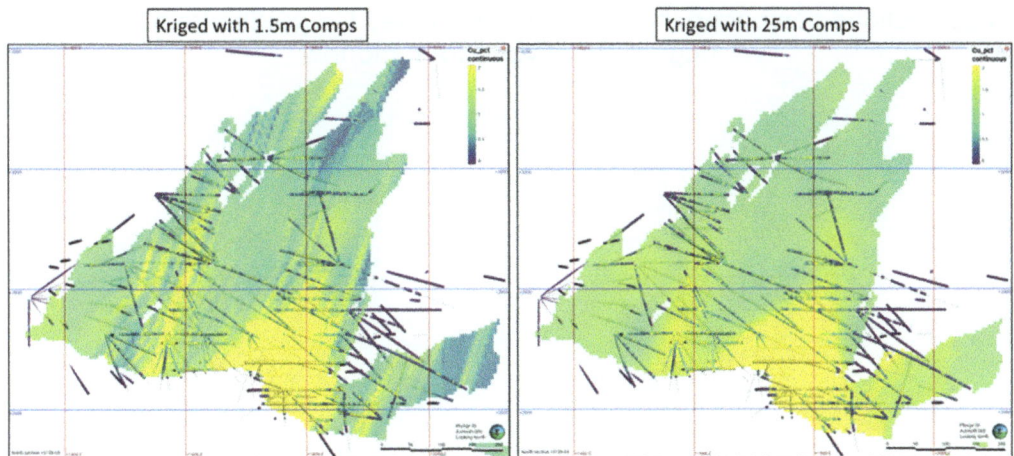

FIG 2 – Cross-sections comparing short (1.5 m) and long composites (25 m) impact on estimated Cu grades.

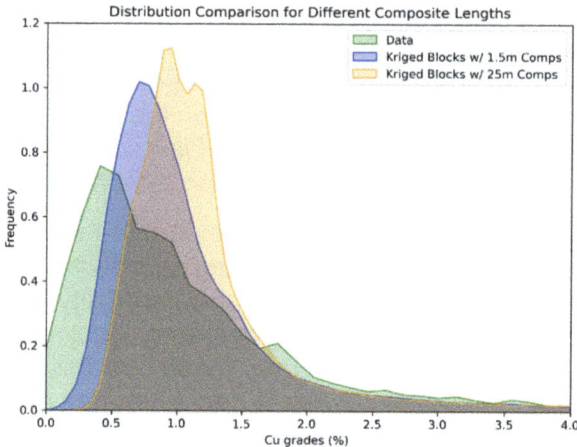

FIG 3 – Distribution comparison between data and kriged estimates using 1.5 m composites and 25 m composites.

For the variogram continuity example, we employed the same block model for testing both continuous and pure nugget variogram models. The continuous variogram model used in resource estimation, exhibits negligible nugget effect, low contributions for the first structure and considerable range in major and semi-major directions. The second variogram model, a pure nugget effect for illustration, is used with the same estimation parameters (composite sizes of 3 m). Results in Figure 4 reveal that estimates with the continuous variogram model represent well the input data, while pure nugget effect estimates are once again oversmoothed. The degree of smoothing in the kriged estimates can be confirmed by evaluating the grade distributions as shown in Figure 5.

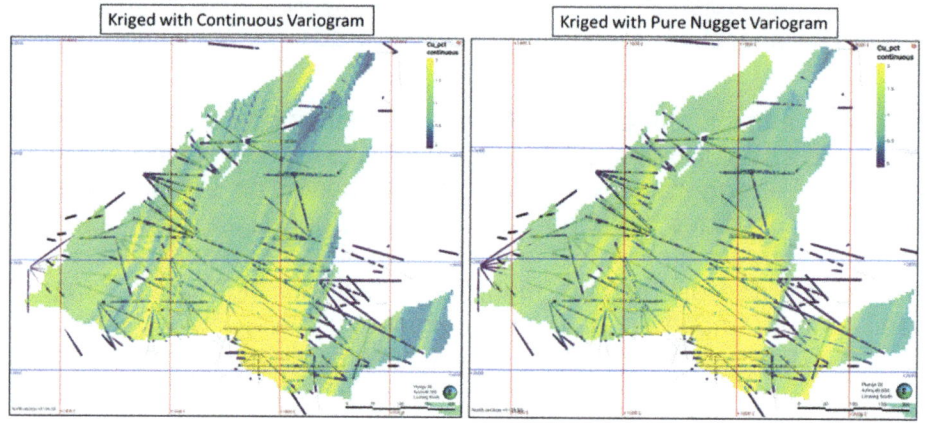

FIG 4 – Cross-sections comparing the impact on estimated Cu grades from continuous and pure nugget effect variogram models.

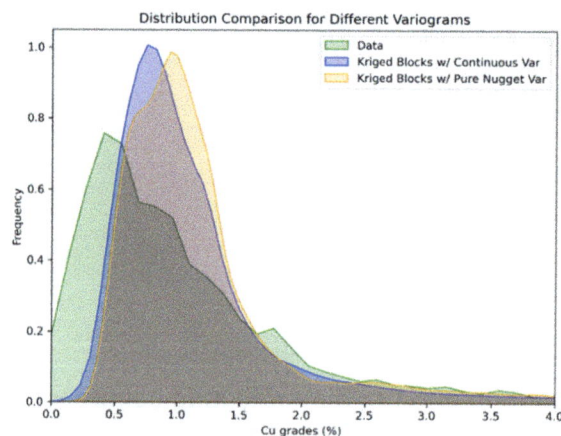

FIG 5 – Distribution comparison between data, kriged estimates using very continuous variogram model and pure nugget effect variogram model.

A table of summary statistics for both composite sizes and variogram continuity tests is shown in Table 1. In the cross-sections and distributions, the variance is considerably lower for the larger support (composites of 25 m) and less continuous variogram (pure nugget). The additivity of the variance explains these two examples. Recall Krige's relation:

$$D^2(SMU/D) = D^2(s/D) - D^2(s/SMU)$$

TABLE 1

Summary statistics comparison between data and different kriged scenarios.

	Data	Kriged w/1.5 m comps	Kriged w/25 m comps	Kriged w/cont. var	Kriged w/pure nugget
mean	1.301	1.315	1.427	1.350	1.380
var	2.804	1.706	1.461	1.757	1.443
min	0.000	0.056	0.187	0.072	0.264
0.25	0.440	0.672	0.860	0.709	0.771
0.5	0.850	0.923	1.089	0.962	1.035
0.75	1.457	1.367	1.398	1.350	1.377
max	26.700	15.131	11.759	14.533	9.552

For the composite test, as the internal variance is the same for both estimates ($D^2(s/SMU)$), it is expected that if composites are bigger (larger support) the dispersion variance of the data ($D^2(s/D)$) will be lower, and so, the dispersion variance of the blocks/estimates ($D^2(SMU/D)$) will also be lower.

In the variogram continuity test, as the dispersion variance of the data ($D^2(s/D)$) is the same for both estimates, it is expected that if the average variogram value is higher ($D^2(s/SMU)$), the dispersion variance of the blocks/estimates ($D^2(SMU/D)$) will once again be lower.

Block sizes and grade control (ore loss and dilution)

For the second case study, the same deposit and data set from the previous study was utilised. A small stope/mining area was selected to demonstrate the influence that different supports, now in terms of block sizes, can have on Cu estimates. The idea is to show how much of the material can be misclassified if unsuitable block sizes are defined. The designed stope has roughly 150 m of length, 110 m of width and 130 m of height. In this study, two different block sizes were used: 5 × 10 × 5 m and 25 × 50 × 25 m. A 'truth' model was used as basis of comparison for material classification. This 'truth' model was estimated on a higher resolution block size (2.5 × 5 × 2.5 m) and used grade control samples. The same estimation parameters and variogram models were used for all estimates.

The results from both Cu estimates and material classification (ore and waste) can be seen in Figure 6. Ore and waste classification was defined on an arbitrary but reasonable cut-off of 1.2 per cent Cu. The image clearly shows how the block sizes can influence the resolution of Cu estimates in that stope area. The larger support estimates do not efficiently capture the mineralisation geometry and the natural gradation between low and high-grade zones, compared to the 'truth' model. The smaller support estimates, on the other hand, are more suitable given the mineralisation characteristics. Moreover, if the selectivity is improved, optimised stopes could be better designed, incurring less ore loss and dilution. Table 2 confirms the visual inspection and shows that the large support model is -19 per cent different from the 'truth', while the smaller support model shows -4.2 per cent difference. Both large and small supports display poor performance on waste definition for this study. Nevertheless, Table 3 confirms once more that the smaller support model performs better for this deposit, given it displays almost 78 per cent less ore loss than the large support model.

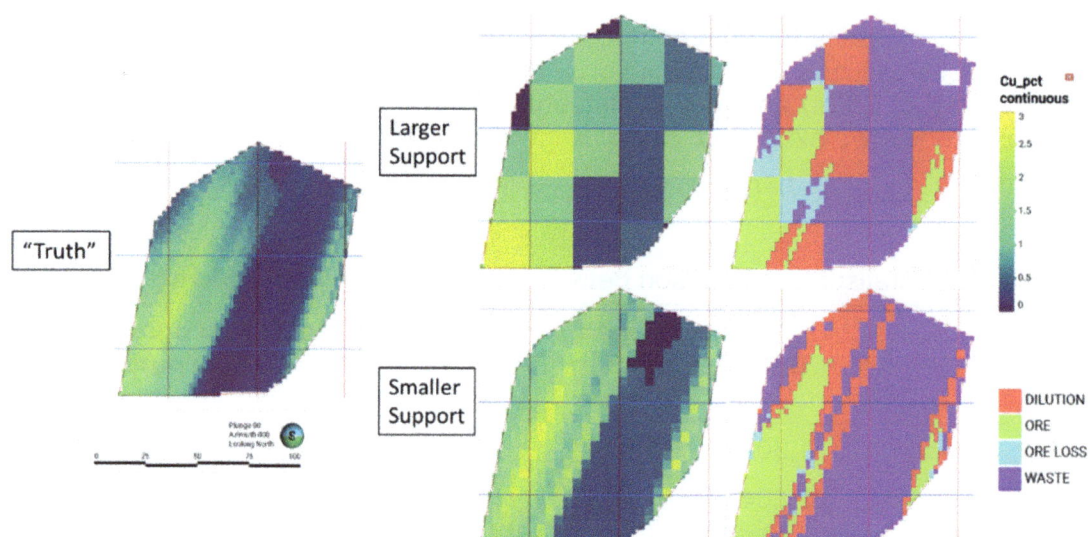

FIG 6 – Multiple cross-sections displaying Cu grade estimates and material classification (ore and waste) comparing the 'truth' model, smaller and larger supports.

TABLE 2

Ore and waste material classification comparison for a given stope, between different supports.

Material type	Block model	Cu content (t)	Relative difference (%)
Ore	'Truth'	24 199	
	Larger Support	19 592	-19.0%
	Smaller Support	23 179	-4.2%
Waste	'Truth'	19 565	
	Larger Support	14 211	-27.4%
	Smaller Support	13 048	-33.3%

TABLE 3

Misclassified material comparison for a given stope, between different supports.

Material type	Larger support metal (CuT)	Smaller support metal (CuT)	Relative difference (%)
Ore loss	4607	1020	-77.9%
Dilution	5354	6517	21.7%

Long-term resource model, block sizes and cut-off

The next case study evaluates how varying block sizes affect grade distribution, smoothing and resource reporting at a cut-off for a Zn-Pb-Ag stratiform deposit. A small portion of the deposit and a single mineralised layer was selected for this exercise.

For the reporting test, Ag grades were estimated for two different block model sizes: 10 × 10 × 2 m and 50 × 50 × 10 m. The first block dimension is closer to actual block sizes used in this operation, while the second block size is a larger cell that could potentially be used in this deposit, given orebody dimensions and geometry and drill hole spacing. Both estimates use the same data, estimation parameters and variogram models. The results from the estimated values for both cases can be seen in Figure 7.

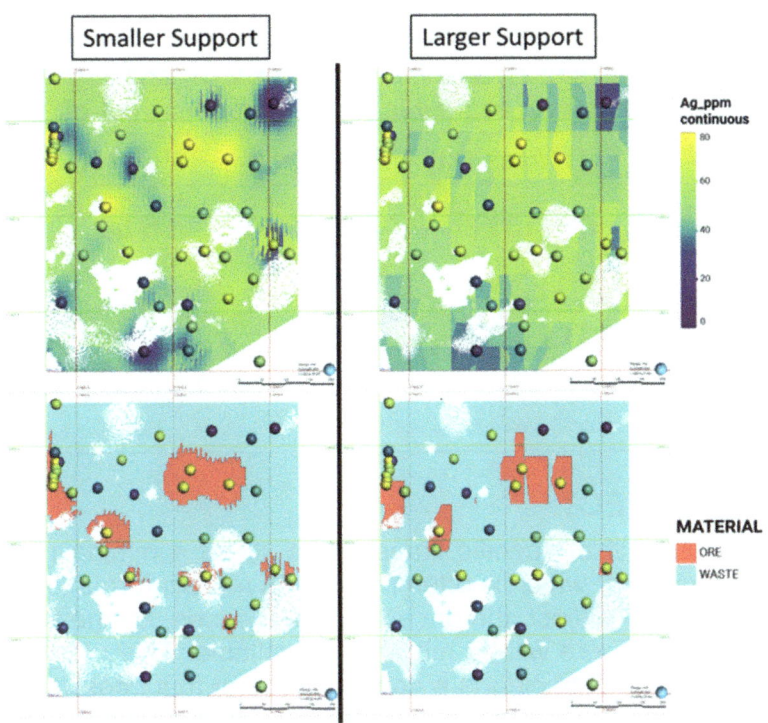

FIG 7 – Plan view comparison between smaller (10 × 10 × 2 m) and larger (50 × 50 × 10 m) supports for Ag estimates and material classification (using 60 ppm as cut-off).

The estimate using the smaller support shows better resolution highlighting low and high-grade zones than the larger support. An arbitrary cut-off of 60 ppm was selected to separate ore and waste for this exercise. Once more, the delineation of the ore boundaries is substantially better defined than the larger support. The degree of smoothing in the kriged estimates can also be confirmed analysing all distributions (Figure 8). From Krige's relation, we can expect that the larger blocks will produce lower dispersion variance and so stronger smoothing is also anticipated. From the interestingly normal-shaped (and real) distribution, we can see how the high-grade tail is absent in the larger support, and so, metal under-reporting is expected. Assuming a silver price of USD 725 706.65/tonne, Table 4 highlights 26 per cent less potential revenue from the larger support model.

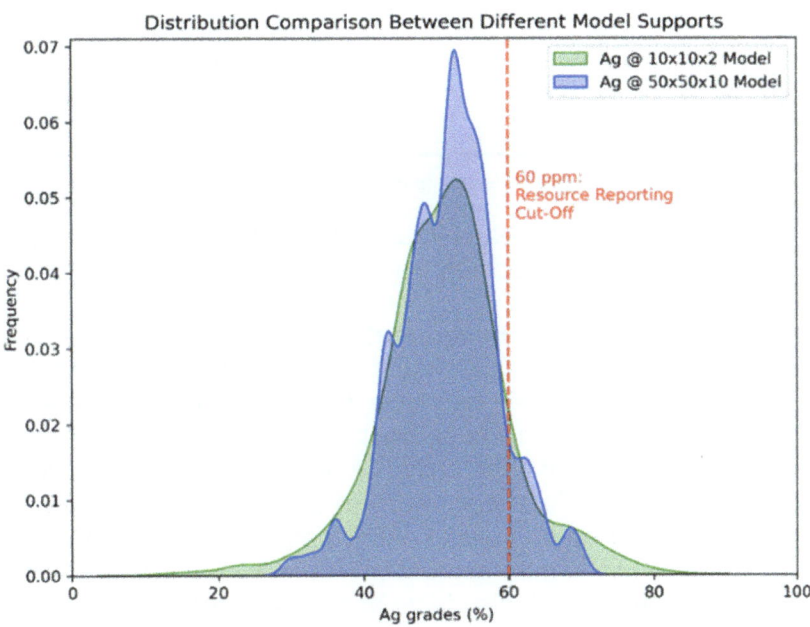

FIG 8 – Distribution comparison between smaller (10 × 10 × 2 m) and larger (50 × 50 × 10 m) block supports.

TABLE 4

Potential revenue from mining based on smaller (10 × 10 × 2 m) and larger (50 × 50 × 10 m) block supports for Ag estimates (using 60 ppm as cut-off).

	Mass (kt)	Ag (ppm)	Metal (t)	Value (USD)
Smaller Support (10 × 10 × 5 m Model)	103.34	67	6.88	4 992 420
Larger Support (10 × 10 × 5 m Model)	79.52	64	5.11	3 707 465
Difference (USD)				-1 284 955
Difference (%)				-26%

THE INFORMATION EFFECT

Long-term models are first built at specific times in the life of a mining project and consider the information available at that time, which is often limited. As more information and data are acquired with the progress of the mining project, the uncertainty decreases. By the time a mid-term model or a grade control model is built, there will be more data available. This reduction in uncertainty from the resource model to the time of mining is a consequence of this information effect. Even though uncertainty decreases from the resource to the grade control modelling time frames, there will always be some remnant uncertainty at the time of mining (Figure 9). This remnant uncertainty is the result of various factors, including the grade control sampling being imperfect, non-exhaustive, the presence of small-scale geological variability, the mining practice and others. There will always be a certain degree of bias or lack of representativeness in sampling methods and no deposit will ever be sampled exhaustively. Typically, only about one billionth of the total volume of a deposit is sampled and we must estimate the grades at non-sampled locations.

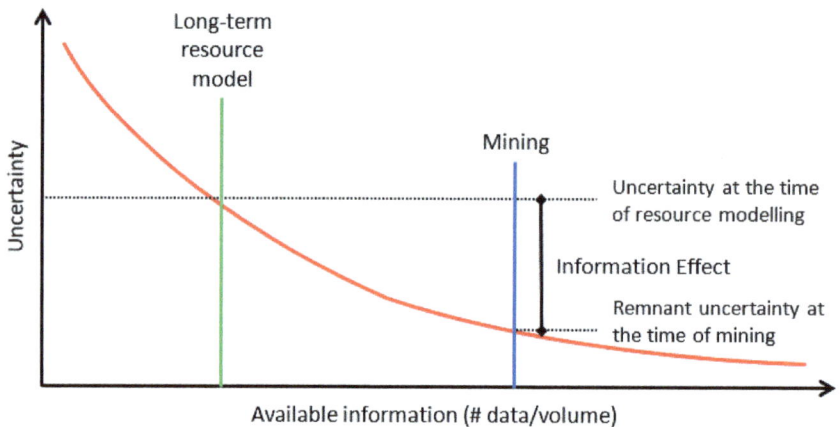

FIG 9 – The information effect – the decrease in uncertainty from the resource model to the time of mining (modified from Chiquini, 2018).

There will always be misclassification of mined material due to incomplete information during the mining process, that is: ore mining blocks estimated and classified as waste and vice versa. There are two types of classification errors: type I and type II. The type I error refers to the material that is classified as ore, but in reality is waste (a false positive). The type II error is a false negative, the material that is thought to be waste but is actually ore (Rossi and Deutsch, 2013) (Figure 10). Both of these errors negatively impact the financial return of a mined area. These classification errors of mined material destination should be minimised. Anticipating future information to correctly predict ore/waste boundaries is critical. In the mining practice, classifying the material being mined is further complicated by having to consider multiple destinations. We typically not only consider ore and waste, but also multiple stockpiles of different grade cut-offs, leach pads, etc. The following example illustrates the impact of the information effect as we move from resource modelling (based on widely spaced drilling) to grade control modelling (based on more closely-spaced data) and its effect on the financial return.

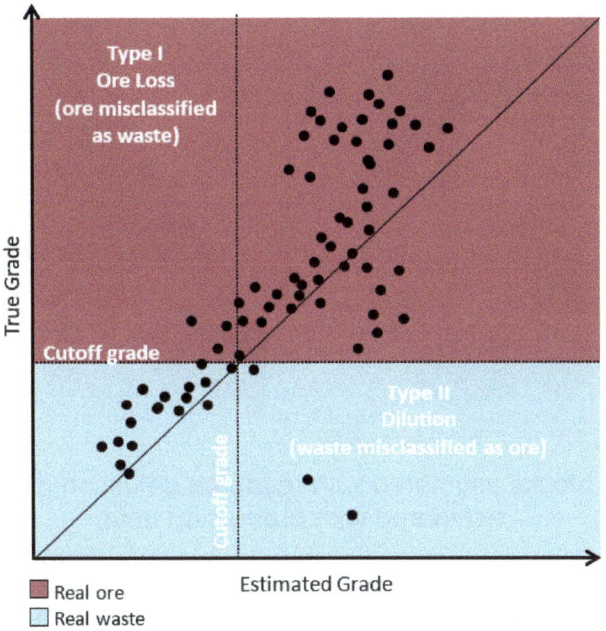

FIG 10 – Type I and type II material classification errors.

Consider a mining area of an open pit copper porphyry deposit located in a single bench and measuring approximately 900 m × 500 m. The drill hole spacing for resource definition is roughly regular at 90 × 90 m and average production data (blastholes) spacing for grade control modelling is 8 × 8 m. Figure 11 shows diamond drilling data available at the time of resource definition and production data available at the time of mining for this specific area of interest.

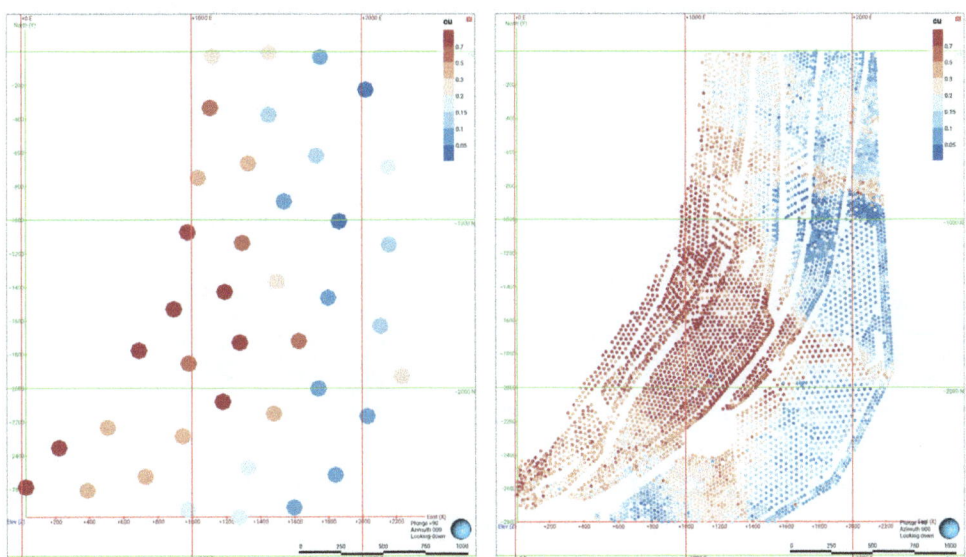

FIG 11 – Left: resource definition drilling spaced 90 × 90 m. Right: production data spaced 8 × 8 m.

For this illustrative example, an inverse distance interpolation is being used to estimate the blocks in the mining area, with an isotropic search volume. The block size utilised for long-term resource modelling is 15 × 15 m, while for grade control modelling it is 8 × 8 m. Figure 12 shows the results of this estimation for the resource model and grade control model.

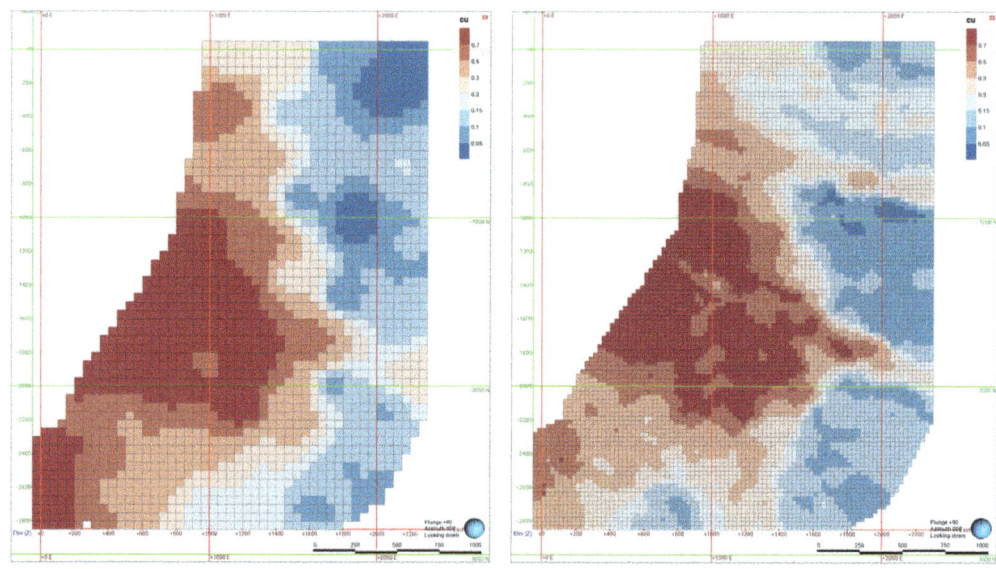

FIG 12 – Left: 15 × 15 m blocks estimated with resource definition drilling. Right: 8 × 8 m blocks estimated with production data.

Now consider the cut-off grade to define the destination for the mined material is 0.2 per cent Cu; blocks above 0.2 per cent will be sent to the mill and blocks below 0.2 per cent will be sent to the waste dump. Figure 13 illustrates the final destination of each block if we were to define ore/waste boundaries on the resource model and the grade control model. Note the areas highlighted as A and B in Figure 13. A shows an area that was underestimated in the resource model and B highlights an overestimated area in the resource model.

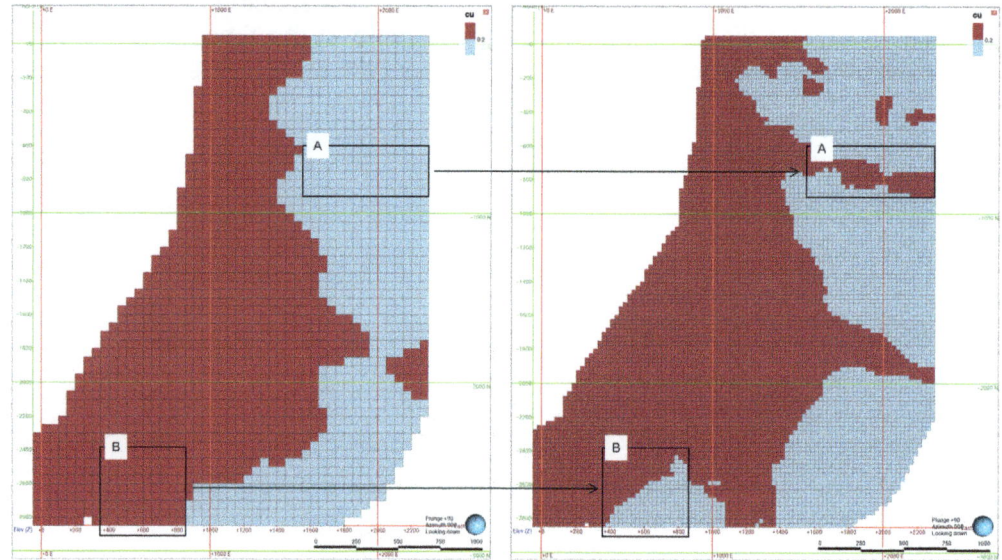

FIG 13 – Ore/waste boundaries defined on the resource model (left) and grade control model (right).

It is not possible to represent the complete variability of the bench at the time of resource modelling due to the information effect. The grade control model represents the local variability of the grades more accurately than the resource model due to closely spaced production data. During the project study phase, 'optimised' mine designs and production schedules are built on the resource model, based on limited information. The reality at the time of mining will be different, positive or negative. Assuming a copper price of USD3.75/lbs, Table 5 compares the potential revenue from carrying out mining activities based on either the resource or the grade control model. In this example, the resource model would predict 11 per cent more revenue than the grade control model would deliver.

TABLE 5

Potential revenues carrying out mining activities based on resource or grade control model.

	Mass (kt)	Cu (%)	Metal ('000 lb)	Revenue (USD)
Resource model	25 613	0.60	337 336	1 265 009
Grade control model	25 694	0.53	299 681	1 123 802
Difference (USD)				-141 207
Difference (%)				-11%

Even though practitioners acknowledge the importance of the information effect for assessment of total resources, accounting for or anticipating the information that will be available at the time of mining is not traditionally embedded in the workflow of long-term models. Most commercial software packages do not offer straightforward ways to include an assessment of the information effect in resource modelling workflows. Chiquini (2018) and Chiquini and Deutsch (2020) proposed a probabilistic (geostatistical simulation-based) workflow to address the information effect at the time of resource modelling. The information effect is accounted for by anticipating the additional production data that will be available at the time of mining. The proposed workflow involves sampling a set of simulation realisations at the anticipated production data spacing. The goal is to mimic the production data planned in the future. Then, it is necessary to estimate all variables that are required for grade control for every set of sampled final data. By sampling each high-resolution realisation at the anticipated final grade control data spacing, we will have a different data set for each realisation, that will be used together with the actual resource definition drilling available at the time to interpolate the grade control variables at a larger block size used for simulation. Sampling the realisations from geostatistical simulation will provide an approximation of the final data that will be acquired at the time of mining and preserve the true grade variability.

Anticipating the information effect at the time of resource modelling will provide a prediction of recoverable mineral resources closer to what will be mined in the future. With that purpose, Emery and Torres (2005) discussed an adaptation of the DGM formalism for change of support to incorporate the information effect in the context of recoverable resources. The DGM predicts the grade distribution (Z_x) at a point of support (x) by the sum of n Hermite polynomials (H_n) associated with transformed Gaussian values of Z (Y_x):

$$Z_x = \sum \phi_n H_n(Y_x)$$

The block support (Z_v) grade is deduced as:

$$Z_v = \sum \phi_n r H_n(Y_v)$$

The coefficient r in the last equation is commonly referred to as the change of support coefficient and serves as a scaling factor for variance (its values decrease as the support v grows). The variances of both the point and block are established as a sum of the expanded polynomial coefficients ϕ_n. Consequently, the change of support coefficient can be ascertained through the following identity:

$$Var\ Z_v = \sum \phi_n^2 r^{2n}$$

The DGM performs support correction by considering the difference in scale from points to blocks on the original point data. In contrast, the adaptation proposed by Roth and Deraisme (2001) calculates the change of support from theoretical block estimates Z_v^*. Therefore, the block grade distribution, accounting for the information effect, can be expressed as:

$$Z_v^* = \sum \phi_n r^* H_n(Y_v^*)$$

Where r^* is the new change of support factor that reduces the original data variance to a theoretical estimated block support. However, the calculation of r^* requires the following assumptions:

$$\begin{cases} Var\ Z_v^* = \sum \phi_n^2\ (r^*)^{2n} \\ Cov\ (Z_v, Z_v^*) = \sum \phi_n^2\ (rr^*r_v)^{2n} \end{cases}$$

Fortunately, the theoretical inference of $Cov\ (Z_v, Z_v^*)$ and $Var\ Z_v^*$ is viable, given their dependence on the modelled variogram and data spacing. Therefore, this adaptation of DGM permits us to predict grade-to-tonnage relationships at a block support for different drill hole spacings and gain a clear understanding of the information effect impacts on recoverable resources.

To illustrate how DGM works with the information effect, the example below shows support correction on the data of a real gold deposit. For the information effect, five different drilling meshes (2 × 2 × 1 m, 5 × 5 × 1 m, 10 × 10 × 1 m, 20 × 20 × 1 m and 40 × 40 × 1 m) were tested (Figure 14). Note that as the mesh resolution decreases, the corresponding DGM model becomes smoother due to reduced $Var\ Z_v^*$ and support correction factor. The DGM lacking information effects provides a less smoothed model, as its variance correction solely accounts for the change of support. On the other hand, the information-effect DGM incorporates a degree of smoothing arising from the estimates with different drilling configurations. In this analysis, assuming a mining cut-off grade of 1.5 g/t of gold, the recovered tonnage ranges from 47 per cent (without information effect) to 56 per cent (with a 40 × 40 m final mesh). Such exercises are valuable for conducting sensitivity analyses on how the final drilling configuration impacts the resources recovered during mining operations.

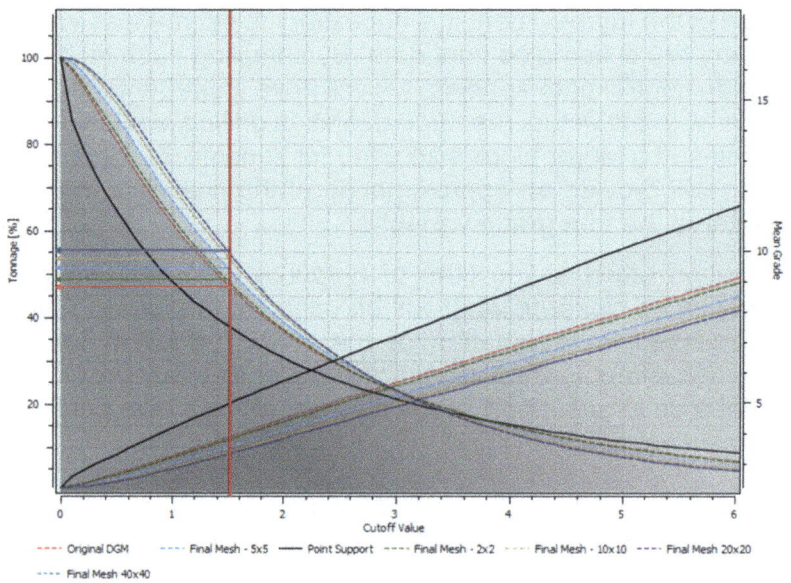

FIG 14 – Grade-tonnage curves comparing the original point-support data to the DGM models with and without information effect.

THE ESTIMATION STRATEGY

Among the subjects discussed in this paper, the estimation strategy or neighbourhood analysis likely has been the focus of the most numerous publications over time. This is relevant when creating fit-for-purpose estimates because how the estimation strategy is defined will define the behaviour of grade distributions on both global and local scales. Therefore, a proper understanding of how the estimation parameters influence the estimates is essential to ensure that the resulting model achieves its intended objective.

While acknowledging the risk of oversimplifying the subject, we can assert that the distinction between resource models and production models is a matter of accuracy, whether local or global, from a geostatistical perspective. Local accuracy is closely linked to selectivity, a crucial aspect of production models as it fundamentally informs the destination of mined material. In contrast, global accuracy pertains to the reproduction of the total grade-tonnage curve. This information holds significant value for strategic decision-making in the initial phases of a mining venture, determining whether the estimated economic tonnes are sufficient to justify further investment in the project. We can employ both theoretical and empirical approaches to tailor the estimates to the accuracy desired.

The ideal scenario would involve globally accurate and locally unbiased estimates simultaneously. Unfortunately, achieving both local and global accuracy is unlikely. Isaaks (2005) demonstrated this in a statistical yet intuitive manner. In a hypothetical scenario, consider a scatter plot with estimated block grades ($Z^*_{(u)}$) plotted on the X-axis and known true block grades ($Z_{(u)}$) plotted on the Y-axis. The relationship between the estimated and true grades can be expressed as follows:

$$Z_{(u)} = a + bZ^*_{(u)}$$

The equation above represents a simple linear regression, where 'a' is the intercept and 'b' is the slope of the regression line. The slope of regression (SOR) parameter is a widely used criterion for defining the search neighbourhood. It is a metric provided by many QKNA (quantitative kriging neighbourhood analysis) tools built into traditional mining software packages. However, the limitation of arbitrarily choosing the estimation neighbourhood to maximise SOR is demonstrated by Kentwell (2022). The following suggests why optimising this parameter may not always be the best approach. The slope of regression b in the previous equation can be expressed as:

$$b = \frac{\sigma_Z}{\sigma_{Z^*}} \rho_{Z,Z^*}$$

Where σ_Z is the dispersion of the true block grades, σ_{Z^*} is the dispersion of the estimated block grades and ρ_{Z,Z^*} is the correlation coefficient between estimated and true grades. Thus, the requirement for a conditionally unbiased estimate is that the SOR equals 1. This implies a perfect correlation between the true block grades and the estimated block grades and a 45-degree regression line. However, perfect correlation between estimates and true values is unattainable in practice and the slope of the regression will always be less than 1. Consequently, the ratio $\frac{\sigma_Z}{\sigma_{Z^*}}$ must be greater than one if we want a perfect SOR. In other words, if the estimated grades are conditionally unbiased (local precision), then the estimated histogram (σ_{Z^*}) will not align with the true grade histogram (σ_Z). But for the resource grade-tonnage curves to accurately predict the tonnes above a cut-off (globally), σ_{Z^*} and σ_Z must be equal (Isaaks, 2005).

With this contradiction (or oxymoron, as characterised by Ed Isaaks) in mind, the following example shows a study aimed at determining the best estimation strategy by assessing how the chosen parameters drive the model more towards local or global accuracy. This assists in making a better-informed decision on the estimation strategy, considering the long-term and resource reporting model's purpose. The following case study was undertaken on an operating open pit gold mine.

The calibration of the kriging strategy involves a combination of simulation and resampling approaches, similar to those typically employed in drill hole spacing studies. For the sake of simplicity, in this example, the primary parameter being examined is the number of composites used in kriging. However, the workflow could be implemented for any combination of parameters. The initial step consists in creating simulated realisations to serve as the truth, to be used as reference against which all kriging scenarios will be compared to (Figure 15b). This simulation should be performed on a grid fine enough to match the sampling resolution of the actual drilling data (Figure 15a). It is also recommended to align the simulation grid with an integer fraction of the SMU ultimately used in resource model estimates to facilitate reconciliation between the reference model and calibration scenarios. After validating the realisations, they should be sampled to mimic the spatial configuration of the original data (Figure 15c). The decision to use simulated reference models over alternatives like uniform conditioning (UC) or discrete Gaussian models (DGM) was driven by the fact that simulation models localise the grades and enable local-scale calibration. In contrast, UC and DGM primarily reproduce the global theoretical grade-tonnage relationship.

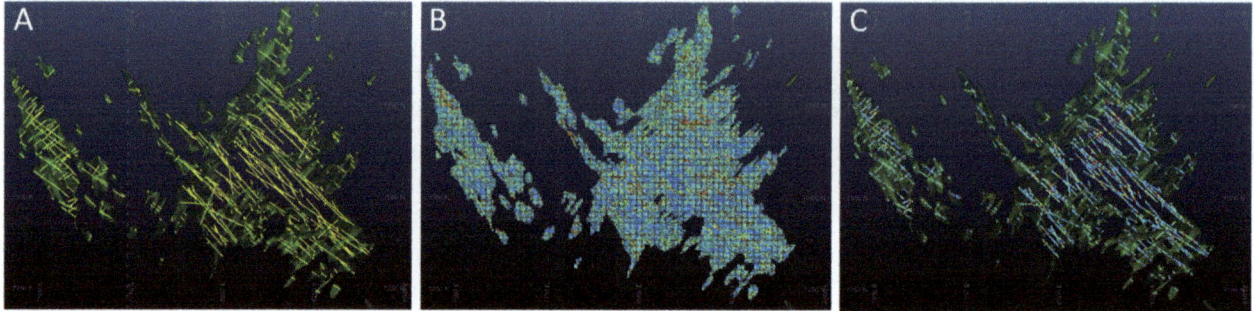

FIG 15 – Plan view showing: (a) the original diamond drilling data, (b) one of the simulated realisations, and (c) the re-sampled realisation mimicking the original data configuration.

The synthetic drilling data was used to create kriged models with different estimation strategies, varying from 4 to 60 samples. The estimates were reconciled against the reference realisation on the reproduction of the global grade-tonnage relationship and average grades (Figure 16).

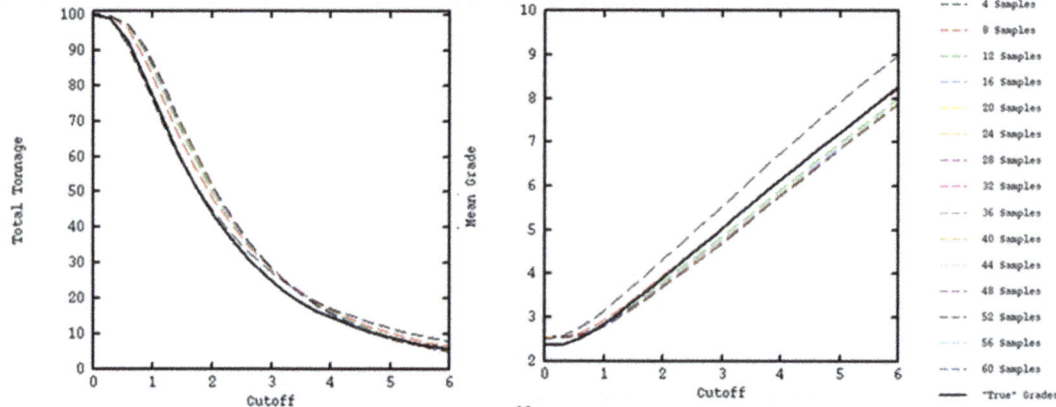

FIG 16 – Relative grade-tonnage and grade-cut-off curves comparing all kriged scenarios (dashed lines) with the reference model (solid black line).

The final analysis involves quantifying discrepancies between estimates with different strategies and the simulated reference model. The graphical representation in Figure 17a illustrates these deviations as a function of root mean squared error (RMSE) on the total grade-tonnage relationship. Higher data quantities lead to increased overall smoothing. Notably, deviations stabilise after 24 samples, indicating that no incremental gain is achieved by using a larger number of samples. Considering global accuracy, the optimal result is achieved with eight samples as it offers the best global match against the reference model. Note that despite producing the lowest RMSE, using only four samples makes the estimates highly unstable and sensitive to local extreme values, as shown by the grade RMSE curve (Figure 17a).

Local accuracy is assessed by comparing estimates to the reference model on a block-by-block basis. To facilitate this comparison, the initially dense simulated grid was upscaled to match the kriging support by averaging grades within the SMU support. The local checks (block by block) for all scenarios were verified by calculating the root mean squared error (RMSE) between the actual and estimated grades (Figure 17b). The local accuracy plot shows no significant improvement in local accuracy when considering more than 16 samples in kriging. Most improvements are observed between the range of 4 to 16 samples.

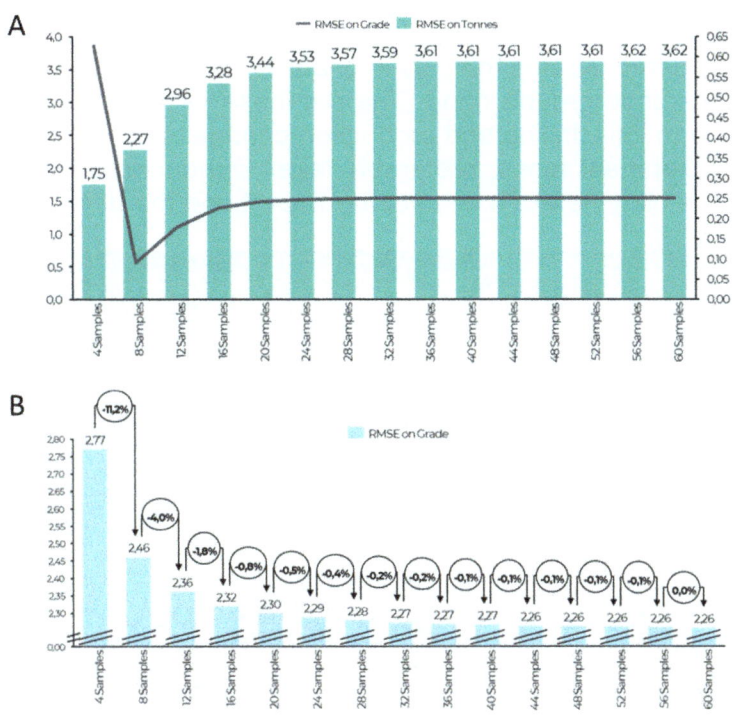

FIG 17 – Reconciliation of the estimates against the reference model. In (a), the bars represent the variation between the kriged tonnes and the reference tones at all cut-off grades. The line represents the variations of the average grade between the estimates and reference models at all cut-off grades. (b) represents the local accuracy check. The bars represent the variation between the estimated grades and the reference model grades.

Hence, the decision regarding the 'optimum' number of kriging samples considers a reasonable balance between local and global accuracy. This consideration is made in a context of mineral resource evaluation and reporting. Therefore, using fewer samples, such as four, would render the estimate unstable, particularly for the higher-grade tail of the distribution, leading to over-smearing of extreme values. The scenario with eight samples best reproduces the 'actual' grade-tonnage relationship albeit with a higher local error. This exercise allows for a better understanding of the effects of parameters on the estimates, enabling effective decision-making and transparency in justifying choices on the estimation strategy.

SAMPLING METHOD AND QUALITY

Managing data from various sources in mineral resource estimation poses a challenge for practitioners. Rarely are there situations where only a single type of data is available for estimation, such as during early exploratory stages. As the mining venture progresses, it is probable that diverse sources of data or even different data vintages get integrated into the geological database. This variety goes beyond drilling types and may include sample preparation methodology, laboratory protocols, different contractors, equipment and other sources of variation. All these factors may result in bias between the different sets of data. This is the case in most mature operations.

Various studies have been conducted on the use of samples with varying quality, including Ribeiro *et al* (2011), who investigated the correlation between samples acquired from diamond drilling and reverse circulation drilling in an iron ore deposit. Paixão (2015) and Donovan (2015) studied the integration of different data sources by employing multivariate methods to address variations in sample quality during estimation. In a different study, Miguel-Silva (2015) examined the effects of sampling errors on final estimates by artificially perturbing the original data.

Biases may occur from many reasons throughout the data acquisition process. This section demonstrates practical aspects of the risk of different sets of data on estimates rather than discussing the reasons behind varying level of sampling quality and best practices in data acquisition. For interested readers, this can be found in the literature dedicated to QA/QC and the theory of sampling, such as Gy (1982).

In the following example, the bias between different sets of data is demonstrated in a real-world deposit. This case study demonstrates the bias in channel samples compared to diamond drilling in an operating and mature underground mine. The database in this deposit comprises exploration diamond drilling, short-term infill drilling and production channel samples. The channel samples are collected along the development faces, preferably taken orthogonally to the lithological contacts. Figure 18 demonstrates a plan view of a development level and the different data types.

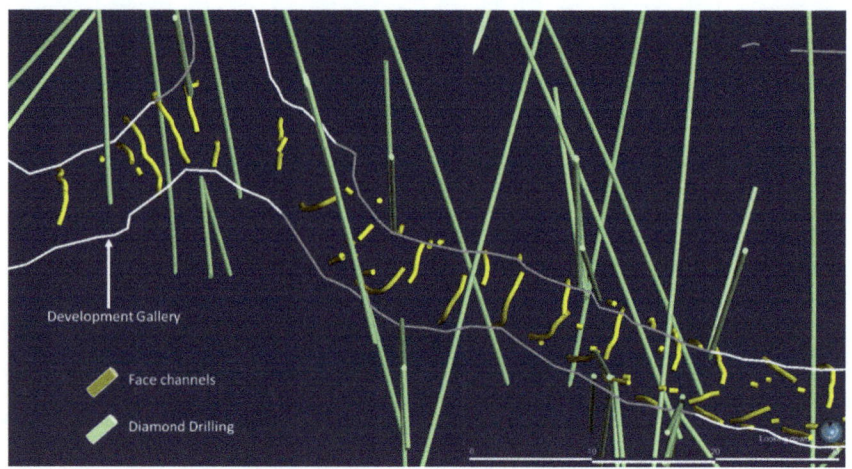

FIG 18 – Plan view of an underground development level showing the face channel samples (yellow) and diamond drilling (green).

The first step is to identify whether bias is present due to different sampling techniques. Twin holes or samples are a traditional technique to assist in assessing and correcting biased data. Abzalov (2016) dedicates an entire chapter to the execution and statistical analysis of twinned samples. However, it can be impractical due to cost constraints and often is equivocal. Redrilling a previously drilled location for exclusive verification and checking purposes is an unpopular decision in practice and is rarely considered a priority when allocating budget for drilling.

A cost-effective and tangible alternative is to statistically analyse the grades of both sampling techniques in locations where they overlap or in an optimal situation where they cross-cut each other. There are considerations to keep in mind when carrying out this exercise. Channel samples are usually collected from areas determined to be economically viable. Consequently, it is expected that channels, on average, will exhibit higher grades than resource definition drilling. There could be an increased internal dilution within channel samples compared to borehole samples, influenced by their respective positions and higher sample density. Moreover, the complete data set encompasses samples generated through various sampling and analytical protocols. Therefore, any observed deviation may reflect other sources of variability.

The paired data for evaluation (channels and diamond drilling) were obtained according to the following criteria:

- Migration of grades from channel samples and diamond drilling to a common target within a volume of influence (by nearest neighbourhood searches).

- Use of a volume no larger than 10 m to select channel and diamond drilling samples. These volumes guarantee that the paired data represent the same geological context.

- Analysis of the statistics of paired data at different distances (from 1 to 10 m). As the distance increases, the number of paired data points also increases. Finding both data types at short distances, such as 1 m or less, is rarely accomplished, and this could lead to statistically insignificant outcomes when working with extremely limited data.

- Preprocessing the data before evaluation to eliminate the influence of any undesired information. The case study's analysis was conducted exclusively with modern drilling data, excluding historical data, to ensure consistent sampling and analytical methodologies.

- Ensuring the paired data is constrained to the same geostatistical domain.

After following the criteria above, a fair comparison of the collocated distribution for both sets of data was possible. The histograms and quantile plots in Figure 19 reveal a bias in channel samples compared to diamond drilling in distinct data sets.

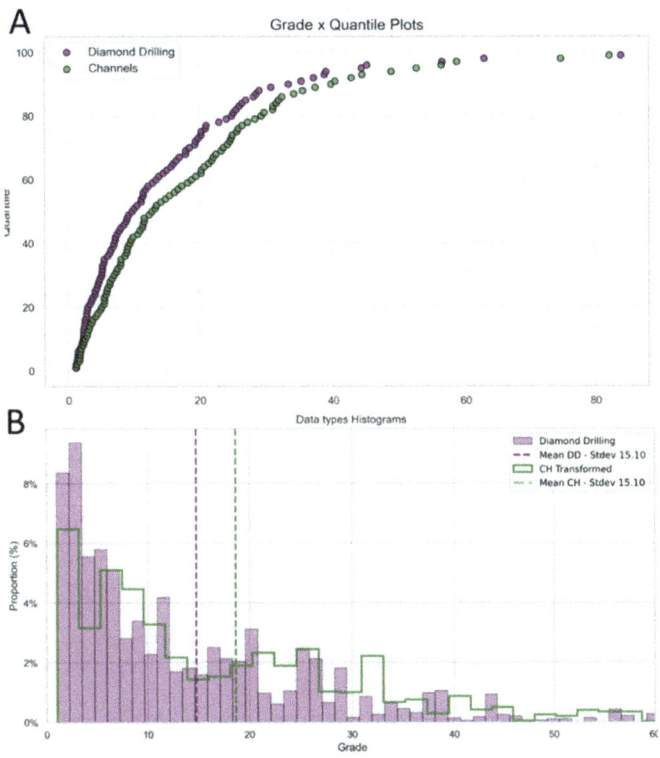

FIG 19 – Statistical analysis comparing the 'collocated' grade distributions of DD and CH grades. The quantiles against the grades are shown in (a) and the histograms with respective statistics are shown in (b).

This bias observed in channel samples, if left untreated, can manifest in the grade estimates. An effective approach to address this issue involves treating the channel samples as secondary variables and correcting the precision bias through statistical transformations. To implement this, the channel grades were standardised using the following calculation:

$$Z_k(u)^{Std} = \frac{Z_k(u) - m_k}{\sigma_k}$$

Where $Z_k(u)^{Std}$ is the standardised channel value, $Z_k(u)$ is the original channel value, m_k and σ_k are respectively the mean and standard deviation of the channel samples distribution. The transformed distribution is finally achieved by:

$$Z_k(u)^{Scaled} = Z_k(u)^{Std} * \sigma_j + m_j$$

Where $Z_k(u)^{Scaled}$ represents the transformed channel value, m_j and σ_j the respective mean and standard deviation of the primary target distribution (diamond drilling histogram). After transformation, the distribution of channel samples, specifically the mean and standard deviation, aligns with the statistical parameters of the distribution of diamond drilling samples. The transformation applied to the channel samples enables the utilisation of both data sets simultaneously without compromising the estimates with the bias previously observed in the channel grades. The transformation's impact on the channel sample set is illustrated in Figure 20. Observing the match in mean and standard deviation is important, highlighting the removal of bias.

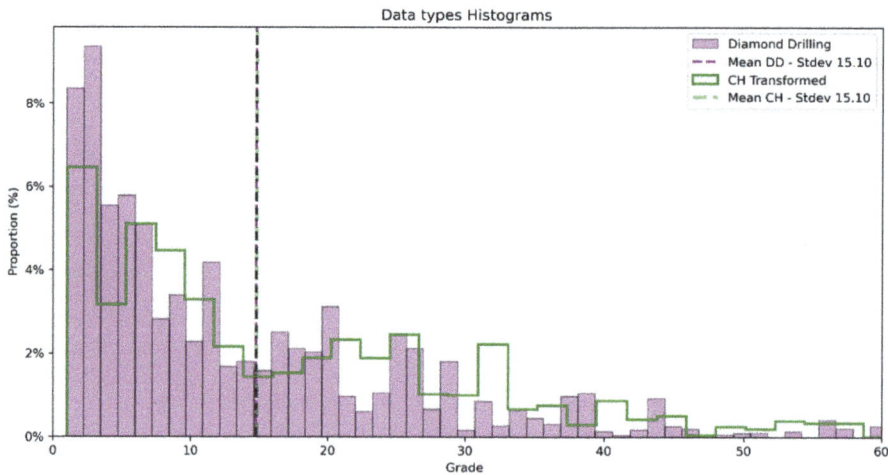

FIG 20 – Histogram of diamond drilling grades and transformed channel grades.

An effective approach for assessing the effects of the data correction on the spatial grade distribution is to generate trend plots for both distributions prior to and after the transformation. Swath plots of both data sets are presented in Figure 21, illustrating the notable shift of grades in channel samples towards the diamond drilling local averages while maintaining their local fluctuations. One should be aware of negative values after rescaling because of the mean of the secondary variable (channels) being greater than the primary (DDH).

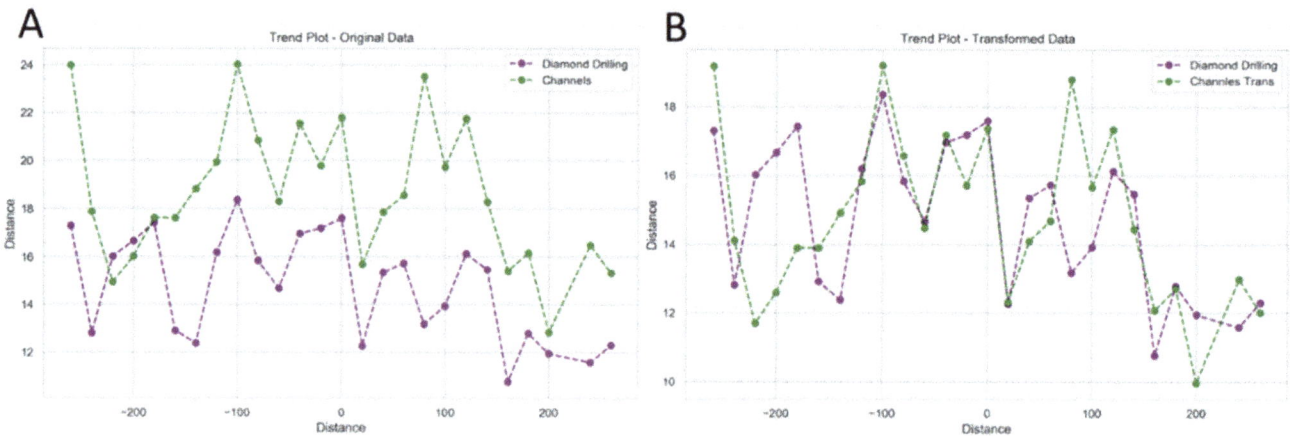

FIG 21 – Trend plots comparing the diamond drilling and channel grades before (a) and after transformation (b).

While literature offers more theoretically sophisticated approaches to address biased data sets, some can be relatively time-consuming and unsuitable for production models. Data transformation has emerged as an efficient and practical solution. This approach can be automated using scripts and can incorporate domain-specific transformation rules prior to grade estimation.

DISCUSSION AND CONCLUSIONS

Based on the authors' experiences in evaluating resources for projects and mining operations, it is reasonable to affirm that practitioners may overlook certain aspects of resource models and grade control as operations evolve. The common practice in many mine sites is to use a predefined estimation script and replicate it on a regular basis for subsequent models, irrespective of the purpose of the later models. This motivated the discussions presented in this paper.

The concept of the support effect in resource estimation highlights how scale and variability impact estimation outcomes. This concept is critical in long-term recoverable resource planning, addressing the spatial variability of ore deposits and the understanding of selectivity in mining operations. Change of support models, including techniques like affine correction, indirect lognormal correction, and the Discrete Gaussian Model (DGM), play a significant role in these evaluations. Case studies from base metal operations illustrate the practical implications of different factors on grade estimates, such as composite length, block dimensions and variogram continuity. These can vary significantly depending on the type of model being generated, that is, grade control for short-term mine planning versus resource reporting.

As discussed previously, the reduction in uncertainty as we acquire more data and move from the resource model to grade control modelling at the time of mining can be referred to as the information effect. The examples in this paper illustrate the impact of the information effect as we move from resource modelling to grade control modelling on the financial return of a mining area. The grade control model represents the local variability of the grades more accurately than the resource model due to closely spaced production data. Disregarding future data that will be available at the time of mining will impact the revenue of a mining project by introducing classification errors on the predicted destination for the mined material. Anticipating the information effect at the time of resource modelling will provide a prediction of recoverable mineral resources closer to what will be mined in the future.

Developing a neighbourhood calibration exercise based on simulation-resampling can be time demanding and computationally intensive. Simulated models are typically not part of the short-term estimation routine in operational mines. However, it is important to note that this type of study does not require regular updates unless material changes happen in geological interpretation, grade continuities, geostatistical workflows or substantial new data is acquired. Some may fairly argue against using a single reference realisation for calibrating the estimates. Using a small subset of realisations can also be argued, and we agree that managing multiple realisations is key for addressing most issues treated by simulation. Optimising estimation parameters based on a reduced number of realisations, especially for global assessments, does not disqualify the results of the method. This is because metrics for error, such as RMSE, are direct error functions and not probabilistic ones. Every realisation represents a fair reproduction of the overall grade variability (same variogram) and data distribution (same grade-tonnage curve). This assumption was empirically tested by rerunning the optimisation workflow using different data subsets and over different realisations. The results are consistent regardless of the chosen reference scenario.

Production sampling methods in most operations have less rigour in ensuring quality and representativeness. Despite that, they provide valuable information on short-scale grade variability as they are closely spaced. In practice, we want to ensure the sampling routine is robust enough to provide good quality samples by monitoring it through dedicated QA/QC programs. It is important to note that correcting eventual biases using the presented approach may only partially mitigate the impact of poor sampling. This is especially true because correcting the bias globally may overlook regionalised trends or even temporal features in the assay values. However, the presented approach is a practical way to benefit from already collected data. Adjusting the distribution statistically is a very simple and straightforward approach. It does not require time-consuming geostatistical techniques such as standardised co-kriging, which could be prohibitive for time-constrained modelling workflows. However, adjusting for bias in a data set is a coarse tool that must be applied with care, transparently and disclosed.

The main purpose of this paper is to discuss some of the key aspects when transitioning from resource modelling to grade control modelling through real targeted case studies. These factors greatly impact the accuracy of model estimates, making them unsuitable for their intended purpose if overlooked. In a mine geology context, correctly understanding and accounting for all of these factors is crucial for creating short-term estimates that allow effective short-term mine planning and successful mining practice. Ignoring them can lead to incorrect estimates, yielding material misclassification and financial losses. The methods presented consist in well-established techniques in the mining industry, ensuring that estimates are tailored precisely to their purpose. The authors believe these to be key considerations and acknowledge it is not an exhaustive list.

REFERENCES

Abzalov, M, 2016. *Applied Mining Geology*, vol 12 (Springer).

Chilès, J-P and Delfiner, P, 2009. *Geostatistics: Modeling spatial uncertainty*, vol 497 (John Wiley and Sons).

Chiquini, A and Deutsch, C V, 2020. Mineral Resources Evaluation with Mining Selectivity and Information Effect, *Mining, Metallurgy and Exploration*, 37:965–979. doi:10.1007/s42461-020-00229-2.

Chiquini, A P, 2018. Mineral resources evaluation with mining selectivity and information effect, Master's thesis, Department of Civil and Environmental Engineering, University of Alberta.

Deutsch, C V and Deutsch, J L, 2015. Introduction to Choosing a Kriging Plan, in *Geostatistics Lessons* (ed: J L Deutsch). Available from: <http://geostatisticslessons.com/lessons/introkrigingplan>

Deutsch, J, Szymanski, J and Deutsch, C, 2014. Checks and measures of performance for kriging estimates, *Journal of the Southern African Institute of Mining and Metallurgy*, 114(3):223–223.

Donovan, A P, 2015. Resource estimation with multiple data types, Master's thesis, Department of Civil and Environmental Engineering, University of Alberta.

Emery, X and Torres, J, 2005. Models for Support and Information Effects: a comparative study, *Mathematical Geology*, 37(1):2005. doi:10.1007/s11004-005-8747-8.

Gy, P M, 1982. *Sampling of Particulate Materials, Theory and Practice* (Elsevier: Amsterdam).

Harding, B E and Deutsch, C V, 2019. Change of Support and the Volume Variance Relation, in *Geostatistics Lessons* (ed: J L Deutsch). Available from: <http://geostatisticslessons.com/lessons/changeofsupport>

Isaaks, E H and Srivastava, M R, 1989. *An introduction to applied geostatistics* (Oxford University Press).

Isaaks, E, 2005. The kriging oxymoron: A conditionally unbiased and accurate predictor, in *Geostatistics Banff 2004* (eds: O Leuangthong and C V Deutsch), second edition, 14:363–374 (Springer: Netherlands).

Journel, A G and Huijbregts, C, 1978. *Mining geostatistics*, 599 p (Academic Press: New York).

Kentwell, D, 2022. Empirical geostatistics #1 – kriging slope of regression: sensitivities and impacts on estimation, classification and final selection, in *Proceedings of the 12th International Mining Geology Conference 2022*, pp .

Machuca-Mory, D F, Babak, O and Deutsch, C V, 2008. Flexible change of support model suitable for a wide range of mineralization styles, *Mining Engineering*, 60:63–72. Available from: <https://www.researchgate.net/publication/243055025_Flexible_change_of_support_model_suitable_for_a_wide_range_of_mineralization_styles>

Miguel-Silva, V, 2015. Análise de sensibilidade das estimativas ao erro amostral, posicional e suas aplicações (Sensitivity analysis of estimates to sampling and positional error and their applications), Master thesis, Programa de pós-graduação em engenharia de minas, metalúrgica e materiais (PPGEM; Postgraduate Program in Mining, Metallurgical and Materials Engineering), Universidade Federal do Rio Grande do Sul.

Paixão, C A, 2015. Uso de informação secundária imprecisa e inacurada no planejamento de curto prazo (Use of imprecise and inaccurate secondary information in short-term planning), Master thesis, Programa de pós-graduação em engenharia de minas, metalúrgica e materiais (PPGEM; Postgraduate Program in Mining, Metallurgical and Materials Engineering), Universidade Federal do Rio Grande do Sul.

Parker, H M, 1979. The volume–variance relationship: a useful tool for mine planning; *Eng and Min Journal*, October, pp 106–123.

Ribeiro, D T, Monteiro, C, Cunha, E, Catarino, M, Augusto, V and Gomes, M, 2011. Correlation study between reverse circulation and diamond drilling in iron ore deposits, in *Proceedings, 5th Conference on Sampling and Blending – WCBS-2011*, Santiago.

Rivoirard, J, 1987. Two Key Parameters When Choosing the Kriging Neighborhood, *Mathematical Geology*, 19(8).

Rossi, M E and Deutsch, C V, 2013. *Mineral Resource Estimation* (Springer Science).

Roth, C and Deraisme, J, 2001. The information effect and estimating recoverable reserves, in *Proceedings of the Sixth International Geostatistics Congress* (eds: W J Kleingeld and D G Krige), pp 776–787.

Sinclair, A J and Blackwell, G H, 2002. *Applied Mineral Inventory Estimation* (Cambridge University Press).

Vann, J, Jackson, S and Bertoli, O, 2003. Quantitative Kriging Neighbourhood Analysis for the Mining Geologist – A Description of the Method With Worked Case Examples, in *Proceedings of the 5th International Mining Geology Conference* (The Australasian Institute of Mining and Metallurgy: Melbourne).

An ever-evolving resource model as new pieces are added to the Avoca Tank puzzle, Tritton Operations

D Beckett[1], A Dimond[2], N Cox[3], O Gonzalez[4], A Jackson[5] and B Cox[6]

1. Core Logging Geologist, Aeris Resources, Brisbane Qld 4000.
   Email: dbeckett@aerisresources.com.au
2. Mine Geology Superintendent, Aeris Resources, Brisbane Qld 4000.
   Email: adimond@aerisresources.com.au
3. Principal Geologist – Structural Geology, Aeris Resources, Brisbane Qld 4000.
   Email: ncox@aerisresources.com.au
4. Principal Geologist – Mine Geology, Aeris Resources, Brisbane Qld 4000.
   Email: ogonzalez@aerisresources.com.au
5. Senior Mine Geologist, Aeris Resources, Brisbane Qld 4000.
   Email: ajackson@aerisresources.com.au
6. General Manager Geology, Aeris Resources, Brisbane Qld 4000.
   Email: bcox@aerisresources.com.au

ABSTRACT

Aeris Resources owns and operates the Tritton Copper Operation in central west New South Wales. Underground mining commenced in late 2004 at the Tritton deposit, with supplementary ore sources providing additional mill feed. Approximately 450 kt copper metal has been produced from the Tritton Operation (circa June 2023), with a further 369 kt contained copper metal in Mineral Resource (Aeris Resources, 2023a). The Tritton deposit has been the major contributor to produced copper; however, over recent years, other deposits have been increasingly relied upon as mining at the Tritton deposit becomes marginal.

The Avoca Tank deposit is a small high-grade copper deposit with a current reported Indicated plus Inferred Mineral Resource of 720 kt @ 3.4 per cent Cu and 1.1 g/t Au (Aeris Resources, 2023b). Despite the deposit's relatively close-spaced drilling (nominal 20 m × 20 m), the complex geological setting poses challenges in confidently correlating sulfide intersections between drill holes.

The geology encountered on the first level of development (4875 mRL level) differed significantly in places from the original interpretation. This situation called for both exploration and production to take place simultaneously.

In response, the Aeris Geoscience Team completed significant work to improve the geological understanding of the deposit. This included understanding the controls on mineralisation, improving the confidence and consistency of data collection, and refining geological interpretation and modelling techniques. These efforts have greatly improved the data collection to grade estimation processes at Avoca Tank, resulting in a more accurate model for mine planning purposes.

The work is far from complete; our processes continue to evolve as we understand a little more with every drill hole and development level advancement.

INTRODUCTION

Aeris Resources owns and operates the Tritton Copper Operation in Central West New South Wales. The closest regional town, Nyngan (population 3000), is approximately 49 km south-east of Tritton. The Tritton Operation includes a copper sulfide processing plant, three current underground operations and several advanced projects coming into production in the coming years.

The Avoca Tank deposit is one of the active underground operations. It is located approximately 27 km north-east of the Tritton processing plant (Figure 1).

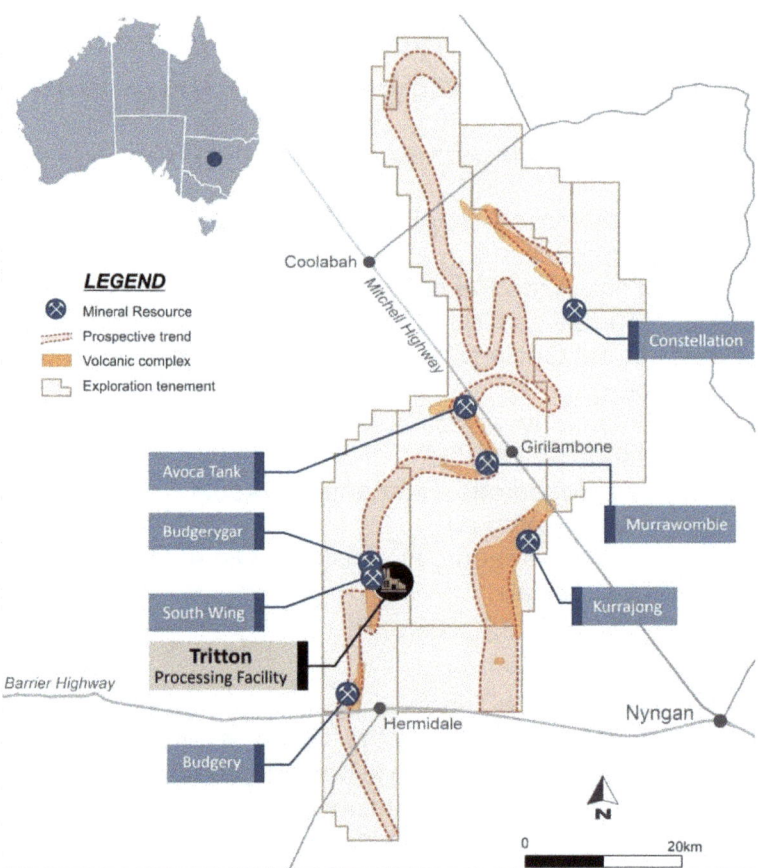

FIG 1 – Plan view showing the location of the Avoca Tank deposit within the Tritton Operation tenement package.

The deposit was first discovered in the 1970s by Australian Selection Trust Pty Ltd (Seltrust). Several shallow drill programs were completed with limited success. It wasn't until 2011 that economic grades of copper mineralisation were discovered during an RC drill program by Straits Resources Limited (Straits). Following the discovery, Straits completed a significant drill program, leading to the release of a maiden Mineral Resource Estimate (MRE) in 2013 (Aeris Resources, 2014).

The 2013 MRE was based on a series of thin, narrow, high-grade sulfide lenses dipping steeply to the north-east from underneath an old mine shaft. The resource estimate detailed the potential complexity of the mineralised lenses, with the possible splitting of the main ore zone into five lenses.

Mineralisation of massive chalcopyrite and high gold grades made this deposit very attractive economically; however, its small size and limited understanding of the local-scale geology, including structural controls, marked it as high risk.

In 2021, Aeris Resources decided to commence underground mining of Avoca Tank, acknowledging the complexity and that the Mineral Resource estimate would likely change as underground mining progressed.

LOCAL GEOLOGY

The Avoca Tank deposit is a copper and gold rich mineralised system occurring within the Girilambone province of central-west New South Wales. Mineralisation displays a strong structural control within chloritic shear zones hosted by Early Ordovician (ca 485–470 Ma) metasedimentary rocks at and around the margins of a metadolerite complex (Burton, 2011; Burton, Trigg and Campbell, 2012; Gilmore, Trigg and Campbell, 2018; Simpson *et al*, 2023).

Structure

Several regional deformation events are evident throughout the Girilambone Group sediments within the Tritton tenement package. Surrounding the Avoca Tank deposit, the most prominent fabric is interpreted to be a bedding parallel S2, strongly spaced to pervasive foliation best identified as differentiated layers within metasandstone horizons. This fabric is associated with compressional deformation and greenschist facies metamorphism during the Late Ordovician to Early Silurian Benambran Orogeny (490–435 Ma) (Simpson *et al*, 2023). The S2 fabric is typically folded (moderately-shallowly inclined, close-tight) and sheared by a D3 event. D3 structures include axial planar foliation (S3) and brittle-ductile shear zones. The S3 fabric is observed as an axial planar weak to strong crenulation cleavage, occasionally observed to accommodate extensional shear strain. A renewed compression event (D4) is observed to fold, reactivate, and fracture previous structures, currently interpreted with a reverse to dextral shear sense. D4 folds trend NNW-SSE are gently plunging to the North, upright and close-open. D4 strain appears to be higher within mineralised zones, producing tighter folds that moderately-steeply plunge (reclined) both North and South.

There are plenty of small-scaled faults (traced across a single level or two) that have currently been mapped with minor offset, but occasionally observed to disrupt mineralisation. More mapping and fault characterisation is required to help build the bigger faulting picture.

Mineralisation

Early-stage magnetite has been overprinted by pyrite and chalcopyrite. Fine-grained pyrite-chalcopyrite is typically observed along D2 penetrative foliations and early magnetite-chlorite shear zones. Overprinting D3 structures, such as brittle-ductile shears and folds, can contain a concentration of coarse pyrite and/or remobilised chalcopyrite. Reactivation during D4 resulted in the milling of brittle pyrite and remobilisation of ductile chalcopyrite. Sheared metadolerite margins are observed to be included among the mineralised reactivated structures.

When compared to other Tritton tenement deposits, Avoca Tank mineralisation has a shorter strike length (≤75 m) and much higher copper grades. Eight discrete sulfide lenses have been defined, averaging 2–3 per cent Cu, with individual drill hole intersections often reporting well above 5 per cent Cu. Gold grades vary from negligible to 1 g/t Au depending on the sulfide lens. Sulfide textures vary from semi-massive to massive, banded, stringer and brecciated. Sulfide textures, mineralogy and copper grades delineate each sulfide lens.

The 2013 interpretation of the Avoca Tank mineralisation has been revised. Previous geological modelling included several short strike-length parallel and *en echelon* lenses. Recent underground diamond drilling, subsequent development mapping, and a greater understanding of the mafic volcanics have significantly improved knowledge of the constraints on mineralisation (Figures 2 and 3).

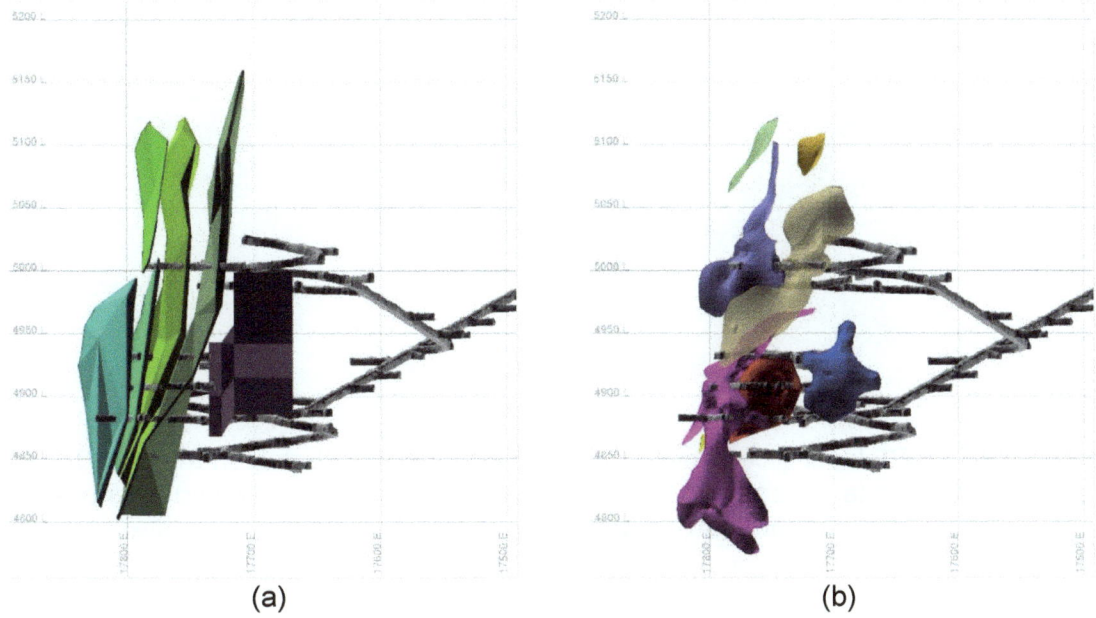

FIG 2 – (a) Cross-section looking north showing interpreted sulfide lenses used for the 2013 Mineral Resource; and (b) the 2023 grade control model.

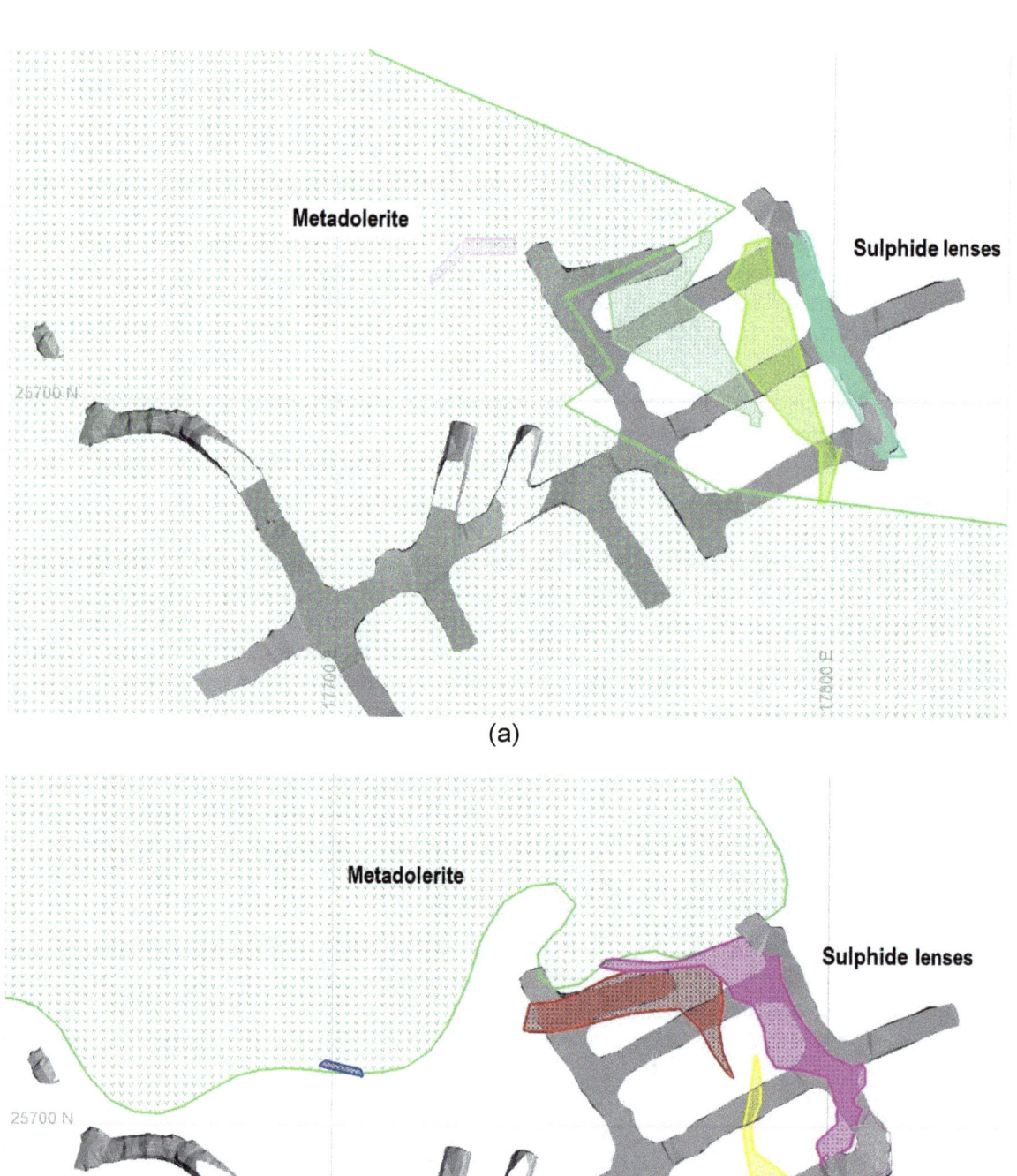

FIG 3 – The 4875 mRL level slice showing the modelled geology between the (a) 2013 resource model and (b) 2023 grade control model. Development is shown by a grey solid wireframe. Areas denoted by white represent metasediments.

Interpretation

Generally, the western half of the Avoca Tank deposit is dominated by metadolerite, and the eastern half is dominated by metasediments. The updated Avoca Tank geological interpretation is based on discrete sulfide lenses located within metasediments and/or at contacts with complexes of metadolerite, with their geometry and continuity generally restricted by their proximity to metadolerite envelopes.

FIRST 24 MONTHS OF MINING – GEOLOGY LEARNINGS

Mining activities began in October 2021 with a 2.5 km decline to access the Avoca Tank deposit. The evident geological uncertainty meant that six additional 500 m resource definition holes were needed before the decline access was completed. The drill program was designed to validate the 2013 geological model around the initial levels of development (4875 mRL and 4900 mRL levels). Results obtained from the infill drill program broadly correlated with the modelled geology, noting several areas where drill results did not correlate. High-grade copper intersections were reported from several massive sulfide lenses (Aeris Resources, 2023c, 2023d), including:

- ATEL001 10.7 m @ 5.48 per cent Cu, 0.30 g/t Au and 9.7 g/t Ag (7.8 m (true thickness))
- ATEL001 2.6 m @ 2.94 per cent Cu, 1.66 g/t Au and 68 g/t Ag (1.9 m^5)
- ATEL003 10.9 m @ 5.83 per cent Cu, 1.03 g/t Au and 17 g/t Ag (7.6 m^5)

Initial level development commenced on the 4875 mRL level. The geology exposed along the access cross-cut and initial ore drives differed from what was expected. The mineralisation in the 2013 geology models based on a nominal 50 m × 50 m drill spacing was classified as an Indicated Mineral Resource. The geological discrepancies between the modelled geology and underground exposures were significant, exceeding differences expected from an Indicated classification.

The increased structural complexity of the sulfide lenses and orientation changes necessitated further drilling to de-risk the initial production front. A close-spaced grade control program was completed at a nominal 15 m × 15 m drill spacing. Two drill rigs positioned in stub drives off the initial access drilled out a 60 m vertical area centred around the 4875 mRL level (25 holes totalling 1711 m). Given the time critical nature of the drill program, the Geoscience team focused on updating geology models quickly once holes were completed.

Significant effort was placed on generating graphical cross-section interpretations whilst the drill holes were being logged. Core logging geologists were empowered to interpret the geology from infill drill holes. In doing so, there were opportunities to adjust the drill program to target further drilling in areas with complicated geology or unresolved interpretations. The graphical interpretations were used with the level mapping to update the 2013 geology model. The model proposed that the massive sulfide mineralisation primarily formed along a metadolerite contact in the northern half of the level. In the southern half, mineralisation was more erratic, difficult to correlate and not associated with a metadolerite contact; thus, mineralisation was interpreted to be more folded and sheared in this vicinity.

An updated grade control model incorporating revised geological interpretations significantly changed the reported Mineral Resource within the revised 60 m vertical window (Figure 4 and Table 1). At the reporting cut-off grade (0.6 per cent Cu), the grade control model reported 47 per cent more copper metal (8996 t versus 6099 t) in comparison to the previous 2013 model. The increased geological information and understanding resulted in improved sulfide lens wireframes.

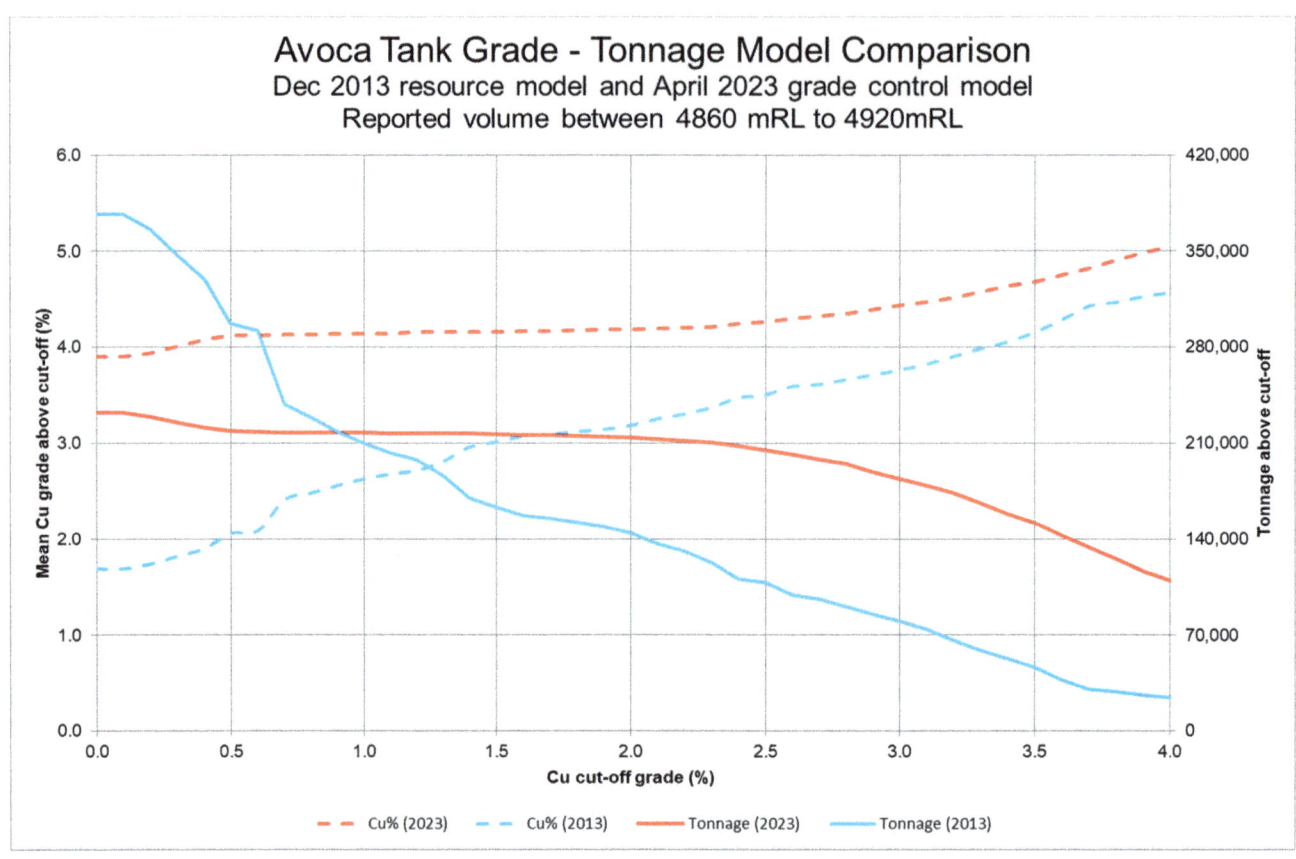

FIG 4 – Grade-tonnage curve comparison between the Avoca Tank December 2013 resource model and the April 2023 grade control model. Reported between the 4860 mRL to 4920 mRL window.

TABLE 1

Reported tonnes, copper grade and contained copper metal above cut-off grade in the December 2013 resource model and April 2023 grade control model.

Cut-off grade (%Cu)	Dec 2013 Resource Model			April 2023 GC Model			% Differences		
	Cu Grade (%)	Tonnage (kt)	Contained Cu metal (kt)	Cu Grade (%)	Tonnage (kt)	Contained Cu metal (kt)	Cu Grade	Tonnage	Contained metal
0.0	1.69	377.0	6.4	3.90	232.1	9.0	131	-38	42
0.6	2.09	291.8	6.1	4.12	218.3	9.0	97	-25	47
1.0	2.63	209.7	5.5	4.14	217.7	9.0	57	4	63
1.5	3.01	163.6	4.9	4.15	216.6	9.0	38	32	83
2.0	3.18	144.5	4.6	4.18	214.0	8.9	31	48	95

It was clear from the early level development and infill drilling that the pre-existing resource definition drilling at a nominal 50 m × 50 m spacing was insufficient to accurately locate the copper sulfide lenses for accurate mine planning purposes. Based on experience at other mined deposits at the Tritton Operation a similar drill spacing would be adequate for detailed mine planning purposes. Geological reinterpretations led to material changes to the planned level development and stope design that resulted in a material change to the reported copper grade and copper tonnes per mine level.

The Geology Department recognised the need to adjust embedded site systems and procedures to update geology models quicker in response to a poorly understood mineral system and evolving knowledge base. The key changes that have been implemented are summarised as:

Upskilling:

- Aeris Resources Principal Structural Geologist provided training and ongoing support, upskilling and empowering the core logging and mine geologists to identify and measure key structural features.

Collaboration:

- Mine site and Brisbane based geologists participated in determining the appropriate geological features used to characterise each sulfide lens.
- Regular information sharing sessions considered the impacts of new data on the existing geological interpretation.

Process improvements:

- Graphic logs:
  - First-pass geological interpretation was completed in the core shed by the core logging geologists.
  - Complex geological areas were quickly identified and flagged for follow-up work (mapping and drilling).
- Orientated drill core:
  - This was mandatory to maximise the collection of structural features from the drill core and assist with interpreting lithology, mineralisation contacts and key structural features.
- 3D Modelling:
  - Implementing new geological modelling software (Vulcan Geology Core) meant wireframes could be updated quickly as new information became available.
- Drill targeting:
  - Geologists reviewed and changed drill programs as new data was captured.
- Grade estimation:
  - Implementation of a grade control system enabled the site team to update grade estimates promptly.

CONCLUSION

The Tritton Operation has a strong history of discovering and mining multiple copper-rich base metal deposits. The deposits have been predictable with manageable geological variance between each data collection stage, ie resource definition drilling → grade control drilling → level development. The Avoca Tank deposit has been an exception. The complex and unpredictable nature of the mineralised system at Avoca Tank has required the Tritton geology team to adapt, upskill and design an efficient and practical process, reducing the time between data collection and updating geology models for mine planning and production purposes.

REFERENCES

Aeris Resources, 2014. Straits Resources Limited, Tritton Operations: Updated Mineral Resource and Ore Reserve Estimate, ASX/media release, 9 April 2014. Available from: <https://clients3.weblink.com.au/pdf/SRQ/01507961.pdf>

Aeris Resources, 2023a. Group Mineral Resource and Ore Reserve Statement, ASX/media release, 18 April 2023. Available from: <https://clients3.weblink.com.au/pdf/AIS/02655216.pdf>

Aeris Resources, 2023b. Avoca Tank Mineral Resource Update, ASX/media release, 25 October 2023. Available from: <https://clients3.weblink.com.au/pdf/AIS/02729821.pdf>

Aeris Resources, 2023c. High-grade copper and gold intersected at Avoca Tank, ASX/media release, 10 January 2023. Available from: <https://clients3.weblink.com.au/pdf/AIS/02619771.pdf>

Aeris Resources, 2023d. Tritton Operations Update, ASX/media release, 30 May 2023. Available from: <https://clients3.weblink.com.au/pdf/AIS/02670778.pdf>

Burton, G R, 2011. Interpretation of whole rock geochemical data for samples of mafic schists from the Tritton area, central New South Wales, Geological Survey of New South Wales Report GS2012/0264, Maitland.

Burton, G R, Trigg, S J and Campbell, L M, 2012. Sussex and Byrock 1:100 000 geological sheets 8135 and 8136, Explanatory notes, Geological Survey of New South Wales.

Gilmore, P J, Trigg, S J and Campbell, L M, 2018. Coolabah 1:100 000 geological sheet 8235, first edition, Explanatory notes, Geological Survey of New South Wales.

Simpson, B, Fitzherbert, J, Moltzen, J, Baillie, I, Cox, B and Huang, H, 2023. Magnetite trace element characteristics and their use as a proximity indicator to the Avoca Tank Cu-Au prospect, Girilambone copper province, New South Wales, Australia, *Mineralium Deposita*.

Leveraging geological logging as pathfinders and proxies in machine learning resource and geometallurgical modelling – benefits, limitations, and pre-screening techniques

D M First[1], D Mogilny[2], Y H Rajcoomar[3] and S Vedrik[4]

1. Chief Geologist, Stratum, Brisbane Qld 4075. Email: david@stratum.ai
2. CTO, Stratum, Toronto ON L4C 0Y3, Canada. Email: daniel@stratum.ai
3. ML Engineer, Stratum, Toronto ON L4C 0Y3, Canada. Email: yush@stratum.ai
4. ML Engineer, Stratum, Toronto ON L4C 0Y3, Canada. Email: sam@stratum.ai

ABSTRACT

Machine learning models are a class of algorithms that learn patterns in data without being given explicit instructions, typically by learning from historical data. These models are creating value in all facets of the mining industry, from exploration to production. The authors provide a high-level overview of machine learning (ML) and deep learning (DL). They discuss feature engineering, a key concept in ML that refers to identifying what data to use as input and the best way to encode such data.

While powerful, ML and in particular DL are not without challenges. Applying ML systems to geological modelling in mining, be it mineral resource, geometallurgical or geotechnical, requires quantitative decisions to be made as to which data should be incorporated into the system. In ML, this is referred to as feature engineering. The naïve answer is to use all the available data. However, this is unfeasible for several reasons; finite computer resources, the concept of overfitting, managing the signal-to-noise (SNR) ratio, false trends due to assay bias and the subjective nature of logging are significant detriments to modelling efficacy.

The authors present and test a pathfinder screening algorithm (PSA) that can be used to identify which logging data is useful for ML modelling. They have applied the technology on three different deposit types: (1) gold modelling at a Western Australian orogenic lode gold deposit; (2) copper modelling at a Chilean manto-type iron oxide copper gold deposit; and (3) sulfur modelling at a Papua New Guinean low sulfidation epithermal gold deposit. In addition, the authors also test different methods of integrating geological logging into machine learning, quantify accuracy improvements, and establish a set of best practices.

INTRODUCTION

Objective

- To create a method of reliably feature engineering geological logging for mine site spatial modelling (mineral resource, geometallurgical and geotechnical). Feature engineering refers to the selection criteria and techniques needed to incorporate data into machine learning.

- To evaluate whether DL models that use geological logging are more accurate than models based entirely on geochemistry and/or kriging models (based on categorical indicator kriging (CIK) and/or mineralised zone (MZ) domaining).

- To create a computationally-efficient method for screening which geological logs are useful for DL modelling prior to model training.

The three deposits where the pathfinder screening algorithm (PSA) system is applied are Northern Star's Jundee gold mine, an orogenic lode gold deposit located at the northern end of the Yandal belt/gold province, Western Australia; SCM Atacama Kozan's copper mine, a manto-type iron oxide copper gold deposit located in the Candelaria – Punta del Cobre district, Region III, Chile and St Barbara's Simberi gold mine, a low-sulfidation epithermal gold deposit.

BACKGROUND

Machine learning and deep learning

Recently, machine learning (ML) has emerged as a powerful tool for revealing complex patterns in data. At its core, ML algorithms learn from historical data to better forecast a future pattern or trend. Although ML and more generally, AI is defined as such above, DL is the term used for one of the most powerful ML algorithms; it uses multiple layers of artificial neurons that are composited into a deep neural network (ie convolutional neural network (CNN)) (O'Shea and Nash, 2015). The concept is loosely modelled on the way neuroscientists believe the brain behaves when it identifies patterns in very large data sets. DL has seen much success in the field of image recognition (eg medical imaging) as well as machine translation, the AI process of automatically translating text from one language to another (Goodfellow, Bengio and Courville, 2016).

ML and DL's inherent advantage over traditional kriging methodologies is its unique ability to leverage non-linear correlation trends, its capacity to model geological logging data and identify high quality data sources by learning from historical data. Geological logging data can be defined as any type of data that is collected through visual inspection of samples, done typically with drill core. While geological logs are inherently qualitative in nature when compared with assays, it nevertheless has significant value as a data set. The biggest advantage of geological logging is its cost-efficient acquisition. However, the major challenges of working with geological logs in spatial modelling include identifying how to best leverage the data while considering its qualitative limitations, interpretation bias (eg a deposit logged by multiple geologists with evolving interpretations) and a relatively low SNR ratio (given that some of the parameters logged may be irrelevant to spatial modelling).

The DL algorithm and nomenclature referenced in this paper by First *et al* (2023) is repeated in this paper. For example, DRC(Au, MZ) ~ DR references a gold resource model that uses diamond drill holes (D), RC drill holes (R), rock-chip (C) assays as inputs into the model, Au assays and MZ geological logging data as a separate input channels and DR (on the righthand side) as drill holes (D) and RC drill holes (R) as a proxy for ground truth or the 'correct' answer, by which the DL model learns the spatial distribution of gold.

The primary advantage of subdividing the inputs from the ground-truth readings is to manage lower quality assays, such as rock-chip samples, which may be useful in spatial modelling but cannot be relied on greatly to teach the model the 'correct' answer. In other words, rock-chipping 10 g/t Au from a block may be useful information for a machine learning model to understand the spatial distribution of gold; however, that sample cannot be used to denote the entire 3 m × 3 m × 3 m smallest mining unit (SMU) as 10 g/t Au given that the chip samples could be collected inside of a 50 cm vein, a highly bias form of data.

The ML algorithms used in this paper are written in Python, a programming language that distinguishes itself from other programming languages with its flexibility, simplicity and large number of available open-source tools required to create modern software, including machine learning algorithms. Python helps software engineers focus on solving logical problems rather than spending time on the basics of the programming language. This is one of the primary reasons that Python is the language of choice for machine learning and data science in general. PyTorch is the ML library that houses the open-source tools used to construct neural network layers. These neural network layers are paired with CUDA (Compute Unified Device Architecture), a computing platform developed by NVIDIA to interact with Graphics Processing Units (GPUs). NVIDIA is a technology company that designs and manufactures GPUs.

Deep learning limitations

While powerful, DL models are not without limitations. A DL model is inherently error-resistant to a certain level of noise within data but are not totally immune. Unfortunately, most geological logging data is noise from the perspective of its usefulness in resource estimation. Most geological logs carry limited value for copper or gold grade modelling. However, a recently developed method, ablation analysis, has been found to be invaluable when selecting which geological data channels are productive inputs into DL models (Meyes *et al*, 2019).

The ablation analysis method individually runs all potential input channels to identify which ones add value to an estimation. For example, if the objective is to build a gold resource model that is enhanced using geological logging, ablation analysis will produce recommendations akin to D(Au, X) ~ D, where every X is a unique geological logging code, whether it be a lithology, alteration or geotechnical code.

However, many particularly larger deposits, such as the Jundee Mine, have a penchant for numerous unique lithological and alteration logging codes, making the analytical process cumbersome and computationally impractical. The techniques to identify which input channels to use and the most efficient way to encode them, collectively feature engineering, is explained in Methods.

Geological logging limitations

Geological logging has two major applications when being applied to spatial modelling at mine sites: (1) a *pathfinder* for mine site exploration and resource estimation; and (2) as a *proxy* for a geochemical assay or mineralogical test.

Its usefulness as a pathfinder in the discovery of new and/or missed ore is evident when drill core is logged as being barren or poorly mineralised but contains a geological logging code that is directly or indirectly indicative of high-grade mineralised zones nearby. A common example discussed below is mineralised zone (MZ) logging.

Geological logs can be used as proxies for assays, alteration mineralogy, rock competency, etc. Most mines only prescribe a full assay suite (ICP-MS), detailed mineralogy/petrology or rock strength tests on a select few mine site samples (as it would be prohibitively expensive to collect thorough assay suites for all inputs used in a resource model).

Mines, particularly underground operations, have significant budgetary constraints on data collection expenditure, therefore the number of samples assayed, especially from third party laboratories, are restricted and/or relegated to mine site laboratory, apart from a few confirmation assays. The total metreage of core assayed is also restricted, often resulting in weak to moderately mineralised core not being assayed at all if it visually appears to not host economic mineralisation, thereby directly impacting modelling accuracy.

However, fortuitously geological logging can serve as a proxy for a geochemical assay and alteration mineralogy. While out of scope for this paper, geotechnical parameters such as RQD, have the potential to be a proxy for rock competency that cannot easily be measured in a laboratory.

Geological logging is already used in spatial modelling in the mining industry in two major ways: categorical indicator kriging (CIK) and MZ domaining (Glacken and Blackney, 2022). In categorical indicator kriging, categorical logs are used to encode whether a sample is oxide, transitional or sulfide and is represented by integer values when applying kriging to the data; for example, oxide ore is represented by 0, transitional ore by 0.5, sulfide ore by 1. The final estimates are rounded based on the mine's error tolerance in each class. Some mines may vary the transition ore estimate (eg 0.7 considered to be sulfide) depending on processing constraints.

Domaining gold with geological logs has some unique challenges, particularly with respect to nuggety gold deposits. Many orogenic and intrusion related gold deposits, geologically log MZ and quartz veins. Essentially, if the core from the geologist's perspective visually looks like it is potentially mineralised, it is logged as such, irrespective of the gold assay collected later. An issue of note is that MZ is often used as a descriptor for many of the pre-mineral lithologies, resulting in extra lithological codes.

The logging data used to assist in the construction of domains by constraining the mineralised zones of these nuggety deposits, thereby critical to the mineral resource estimation process as there is extreme variation in gold distribution within a small volume. Frequently within these orebodies, unrepresentative (barren or subeconomic) rock-chip or drill core assays can be sampled in very close proximity to well-mineralised drill core; as such, RC chip or rock-chip samples make grade estimation of ore blocks very challenging.

While geologists are proficient at handling the non-linear nature of geological logging, there is substantial risk, as the resource model becomes beholden to the subjective interpretation of the geological logging team and potentially an overreliance on categorical indicator kriging and/or mineralised zone domaining (Glacken, Rondon and Levett, 2023; Sims, 2023).

While these two methods demonstrate that geological logging has inherent value in spatial estimation, they are limited in its usefulness due to kriging's inherent linear interpolation-based algorithm. The two methods are also incapable of accurately modelling mixed data types, like unassayed core where it could be interpreted as either barren or weakly mineralised even though it visually appears barren. There are non-linear geostatistical methods that have been applied, like multiple indicator kriging and localised uniform conditioning; however, they have proved challenging to implement (Zhang and Glacken, 2023).

As discussed, geological logs are critical when manually domaining an orebody, but have proved imperfect when modelling due to the subjective nature in the logging process. There is a natural tendency for geologists to record too much detail, effectively subdividing or splitting channels, instead of looking holistically for commonality within the data such that productive modelling inputs are derived by lumping channels together.

This results in situations where a large component of the 'signal' is lost such that the detailed information is not incorporated into the kriging model, even though there may be good MZ logging of an orogenic or intrusion related gold deposit. To circumvent this issue, many mines create ever smaller domains with the aim of capturing the geological complexities of the deposit. This results perversely in the domains guiding mine planning and mine site exploration, rather than kriging estimation.

A naïve solution would be to undertake an ablation analysis on the ten most common logging codes. Regrettably, the most common codes are not necessarily the most useful ones for resource modelling as the economic mineral resources is invariably restricted to anomalous geologically zones. Therefore, it has become necessary to derive a method that can screen geological logs with relatively high degree of accuracy for their usefulness in spatially estimating a parameter (eg gold or copper value).

METHOD

Machine learning models are trained using geological logs at three deposits to test the applications of proxy and pathfinder logging under different geological environments. This protocol derives a general solution with wide applicability for feature engineering of geological logging.

Jundee Mine – introduction

The deposit is an Archean orogenic lode gold deposit hosted within the northern Yandal greenstone belt, Yilgarn Craton (Vearncombe and Elias, 2017). The host rocks are dominated by a succession of diabase sills and tholeiitic basalt units. The mine sequence is composed of two basalt units separated by a sediment formation intruded by the sills. Locally the entire mine sequence is intruded by dacite, granodioritic porphyries and lamprophyres (Phillips, Vearncombe and Eshuys, 1998; Phillips, Vearncombe and Murphy, 1998; Smith, Haese and Grigson, 2017).

Gold is deposited and controlled by brittle-ductile shears within an array of transverse or oblique faults, primarily developed within the diabase and tholeiitic and to a lesser extent the felsic porphyries. The major oblique structures typically contain fault gouge with variable displacement up to 120 m with the granodioritic porphyries preferentially exploiting these faults. The faults and shear zones locally have variable orientation and dip (vertical to 20° in either direction) as conjugate subparallel shears, resulting in mineralised zones hundreds of metres wide.

The mineralisation is hosted in numerous discrete high-grade tabular, narrow and discontinuous fault-shear zones. High-grade mineralised zones typically range from <0.5 to 1.5 m and locally can be up to 5 m wide, with a vertical and lateral extent up to 500 m and 1000 m respectively, although within these zones, individual mineralised structures are highly discontinuous.

The host alteration is characterised by quartz-sericite-ankerite-sulfide (pyrite ± arsenopyrite ± chalcopyrite) assemblage. Gold mineralisation has a heterogenous distribution pattern, invariably restricted to the mineralised structure, with the local country rock barren. Course visible gold is a common occurrence, although the relative distribution of visible gold between areas is quite variable, resulting in a pronounced nuggety grade distribution pattern with grades ranging up to several thousand g/t Au.

At the Jundee Mine, the mineral resource team are in part tasked with identifying and predicting the direction of the relatively narrow high-grade veins for mine planning ore control, ore reserve estimation, which has proved challenging. It has been particularly challenging for the team to rely on the resource model to guide underground drilling or for mine planning in areas where underground drill spacing >20 m.

Jundee Mine – methodology

Gold mineral resource modelling at the Jundee Mine with lithology logging is essentially a pathfinder logging application; it is critical to identify which lithological logs are pathfinders to high-grade veins and allow the DL to leverage those logs to better identify (new and missed) areas of mineralisation.

As stated above, it is impracticable to test every potential lithological log channel via DL for its ability to enhance the gold model accuracy. To identify pathfinder logs, there are three potential options:

1. Substantially simplify the logging by consolidating logs based on inherent commonality to reduce the number of categories/channels.
2. Leverage the institutional knowledge of the mine site geologists for known or suspected geological pathfinders.
3. Create a fast and accurate screen for geological pathfinders to determine the primary candidates for incorporation as inputs into DL models.

The principal lithology and/or alteration logging codes used by the Jundee Mine geologist are found in Table 1. Miscellaneous codes referencing minor and post-mineral lithologies, geotechnical parameters etc, were excluded as their association with the spatial distribution of gold mineralisation was believed to be absent or at best tenuous. More than 200 lithological related codes are currently in use at the Jundee Mine.

TABLE 1

Jundee Mine – 28 lithological logging codes with a frequency >1000.

Code	Description	Code	Description
Abt	Tholeiitic basalt	**Aog**	Gabbro
AbtMZ	Mineralised basalt	Ash	Shale
Abv	Undifferentiated basalt	Ashb	Carbonaceous shale
Ach	Chert	**Asp**	Sandstone
Agc	Dacitic porphyry	**Ast**	Siltstone
AgcMZ	Mineralised dacitic porphyry	Asu	Undifferentiated sediment
Agd	Granodiorite	**AsuMZ**	Mineralised sediment
Agg	Granite	Asx	Polymictic breccia
AggMZ	Granite MZ	**Auu**	Undifferentiated ultramafic
Agp + Aip	Porphyry	Avn	Quartz – carbonate
Alp + Alu	LAMP Lamprophyres (undifferentiated)	**MZ**	Mineralised zone
Aod	Dolerite	**Pod**	Proterozoic dolerite
Aodm	Dolerite (magnetic)	**QV**	Quartz vein
AodMZ	Mineralised dolerite		

Bold geological codes have an elevated signal with the barren pathfinder screening algorithm (PSA).

The first potential solution was to simplify the logs based on geological commonality, referred to as 'LITHO'. This involves grouping codes based on general rock type; felsic volcanic/intrusive, mafic/ultramafic, sedimentary, etc. It is believed this method may be effective with disseminated deposits, but carries substantial information loss with orogenic lode gold deposits.

The second potential solution was to leverage known geological pathfinders. The best two lithological pathfinders at the Jundee Mine are suffixed by 'MZ' (mineralised zone) and the lamprophyre code 'LAMP'. As discussed earlier, MZ is the suffix on a lithological code that the geologist logs when they believe the rock to be mineralised, irrespective of actual gold grade. The concept is that the geologist believes the core could be auriferous based on its visual appearance (primarily alteration and sulfide mineralisation), but may be barren or weakly auriferous due the nuggety grade distribution.

To ensure the DL model captures all permutations, samples are assigned a 1 if a lithology log has an 'MZ' suffix, 0.5 if the 'MZ' suffix is absent but this lithology can have a 'MZ' suffix in other core samples and 0 if 'MZ' is absent from a lithology log and that lithology is never logged with 'MZ' (such as lamprophyres, and other post-mineral intrusives emplaced after the primary mineralising event).

The third solution is to create a statistical screen for the geological pathfinders. A novel pathfinder algorithm is developed that is described and referenced below as a Pathfinder Signal Analysis (PAS).

1. Remove samples with lithology logging code than occur less than a predetermined minimum frequency and cap individual samples at 50 g/t Au.

2. Match each drill hole sample to its assay (Au_{self}) and its lithological log (l).

3. For each drill hole sample (no composite) find the highest grade within 10 m denoted $Au_{max\,(10m)}$ excluding the sample itself.

4. Group samples together based on their (Au_{self}, L) and calculate the average $P(Au_{max(10m)}|L = l, Au_{self} = b)$ for each lithology l. To group Au_{self}, group based on grade bucket b that contains Au_{self} as barren (<0.1 g/t Au), low-grade (0.1–0.5 g/t Au), mid-grade (0.5–2.2 g/t Au) and HG (>2.2 g/t Au). For context, the cut-off grade across an SMU is 2.2 g/t Au.

5. Calculate the average $P(Au_{max(10m)}|Au_{self} = b)$ maximum nearby gold grade for a sample in each of the four grade buckets.

6. Calculate signal strength $S(L = l, Au_{self} = b)$ as $P(Au_{max(10m)}|L = l, Au_{self} = b)/P(Au_{max(10m)}|Au_{self} = b)$ for each grade bucket b, lithological log l. Rank lithology logs for each grade bucket based on the signal strength.

The most important pathfinders are those that indicate nearby high-grade mineralisation while themselves being barren (<0.1 g/t Au) or low-grade (0.1–0.5 g/t Au). The logic is that a DL model does not need to rely on lithology logging if the drill core is already HG because the assay itself will be the HG gold signal. The signal strength refers to the ratio between the average maximum gold grade found nearby with a specific lithology log and the gold assay $P(Au_{max(10m)}|L = l, Au_{self} = b)$ relative to the nearby maximum gold grade found nearby as expected by purely the gold assay irrespective of the lithology log $P(Au_{max(10m)}|Au_{self} = b)$. The top pathfinders are reviewed in results section.

After selecting a short list of potential pathfinders (LITHO, MZ, LAMP) as well as pathfinders discovered by the PSA, a series of DL models were created with slightly different inputs to evaluate the usefulness of the respective input channels. Each DL model has access to the gold channel from drill hole, RC drill hole, chip data and a single lithology channel. The goal is to test which lithology channels are most useful when logging the core. Pathfinders discovered by PSA have access to the MZ channel, should the proposed pathfinder code co-exist with a MZ suffix.

Atacama Kozan Mine – introduction

The mine is located about 16 km SSE of Copiapo, Region III Chile. The deposit lies at 300–400 m below the surface and does not outcrop. Exploration commenced in 1990. Mine construction was completed in 2003 based on a mineral resource of 30 Mt (recoverable 20 Mt) at 1.5 per cent Cu (Ichii et al, 2007).

The district, is characterised by an early-Cretaceous volcanic-sedimentary arc sequence with mineralisation hosted primarily in the upper part of the Lower Andesite member Punta del Cobre Formation, which is overlain by volcano-sedimentary and dacite members. This is overlain by the marine-sedimentary Chañarcillo Group. To the west the Copiapó batholith (diorite to quartz monzonite) was emplaced during a period of regional tectonic reversal from extensional to transpressional. Geochronological studies infer that the main phase of mineralisation overlaps with the two major early phases of the Copiapó batholith emplacement, although there is no conclusive evidence to indicate from the exposed phases of the batholith that it was the source of mineralising fluids (del Real, Thompson and Carriedo, 2018).

The orebodies are mineralised with magnetite, chalcopyrite, and pyrite, with lesser pyrrhotite and sphalerite as veinlets and disseminations (locally semi-massive sulfide bodies) and is hosted within highly altered favourable lithological units, fault zones and breccias. The stratigraphically controlled replacement mineralisation forms extensive stratabound orebodies that are locally termed 'mantos'. Textural studies indicate that the hydrothermal system evolved and progressed outwards and upwards from sub-vertical feeder structures as the replacement occurred.

A distinctive early sodic-calcic alteration (actinolite, albite, scapolite, epidote) characterises the district, which is locally overprinted by potassic ± calcic alteration (actinolite–biotite (green–high Mg?) –K-feldspar) alteration associated with the manto mineralisation (Ichii *et al*, 2007).

Atacama Kozan Mine – methodology

Copper resource modelling at Atacama Kozan integrated with lithology logging is another example of a proxy logging application. Roughly half the Atacama Kozan drill hole data set is visually deemed to be barren and remains unassayed for copper or any other element. However, irrespective as to whether the core remains unassayed, it cannot be assumed to be barren (~0.0 per cent Cu) from a modelling perspective. Although the underground mine has a relatively high cut-off grade when compared to an open pit mine it is critical to note that a weakly mineralised (eg 0.2 per cent Cu) assays are fundamentally different from barren (0.05 per cent Cu) assays, as the former sample may indicate mineralisation in close proximity, whereas the latter is likely to have little significance and be indicative of a barren zone.

Two solutions are proposed to resolve the issue using a similar concept to that trialled at the Simberi Mine (detailed in the next section):

1. D(Cu, ZFCU) ~ D(Cu), utilise an independent input channel. Rather than assuming that unassayed core can be assigned a 0.0 per cent Cu value, use the ZFCU (zero filled copper) as an extra channel into the model to indicate material that has been visually logged to be barren but is unassayed.

2. D(Cu, ZFCU) ~ D(Cu, ZFCU), utilise both as an independent input channel and as a measure of ground truth. In addition to approach 1, for samples that are unassayed and logged as barren, assign 0.0 per cent Cu and use it to teach the model the correct answer for the copper grade for a certain block.

It is impossible to sample the true copper distribution for unassayed core without additional data collection (assaying) and it is improbable the mine will assay significant quantities of core previously logged as barren or weakly mineralised and excluded from their mineral resource model.

Three DL models are created: (1) using available Cu assays only; (2) using the Cu assays and unassayed drill core (ZFCU) as an independent input; and (3) using the second method as input but with a measure of ground truth for areas that are geologically logged as barren (ie unassayed).

Figure 1 illustrates it is not easy to visually differentiate barren core (<0.05 per cent Cu) from low-grade (weakly mineralised) core (0.05–0.25 per cent Cu). Visual observations indicate that assigning a 0.0 per cent Cu grade to unassayed core carries a substantial risk to the DL model as the algorithm is likely to determine an area is barren to economic mineralisation due to the predominance of unassayed core.

FIG 1 – Visual inspection of low-grade material from Atacama Kozan.

Simberi Mine – introduction

Simberi Island, New Ireland Province, north-east Papua New Guinea is part of the Tabar island group and geologically lies at the north-west extension of the Tabar–Lihir–Tanga–Feni (TLTF) a Pliocene–Holocene volcanic island chain (younging to the south-east). This island chain is renowned for hosting the world-class low-sulfidation epithermal gold mine on Lihir, the Simberi gold mine and significant sub-economic gold mineralisation on Tabar and Ambitle (Feni) islands. The mineralisation is believed to coeval with magmatism and hosted in porphyritic high-K calc-alkaline (transitional to shoshonites) trachytic and monzodioritic volcanics and hypabyssal intrusives. There is no agreed consensus on the tectonic and petrogenetic evolution of the island arc, with several hypothesis postulated (Ponyalou, Petterson and Espi, 2023).

The open pit mine commenced production in February 2008 based on a mineral resource 4.7 Moz Au (measure, indicated and inferred) (Allied Gold Limited, 2009), subdivided into oxide (1.28 Moz Au), transitional (0.145 Moz Au) and sulfide (3.285 Moz Au) from several orebodies, the primary economic ones being Sorowar, Pigiput and Pigibo. The mineralisation is controlled and emplaced as irregularly distributed breccia zones within altered volcanics. Gold mineralisation occurs as an irregular <50 m thick oxide cap, which is frequently located at higher points within the topography and invariably overlies a refractory sulfide (predominantly pyrite and arsenian pyrite). Typically, the grade of the mineralisation correlates well with the degree of host rock fracturing.

Within the auriferous oxide zones gold is associated with arsenic and antimony, whereas silver is invariably depleted together with base metals, due to tropical weathering (leaching). The auriferous sulfide mineralisation is associated with disseminated pyrite mineralisation. Earlier petrographic studies reportedly indicate there are two pyrite dominated mineralisation events, an early, pervasive, pre-auriferous event and a later refractory auriferous arsenian event hosted by pyrite and marcasite and lesser arsenopyrite at depth (Porter, 2008; Godfrey, Battista and Hearse, 2011).

Simberi Mine – methodology

Sulfur geometallurgical modelling at the Simberi Mine utilising geological logging is a proxy logging application. Sulfur is the strongest indicator of gold recovery at the mine as it correlates with degree of oxidation of the sulfide mineralisation. The challenge is determining the gold recovery of ore prior to processing.

Unfortunately, most grade control RC drill holes are not assayed for sulfur due to inconsistent availability of LECO sulfur assays. In its place, visual logging of the ore oxidation (weathering) state is undertaken; ie orange–red for oxide ore, red–grey or white for transitional ore, grey for sulfide ore). Therefore, the goal is to leverage oxide logging (OX) as a substitute for sulfur assay in RC drill holes, which are unassayed for sulfur, but have oxide logging.

To best model by proxy, the authors first needed to determine the correlation between visual logging (OX) and sulfur (S). This is critical given that drill hole data and recently collected grade control data was visually logged for oxide ore and assayed for sulfur. Therefore, this data was used to understand the relationship between S and OX. Geochemically <0.4 per cent S is considered oxide ore, 0.4–2.0 per cent S transitional ore and >2.0 per cent S sulfide ore. Figure 2 shows the sulfur grade for material visually logged as oxide, transitional and sulfide.

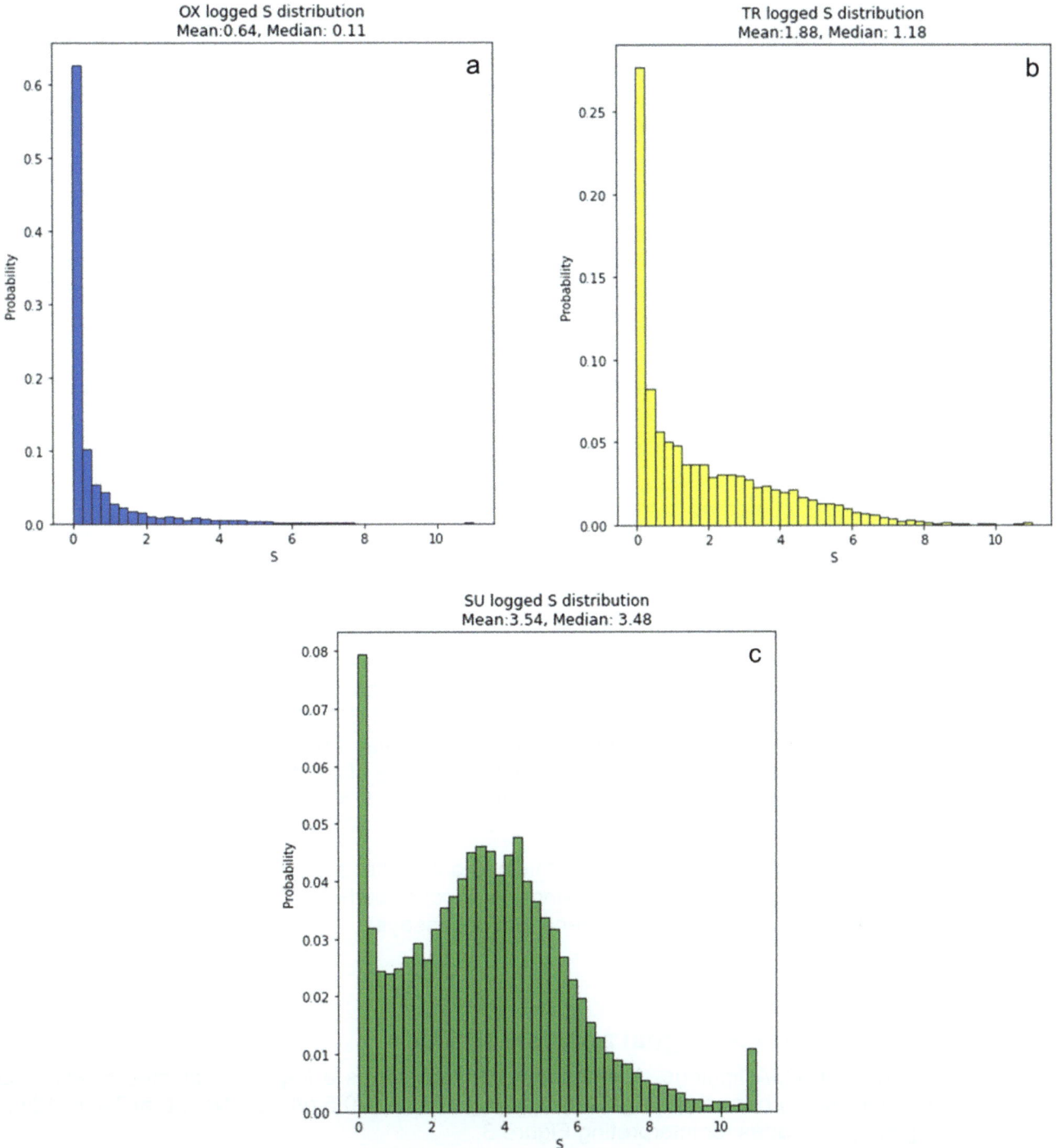

FIG 2 – Sulfur distribution in RC samples visually logged as: (a) oxide ore (OX); (b) transitional ore (TR); and (c) sulfide ore (SU).

The median S grade for material logged as $OX = OX$ is 0.11 per cent S, $OX = TR$ is 1.18 per cent S, and $OX = SU$ is 3.48 per cent S, with 69.7 per cent of material logged as OX is geochemically classified as an oxide (<0.4 per cent S), 32.9 per cent of material logged as TR is geochemically classified as an oxide (0.4–2.0 per cent S), 73.2 per cent of material logged as SU is geochemically a sulfide (>2.0 per cent S). The major issue that arises is that while OX logging is correlated with sulfur grade, it is not a perfect representation. Even if misclassification of transitional ore is ignored (ie transitional ore geochemically being logged as oxide or sulfide or sulfide ore logged as transitional), there are statistically significant occurrences of sulfide ore geochemically being logged as oxide (9.8 per cent of total oxide logs) and oxide ore geochemically logged as sulfide (10.0 per cent of total sulfide logs).

The mine suspects that visual oxide logs with sulfide geochemical signature could be due to higher than normal concentrations of oxide related sulfates, such as jarosite or gypsum. Another common inconsistency is the presence of weakly mineralised (<0.3 g/t Au) visually logged as sulfide ore (SU) but with an oxide geochemical signature, likely indicating the presence of weak to unmineralised country rock within the orebody. Overall, visual logging acts as an imperfect, but important channel with respect to sulfur distribution within the orebody.

Given the discrepancy between OX logging and S grade there are two promising options on how to encode the data:

1. D(S,OX) ~ D(S), exclusively an independent input channel. Rather than assuming that OX logging can be interchangeable and filled in for a missing sulfur value, use the OX logging as an extra channel into the model (OX=0, TR=0.5, SU=1.0) that the DL can choose to interpret as it sees fit.

2. D(S,OX) ~ D(S,OX), both as an independent input channel and a measure of ground truth. In addition to approach 1, for samples that lack sulfur assays but contain visual logging, map the visual logging to a S assay value and use it to teach the model the correct answer for the sulfur grade for a certain block.

The advantage of the first method is that data quality issues associated with visual logging can be well constrained as the model can determine how much to rely on OX logging in S estimation. The advantage of the second approach is that it increases three times the number of 'correct' examples that be given to the model to teach it to understand sulfur distribution at the mine (there is three times more grade control than exploration drill hole data).

For the second method, it is necessarily to assign a sulfur proxy value to each visual log. Given that the visual data is somewhat biased, the authors assign mapping based on the distribution in Figure 2. That is an $OX = OX$ log does not necessarily indicate 0 per cent S, but instead maps to the median sulfur grade found on samples that have both a known S assay and an $OX = OX$ log. Using this methodology, $OX = OX$ maps to S = 0.1 per cent in the absence of a sulfur assay, $OX = TR$ maps to S = 1.18 per cent in the absence of a sulfur assay and maps to S = 3.48 per cent in the absence of a sulfur assay.

Three deep learning models are then created: (1) using only sulfur assays, (2) using the first method (OX as an independent input only); and (3) using the second method (OX as an independent input and measure of ground truth in absence of geochemical assays).

RESULTS

Jundee Mine – pathfinder signal analysis (PSA)

Figure 3 illustrates the PSA findings. A pathfinder lithology log is a log that indicates nearby HG mineralisation while being barren (<0.1 g/t Au) or low-grade (0.1–0.5 g/t Au). Note that Table 1 lists the principal geological codes for interpreting Figure 3.

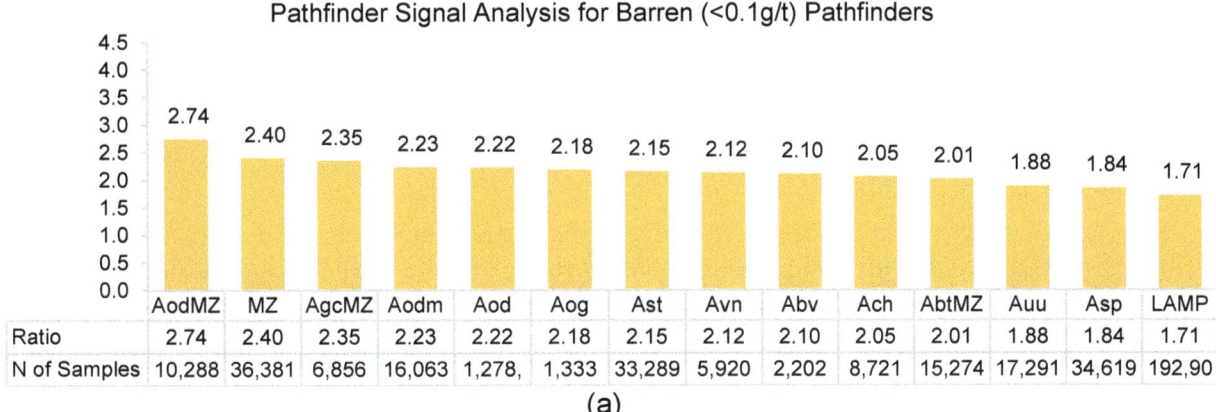

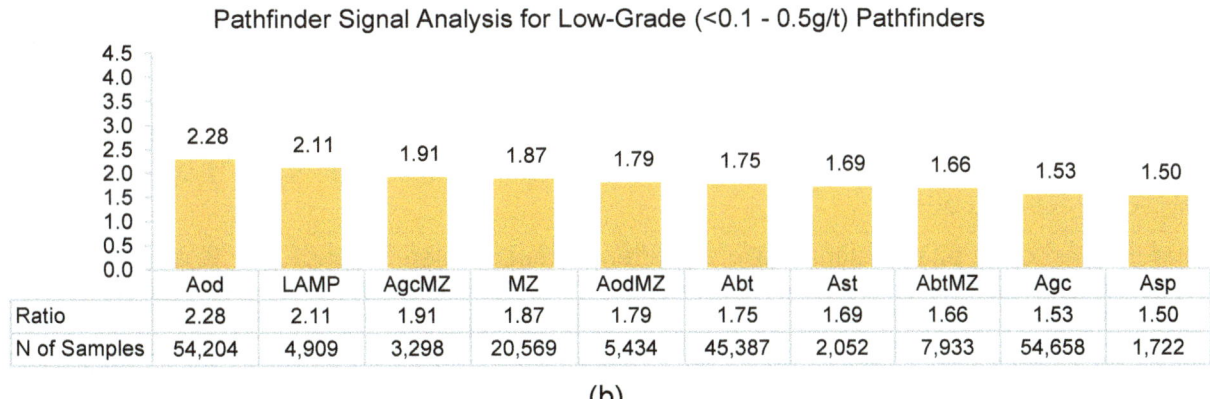

FIG 3 – (a) Pathfinder signal analysis (PSA) for barren pathfinders at the Jundee Mine.
(b) Pathfinder signal analysis (PSA) for low-grade (LG) pathfinders at the Jundee Mine.

The barren pathfinders with the strongest signal strength are AodMZ (mineralised dolerite), MZ (mineralised zone), and AgcMZ (mineralised dacite porphyry) at 2.74×, 2.40× and 2.35× signal strength respectively.

The low-grade (LG) pathfinders with the strongest signal strength are Aod (diorite), LAMP (lamprophyres), AgcMZ (mineralised dacite porphyry) at 2.28×, 2.11× and 1.91× signal strength respectively.

Note that signal strength tends to be stronger for barren pathfinders because signal strength logic is calculated relative to the base signal strength of the sample irrespective of the lithological log. For example, LG 0.1–0.5 g/t Au samples have a higher average gold assay near HG gold, than barren <0.1 g/t Au samples.

DL MODEL EVALUATION

It is necessary to create an accurate method by which to test and evaluate DL models against one another to establish the value of incorporating lithology logs for each deposit into the DL modelling process.

Overview

As Jundee and Atacama Kozan mines are underground operations it was best to evaluate the quality of each model by their forward-facing precision and recall; ie the Jundee Mine DL model is created using data collected prior to 2023 and compared against drilling conducted in 2023.

Similarly, the Atacama Kozan DL model is created using data collected prior to 2021 and compared against drilling data collected in 2021–2022. To ensure enough data was used to evaluate Atacama

Kozan DL model, two years of data collection were used for comparison instead of one. The models were evaluated based on precision and recall.

- Precision is the percentage of blocks predicted as economic high-grade (HG) that are reconciled as HG in forward-facing diamond and RC drilling. It tracks the frequency of false occurrences, as in incidences when a HG block or vein projected in the mine plan reconciles as waste.

    Precision can alternatively be interpreted as the false positive rate, denoted in Figures 4a and 5a. A model with a precision of 100 per cent reconciles HG in all blocks predicted as HG while a model with a precision of 0 per cent exclusively reconciles waste inside of HG blocks.

- Recall is the percentage of reconciled HG that is predicted as HG. It tracks the frequency of false negative occurrences, as in occurrences that veins exist, but were missed by the resource model.

    Recall can alternatively be interpreted as the missed mineralisation rate, denoted in Figures 4b and 5b. A model with a recall of 100 per cent misses no mineralisation while a model with a recall of 0 per cent does not predict any block drilled as HG.

The objective of the resource model is to model additional gold mineralisation at the Jundee Mine and copper at Atacama Kozan Mine, that can be incorporated into the resource definition drilling targeting program, without lowering the sensitivity of HG misclassification beyond a minimum threshold (ie cut-off grade).

For the Simberi open pit operation a slightly different methodological approach is required as it is an ore type classification objective. As discussed, the mine geochemically classifies <0.4 per cent S ore as oxide, 0.4–2.0 per cent S as transitional and >2.0 per cent S as sulfide.

The objective of improving the ore type DL modelling is to find additional oxide material without lowering the sensitivity of misclassifying it beyond a minimum threshold. This can be measured by tracking the oxide recall and oxide precision:

- Oxide precision is the percentage of blocks predicted as oxide (OX) that are reconciled as oxide (as measured by sulfur assay being <0.4 per cent S) in diamond and RC drilling. It tracks the frequency of false occurrences, as in instances when an oxide block or vein projected in the mine plan reconciles as transitional or sulfide. The oxide precision can be interpreted as false oxide rate, as denoted in Figure 6a.

- Oxide recall is the percentage of reconciled oxide that is predicted as oxide. It tracks the frequency of false negative occurrences, as in instances where additional oxide ore exists, yet was missed by the resource model. Oxide recall can be interpreted as missed oxide rate, as denoted in Figure 6b.

The alternative terms of missed mineralisation rate and false positive rate, rather than precision and recall are used when discussing the results as they are more commonly used by resource geologists.

Jundee Mine

Figure 4 illustrates the false positive and missed mineralisation rates, respectively, between different DL[AI] models. The 2023 kriging model is also included which is created using ordinary kriging in narrow well-constrained domains, provided by the site team. Additional DL models will be added in the final revision that reflect additional lithology logs.

The high forward-facing false positive rate in reconciliation is common in resource definition drilling of narrow vein gold deposits. This is due to the nature of narrow vein gold deposits being challenging to model due to nuggety gold distribution and geological complexity, thereby requiring intensive resource definition and grade control drilling to block-out stopes.

The DL[AI] (Au, LAMP) lithology model has the lowest missed mineralisation at 61.1 per cent when compared to kriging model at 88.5 per cent and the basic DL[AI] (Au) model at 68.5 per cent. The DL[AI] (Au, LAMP) missed mineralisation rate of 61.1 per cent indicates that 38.9 per cent of material reconciled as HG (>2.2 g/t Au) in 2023, was predicted as HG in the DL model using pre-2023 data.

Likewise, the kriging model missed mineralisation rate of 88.5 per cent indicates that 11.5 per cent of material reconciled as HG in 2023 was predicted as HG by the kriging model using pre-2023 data. This translates to a 3.38× increase in reconciled mineralisation while having a higher sensitivity (ie less false positives) than kriging. DL[AI] (Au, LAMP) and DL[AI] (Au, MZ) models also outperform the DL[AI] (Au) model, indicating the machine learning architecture is directly leveraging the LAMP and MZ lithology codes to identify areas of missed mineralisation.

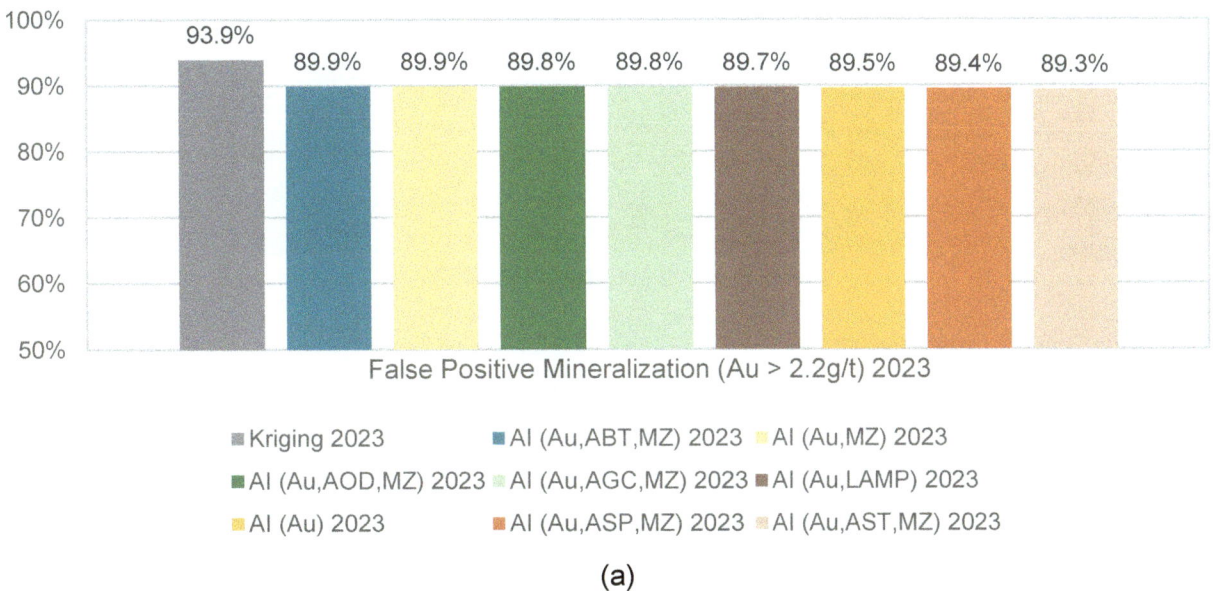

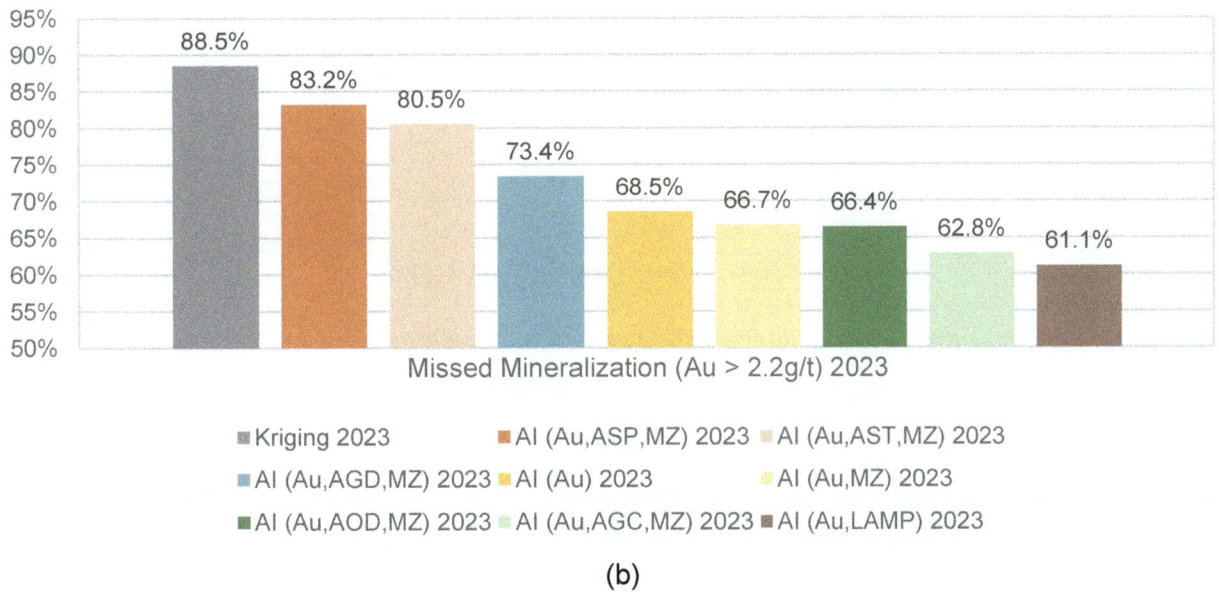

FIG 4 – (a) False positive mineralisation rate (2023 DDH) for the Jundee Mine; (b) Missed mineralisation rate (2023 DDH) for the Jundee Mine.

Atacama Kozan Mine

Figure 5 illustrates false positive and missed mineralisation rates, respectively, between different DL models. The 2021 kriging model is also included which is created using ordinary kriging in mineralisation domains.

The elevated forward-facing false positive rate in reconciliation is common in resource definition drilling of underground base metals and precious metals deposits. This is because resource definition drilling tends to be drilled in areas with less data than grade-controlled regions in underground deposits, which tend to have a higher cut-off grade. This invariably results in less HG blocks defined as a whole, when compared to waste. The false positive rate generally decreases to <50 per cent for grade control drilling.

FIG 5 – (a) False positive mineralisation 2021–2022 for the Atacama Kozan Mine; (b) missed mineralisation 2021–2022 for the Atacama Kozan Mine.

For the D(Cu, ZFCU) ~ D(Cu, ZFCU) model, it has the lowest missed mineralisation rate of 71.8 per cent compared to the kriging model rate of 83.2 per cent, whereas the D(Cu) ~ D(Cu) model is 82.8 per cent. The D(Cu, ZFCU) ~ D(Cu, ZFCU) model missed mineralisation rate of 71.8 per cent indicates that 28.2 per cent of material reconciled as HG in 2021–2022 was predicted as HG by the DL model using pre-2021 data. Similarly, the kriging model missed mineralisation rate of 83.2 per cent indicates that 16.8 per cent of material reconciled as HG in 2023 was predicted as HG by the kriging model using pre-2021 data. This translates to a 1.67× increase in reconciled mineralisation while having the equivalent sensitivity (ie false positive rate) as kriging. D(Cu, ZFCU) ~ D(Cu, ZFCU) uses assayed copper and visually inspected (ie barren versus not barren) copper as two separate channels into the model.

The model inputs use unassayed visually barren core as an example of ground truth to teach the model that visually determined barren material has a copper grade of 0.0 per cent. The D(Cu, ZFCU) ~ D(Cu, ZFCU) model has a lower missed mineralisation rate than both D(Cu) ~ D(Cu) model (which ignores all visually inspected core) and the D(Cu, ZFCU) ~ D(Cu) model(which uses logged core as an input without ground truth to verify the accuracy of the log).

Simberi Mine

Figure 6 shows the false oxide and missed oxide rates, respectively, between different DL models. The 2022 kriging model is also included which is a categorical indicator kriging model based on visual logging of OX, TR, SU where TR is halfway between OX and SU. Categorical indicator kriging (CIK) is used for the JORC resource model at the mine as it has historical reconciliation advantages over ordinary kriged sulfur estimations. The sulfur ordinary kriging estimate had a false oxide rate of 34.8 per cent with a missed oxide rate of 89.8 per cent due to its tendency to overestimate oxide ore as transitionary ore, due to smoothing.

D(S, OX) ~ D(S, OX) and D(S, OX) ~ D(S) models have a lower false oxide rate when compared to kriging, while the D(S) ~ D(S) model has a slightly higher false oxide rate. The decrease in performance D(S) ~ D(S) is likely attributed to the D(S) ~ D(S) model does not have access to the visual logging data set that both the DL and CIK kriging models can access.

All DL models have a lower false oxide rate than kriging, likely due to DL's ability to leverage historical data to interpret and extend oxidisation patterns more accurately.

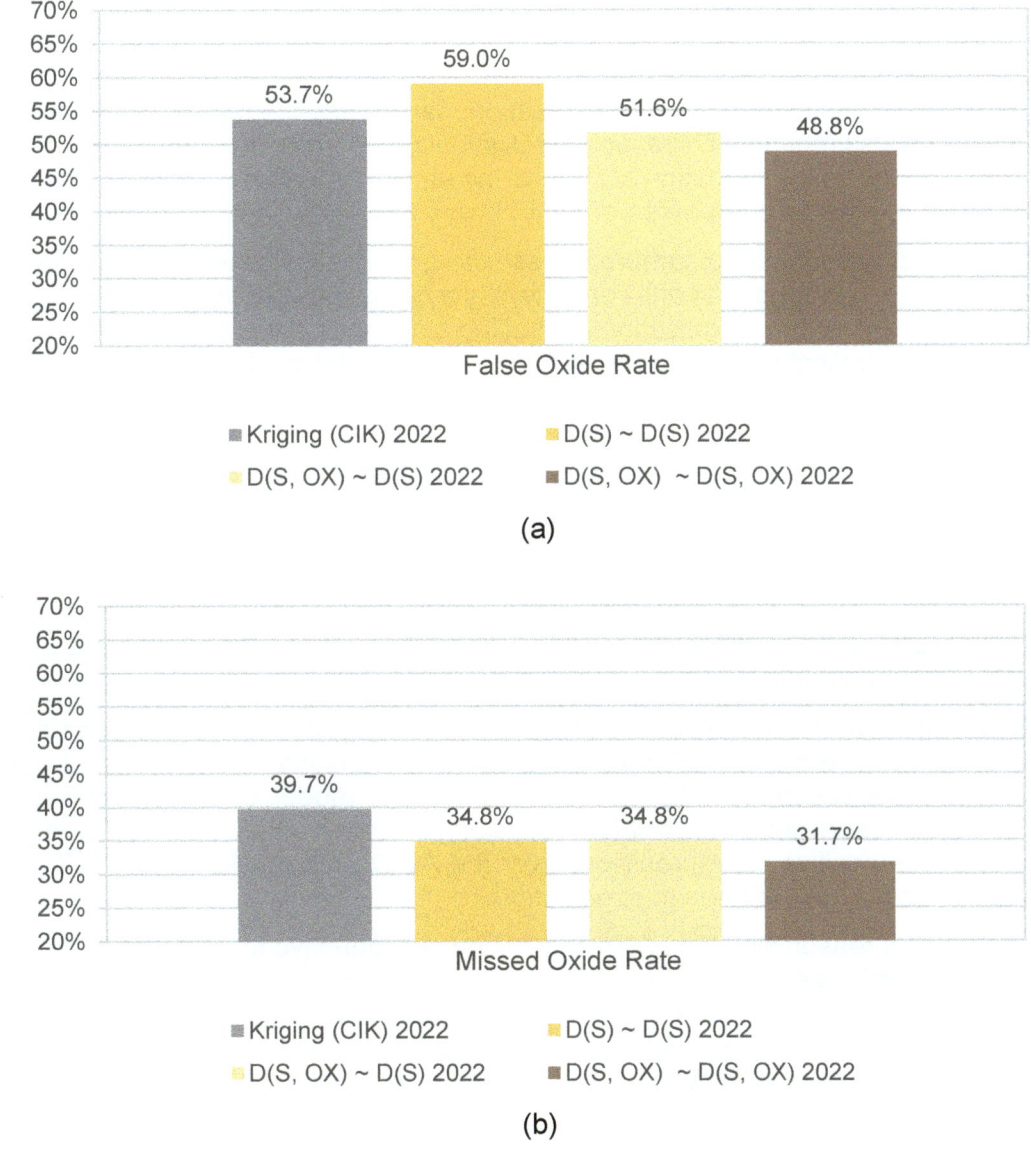

FIG 6 – (a) False oxide rate 2022 for the Simberi Mine; (b) Missed oxide rate 2022 for the Simberi Mine.

The model with the lowest false oxide rate and missed oxide rate is D(S, OX) ~ D(S, OX). This model uses the visual logging and sulfur assays as two separate data channels. It also uses visual logging as a measure of ground truth to teach the model correct answers for samples where a sulfur assay is not unavailable. This model misses 20.2 per cent ((31.7–39.7)/39.7 = 20.2 per cent) less oxide over CIK kriging while predicting 9.1 per cent ((53.7–48.8)/53.7 = 9.1 per cent) less false oxide over CIK kriging.

The second best model Is D(S) + GC(OX, S) ~ D(S), which uses the visual logging and sulfur assays as two separate data channels. However, it does not use visual logging as a measure of ground truth with which to teach the DL model.

DISCUSSION

Lamprophyres

Integrating LAMP logging as a channel into the machine learning Au resource model at the Jundee Mine created a substantial reduction in missed mineralisation rate over DL models that only use Au as an input. It is a well-known observation that lamprophyres are spatially and temporally associated with gold mineralisation, although the relationship is interpreted as indirect. Geologists have noted that lamprophyre emplacement is often contemporaneous with orogenic lode gold mineralisation and commonly occur in proximity of major crustal faults and shear zones, that controlled the hydrothermal systems that generated the gold deposits. These major structures are thought to represent the deep magmatic and fluid conduits (Müller and Groves, 2019).

At the Jundee Mine the multi-stage gold mineralisation, particularly the late-stage bonanza event, is controlled by a late stress regime with associated dilation zones. This deformation event appears to have a significant control on the formation of MZ and the emplacement of post-mineral lamprophyres. However, it is observed that lamprophyres are not always associated with the major gold lodes.

Based on this indirect association of lamprophyres with gold mineralisation, it is logical to assume that when lamprophyres are logged in drill core, there is a higher probability that they were emplaced near to or within the lode gold mineralisation controlling structures. Although not a direct pathfinder, to high-grade gold mineralisation lamprophyres (LAMP) input informs the DL model by significantly reducing the probability of missed mineralisation (recall) and, to a minor degree, false positives (precision).

Mineralised zone

Integrating mineral zone (MZ) logging as a channel into the machine learning Au resource model at the Jundee Mine created a moderate reduction in missed mineralisation rate over DL models that only use Au as an input. Logged mineralised zone is used to better inform manual domaining for the kriging models but also is an invaluable input channel (MZ) informing the DL models interpretation of the mineralisation constraints.

MZ logs indicate the core is potential mineralised as visually it hosts the prospective gangue mineralisation that elsewhere can host high-grade gold mineralisation. As an indirect pathfinder to high-grade gold mineralisation, the MZ input informs the DL model by significantly reducing the probability of missed mineralisation (recall) but less so than LAMP as an input. MZ does not impact positively on the false positive rate (precision) over the Au DL model. As to why MZ input do not improve the false positive rate is open to speculation.

It is speculated that the reason why the LAMP input is more productive than MZ as a DL model input, is because the lamprophyres are more commonly proximal to the major crustal structures. These structures are controlled by the hydrothermal system and therefore have a great influence on high-grade gold distribution. MZ is more widely distributed within the deposit and is not controlled solely by major crustal structures; therefore, it has a lesser impact on the DL model for missed mineralisation compared to LAMP.

Field testing

Extensive testing has been undertaken at the three mine sites.

Jundee Mine

The DL models were used to guide a 5000 m drilling campaign within at the particularly complex Deakin zone. Full results are not yet available.

Atacama Kozan Mine

The DL models were used to guide a successful 2000 m underground drilling program that successfully identified three new zones of additional economic copper ore, where waste was previously predicted by the kriging model. The newly discovered ore will likely be incorporated into the mineral resource estimate. A total of 11 holes were drilled, all intersected significant mineralisation, above the cut-off grade. The best intercepts are listed in Table 2. They were drilled in areas previously interpreted as uneconomic, whereas the DL models, with Cu assays inputs and leveraging geological logs predicated new ore zones.

TABLE 2

Atacama Kozan Mine – eight best economic intercepts, from the guided drilling program.

Target	HG intercept length	Average grade	Max grade (1 m length)
A	24 m	0.81% Cu	3.90% Cu
A	36 m	1.19% Cu	3.62% Cu
B	35 m	2.24% Cu	7.98% Cu
B	59 m	1.58% Cu	9.20% Cu
B	28 m	1.84% Cu	8.08% Cu
C	55 m	0.63% Cu	2.32% Cu
C	17 m	1.88% Cu	3.11% Cu
C	17 m	1.93% Cu	5.91% Cu

Simberi Mine

The technology has been incorporated into operational usage. Guided by the geological logging model, the sulfur model has been able to consistently increase routing of additional economically recoverable material to the mill.

CONCLUSION

The machine learning application at the Jundee Mine demonstrates that integrating by feature engineering geological logging into DL Au mineral resource models can substantially reduce the missed mineralisation rate of the models in forward facing reconciliation, effectively being used as pathfinders to gold mineralisation. Many of these logs such as lamprophyres (LAMP) and mineralisation zone (MZ), can be effectively pre-screened with the use of the pathfinder signal analysis (PSA) algorithm, removing the need to test every lithological log in the geological database.

ML applications at Atacama Kozan and Simberi mines demonstrate the efficacy and best practices surrounding the use of proxy logging, the use of visual logging as a proxy for geochemical assays in ML geometallurgical and resource modelling. Visual logging is best used as a supplementary input channel into DL models, including but not limited to binary codes for core unassayed but visually assumed to be barren for Cu modelling or visual logging OX, TR, SU for sulfur modelling. Additionally, for coordinates that do not have geochemical assays but do have logging; logging, despite its limitations, is a beneficial example of 'ground truth' by which to each the model correct answers by converting geological logs to their respective most likely geochemical proxy. For example, converting visually barren core to 0.0 per cent Cu or visually OX core to 0.1 per cent S (the median value of material in data that is logged as OX and assayed for S).

The results have determined that DL models that use geological logs as input are more accurate than DL models based exclusively on geochemistry and/or kriging models, utilising categorical indicator kriging (CIK) and/or mineralised zone (MZ) domaining in a wide range of deposit classes. Both pathfinder and proxy logging proved to be highly applicable in machine learning resource modelling and can be used both to reduce the missed mineralisation rate, false positive rate and accuracy of ore type oxidisation modelling.

ACKNOWLEDGEMENTS

The authors wish to acknowledge the anonymous reviewers who approved the contents and structure of the paper. SCM Atacama Kozan, for permission to access the Atacama Kozan Mine database and discussions with Katsuhito Terashima; Northern Star Resources Ltd for permission to access the Jundee Mine database and interactive contributions with Heath Anderson, Andre Ferreira, Leon Griesel, Simon Davies and others. St Barbara Ltd for permission to access the Simberi Mine data sets and discussions with Bruce Robertson, David Plowman, Jane Bateman, Brett Ascott and others. A special thanks to Ms Ady Aguilar for drafting several of the figures.

REFERENCES

Allied Gold Limited, 2009. Annual Report 08 | 09, Allied Gold Limited. Available from: <https://announcements.asx.com.au/asxpdf/20091009/pdf/31l7ynbd3qjw5t.pdf> [Accessed: 12 January 2024].

del Real, I, Thompson, J F H and Carriedo, J, 2018. Lithological and structural controls on the genesis of the Candelaria-Punta del Cobre Iron Oxide Copper Gold district, Northern Chile, *Ore Geology Reviews*, 102:106–153.

First, D M, Sucholutsky, I, Mogilny, D and Yusufali, F, 2023. Introducing deep learning and interpreting the patterns – a mineral deposit perspective, in *Proceedings of the Mineral Resource Estimation Conference 2023*, pp 2–14 (The Australasian Institute of Mining and Metallurgy: Melbourne).

Glacken, I M and Blackney, P C J, 2022. Categorical and multiple indicator kriging – are we ignoring the geology?, in *Proceeding 12th International Mining Geology Conference 2022*, pp 67–77 (The Australasian Institute of Mining and Metallurgy: Melbourne).

Glacken, I, Rondon, O and Levett, J, 2023. Drill hole spacing analysis for classification and cost optimisation – a critical review of techniques, in *Proceedings of the Mineral Resource Estimation Conference 2023*, pp 179–191 (The Australasian Institute of Mining and Metallurgy: Melbourne).

Godfrey, S, Battista, J and Hearse, P, 2011. Simberi Gold Project, Simberi Island, Papua New Guinea, Competent Person's Report, in *Allied Gold Mining PLC Prospectus*, pp 290–475. Available from: <https://announcements.asx.com.au/asxpdf/20110620/pdf/41z9hjy1rhc5k6.pdf> [Accessed: 12 January 2024].

Goodfellow, I, Bengio, Y and Courville, A, 2016. *Deep Learning*, 801 p (The MIT Press: Cambridge).

Ichii, Y, Abe, A, Ichige, Y, Matsunaga, J, Miyoshi, M, Furuno, M and Yokoi, K, 2007. Copper exploration of the Atacama Kozan Mine, Region III, Chile, *Shigen-Chishitsu*, 57(1):1–14.

Meyes, R, Lu, M, Waubert de Puiseau, C and Meisen, T, 2019. Ablation studies in artificial neural networks, arXiv preprint. Available from: <https://arxiv.org/pdf/1901.08644.pdf> [Accessed: 12 January 2024].

Müller, D and Groves, D I, 2019. Indirect Associations Between Lamprophyres and Gold-Copper Deposits, in *Potassic Igneous Rocks and Associated Gold-Copper Mineralization* (eds. D Müller and D Groves), fifth edition, ch 8, pp 279–306 (Mineral Resource Reviews).

O'Shea, K and Nash, R, 2015. An introduction to convolutional neural networks, arXiv preprint. Available from: <https://arxiv.org/pdf/1511.08458.pdf> [Accessed: 12 January 2024].

Phillips, G N, Vearncombe, J R and Eshuys, E, 1998. Yandal greenstone belt, Western Australia: 12 million ounces of gold in the 1990s, *Mineralium Deposita*, 33:310–316.

Phillips, G N, Vearncombe, J R and Murphy, R, 1998. Jundee gold deposit, in *Geology of Australian and Papua New Guinean Mineral Deposits* (eds: D A Berkman and D H Mackenzie), pp 97–104 (The Australasian Institute of Mining and Metallurgy: Melbourne).

Ponyalou, O L, Petterson, M G and Espi, J O, 2023. The Geological and Tectonic Evolution of Feni, Papua New Guinea, *Geosciences*, 13(9):257. doi:10.3390/geosciences13090257 [Accessed: 12 January 2024].

Porter, T M, 2008. Simberi, Tabar – Sorowar, Pigibo, Pigiput, Botlu, Samat Papua New Guinea, in *PorterGeo Database* (PGC Publishing). Available from: <https://portergeo.com.au/database/mineinfo.asp?mineid=mn1328> [Accessed: 12 January 2024].

Sims, D A, 2023. An estimation error; in *Proceedings of the Mineral Resource Estimation Conference 2023*, pp 246–249, (The Australasian Institute of Mining and Metallurgy: Melbourne).

Smith, S, Haese, R and Grigson, M W, 2017. Jundee gold deposit, in *Australian Ore Deposits* (ed: G N Phillips), pp 273–278 (The Australasian Institute of Mining and Metallurgy: Melbourne).

Vearncombe, J R and Elias, M, 2017. Yilgarn Craton – mineral deposits and metallogeny, in *Australian Ore Deposits* (ed: G N Phillips), pp 95–106 (The Australasian Institute of Mining and Metallurgy: Melbourne).

Zhang, G and Glacken, I, 2023. Best practise in Multiple Indicator Kriging (MIK) – importance of post-processing and comparison with Localised Uniform Conditioning (LUC), in *Proceedings of the Mineral Resource Estimation Conference 2023*, pp 76–85 (The Australasian Institute of Mining and Metallurgy: Melbourne).

Adaptive mine planning in open pits – the role of near-real-time orebody characterisation in grade control modelling

S O'Brien[1] and S Maguire[2]

1. Geoscientist Manager, Veracio, Adelaide, SA 5950. Email: shaun.obrien@veracio.com
2. Lead Geologist, Veracio, Adelaide, SA 5950. Email: shauna.maguire@veracio.com

ABSTRACT

In addressing the widely acknowledged operational inefficiency within the mining sector, a pressing challenge emerges. Mine Geologists often find themselves powerless, with focus primarily shifted towards throughput and cost. Consequently, many opportunities for incremental value creation end up relegated to waste. This inefficiency is not merely a traditional challenge but one that demands a novel approach. Although every mining operation strives for optimisation, they remain shackled by throughput constraints. In the absence of optimal conditions, efforts to innovate and improve can become overshadowed by the looming threat of productivity loss.

Recent advancements in data acquisition, specifically accelerated acquisition of reliable data, have posited new avenues to tackle these inefficiencies. The methodology employed in this study leveraged a combination of high-speed data acquisition and conventional sampling techniques, *in situ* geochemical scanning using high-density X-ray fluorescence (XRF), immediate data processing and a strategic approach to controlled sampling. The end goal? Delivering accurate, timely results to mine planners, expediting decision-making processes.

The novelty of integrating these methodologies cannot be underestimated. Offering a fresh perspective, this approach accentuates the importance of timely, precise planning in enhancing the economic viability of mining operations. It is especially pertinent in open pit scenarios where strategic decisions on bench sequences can significantly influence operational economics against the plan.

Findings suggest that this methodology offers a robust alternative to traditional approaches, promoting not only increased efficiency but also heightened value recovery. Opportunities have been identified for sampling automation and next-generation sensors to for the Mine Geologist, an enriched, data-driven toolkit, promising more nuanced orebody characterisation and enhanced decision-making capabilities. As for the broader mining industry, adopting such practices could pave the way for more sustainable operations, minimising waste, optimising resource allocation, and, most importantly, bolstering the economic viability of mining projects.

Proof of concept study framework at 'Site A'

At 'Site A' the on-site labs are ill-equipped to handle the projected sample volume generated in-pit operations, leading to triaging of samples.

The study's proof of concept framework involves two components:

1. Determine representivity and repeatability of results.
2. Laboratory analysis if samples to validate correlation and accuracy of results.

The first component of the study was designed to assess the amenability of the deposit to this sampling methodology. This was assessed by analysing ~500 sample pairs. Representative bulk samples were collected at blasthole cones which represented two mineralogical domains in the orebody, the first being a bornite rich zone with high Au and Hg and the other a zone of chalcopyrite mineralisation. Specific mineral domain details for each sample have been kept blind in the study. If the analysis of duplicate pairs was highly variable (ie $R^2 < 0.5$) it would indicate that sampling variability may be a fatal flaw.

The second component was to assess the accuracy of high-speed data acquisition through XRF in-field sampling methodology. Bulk samples were sent to the on-site lab for analysis in parallel to the sub-samples being analysed with XRF. Additionally, samples were sent to an external lab for expanded multi-element analysis, providing a comprehensive understanding of the suitability of this methodology.

Blasthole samples were collected and brought to the site sample preparation facility where they were split and fed through a 5 mm crusher. These samples were then loaded into labelled trays with the capacity for 30 samples each as demonstrated in Figure 1. Each tray was duplicated (primary and duplicate set of 30). The trays were loaded into the TruScan unit, equipped with XRF instruments as displayed in Figure 2, which enabled high-speed scanning, with results available within minutes of the scan.

FIG 1 – Process of removing sub-sample from plastic bag into Chip Tray Assay Pod module ready for field analysis.

FIG 2 – Prepared tray of samples being analysed under XRF sensor.

Primary versus duplicate analysis

The primary versus duplicate scanning confirmed the representativeness of the sampling protocols with an R^2 of 0.89, results displayed in Figure 3. This suggests a low variability in Cu concentrations between the primary and duplicate samples. What variability there was increased with concentration, aligning with typical patterns. The variability at high concentrations is most likely due to a nugget effect as small (<1 mm) grains of chalcopyrite were visible in some of the high-grade samples. In the case of the low-grade (<0.5 per cent Cu) samples there is slight variation between primary and duplicate Cu concentrations due to slight variation in grain size between the samples. Although higher variability exists at greater concentrations, absolute accuracy of values was secondary to the aim of defining samples above a nominal cut-off.

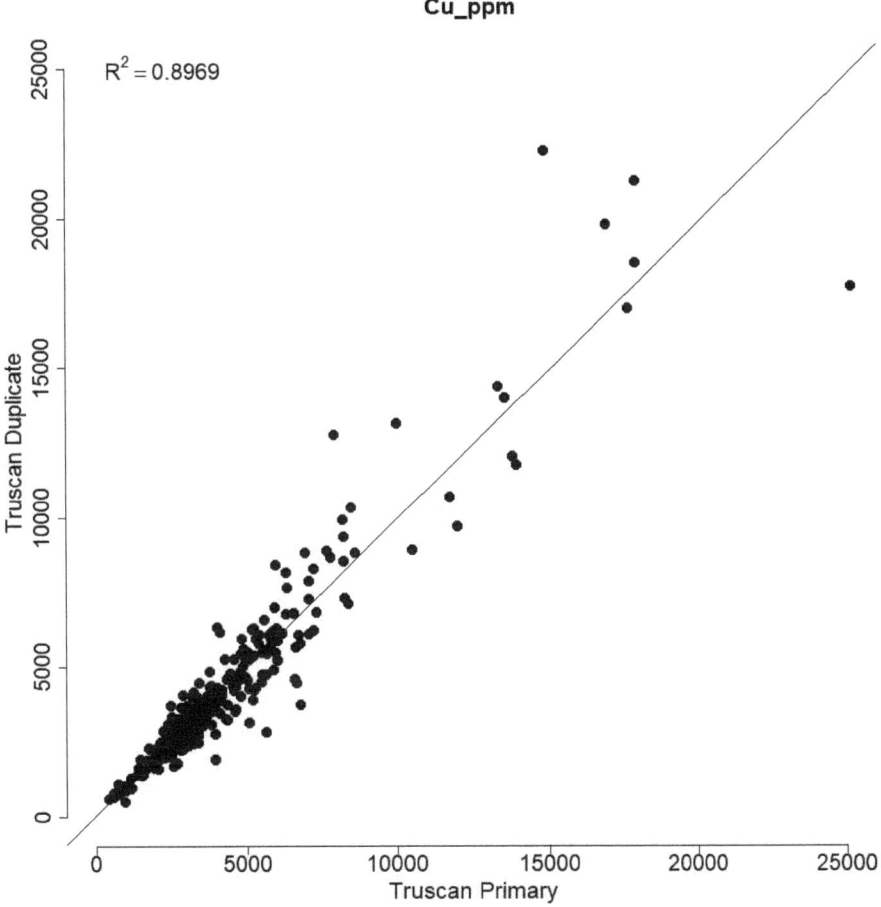

FIG 3 – Primary versus duplicate result for all samples (Cu).

In-field versus lab

The in-field sampling and analysis demonstrated high effectiveness in discriminating ore grades, particularly for copper. I positive bias was identified in the results, resulting in a systemative overestimation of copper concentration and results are displayed in Figure 4. This trivial issue can be overcome through re-modelling of XRF calibration and regression corrections.

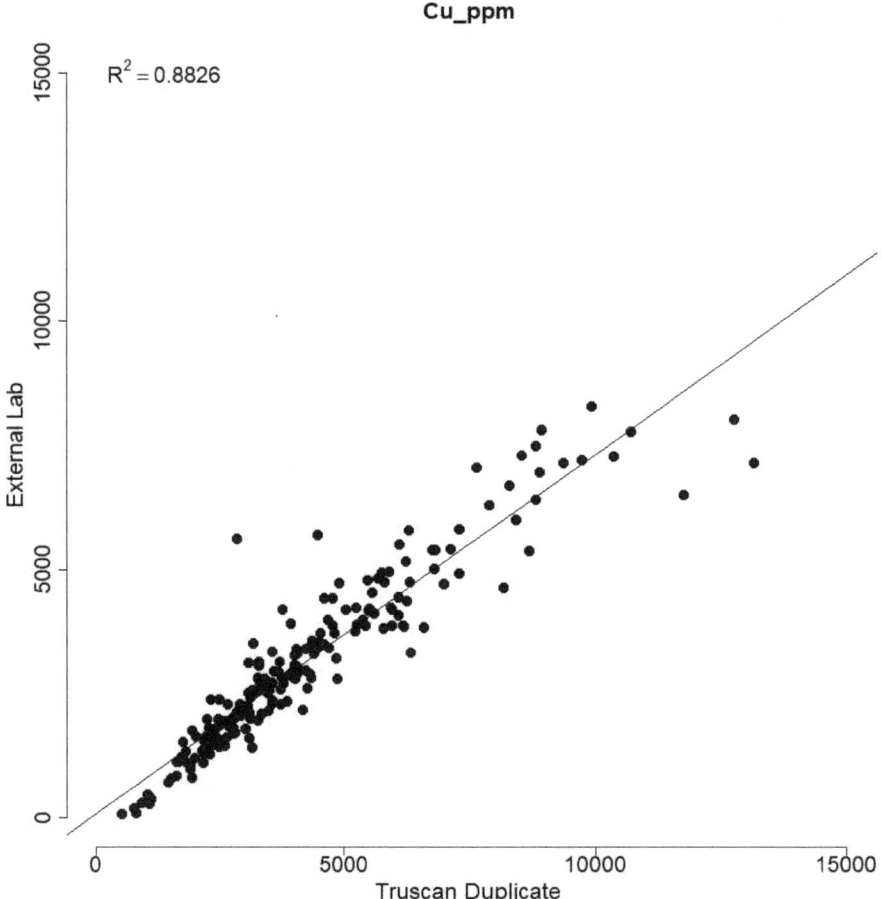

FIG 4 – XRF results versus external laboratory results for copper.

CONCLUSIONS

This proof of concept has demonstrated the effectiveness of the methodology to analyse blasthole samples in the field and triage samples effectively. This method helps to identifies priority samples for gold assays on the day-of-drilling and provides copper concentration data before firing, enabling a responsive approach to Ore Control. At the site highlighted in this study, there is a strong correlation between Cu concentration and Au concentration which follows a ~ 1:1 correlation. Based on this a threshold of 0.8 per cent Cu and greater is used to separate samples out for further Au analysis. Despite the correlation between Cu and Au concentrations, a greater variance is expected for Au due to nugget effect.

When assessing these results, it is important to evaluate what trade-offs are being proposed. In the classical sampling regime, analytical sample preparation is an intensive and time-consuming process. The material will be pulverised and homogenised before being subject to acid digestion to liberate elements from the minerals before being analysed with highly precise analytical equipment. There is a reasonable argument that the additional accuracy and precision gained through these processes does not adequately offset the opportunity cost of delaying decisions. With the in-field approach, the reasoning is that we can afford to sacrifice additional preparation stages that are time consuming, to realise the benefits of high-quality information sooner that can populate coarse bins (high/low-grade and waste), supposing that what is produced is within a reasonable tolerance.

The fact that this method is non-destructive allows classical analysis to provide robust QA/QC. With the potential to process over 200 samples per 12-hour shift, the methodology is scalable with increased mine production with subjective components of ore grade decisions being mostly eliminated. Automated sampling from the blasthole is also an active area of our development which aims to improve operational efficiencies and to reduce manual processes and the uncertainty that accompanies them. Further work is needed to explore ore type characteristics through multi-element chemistry generated with the XRF, or additional sensors such as co-registered hyperspectral with XRF to identify mineralogy that affect recoveries including clay and carbonate species.

Advancing mine geology through optimised sampling strategies – a case study from Porphyry Underground Mine, Western Australia

R Snell[1]

1. Senior Geologist, Northern Star Resources, Perth Western Australia. Email: rsnell@nsrltd.com

ABSTRACT

The Porphyry Underground Mine, located in Western Australia, poses unique challenges to mine geology, particularly in obtaining representative samples for grade and tonnage estimation. The main ore lode within the mine dips approximately 20° to the east with an average width of 9 m. When underground sampling of a wide, shallowly dipping orebody with varying grade subdomains, traditional face sampling may introduce sample bias as the observed domain in the face sample does not often represent the entire lode, leading to biased grade and tonnage estimates. The shallow dip of the orebody also makes underground drilling from dedicated drill drives less feasible and increasing surface drilling may potentially negatively impact the project's economics. In response to these challenges, this paper aims to develop a robust methodology that mitigates sample bias whilst still providing the most accurate and representative data collection in this challenging geological setting, without increasing the operational costs.

This has been undertaken through a comprehensive analysis of various sampling approaches, including face samples, wall samples, diamond drilling, RC drilling, face chip samples and downhole surveying. This paper will explore different methodologies, encompassing both technological advancements and traditional mine geology techniques. Geostatistical estimates will be employed to compare these methods of data collection, aiming to optimise the sampling strategy for the Porphyry Underground Mine.

The proposed methodology's significance lies in its potential to improve the reliability of geological characterisation and subsequent decision-making processes. By addressing the issue of partial lode sampling and sample bias at Porphyry Underground Mine, this research provides valuable insights that disrupt the established norms of mine geology practices.

INTRODUCTION

The Porphyry mining centre is located within Northern Star Resources Carosue Dam Operations (CDO) approximately 135 km north-east of the City of Kalgoorlie Boulder. The deposit is approximately 42 km north of the main administration centre and processing plant of CDO. The mining centre consists of both open pits and underground mines that have been intermittently mined since 1984. The current Porphyry Underground commenced mining in October 2022 and has a current mine life until 2028 and is the basis for this study.

All mines, regardless of mineralisation style or mining methods pose challenges to the geology teams mining them. These challenges can range from geological and estimation complexity to commodity prices or government regulations. Porphyry Underground is no different. This paper will focus on the unusual but not unique challenges for the mine geologists around potentially biased data capture due to partial lode sampling. The paper will present an analysis of methods employed to mitigate bias in estimation, focusing particularly on the data capture stage.

BACKGROUND

Regional geology

The Porphyry Mining Centre deposits are located in the Eastern Goldfields Province of the Archaean Yilgarn Block, within a belt of greenstone-granites. The geological setting includes alternating mafic-ultramafic and felsic-clastic sequences, interpreted as concurrent volcanic episodes. The gold deposit is found in a syenomonzonite intrusion within the Murrin-Margaret sector, delineated by the Keith-Kilkenny tectonic zone to the west and the Laverton tectonic zone to the east. Alkaline syenitic intrusions near the Keith-Kilkenny zone are linked to north-north-west trending faults (Roberts, Witt and Westerway, 2004).

Local geology

The Porphyry Mining Centre, as shown in Figure 1, comprises two main lithological domains, with the primary host for Porphyry mineralisation located in an oval-shaped syenomonzonite granitoid body measuring 4.5 km by 2.5 km. This coarse-grained, cream/grey intrusion contains microperthite phenocrysts in a groundmass of plagioclase, quartz, orthoclase, microcline, biotite and epidote. Aplite veining and tourmaline breccias are common near the intrusion margins and mineralisation is currently confined to the granitic contact margins (Allen, 1987).

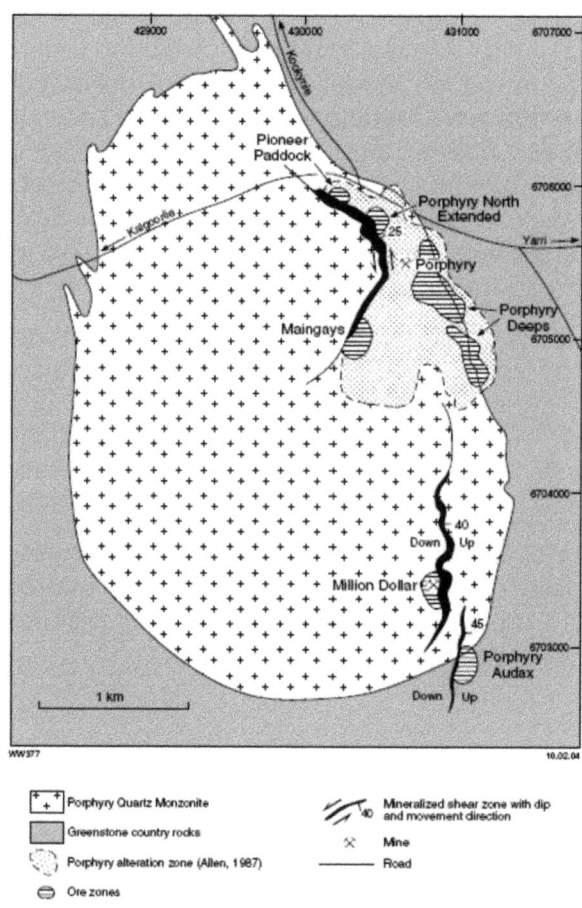

FIG 1 – Schematic geological map showing the Porphyry Quartz Monzonite and relevant gold deposits (Roberts, Witt and Westerway, 2004).

The syenomonzonite intrudes Archean greenstones characterised by tholeiitic basalt, andesite and clastic sediments trending NNW with a steep easterly dip. Within the greater Porphyry shear, this greenstone package exhibits a secondary style of mineralisation, extending into the andesitic unit with shear intensity diminishing further from the contact. A reduction in grade tenure is observed as mineralised lodes cross this lithological boundary (Swager, 1995).

Gold mineralisation

Mineralisation is situated within a north–south-trending shear zone, dipping eastward at 20° to 25°, and is primarily concentrated along the eastern margin of the Porphyry monzonite intrusion. The quartz monzonite has undergone significant shearing, resulting in a well-defined layered fabric containing biotite, muscovite, quartz and carbonate. Localised mylonite has formed due to this shearing process. The outer, weakly-sheared zone exhibits calcite alteration, while the inner, strongly sheared and mineralised zone typically contains pyrite, hematite and narrow quartz veins. The highest gold grades are consistently associated with pronounced deformation (shearing), intense hematite-carbonate alteration, the presence of euhedral pyrite and an increase in quartz veining.

The majority of contained ounces have been sourced from the higher-grade shear zone known as the Apollo shear, with several other shears on the hanging wall providing less than 10 per cent of the overall ounces. The mineralisation in the Apollo shear shows patchy low-grades, with sometimes continuous thin high-grade zones. This is believed to be caused by the anastomosing nature of the shear. It is this heterogeneous distribution of grade within the main lode at Porphyry Underground that introduces the challenge for the mine geologist.

Current data set

The Porphyry geological data set predominantly incorporates diamond drilling, RC drilling, and face chip sampling and stores lithology, grade, density, veining etc. All drilling to date has been conducted from the surface; due to the shallow-dipping nature of the orebody. The initial drilling, conducted in the late 1970s and 1980s had hole identifiers with a prefix of 'PD' occurred before mining the initial pit in 1984. This early data is of comparatively poor quality when compared to the more modern drilling conducted across the project since the 2000s. This older generation of holes typically have incomplete surveys, inconsistent logging and selective sampling.

Since the resumption of mining activities in 2010, additional drilling has taken place, enhancing the data set in both spatial extents and quality. In the years 2021 and 2022, Northern Star has completed 178 exploration holes, further contributing to data accuracy and confidence. Presently, efforts are underway to bolster the data set specifically for the purpose of underground mining grade control, this is involving diamond drilling and underground face sampling. Similar to most selectively mined gold operations, the mine's geology team endeavours to enhance both the local geological and tonnage and grade models through closely spaced, high-quality diamond drilling, as well as underground mapping and sampling.

Given the industry wide pursuit of cost efficiency, surface drilling remains a favourable alternative to underground diamond drilling at the Porphyry site. The shallow geometry of the orebody facilitates the utilisation of surface drilling methods over dedicated underground drilling platforms. By aligning drilling techniques with the geological characteristics of the ore deposit, the mining operations at the Porphyry site can optimise drill efficiency and minimise financial outlays.

Extraction

Due to the shallow dipping nature of the Apollo lode, Porphyry Underground is lateral development intensive with a 10 m lateral and 7 m vertical level spacing as shown in Figure 2. Production ore extraction is conducted via top-down central retreat longhole open stoping with rib pillars designed *in situ* for stope and regional stability.

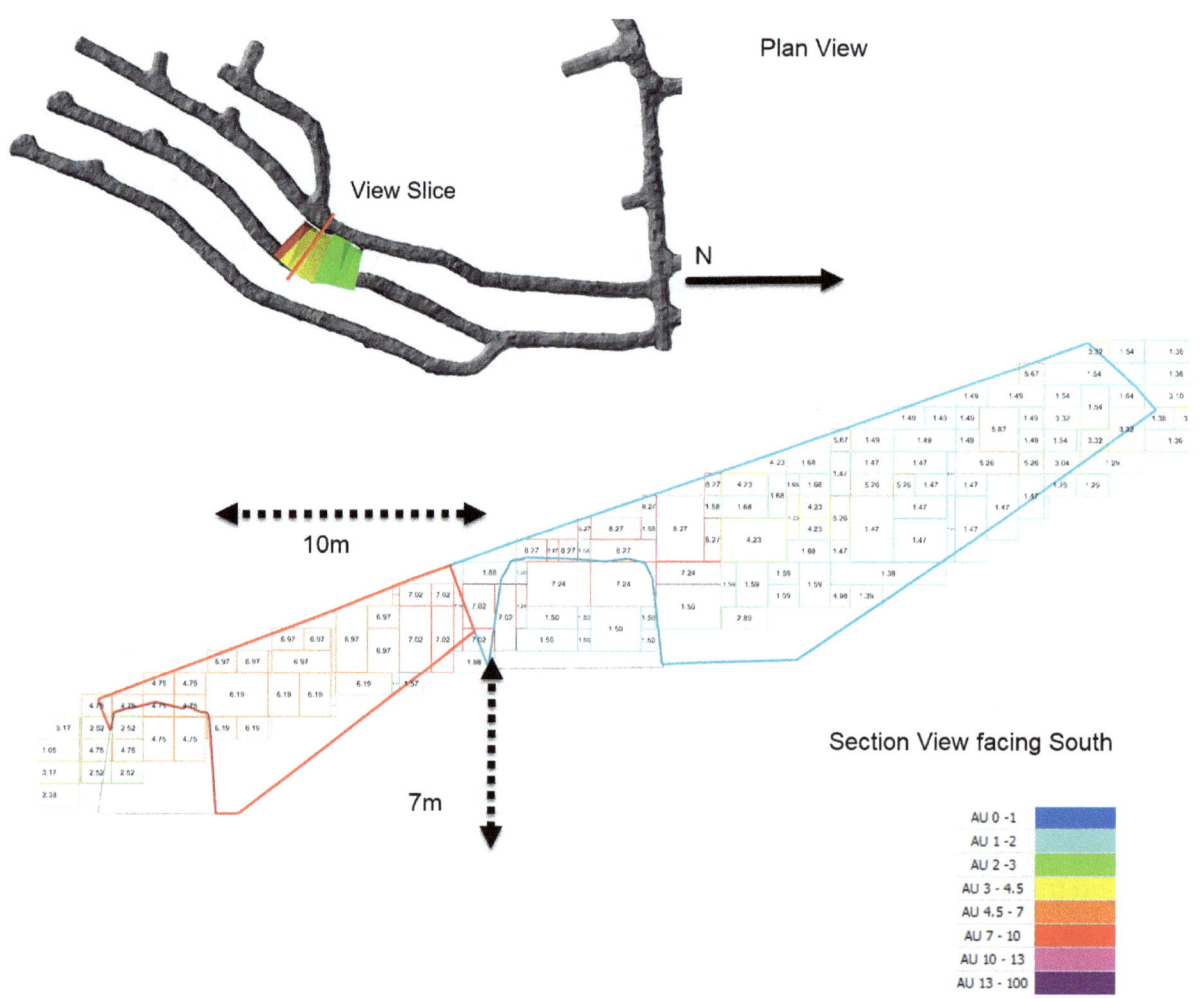

FIG 2 – Typical stope design at Porphyry Underground.

SAMPLE BIAS

There are many different types of bias that impact resource estimations throughout the value chain from sample collection to assaying. This paper will focus on under coverage bias in sample collection and the impact that it has on Porphyry Underground.

Under coverage bias in the context of underground geological face sampling refers to the potential distortion or limitation in the representativeness of collected samples, leading to an incomplete or biased portrayal of the geological characteristics of the deposit. This bias may arise when certain areas or types of rock within the deposit are inadequately sampled or excluded from the sampling process effecting estimation (Dominy et al, 2009). Several factors can contribute to under coverage bias in underground geological face sampling:

- **Selective sampling**: if geologists favour certain rock types, structures, or mineralisation zones over others during the sampling process, it can result in under coverage bias. This may occur due to subjective judgments or preconceived notions about where valuable minerals are most likely to be found, for example, when samplers collect samples that are 'shiner' or more altered than the average sample.

- **Sampling equipment limitations**: the choice of sampling equipment and methods can influence the coverage of different geological features. This may occur if the equipment used is not suitable for obtaining samples from specific rock types or structures, those areas may be underrepresented in the data set, for example, if the mineralisation occurs on the contact of a hard and soft boundary, and an aircore rig is used to collect samples but cannot penetrate the hard rock.

- **Spatial variability**: geological deposits often exhibit spatial variability, meaning that the composition and characteristics of the rock can vary across different locations. If the sampling locations are not strategically chosen to capture this variability, certain geological features may be under covered.

- **Accessibility issues**: underground mining environments can be challenging to navigate, and some areas may be less accessible than others. If accessibility issues lead to the omission of certain regions from the sampling process, under coverage bias may occur. For example, ore drive development is typically 5 m in width, but if the orebody is greater than that, the face sample will not represent the whole orebody.

To address under coverage bias, it is crucial to design comprehensive and systematic sampling protocols that account for the geological heterogeneity of the deposit. This involves selecting sampling locations based on a thorough understanding of the deposit's geology, using appropriate sampling techniques for different rock types, and ensuring a representative distribution of samples across the mining face.

BIAS' OBSERVED AT PORPHYRY UG

Unrepresentative sampling

In general, the mine geology team does not design the ore drives, so with little control on drive placement this sampling control is taken away from the mine geologist and inaccessibility to obtain representative sampling occurs. As seen in Figure 3 the drive placement at Porphyry is best designed for stope extraction and not capturing an entire snapshot of the Apollo lode. The drives are regularly spaced 10 m laterally apart but irregularly distributed in their placement in either the centre, footwall or hanging wall of the ore. This means that although the face sample is not consistently over sampling one portion of the lode; each sample is not capturing the lode in its entirety. The implications of this on the Porphyry Underground was investigated by the on-site geology team.

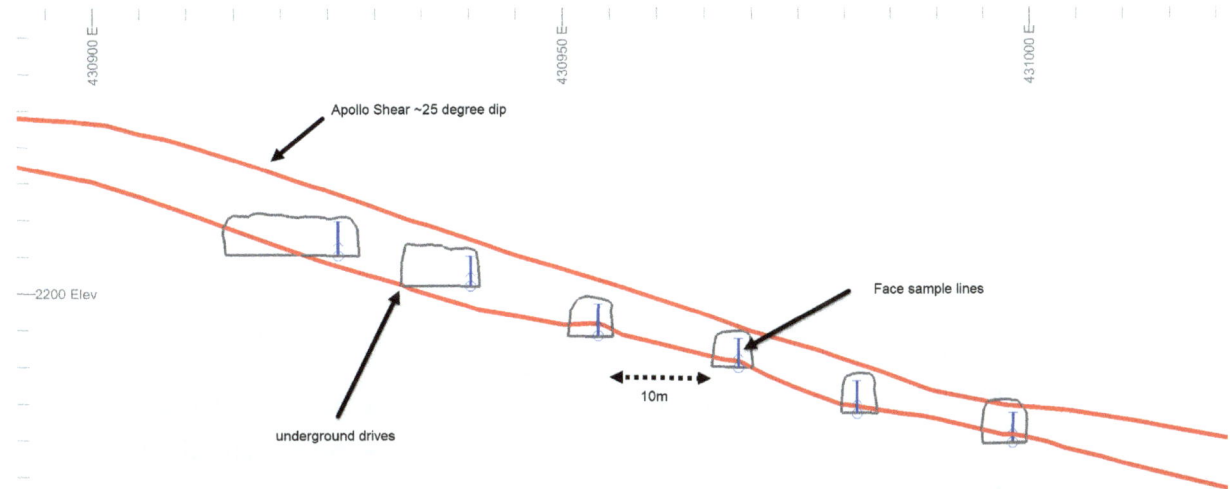

FIG 3 – A schematic from the 2190 ore drive souths at Porphyry Underground showing partial lode sampling, section view facing north.

Sample collection method

Development of sampling protocols on a deposit must be specific to the geology of the mineralisation. The underground mine geologists' methodology of collecting the sample either at the face or otherwise can differ between techniques of varying quality such as face sampling, wall sampling, sludge sampling and muck pile sampling. The sample obtained must be fit for purpose, whether that be for daily production tracking; ore or waste classification; or to aid an estimation. Each sampling method considered introduces specific advantages and limitations that impact the accuracy and reliability of geological data. For instance, chip sampling, commonly employed for its simplicity and rapid execution, may offer real-time insights into the mineralisation trends at the face. However, it may lack the representativeness needed for detailed geostatistical analysis. On the other hand, diamond core drilling provides more comprehensive and continuous samples, offering valuable information on the rock's structure and composition but is time-consuming and not always cost-effective. Careful consideration of factors such as geological heterogeneity, ore variability, and the spatial distribution of mineralisation is essential when selecting the most appropriate sampling method. Additionally, rigorous adherence to standardised protocols and quality assurance measures ensures that the collected samples accurately reflect the geological characteristics of the deposit, thereby enhancing the reliability of subsequent analyses and estimations in the mining process (Roberts, Dominy and Nugus, 2003). Table 1 summarises various techniques and their associated advantages and limitations.

TABLE 1

Review of sampling techniques at Porphyry (adapted from Roberts, Witt and Westerway, 2004).

Collection method	Collection stage	Advantages	Limitations
Face sampling – rock hammer	Within development cycle	• cheap • quick • real time data for production tracking	• potential for operator bias • partial sampling of the lode due to height of the sample line
Face sampling – jumbo drill	Within development cycle	• cheap • quick • real time data for production tracking	• ties up the use of expensive machinery/production delays • potential for operator bias
Sludge sampling	Post-development Post-stope design	• no extra costs for the drill hole if the hole is required for production	• time consuming for a geologist • loss of fines could cause a low-grade bias in a shear hosted orebody • highly contaminated sample • delayed data, not available for development tracking
Underground reverse circulation drilling	Post-development Pre-stope design	• uncontaminated sample • captures lode entirety • provides data for accurate stope extraction	• expensive • additional step outside of the production cycle • not in real time • no structural data
Underground diamond drilling (within ore development)	Post-development Pre-stope design	• uncontaminated sample • captures lode entirety • provides data for accurate stope extraction • can also be used for geotechnical analysis of stope design	• the most expensive option • additional step outside of the production cycle • not in real time
Muck pile sampling	Post-development Post-stope firing	• cheap • has no impact on production cycle	• representative and unbiased samples are hard to obtain • sample taken after dirt is already on the ROM

In choosing face sampling as the physical collection methodology of the sample at Porphyry Underground the influence of bias arising from partial sampling of the ore lode still exists; the only variable risk is to the quality of the individual sample. This approach was chosen as it was deemed the most cost-effective way to provide in-time data that did not have contamination or potential machine downtime.

Face sampling process at Porphyry Underground

Face sampling in underground mining involves collecting samples directly from the face of the ore or host rock during ore drive development. This method of data collection has both advantages and disadvantages, which are important to consider in the context of ore grade control, operational efficiency and resource estimation.

Through geological mapping and sampling directly from the exposed rock face, geologists can capture the subtle variations in mineralogy, texture and structure that may be missed by other sampling methods. This granularity of data is particularly crucial in heterogeneous deposits where ore distribution is irregular. Development face sampling and mapping allows for the identification of distinct geological domains, alteration zones, and unmineralised structures, contributing to a more refined geological model, identifying subtle changes missed by wide spaced drill holes. The collected data aids in characterising the spatial heterogeneity of the deposit, facilitating more accurate resource estimation. In essence, data collected from the development face adds a finer level of detail to the data, enabling a comprehensive understanding of the complexities inherent in the deposit.

When mapping and sampling a typical face at Porphyry it's crucial to follow a systematic and precise procedure to ensure representative samples and the safety of our team. The process begins by selecting a sample line in the face that cover as much width of the ore lode as possible as seen in Figure 4a. The line is 'logged' into domains based off a visual estimate of the alteration intensity, strain intensity, and sulfide percentage, this sample line should be as perpendicular as possible to the dip of the ore lode. A geology map is then drawn onto a template as seen in Figure 4b that captures all relevant geology information including lithology, alteration, and mineralisation across the face. Samples are collected manually from the line using a rock hammer and chisel with the sample being chipped directly into a calico sample bag mounted to a specially designed sampling ring taking care to avoid contamination and ensure the samples are representative of the *in situ* conditions. These sample lines are then converted to drill holes with a collar, azimuth and dip in the geological database. Adhering to these guidelines will help generate a comprehensive data set that accurately reflects the characteristics of the shallow orebody, facilitating more informed geological assessments and mining decisions.

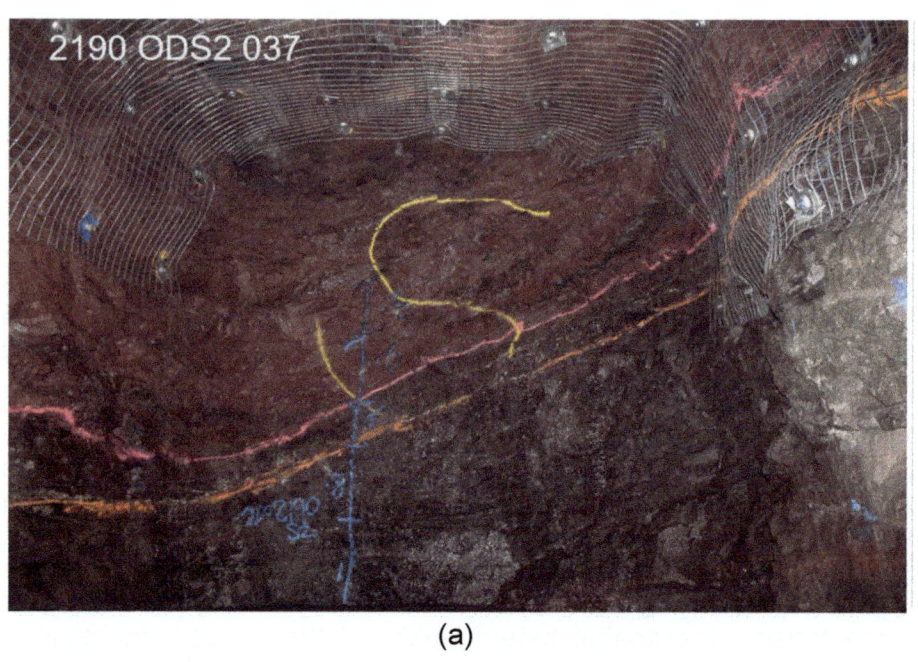

(a)

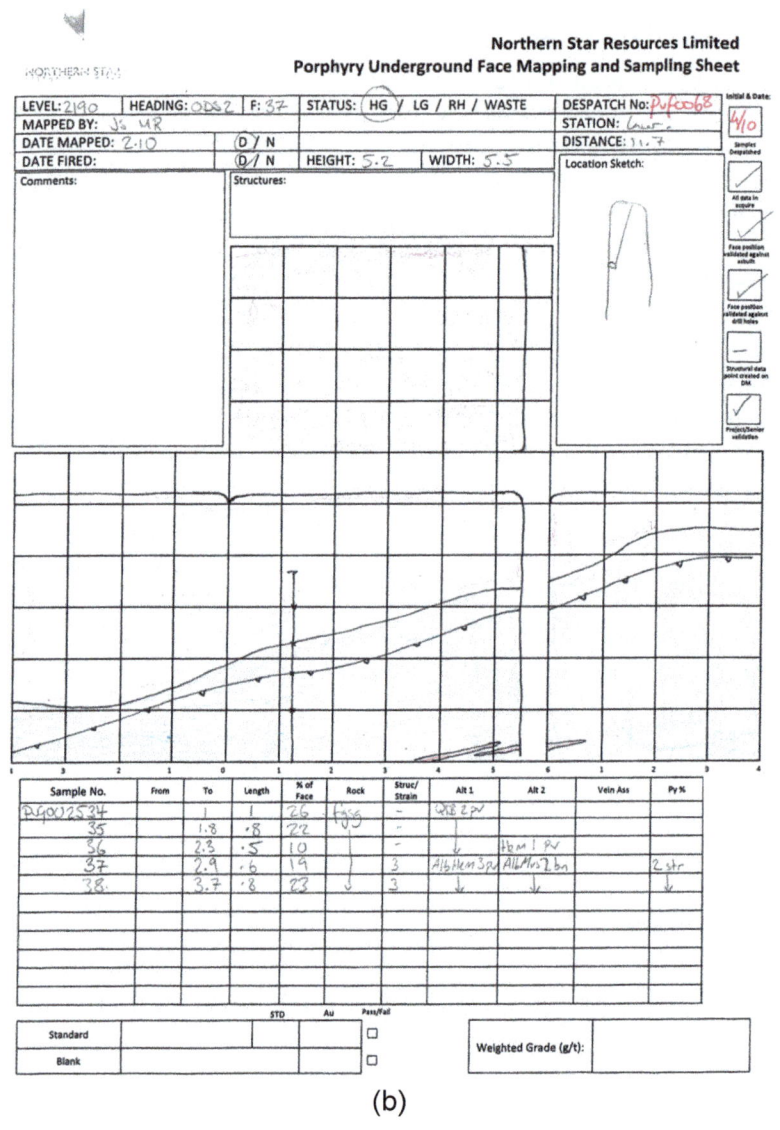

(b)

FIG 4 – 2190ODS_037 face: (a) face sample line in underground photograph; (b) mapping and sampling sheet completed by underground mine geologist.

Trial 1 – face sample versus wall sample

Spatial variability in sampling refers to the natural diversity of geological characteristics within a given area and understanding this variability is crucial in obtaining representative data. Failure to account for spatial variability can introduce bias into sampling results, potentially leading to inaccurate assessments of ore deposits. In mining, orebodies often exhibit differing characteristics at the local scale. If sampling locations are not strategically chosen to capture this variability, the data may be skewed, and the resulting model or estimation may not accurately reflect the true nature of the deposit. Biases may arise when, for example, samples are predominantly taken from areas with higher or lower mineralisation, overlooking the overall diversity within the deposit.

The mine geology team needs to stay attuned to the production cycle, ensuring they contribute value to operations. This involves optimising their time at the site, collaborating with the operational team harmoniously. Approximately 70 per cent of the development cuts made at Porphyry Underground occur in an ore drive. However, due to the mine geologist's limited working hours (06:00 to 18:00) and the necessity to take a sample with each development advance, the pace of underground mining was significantly affected, this was deemed unacceptable and a change was required. Given the inherent characteristics of the Apollo lode, the geology team investigated whether the development drives' walls could serve as a representative sample face. Their theory, as illustrated in Figure 4a, was that the ore intersection, situated on the left-hand wall of the ore drives when facing south, manifests an equally accessible sample selection that would not impact the production cycle.

A trial was conducted in a pre-determined area to take both face and wall samples of every development advance to determine if a bias would be introduced when taking wall samples opposed to traditional face samples. A series of models were then run under the same parameters as a regular grade control model would be and then compared back to the resource model. It should be noted that extraction designs were not considered as part of this trial, the tonnes and ounces evaluated were the whole value of the Apollo lode within the trial area with no modifying factors.

Findings

As with most narrow vein gold mines, the mineralisation is considerably more variable at a local scale during mining than modelled or estimated using broad spaced data points during feasibility. When mapping the ore/waste boundary without the addition of sample data there was significant local change at a 5 m scale in the spatial location of the mineralisation to the degree of up to a metre difference. It was also found that the dispersion of known high-grade lenses were ±20–25 per cent different in size from initial predictions by 40 m spaced drill holes. As can be seen in Table 2, by not including any additional assay data into the estimate we have a reduction in ounces by 6 per cent compared to the initial resource model. However, with the addition of assay data from either face, wall, or both into the estimate we see a small positive change in the overall ounces. This can partially be attributed to no data being present at a local scale for the areas now included within the domain, with the nearest composites being in lower grade the model under-called the newly added blocks.

TABLE 2

Results summary from Trial 1.

Conditions	Ounces total (oz) (in trial area)	Variation from Resource model (%)
Resource model	17 206	0
With face and wall samples	17 895	+4%
With only face samples add	17 551	+1%
With only wall samples added	17 723	+2%
With neither face nor wall samples added, but backs mapping included	16 174	-6%

On a local scale however, as shown in Figure 5 there were significant changes of up to 20 per cent at a 5 to 10 m scale that was captured in the face sample data. These changes whilst globally minor have had material implications to the mine planning and stope extraction processes on a local scale and could have resulted in misallocation of ore and waste and negatively impacted the economics of the mine on a short-term basis.

These findings show the importance of geological mapping and sampling and that they cannot be overlooked by the mining schedule. However, whether that data is captured from the mining face or post-cycle in the wall has little impact on the quality of the estimation. Since the investigation, every alternate face is scheduled into the mining cycle for mapping and sampling, while the alternate wall sample is mapped and sampled outside of the mining schedule. This reduces potential delays caused by Geology team on the development cycle whilst still being able to capture data necessary information prior to stope extraction.

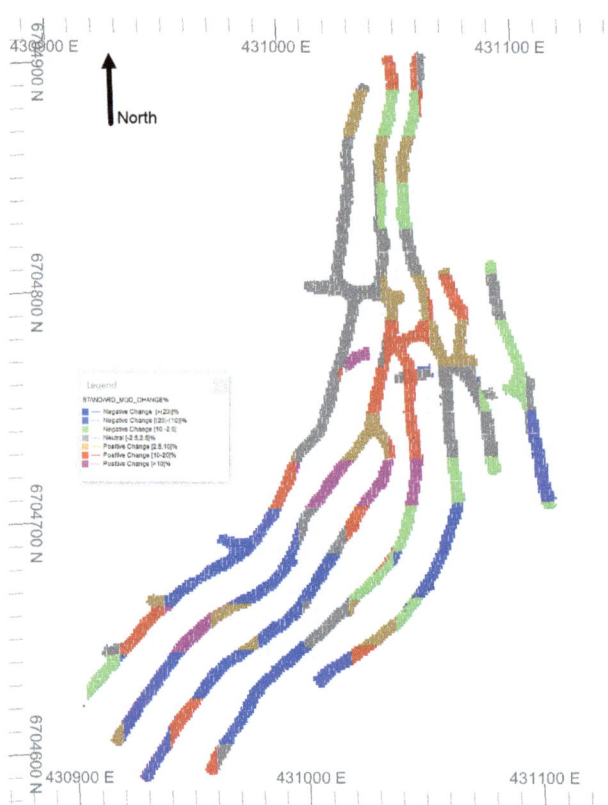

FIG 5 – Model change by per cent within trial area with the addition of faces and walls.

Trial 2 – visual indicators trial

As part of the investigation to minimise the impact of geology to the mining cycle, a trial was conducted in 2023 to eliminate numerical assay data from face samples and instead categorise the sample intervals into grade bins based on geological observations. The intention of the trial was to replace the face assay with an attribute based on a geological observation and then to use that assigned value to support the CIK estimation. Essentially, the goal was to widen the data collection scope beyond standard drill holes without introducing additional numerical data. Consequently, the entire sample was removed to mitigate the bias it introduces.

This was done by a simple numeric model built in Leapfrog Geo (by Seequent), with each grade category being assigned a number from zero to four. Zero being waste rock and four being the highest-grade interval. These numeric isoshell wireframes were then used as a hard boundary for our four subdomains within the Apollo domain during the estimation.

For the trial the mine geologists were tasked with assigning a number from zero to four for a georeferenced scaled photo of each face within the selected trial area. Several other members of the technical services team, with a variety of experience levels, were also involved at this stage to provide an unbiased sample set. The total sample set included 204 sample lines for a total of 909 individual samples and resulted in 406 data points for the numeric model. Figure 6 shows a typical example of how the faces were allocated into their potential grade categories.

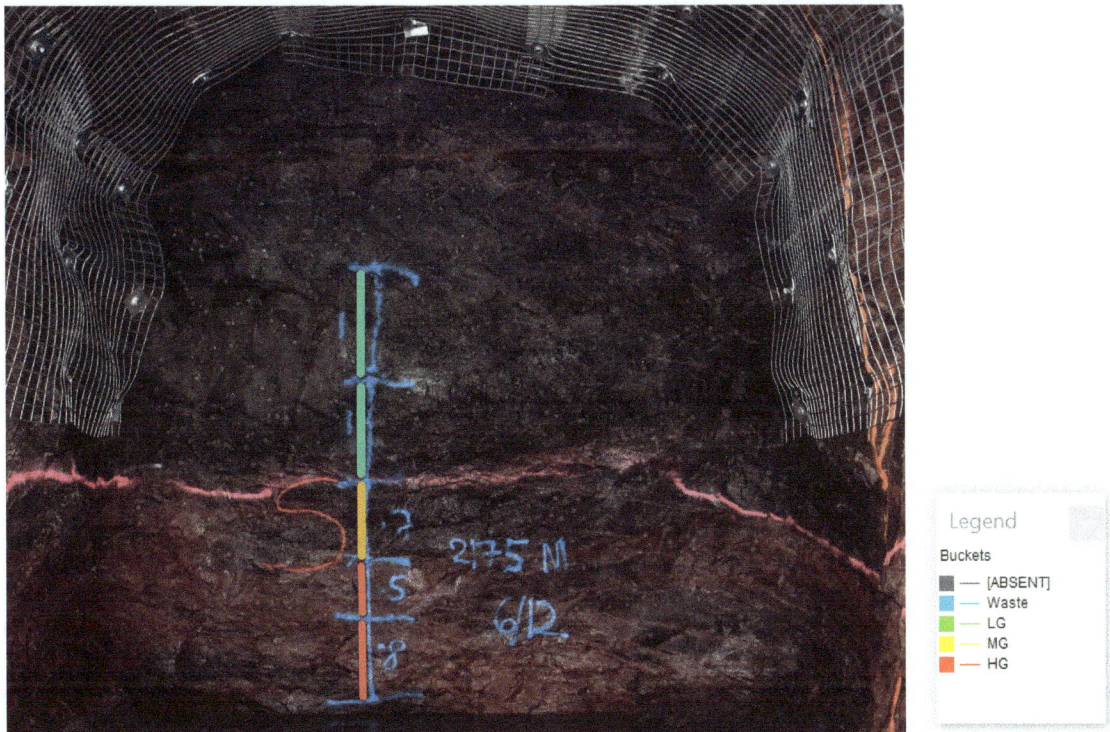

FIG 6 – Typical face sample line with category designations.

Table 3 compares the trial models to the grade control model across different grade bins in the Apollo domain. It evaluates the model's performance in tonnes, gold grade (AU g/t), and ounces against both grade control tonnage and grade estimation. The Δ Grade Control column shows differences between the trial model and the currently interpreted actual values. Discrepancies in tonnes (ΔT) and ounces (ΔOz) are highlighted to indicate estimation accuracy variations. Data is grouped into grade bins from 0 to 3.5, representing different intervals of gold grades. The Grand Total row summarises overall outcomes, including cumulative tonnes, gold grade, and ounces for the trial model, grade control, and budget estimates. This analysis aims to understand how effectively the trial model predicts and controls gold grades in the Apollo domain.

TABLE 3

Results table from trial model run compared back to the most recent grade control model split by grade.

	MODEL						Δ GRADE CONTROL			
	TRIAL MODEL			Grade Control						
Grade Bin	TONNES	AU g/t	OUNCES	TONNES	AU g/t	OUNCES	ΔT		ΔOz	
⊟ APOLLO	2,359,578	1.64	124,759	2,363,019	1.65	125,577	⬇-	3,442	⬇-	818
0	712,156	0.25	5,661	664,101	0.26	5,539	⬆	48,055	➡	122
0.5	5,778	0.75	140	24,663	0.63	497	⬇-	18,885	➡-	357
0.8	311,592	1.07	10,710	343,690	1.07	11,806	⬇-	32,099	⬇-	1,096
1.2	669,850	1.34	28,939	625,448	1.34	26,976	⬆	44,402	⬆	1,964
1.5	360,413	1.66	19,292	355,655	1.67	19,121	⬆	4,758	➡	171
2	20,479	2.15	1,417	29,140	2.17	2,038	⬇-	8,661	⬇-	621
2.5	569	2.61	48	9,612	2.80	866	⬇-	9,043	⬇-	819
3	224	3.37	24	12,517	3.22	1,297	⬇-	12,293	⬇-	1,272
3.5	278,518	6.54	58,529	298,194	5.99	57,438	⬇-	19,675	⬆	1,091
Grand Total	2,359,578	1.64	124,759	2,363,019	1.65	125,577	⬇-	3,442	⬇-	818

The trial model exhibits diminished granularity within each of the four categories as perceptible in the delta column of Table 3, where the gradations of gold concentrations tend to concentrate near the peripheries of each grade bin. Notably, material falling within the 0.2 g/t to 0.8 g/t range, following the model's selection process, are now inaccurately categorised as waste, thereby depleting the modelled ounces. Materials initially deemed mid-grade by mine geologists or category 3, which had been assayed to be high-grade (category 4), find themselves reassigned to the 1.2 g/t to 1.5 g/t grade bin. This shows us that geological observations are not quantitative, particularly when being applied under time pressure or from a distance.

The observed disparities in the trial model's classification of gold grades within distinct bins bear significant implications for resource estimation and decision-making processes within the mining operation. The inaccurate assignment of material, particularly the misclassification of mid-grade substances as waste and vice versa, has the potential to skew the overall understanding of the deposit's composition. This misalignment between the trial model and established grade control methods could lead to suboptimal strategic decisions and resource allocation. Geological observations are difficult to quantify, particularly when estimating the percentage abundance of a particular mineral in a sample. This can have major influence when determining what grade bin to allocate a sample to and plays a significant role in the observed results of this trial. It is imperative to critically assess the factors contributing to these discrepancies, such as the interplay between geological features and numerical values, to enhance the accuracy and reliability of future modelling efforts in mining operations.

Findings

The trial conducted in 2023, aimed at substituting numerical assay data with qualitative categorisation based on geological observations, did not yield the anticipated outcomes. The objective was to mitigate reliance on drill hole data by introducing a categorical attribute derived from geological features, intending to enhance the spatial range of data collection. However, the results indicate that the assignment of grade categories by mine geologists proved challenging. Particularly, the differentiation between low-grade and waste classifications proved unreliable. While high-grade classifications exhibited a notable correlation with assay data, the overall effectiveness of the trial fell short of expectations.

CONCLUSIONS

While face sampling offers valuable real-time information for grade control and operational decisions, its limitations in terms of representativeness, logistics and potential biases should be carefully considered. A combination of face sampling and other sampling methods, along with rigorous quality control measures, is employed at Porphyry Underground to obtain a more comprehensive understanding of the orebody. This author has found that a combination of data collection methods and an increased understanding of the geological characteristics of the deposit has proven crucial in mitigating the challenges encountered during the trial. Despite the deviations from the initial aim of this paper, the amalgamation of various sampling techniques has revealed a pragmatic solution that sufficiently fulfills the purpose of the mine geology department. This adaptive approach not only underscores the complexity of the underground mine geologist's role but also highlights the importance of flexibility in deviating from the standard practice to form an approach best fitting the specifics of the deposit.

ACKNOWLEDGEMENTS

The authors would like to acknowledge Northern Star Resources for their permission to use the data from Porphyry in this publication. The data used has been gathered over 2023 by the Porphyry Underground Geology team, the author is sincerely grateful to the team members who have given their time to collecting this data and the support they have provided during the time writing this paper. Finally, to Brett Thomas and Emma Murray-Hayden for their review of the manuscript.

REFERENCES

Allen, C A, 1987. The nature and origin of the Porphyry gold deposit, Western Australia, in *Recent advances in understanding Precambrian gold deposits* (eds: S E Ho and D I Groves), University of Western Australia, Department of Geology and University Extension, publication #11, pp 137–146.

Dominy, S C, Platten, I M, Fraser, R M, Dahl, O and Collier, J B, 2009. Grade control in underground gold vein operations: The role of geological mapping and sampling, in *Proceedings of the International Mining Geology Conference*, pp 291–307 (The Australasian Institute of Mining and Metallurgy: Melbourne).

Roberts, F I, Witt, W K and Westerway, J, 2004. Gold mineralisation in the Edjudina-Kanowna Region, Eastern Goldfields, Western Australia, Western Australia Geological Survey, Report 90, 263 p, ch 10.

Roberts, L S, Dominy, S C and Nugus, M J, 2003. Problems of sampling and assaying in lode-gold deposits – Case studies from Australia and north America, in *Proceedings of the International Mining Geology Conference*, pp 387–400 (The Australasian Institute of Mining and Metallurgy: Melbourne).

Swager, C P, 1995. Geology of the Edjudina and Yaboo 1:100,000 sheets, *Western Australia Geological Survey*, 1:100,000 Geological Series Explanatory Notes, 35 p.

SBRE workflow development for Nickel West – enabling decisions under uncertainty

C Williams[1], I Minniakhmetov[2], G Merello[3] and R Finch[4]

1. Principal Resource Modelling, BHP, Brisbane Qld 4000. Email: craig.williams1@bhp.com
2. Principal Global Modelling and Data, BHP, Perth WA 6000.
   Email: ilnur.minniakhmetov@bhp.com
3. Principal Resource Geologist, BHP, Perth WA 6000. Email: gianpiero.merello@bhp.com
4. Superintendent Geoscience Technical, BHP, Perth WA 6000. Email: richard.finch1@bhp.com

EXTENDED ABSTRACT

An understanding of the uncertainty related to geological variables, particularly those that are key drivers for grade control strategy and mineral project economics, is required to make informed project investment decisions. Conditional simulation is recognised as best practice for the quantification of geological uncertainty (Journel, 1974) which can in turn be used in optimisation studies, inclusive of grade control practices. BHP has developed the Simulation Based Resource Evaluation (SBRE) Python package as a standardised platform for the design, testing and running of conditional simulation workflows for assets within the Group. SBRE incorporates a modular workflow approach, allowing for assets to use existing software platforms as required for any stage in the workflow, whilst also providing a standard code base with access to the latest geostatistical techniques. Once adapted to the data and commodity specific requirements of each asset, workflows allow for automation and scalability through cloud computing, thereby providing wider access to conditional simulation for use in uncertainty quantification studies, within each asset. A case study is presented outlining the development of a SBRE workflow for Nickel West, using data from the Jericho Nickel Project.

Jericho Project geological setting

The Jericho Nickel Project is located in the northern portion of the Agnew-Wiluna greenstone belt of Western Australia. It is situated along a prominent north-west-south-east magnetic lineament, which connects with the Honeymoon Well Nickel Project to the north and the world-class Mount Keith nickel operation to the south (Figure 1). Nickel mineralisation within all three of these locations is hosted within steeply dipping Archean aged, ultramafic volcanic units (komatiites).

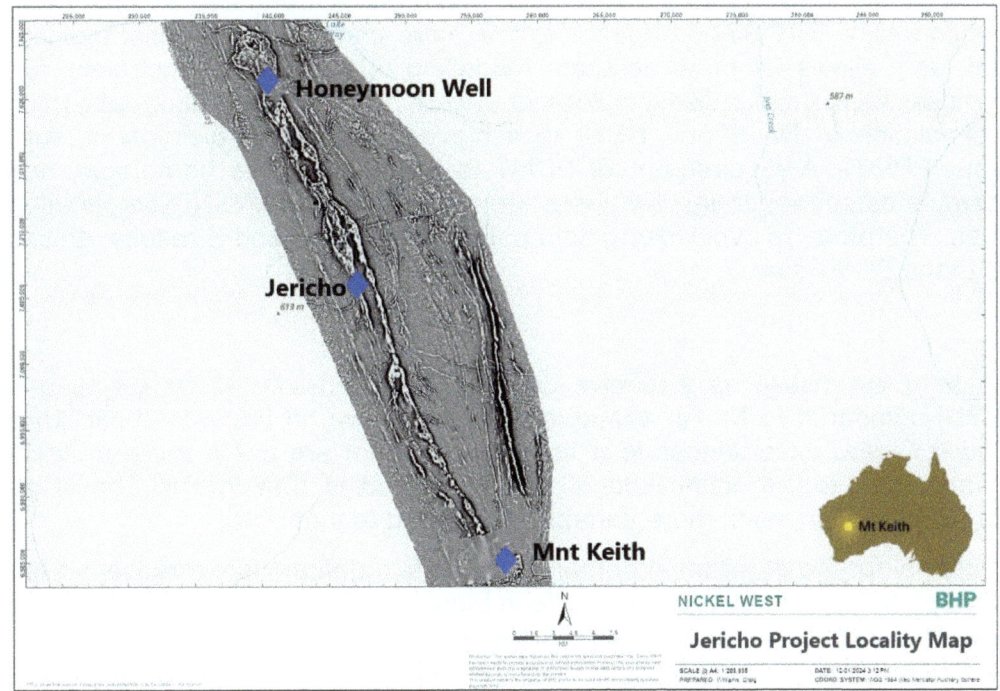

FIG 1 – Location of the Jericho Project.

The steeply dipping ultramafic units, as seen by the light blue wireframe in Figure 2, host nickel sulfide mineralisation (dark blue wireframe) and talc alteration (white wireframe), which are typically situated along the margins and within major structures of these ultramafic units. There are three weathering domains which have been simulated, one realisation of which is shown in the block model in Figure 2. An oxidised domain (yellow) exists at surface, followed by a transitional domain (red) and fresh domain (blue). Characterising the transitional domain is critical for stockpile hygiene, as well as to determine project sequencing, strategy and fundamental economics, as it contains a complex assemblage of primary, secondary and oxidised nickel bearing minerals. Predicted and realised processing outcomes (namely recovery and concentrate quality) are particularly sensitive to the characteristics of the weathering and mineralisation systems of nickel sulfide deposits with stockpiling strategy playing a key role to produce a stable supply chain. In lieu of quantitative mineralogy data, non-sulfide Ni (NSNI) analysis provides a valuable and readily available semi-quantitative proxy for determining mineral assemblage and degree of oxidation.

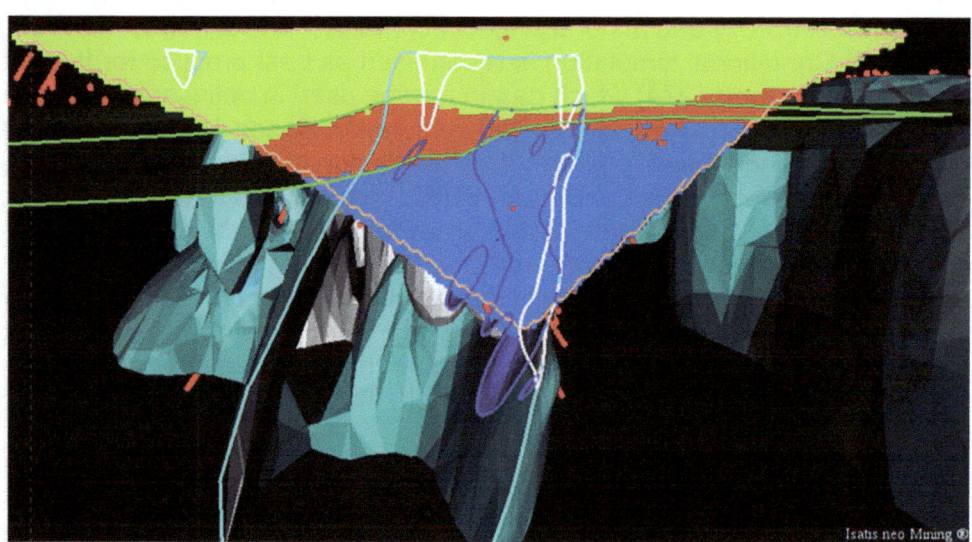

FIG 2 – West–east cross-section showing geological domains inside the economic pit limit.

SBRE workflow development for the Jericho Project

The SBRE conditional simulation workflow developed for Nickel West incorporates categorical domain simulation, which employs the advanced hierarchical truncated pluriGaussian simulation (HTPGS) methodology. HTPGS is superior to the more common sequential indicator simulation (SIS) method, as it allows for more accurate modelling of geological constraints, for example at Jericho, the nickel sulfide mineralisation does not occur outside of the ultramafic unit. Multivariate conditional simulation of Ni, S and NSNI is achieved by using projection pursuit multivariate transformation (PPMT). A requirement of PPMT is that the data is homotopic, however in the composite sample database for Jericho, there are significantly less NSNI samples compared to Ni and S samples. Therefore, to avoid losing data points with only Ni and S results, an imputation step is used prior to the PPMT step.

Results

Grade uncertainty, expressed as a relative percentage uncertainty of the e-type grade, is much higher for NSNI compared to Ni. An example of this is shown in the west–east cross-sections in Figures 3 and 4, where uncertainties less than 100 per cent are much more restricted for NSNI, found only very close to drill sample locations, compared to that for Ni. The single simulation realisation for NSNI is also much more variable, compared to that of Ni.

Volume uncertainty for the transitional domain and the talc domain is expressed in terms of the e-type probability of the domain at each grid node in Figures 5 and 6. The red 100 per cent probability zone is narrower than the deterministic wireframe, especially towards the edge of the planned pit, where there is less drilling. This indicates that the deterministic wireframe volumes are likely to overestimate the domain volume, compared to the e-type volume.

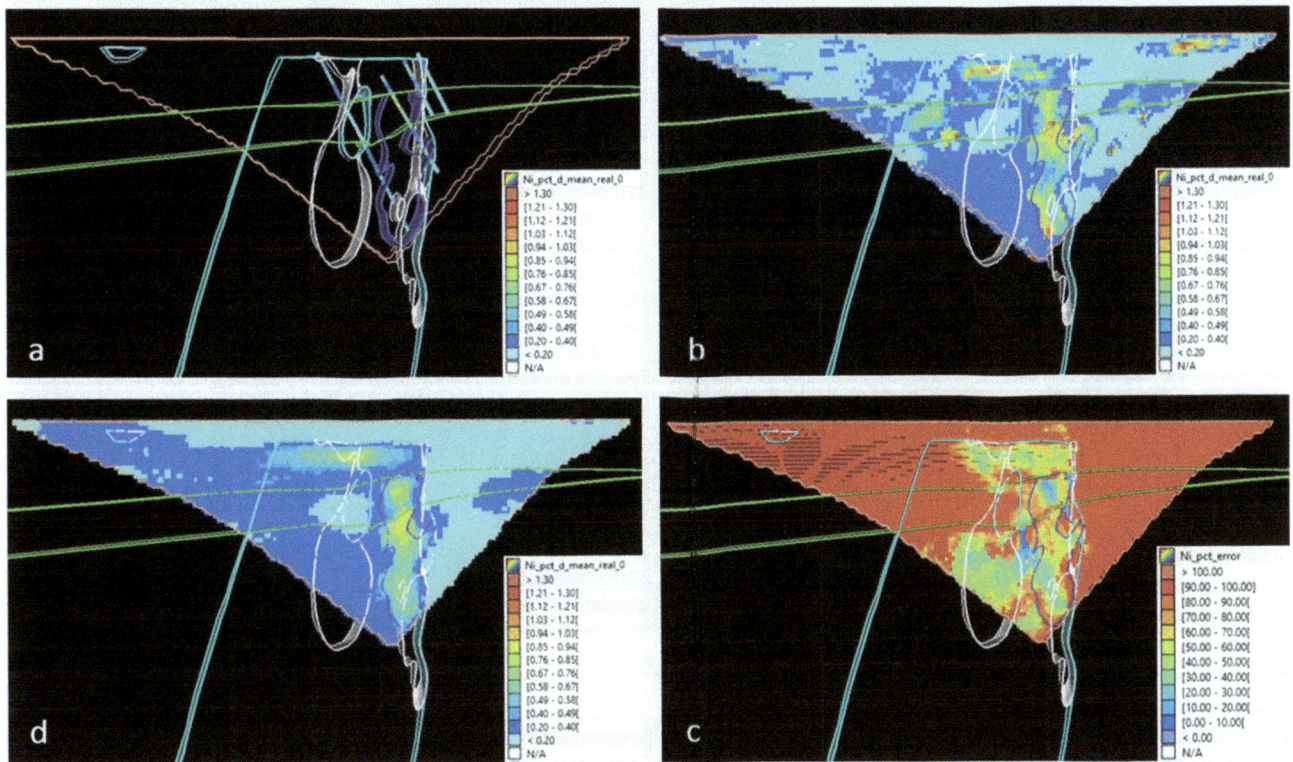

FIG 3 – (a) Ni drill samples, (b) single realisation of Ni grade, (c) Ni percentage uncertainty, (d) e-type Ni grade green line = transitional domain, blue line = ultramafic, dark blue line = nickel sulfide, white line = talc.

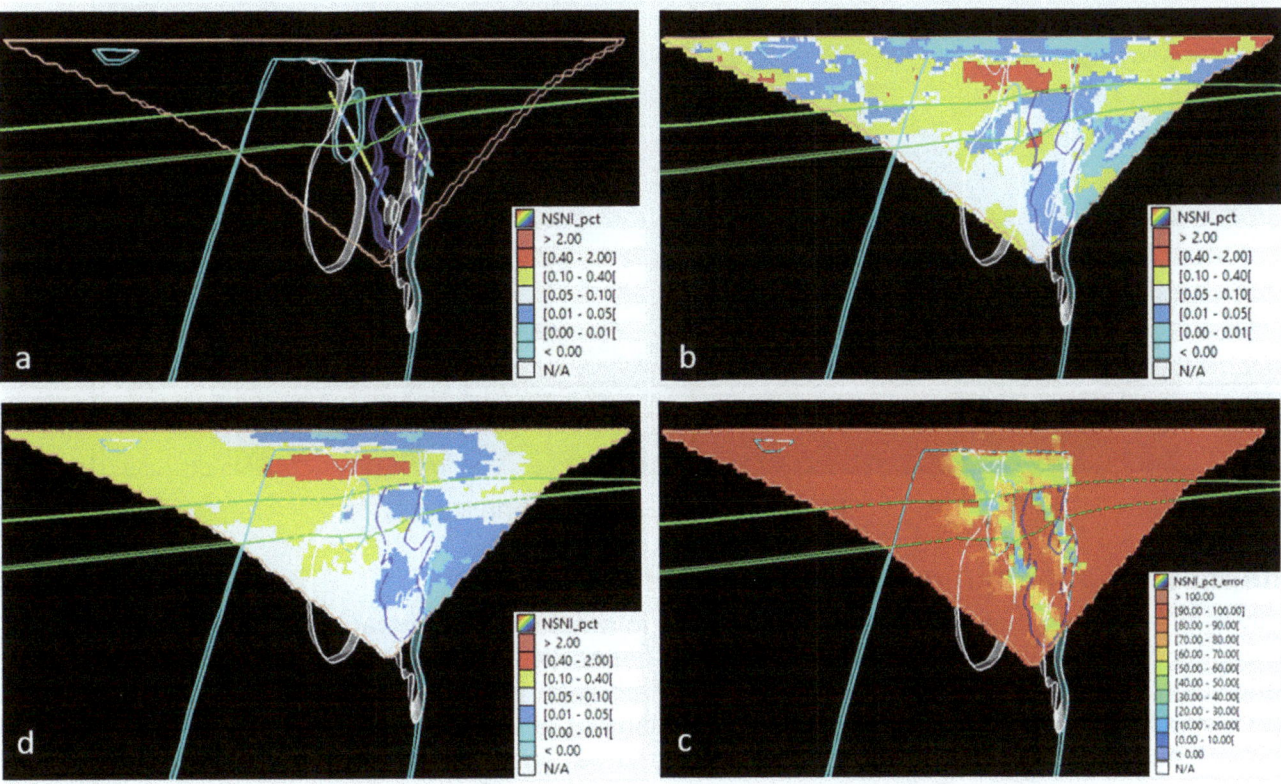

FIG 4 – (a) NSNI drill samples, (b) single realisation of NSNI grade, (c) NSNI percentage uncertainty, (d) e-type NSNI grade, green line = transitional domain, blue line = ultramafic, dark blue line = nickel sulfide, white line = talc.

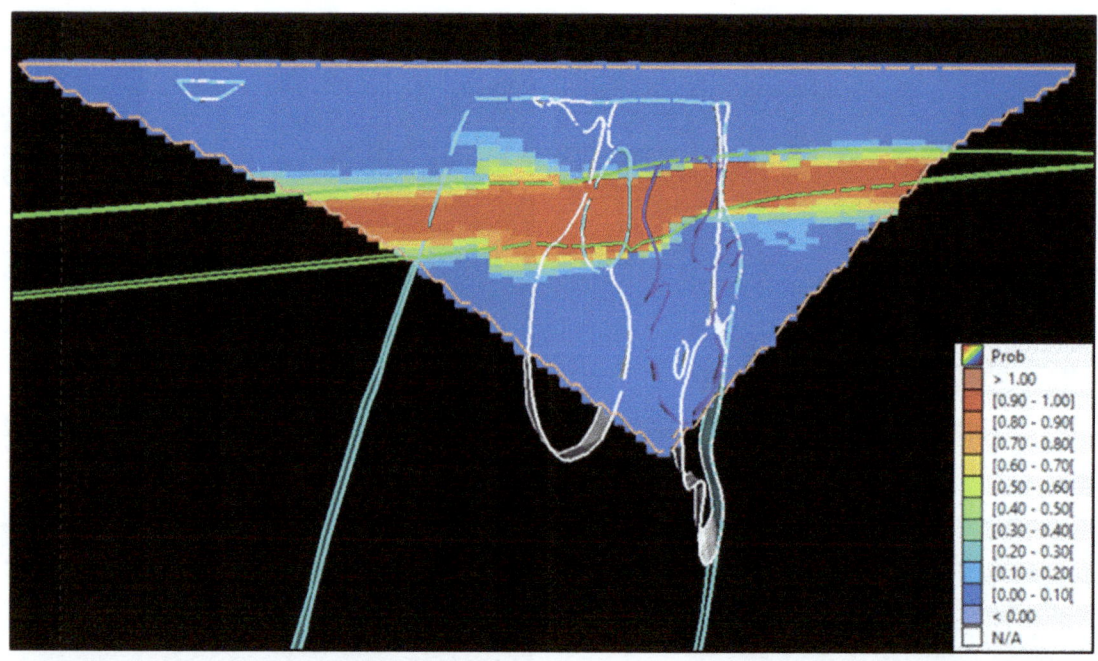

FIG 5 – Transitional domain probability.

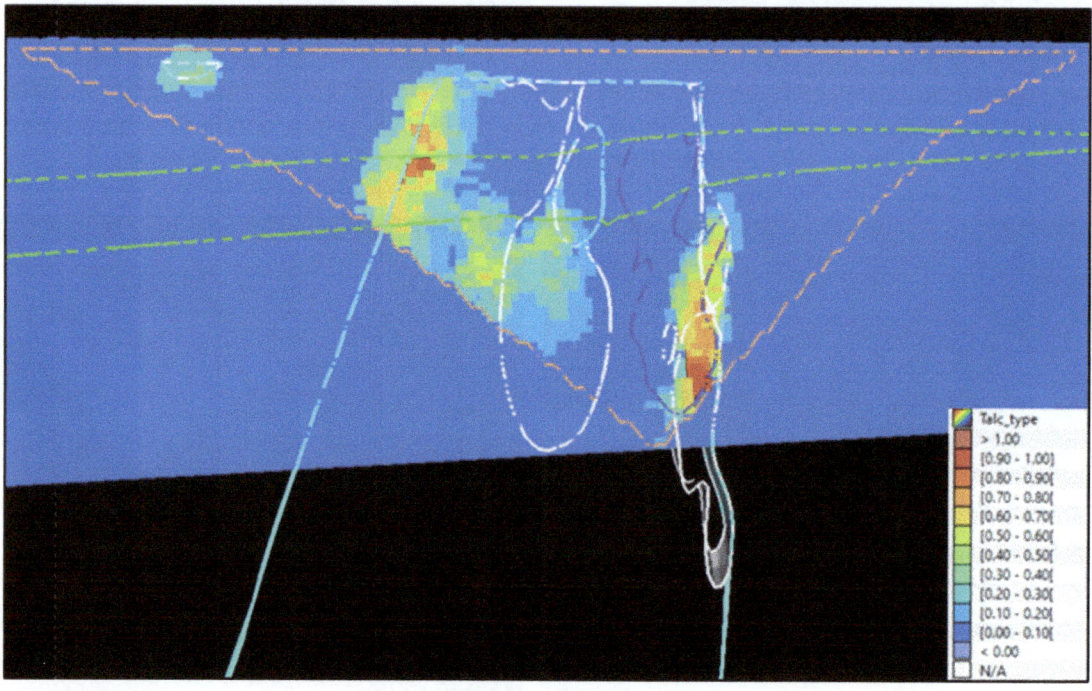

FIG 6 – Talc domain probability.

Implications for mining and exploration

Given the key role that the NSNI estimate plays in determining mineral assemblage and degree of oxidation for mine planning and grade control purposes, the quantification of the uncertainty in grade and volume for the Jericho Nickel Project, particularly with respect to the highly variable NSNI, allows for a much more robust techno-economic assessment of the project. An example of the potential project sensitivity to NSNI, compared to Ni and S, is illustrated in the transitional domain grade tonnage curves in Figure 7. The grade tonnage curves for different NSNI realisations show that the range in potential NSNI grades for a given Ni cut-off grade widens significantly with increasing Ni cut-off grade.

In addition to allowing for a more robust assessment of project economics, the Jericho Nickel Project conditional simulation study clearly shows that the current level of confidence in volume and grade is too low for detailed mine planning purposes. This provides an important motivation to increase drilling density and thereby reduce the project risk.

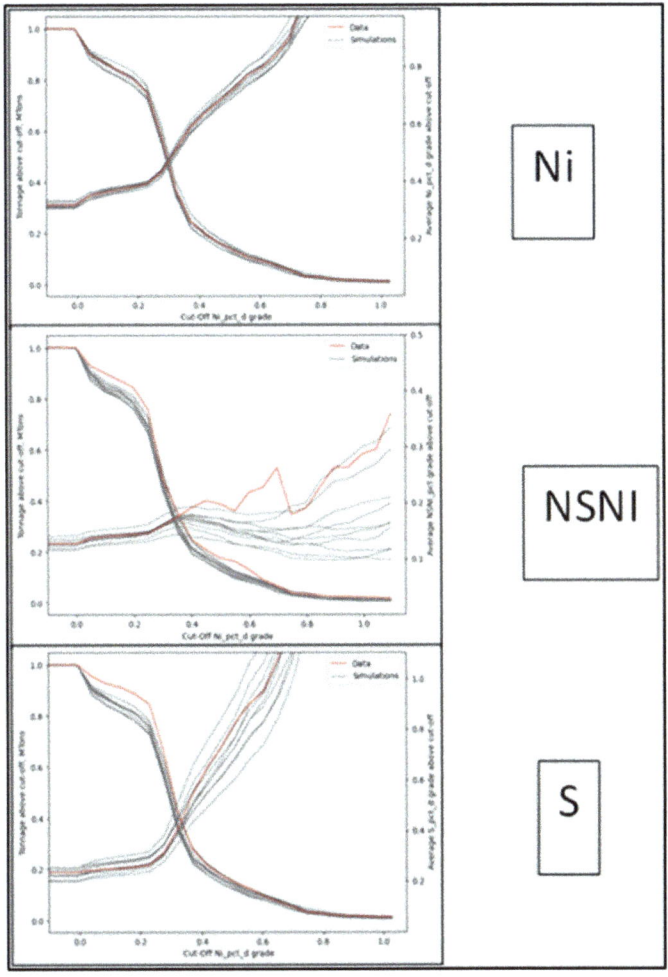

FIG 7 – Transitional domain grade tonnage curves at a Ni cut-off grade for Ni, NSNI and S.

ACKNOWLEDGEMENTS

The authors would like to acknowledge BHP for the permission to publish this paper. We would also like to thank Kerry Turnock and Ranjan Saha for enabling this study, by allowing for the time and resources to complete the SBRE Nickel West project.

REFERENCES

Journel, A G, 1974. Geostatistics for Conditional Simulation of Ore Bodies, *Economic Geology*, 69:673–687.

Excellence in mine geology – delivering value

Mineable Shape Optimiser and ore wireframing, analysis of tools for scheduling open pit mining at KCGM

J Ball[1]

1. Geology Superintendent Operations, Northern Star Resources, Subiaco WA 6008.
   Email: jball@nsrltd.com

ABSTRACT

The ability to quantify mineable tonnes and grade at different stages of the mining process, ranging from resource evaluation to final ore block design, is of paramount importance in mining operations. Ideally, these predictions should be generated quickly, consistently and should reconcile easily throughout each phase of the mining cycle. Without this it would not be possible to adhere to mining or processing plans. This research evaluates a new method of prediction being trialled at Northern Star's Fimiston Open Pit.

The Fimiston Open Pit is part of the Kalgoorlie Consolidated Gold Mine operations (KCGM), in Western Australia's Golden Mile. Mining on the Golden Mile has been continuous for over 125 years, while the pit is currently 700 m deep, 4 km long and comprises of over 500 discrete ore lodes. The pit has been in operation for over 30 years and currently has over ten years of mine life remaining.

Due to the complexity and number of ore lodes at Fimiston combined with the use of large mining equipment, the quantification of mineable tonnes and grade is challenging. Although existing methodologies such as manual digitisation of ore blocks are suitable for large, continuous, simple lode systems, at KCGM a shift in geological characteristics and lode structure in the current mining areas has resulted in greater variance to quantifiable tonnes and grade. To create a more predictive and consistent result that maximises mineable high-grade ore an alternate process was required.

Mineable Shape Optimiser (MSO) a traditionally underground tool, has been analysed as a tool to create mineable shapes in the pit at all mining stages, from resource evaluation to final ore block designs. Utilising this tool within an open pit setting could facilitate faster scheduling and ensure consistent ore block designs across diverse mining parameters, accommodating various lode geometries while addressing historical mining voids challenges.

INTRODUCTION

Since the discovery of gold in 1893, the Golden Mile has produced in excess of 60 million ounces. And since its discovery, a diverse range of mining methods have been employed, ranging from small-scale handheld operations to extensive stoping and open pit mining. Before the consolidation of mining operations along the Golden Mile in 1989 and the commencement of the Fimiston Open Pit also known as The Super Pit, mining practices were more selective, reflecting the higher head grade requirements of that era.

In recent times, a combination of robust commodity prices and an approach focused on high volume and low costs has enabled KCGM to extract lower-grade ore profitably. While traces of high-grade lode mineralisation still exist, the predominant source of mined ore comes from the lower-grade halos/stockworks surrounding the high-grade core. This has become the cornerstone of a highly successful open pit business over the past 30 years and it is anticipated to remain a profitable operation for the remaining mine life. The total Mineral Resources and Ore Reserves for the KCGM operations as of 4 May 2023 are: 565 Mt @ 1.6 g/t for 28 Moz; and 286 Mt @ 1.3 g/t for 12 Moz respectively. With Fimiston open pit Mineral Resources comprising of 320 Mt @ 1.6 g/t for 16.9 Moz and Ore Reserves of 146 Mt @ 1.7 for 8.1 Moz.

Quantifying mineable tonnes and grade at Fimiston, given its complex ore lodes and use of large mining equipment, poses a considerable challenge to mine geologists and mine planners. Traditional methods, such as manual digitisation of ore blocks, suits simpler lode systems. However, the recent addition of new cutbacks, with more challenging mineralisation, demands a more accurate predictive approach. In response, the Mineable Shape Optimiser (MSO), traditionally an underground tool, was investigated for open pit use. This tool offers accelerated scheduling and ensures consistent ore block designs, accommodating diverse mining parameters and addressing challenges posed by historical mining voids.

GEOLOGY

The Golden Mile, part of the Kalgoorlie Goldfields in the Norseman-Wiluna Greenstone belt of the Yilgarn Craton, is situated within the Kambalda Sequence of the Kalgoorlie Terrain. Bounded by the Ida and Mount Monger faults to the west and east, respectively, the terrain consists of mafic volcanic and volcaniclastic sediments. Figure 1 illustrates the NNW trending Archaean greenstone belt, which has undergone successive deformation and regional metamorphism, ranging from lower greenschist to amphibolite facies (Clout, Gleghorn and Eaton, 1990).

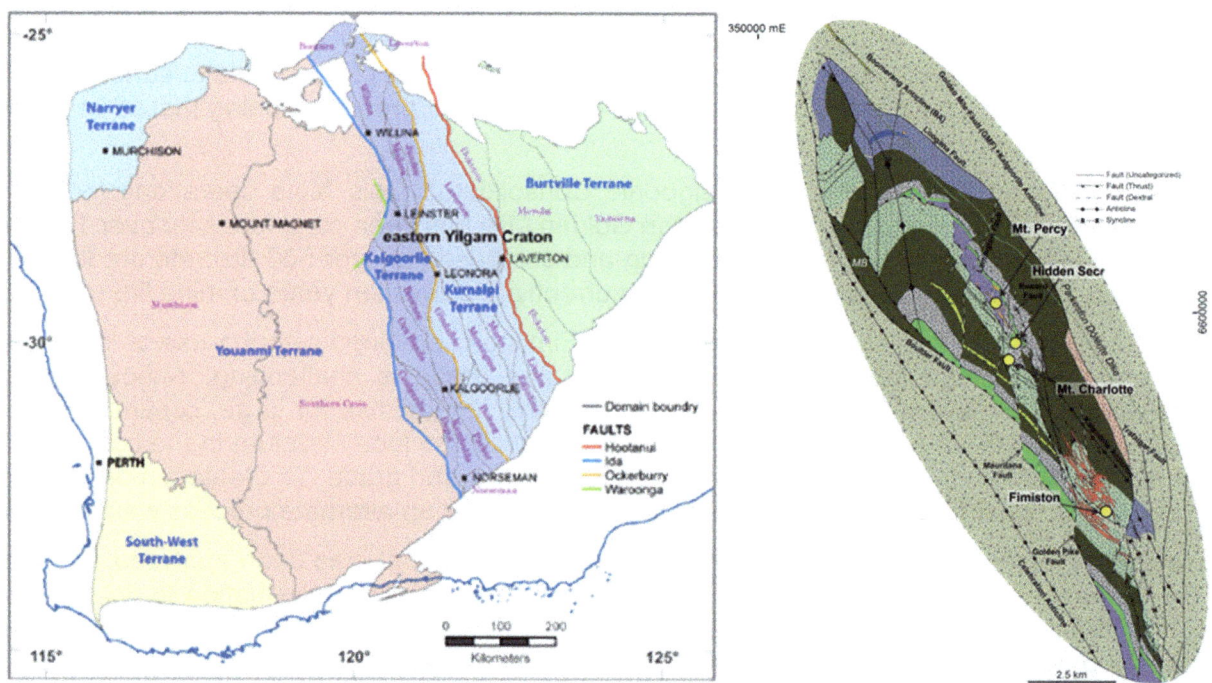

FIG 1 – Regional geological setting of Fimiston (Czarnota, Blewett and Goscombe, 2010). (left) Camp-scale geologic map of Golden Mile with stereonets of bedding, foliation and litholigic contacts. (right) Stratigraphic column of Golden Mile (McDivitt *et al*, 2020).

LOCAL GEOLOGY, GOLD MINERALISATION AND FIMISTON LODES

The stratigraphy of the Kalgoorlie Terrane consists of a lower mafic-ultramafic volcanic sequence with associated sub-volcanic sills overlain by a thick sequence of clastic sedimentary rocks and intermediate to felsic volcaniclastic rocks (Swager, 1997). This stratigraphic sequence was broadly deposited between 2708 ± 7 Ma and 2681 ± 5 Ma (Nelson, 1997) and was later intruded by mafic tholeiitic and komatiitic sills (Witt, 1995).

A robust stratigraphic model has been defined based on field and diamond core observations, supported by extensive geochemical data sets (Gauthier *et al*, 2004). The Hannans Lake Serpentine (HLS) underlies the Kalgoorlie Sequence and is overlain by the Devon Consols Basalt (DCB). This is overlain by Kapai Slate and Williamstown Dolerite. The Paringa Basalt overlies the Williamstown and can be divided into a lower, Cr-rich basalt and upper, pillowed basalt. Oroya Shale separates the Paringa from the Golden Mile Dolerite (GMD). The GMD has been separated into 10 sub-units based on textural observations and geochemistry and intrudes the base of the Black Flag Group near the contact with the Paringa Basalt.

Hydrothermal alteration associated with the emplacement of auriferous veins also took place along brittle–ductile structures. Gold mineralisation has been subdivided into three styles: an older and protracted Fimiston-style vein-forming event, an Oroya-style vein-forming event and a younger Charlotte-style stockwork vein-forming event (Clout, Gleghorn and Eaton, 1990). All mineralisation styles exhibit strong proximal hydrothermal alteration envelopes containing variable amounts of muscovite, carbonate, pyrite/pyrrhotite, quartz, hematite, magnetite and albite (Clout, Gleghorn and Eaton, 1990).

Granophyric Unit 8, within the GMD is the main host rock for mineralisation in the Fimiston deposit and hosts two styles of mineralisation, Fimiston-style Lodes and Charlotte-style stockwork veins. Fimiston-style mineralisation is complex and is characterised by steeply west-dipping lodes. These lodes are associated with brittle-ductile shears forming broad low-grade halos that extend >1 km both laterally and vertically. There are four main lode types within the Fimiston Style of mineralisation (Figure 2), these are separated into; Main (350°–10°), Cross (85°–135°), Caunter (315°–350°) and Oblique (010°–030°) (Gauthier *et al*, 2004).

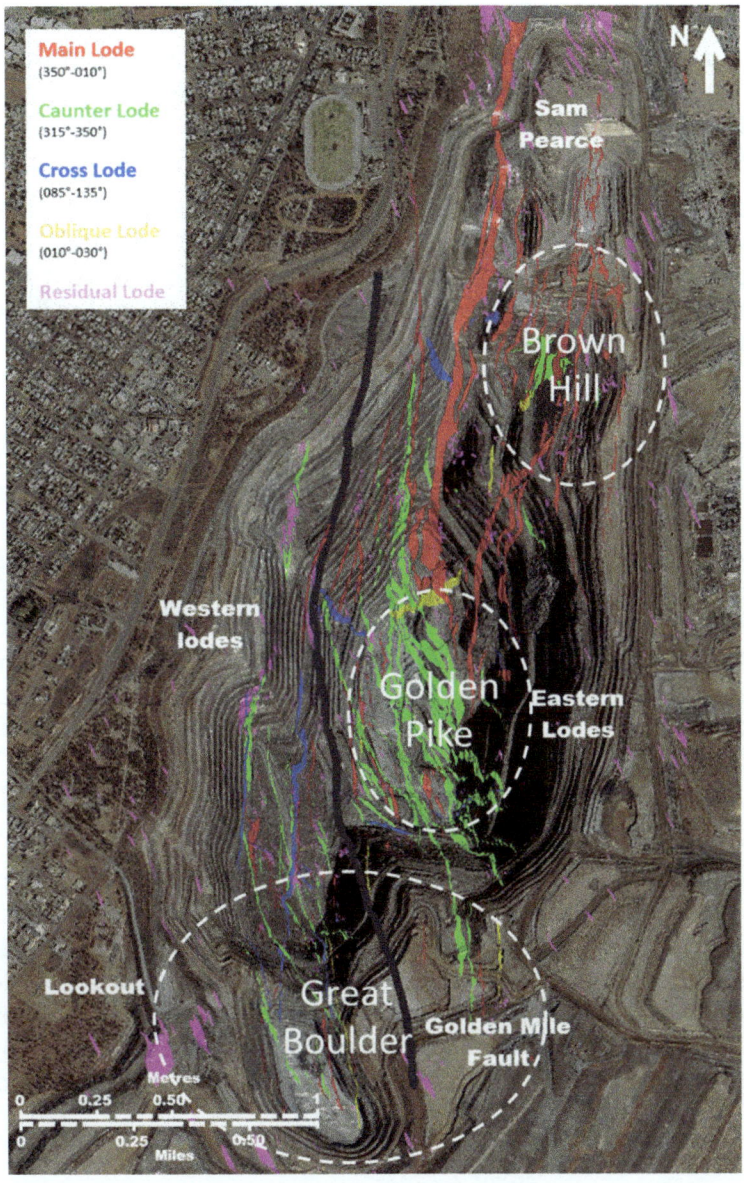

FIG 2 – Fimiston Style mineralisation lode types, the Sam Pearce decline and public lookout are shown for reference.

Mining history

In June 1893, Irish prospectors Paddy Hannan, Thomas Flanagan and Dan Shea made the discovery of gold near Kalgoorlie. Within five years, the Golden Mile had produced half a million ounces of gold, leading to the establishment of the twin towns of Kalgoorlie and Boulder. In the decade following the initial find, there were 49 operational mines, 42 mining companies, 100 headframes and over 3.5 km of underground workings. In the 1980s the mines and leases were purchased by Alan Bond and Homestake Gold, who formed a joint venture, Kalgoorlie Consolidated Gold Mine (KCGM). After numerous further transactions, in February 2021, a merger of equals was finalised between Saracen Mineral Holdings and Northern Star Resources, resulting in the Golden Mile coming under a single owner for the first time in its history.

OPEN PIT SCHEDULING

Current status

Open pit mining at Fimiston occurs amid the once 49 mines and 3500 km of underground development and therefore not only poses challenges to the geologist but to all personnel within the mine to ensure safe production. To mitigate the risk of voids and mining around historic workings, current mining is conducted using large scale equipment; four Komatsu PC8000 face shovels, one Liebherr 9400 excavator and Cat 793F trucks with a payload of 250 t. Again, due to the hazardous nature of mining around the 125 years of void left beneath the current surface mining takes place on 10 m high benches. The use of this large equipment and safety constraints results in a Selective Mining Unit (SMU) of 10 m height, 4 m width, and 8 m strike resulting in a scale and therefore tonnage that is relatively large when compared with other lode hosted open pit gold operations.

There are two types of ore blocks generated at Fimiston, high-grade blocks are those that are above a diluted grade of 0.9 g/t and subgrade blocks that are between 0.5 g/t and 0.9 g/t. High-grade blocks are sent directly to the processing plant while subgrade blocks are sent to long-term stockpiles. These subgrade stocks will be processed in the future when the current mill expansion is completed and KCGM becomes mine constrained.

P-CLASS

The SMU of 8 m × 4 m × 10 m sets the basis of constraints of ore block design for the geology team within Fimiston and when combined with the complex and varied lode geometry presents a variety of challenges for the mine geologist. For the use of scheduling and forecasting mining tonnes and grades a dilution macro named 'Product Class' (P-CLASS) has been created. Developed in 1999, within Datamine, the P-CLASS macro generates a re-blocked block model using SMU parameters, cut-off grades and lode type to create mineable ore blocks. However, when comparing the tonnes and grade between the P-CLASS macro-output, and the reconciled actual mined tonnes and grade, a discrepancy has been observed in the narrower lodes within the recent cutbacks of Brown Hill and Great Boulder. This discrepancy, as seen in Figure 3 shows the actual mined tonnes are generally lower than predicted, but the grade is higher for cumulatively less ounces. This discrepancy has impacts not only on the processing plants schedule waste dump management.

P-Class works by generating mineable shapes using the SMU (with a 1 m skin applied). This is effective in large, bulk mining areas, such as Golden Pike, which is dominated by continuous 'Main Lodes' that are up to 40 m in thickness. Figure 3 shows comparatively much improved reconciliation of reserve to ore mined tonnes from November 2021 to May 2023. During this time Golden Pike accounted for an average of 52 percent of open pit mining. The months prior to this, from July 2020 to January 2021 Golden Pike accounted for an average of 28 percent of open pit mining. The sub graphs demonstrate the results from mining Brown Hill and Golden Pike individually, showing a roughly 30 per cent variance of P-Class to actual mined, with Brown Hill displaying far greater variances.

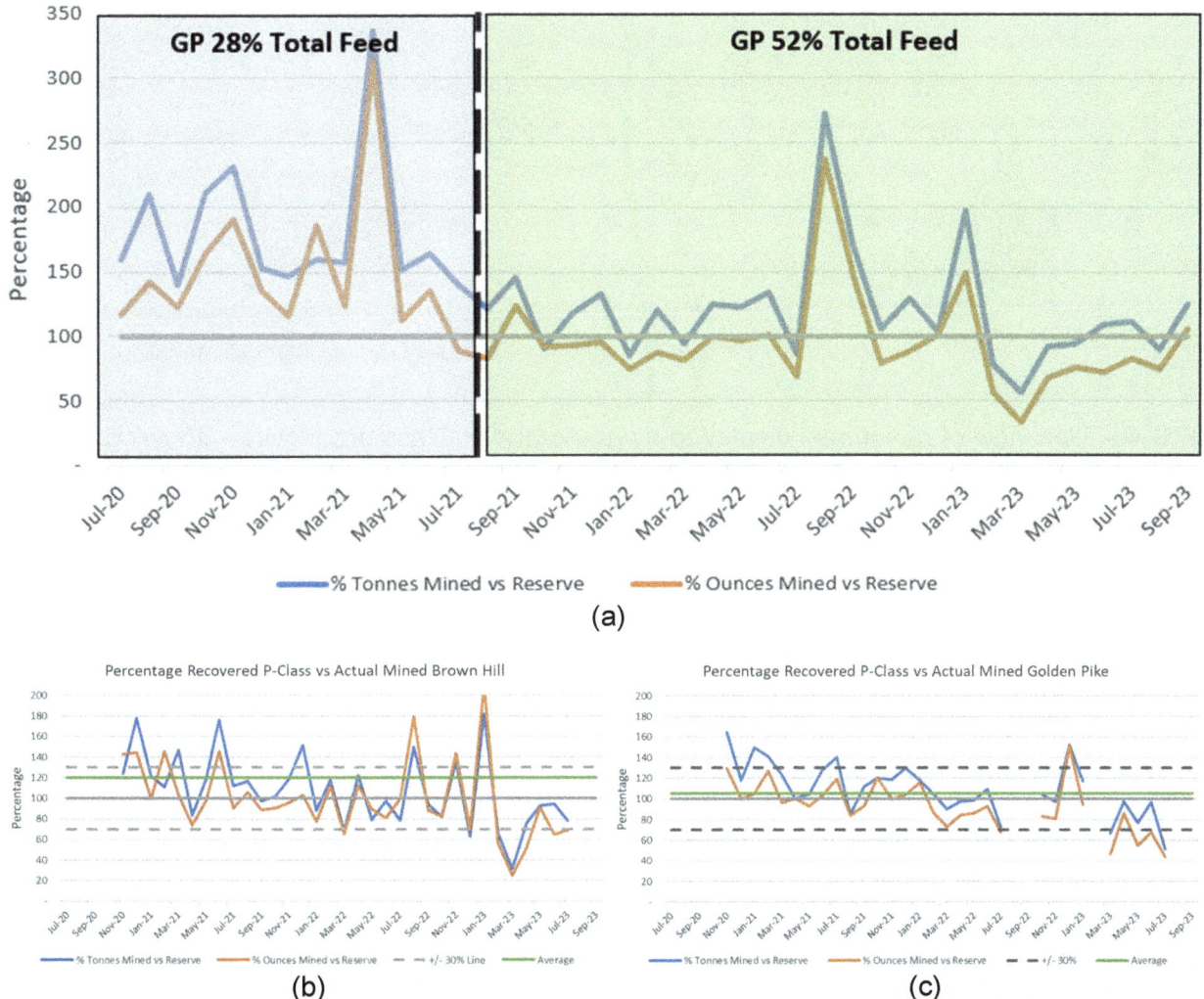

FIG 3 – Percentage recovered P-Class versus Actual Mined. When Golden Pike (GP) is a small proportion of the ore mined, the PCLASS macro is less accurate. (a) Total Reserve versus Actual Mined; (b) Brown Hill Total Reserve versus Actual Mined; (c) Golden Pike Reserve versus Actual Mined. Gaps present in Golden Pike data where no mining occurred.

Where there are smaller narrow lodes or stacked lodes, such as in Brown Hill or Great Boulder, P-Class struggles to make practical mining shapes. It will often generate small 'islands' of sub-grade blocks, within a larger high-grade area or *vice versa*. This is not practical to mine in real-world scenarios and is unlikely to be reproduced by a geologist when designing an actual ore block to be dug. Figure 4 shows an example of where the P-Class has generated isolated blocks of grade within the Brown Hill Cutback. Although these may be accurate, small blocks tend to be diluted resulting in high-grade blocks becoming low-grade or sub-grade.

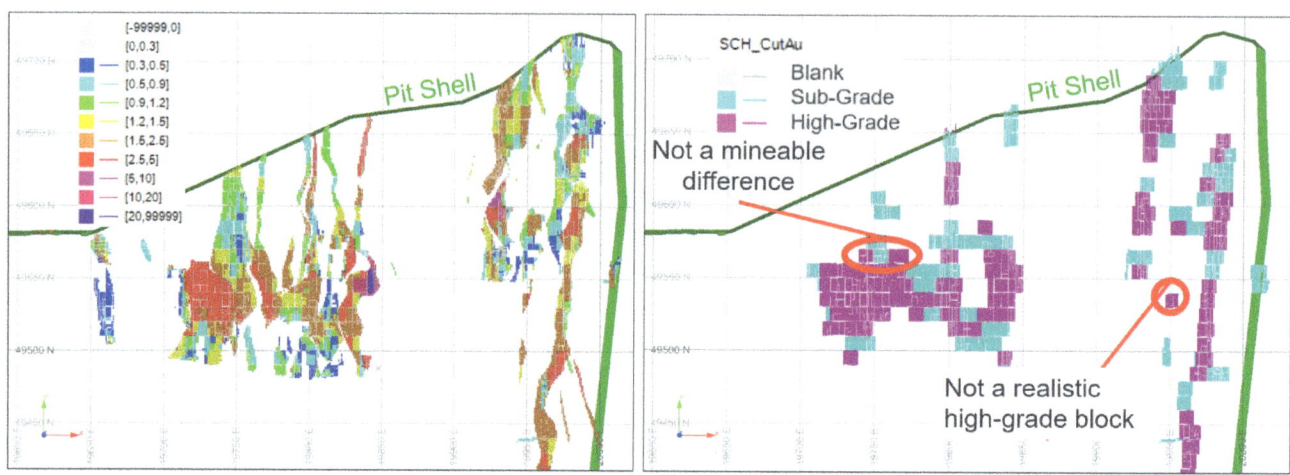

FIG 4 – Plan view of raw model displayed by grade and P-Class ore blocks – showing false or unrealistic mining shapes.

ORE BLOCK GENERATION

At KCGM, ore block design involves a manual digitisation process for each blasted shot. This occurs on a 'moved' or 'translated' block model derived from pre-blast and post-blast scans of the muck pile. These marked-out blocks serve as the final blueprint for mining operations. However, long-term planning integrates both manual geologist designs and computer-aided designs. When creating ore blocks, factors such as SMU (Selective Mining Unit), ore lode characteristics (including dip, strike, grade homogeneity, thickness) and dig direction are considered. The final consideration for the geologist is to understand the likelihood of dilution occurring within each block. The strategy employed at Fimiston is known as the 'tight podding strategy', or the design of ore block boundaries within the ore wireframe, rather than on the boundary. This ensures that blocks when blocks are diluted, as they invariably are due to the machinery used, the dilution is of economic grade. The extent of the strategy applied is based on several factors, the main one being digging direction.

Digging direction is of vital importance when mining to maximise recovery of ore material, as well as understanding the dilution impacts when mining. The Fimiston lodes typically exhibit a westward dip in a mine grid direction (magnetic south-west), ranging from 65° to 90°. The decision to mine from west to east in the mine grid was based on the belief that this approach would minimise dilution due to the shovel's capabilities and bucket movement, enabling more selective extraction of mineralised lodes. The digging direction of individual grade control ore blocks was documented and subsequently compared to the grade control data. A significant correlation was identified between the digging direction and the percentage of dilution (over mining) and the results can be seen in Figure 5 (Cook and Fitzgerald, 2011). As a result, during the ore block design process the digging direction is determined and ore blocks are designed either dipping, if mining from west to east, or vertically if mining from north–south. Mining from the east to west is avoided wherever possible. Occasionally unfavourable digging directions must be employed due to several constraints such as historic workings or highwall proximity.

FIG 5 – Summary of the over-mining of ore block designs within the Fimiston Pit classified by the direction in which the blocks were mined (known in open pit operations as digging direction). Mining from west to east was the most favourable orientation in which to extract ore. Conversely east to west mining caused the most over-mining because of the extra dilution extracted to minimise ore loss (Cook and Fitzgerald, 2011).

HUMAN RESOURCES

Relying on a people and process driven systems for designing ore blocks can prove challenging. As described in detail by Cook and Fitzgerald (2011), KCGM has a high staff turnover rate, with varying degrees of experience. The contrast of experience in geologists is clearly highlighted when ore blocks are reviewed and compared. Figure 6 shows potential variance when solely relying on geologists to design and create ore blocks.

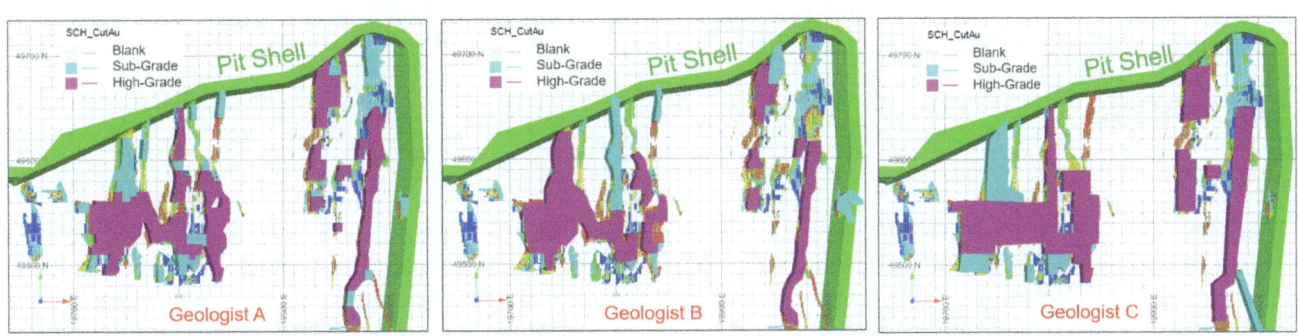

FIG 6 – Plan view displaying blockouts generated by three different geologists displaying high-grade and subgrade ore block designs, pit shell and raw grade control model.

Each of the three geologists of varying experience and tenure, created ore block designs for one blast in the Brown Hill area of Fimiston pit. This area is known for a variety of lode types, with a mix of Cross, Caunter, and Main lodes, which results in a complex ore block design as orebodies with differing strike and dip must be evaluated together. Each of these three block designs conform to the parameters governed by the SMU and are technically all correct, despite the obvious difference in quantifiable ounce recoveries (Table 1).

TABLE 1

Resultant tonnes, grade and ounces derived from ore block designs from a mining bench in the Brown Hill cut back.

	HG (>0.9 g/t)			Subgrade (0.5–0.9 g/t)		
	Tonnes	Au g/t	Oz	Tonnes	Au g/t	Oz
Geologist A	285 313	2.05	18 805	78 819	0.70	1774
Geologist B	255 958	2.06	16 952	45 506	0.81	1185
Geologist C	345 283	1.68	18 650	81 932	0.66	1739

Each geologist's ore block design demonstrates varying levels of complexity and unique block shape concepts, influenced by individual interpretation and human factors. With a tonnage variation of almost 90 kt and over 1800 ounces of high-grade ore between the three geologists, the impact on processing schedules, stockpile and waste dump design as well as overall cash flow cannot be underestimated. However, it is these geologists who currently generate designs for scheduling purposes within Fimiston, emphasizing the need for repeatability and applicability across all pit areas in any scheduling tool utilised. Relying on geologists to design ore blocks is also an intensely long and time-consuming process. This is not feasible for Life of Asset (LOA) planning time frames, nor would there be the numbers of sufficiently skilled geologists to complete such work as a single 10 m bench can take up to several hours to complete. Therefore, in practice, a combination of computer generated (P-Class) and manual generated ore blocks are used.

FORECASTING

With a repeatable methodology, run by computer, it would be possible to generate the same ore block product on a Life of Asset, Long (two-year), Medium (quarterly) and Short (weekly) time frame. The ability to forecast a repeatable result also aids the post-blast ore block process whereby a 'moved' block model can be used to generate post-blast ore blocks for mining.

Comparing various schedules relative to the same parameters greatly enhances reconciliation functions, enabling more accurate assessment of change. This is achieved by eliminating variations stemming from inconsistent application of rules, such as those highlighted among different geologists. By employing consistent and repeatable parameters across models for Resource, Reserve and Grade Control, one variable is removed when comparing results. This facilitates the identification of genuine mineable changes in models resulting from drilling or interpretation adjustments. This enables changes between models, impacts of blast dilution and mine design to be interrogated more thoroughly and highlight areas for improvements.

SOLUTIONS

MSO

The creation of a more predictive and consistent result that maximises mineable high-grade ore in a manageable time frame, with repeatability, lead the geology team to the investigation of the MSO function within Datamine. Traditionally, this tool is used to design stope shapes in underground mines for economic extraction and relies on given parameters. Similar to P-Class, this tool also uses minimum dimensions for ore block strike, height, etc. However, the MSO has allowances for changes in dip angles, strike angles and apparent widths. MSO also allows the integration of a preferred dig direction, or ore block orientation. The results are visual and are easy to compare to the original block model.

A variety of MSO scenarios were completed and compared to manual design in various locations throughout the pit. The scenarios differed by changing the minimum block widths and strike lengths, the distance between blocks and other parameters. The aim of each scenario was to replicate the high-grade tonnes and ounces designed by a geologist. The results of a selection of the most suitable scenarios are displayed in Table 2.

TABLE 2

Results of various MSO runs on Brown Hill pit, compared with geologist ore block designs and P-Class, with Scenario 2 highlighted as the best comparison found.

	Category	Tonnes	Au g/t	Oz
Scenario 1	HG Total	332 692	2.54	27 127
	Waste	49 283	0.08	121
Scenario 2	HG Total	703 214	1.90	43 034
	Waste	175 463	0.18	1016
Scenario 3	HG Total	685 406	2.01	44 185
	Waste	141 491	0.19	877
Scenario 4	HG Total	664 957	2.12	45 250
	Waste	102 561	0.21	684
Scenario 5	HG Total	684 859	2.00	44 055
	Waste	140 614	0.15	662
Scenario 6	HG Total	590 225	1.94	36 883
	Waste	127 476	0.14	561
Geologist	HG	1 011 056	1.51	48 937
	SG	112 934	0.46	1672
	Waste	3 793 438	0.09	10 991
P-Class	HG	714 224	1.81	41 477
	SG	457 344	0.62	9061
	Waste	11 080 611	0.03	10 721

As seen in Table 2, the MSO solutions were unable to replicate the results of a geologist's ore block design and reproduced similar results to the P-Class process. Unlike humans, the MSO lacks the flexibility to deviate from set rules. While it can adjust the prescribed limits of ore blocks, it cannot intuitively adapt like a human can. For instance, where a human might determine an ore block should match the width of the lode, the MSO might design a subgrade block within that area, even if it's the minimum width allowed by the parameters. In practical mining scenarios, however, this subgrade block would typically be mined along with the high-grade material, reflecting the human approach.

ALTERNATIVES

The mining industry has already embraced artificial intelligence (AI) and machine learning, for example in core logging, and the challenge of creating mineable shapes in a pit presents an opportunity for their application. Various AI software solutions have been developed, some of which employ a pixel-based approach to analyse block models. These algorithms assign data to individual pixels, representing block model grades and then combine them to generate target sizes and grades, similar to an MSO. However, unlike traditional methods where ore block size governs the design, software like 'Ozone' operates inversely. It expands or combines pixels (individual blocks) to achieve target size and cut-off grade. Further research and validation are needed to effectively adapt and trial this software for practical use in mining operations.

CONCLUSION

MSO is traditionally an underground mining tool and has been analysed as a tool to create mineable shapes in the pit at all mining stages, from resource evaluation to final ore block designs. The use of this tool in an open pit brings accelerated scheduling capabilities, as well as repeatability of ore block designs on all planning time scales and potentially improved reconciliation. This works to fit varying lode geometries, as well as navigating the challenge around historic workings. After many runs and scenarios of MSO trials on Fimiston the same problems occur to what is seen using the historic P-Class process. More work is needed to investigate if alternative MSO products are available to challenge this, or if artificial intelligence is the future. However, when reviewing results of various computer generated ore blocks with those of geologists it poses the question, is the human result not adhering to SMU rules as it should be and therefore inherently flawed? And are we therefore attempting to find a computer methodology to match biased and impossible target tonnes and grade?

The reality, going forward, is that the Fimiston geology team will aim to reduce the variability between geologists through ongoing coaching and ore block design review, but will also aspire to find a computer aided methodology that works in tandem with humans. The reality of orebodies, and geology in general, is that it cannot be a one size fits all approach that a computer can learn or be instructed. Fimiston is currently made up of four cutbacks within in a single pit, each with subtly different lode characteristics and mineralisation styles. Therefore, an algorithm that can generate ore block designs, under the supervision of a quality mine geology team to ensure economically viable and mineable shapes, is the ultimate goal going forward.

ACKNOWLEDGEMENTS

The author expresses their gratitude to Northern Star Resources for granting permission to utilise the KCGM data in this publication. This data, accumulated over 125 years by numerous geologists and mining personnel, represents a collective effort and the authors sincerely appreciate the dedication of everyone involved in its collection. The author would like to acknowledge the KCGM mine geology team for ore block design evaluations and Datamine Australia for support with MSO work. Finally, to Emma Murray-Hayden for ongoing support throughout.

REFERENCES

Clout, J, Gleghorn, J and Eaton, P, 1990. Geology of the Kalgoorlie gold field, in *Geology of the Mineral Deposits of Australia and Papua New Guinea* (ed: F E Hughes), pp 411–431 (The Australasian Institute of Mining and Metallurgy: Melbourne).

Cooke, D and Fitzgerald, M, 2011. The value of geological human capital in the improvement and maximisation of ore deposit value – the Kalgoorlie Consolidated Gold Mines experience, in *Proceedings of the Eighth International Mining Geology Conference*, pp 59–74 (The Australasian Institute of Mining and Metallurgy: Melbourne).

Czarnota, K, Blewett, R S and Goscombe, B, 2010. Predictive mineral discovery in the eastern Yilgarn Craton, Western Australia: an example of district scale targeting of an orogenic gold mineral system, *Precambrian Research*, 183:356–377.

Gauthier, L, Hagemann, S, Robert, F and Pickens, G, 2004. Structural architecture and relative timing of Fimiston gold mineralization at the Golden Mile deposit, Kalgoorlie, GSWA Record, pp 53–60.

McDivitt, J A, Hagemann, S G, Baggott, M S and Perazzo, S, 2020. Geologic Setting and Gold Mineralization of the Kalgoorlie Gold Camp, Yilgarn Craton, Western Australia, *SEG Special Publications*, 23:251–274.

Nelson, D R, 1997. Evolution of the Archaean granite-greenstone terranes of the Eastern Goldfields, Western Australia: SHRIMP U-Pb zircon constraints, Geological Survey of Western Australia, pp 57–81.

Swager, C P, 1997. Tectono-stratigraphy of late Archaean greenstone terranes in the Eastern Goldfields, Western Australia, Geological Survey of Western Australia, pp 11–42.

Witt, W K, 1995. Tholeiitic and high-Mg mafic/ultramafic sills in the Eastern Goldfields Province, Western Australia: Implications for tectonic settings, *Australian Journal of Earth Sciences*, pp 407–422.

Overcoming practical challenges to honour geological controls in short-term modelling of the Kamoa-Kakula copper deposit

H Bananga[1], G Gilchrist[2] and J Chitambala[3]

1. Superintendent I, Mine Geology, Kamoa Copper, Kolwezi, DRC.
   Email: herveb@kamoacopper.com
2. Vice President, Resources, Ivanhoe Mines, Johannesburg, South Africa.
   Email: george.gilchrist@ivanplats.com
3. Resource Manager, Ivanhoe Mines, Johannesburg, South Africa.
   Email: joshuac@ivanhoemines.com

ABSTRACT

Kamoa-Kakula is a world-class, high-grade stratiform copper deposit recently discovered by Ivanhoe Mines on the western edge of the Central African Copperbelt in the Democratic Republic of Congo (DRC). Two main rock types are present at Kamoa-Kakula; diamictite and an interbedded, often highly pyritic siltstone. The copper mineralisation is controlled by the hydrological framework and the redox boundary at the base of the diamictite.

Surface drilling broadly defined the geometry of the orebody and allowed construction of the resource model. Underground exposure allows surface holes to be supplemented with underground data (geology profiles and channel samples) to refine the stratigraphic surfaces and inform the grade control model. Kakula is characterised by a bottom-loaded grade profile with a very well-defined vertical gradation. The orebody is thicker than a single mining cut can expose. This creates a challenge for underground profiles and channel samples to fit in with surface drill holes that have full stratigraphic intersections. A robust resource estimate relies on data positioned in the correct stratigraphic context.

Additional challenges faced include capturing accurate coordinate positions (particularly elevation) of underground data. Significant demand on the survey department prevents them from recording data locations for hundreds of profiles and channel samples. Monthly survey scans are used through a scripted Datamine macro process to correct the elevation to allow the data to be represented as drill holes in the geological modelling.

To determine the spatial position relative to the key stratigraphic markers, profiles and channel samples are artificially expanded downward and upward to represent missing stratigraphic units. The process relies on knowing the thickness trends for each stratigraphic unit. This is made possible by a two-pass vertical thickness estimate to account for partial intercepts of stratigraphic units in the underground data. The final step applies a coordinate transformation prior to data analysis and grade estimation.

The grade control modelling routine applies innovative approaches to overcome practical challenges whilst honouring key geological controls.

INTRODUCTION

The Kamoa-Kakula Project is situated in the Lualaba Province in the Democratic Republic of Congo (DRC). The deposit is located within the Central African Copperbelt, approximately 25 km west of the provincial capital of Kolwezi and about 270 km west of the regional centre of Lubumbashi. The Project includes the stratiform copper deposits of Kamoa (discovered in 2008) and Kakula (discovered in 2015). After 16 years of exploration, nearly 2200 holes have defined the Mineral Resources of the largest copper deposit ever discovered in Africa. The Indicated Mineral Resource includes 1387 Mt at 2.64 per cent Cu at 1 per cent cut-off. The Kamoa-Kakula project is currently in production, operating from two underground mines and two concentrator plants with a processing capacity of 9.2 Mt of ore per annum.

Once in development and production, an infill drilling campaign, combined with underground mapping and sampling have contributed to enhancing the geological and grade control models and allowing a better understanding of the geology of the deposit.

At Kamoa-Kakula, the short-term model is used for short-term planning, grade control, monthly production (tonnes-grade) reporting and reconciliation. The short-term model is informed by the exploration drill holes augmented by geological profiles and channel samples from the grade control process. Several challenges exist to incorporating diverse data sets into a single estimation model, in particular:

- Establishing accurate underground coordinates (especially elevation) due to the high demand on surveyors.
- Incorporating underground data that represents partial intercepts of stratigraphic units.
- Ensuring estimation consistently honours corresponding parts across grade profiles.
- Producing a high-quality short-term model on time to comply with the month-end production reporting deadline.

GEOLOGY

Regional geology

The metallogenic province of the Central African Copperbelt is hosted in metasedimentary rocks of the Neoproterozoic Katanga Basin (Kampunzu and Cailteux, 1999). The lowermost sequences were deposited in a series of restricted rift basins which were then overlain by laterally extensive, organic-rich, marine siltstones and shales. The Katanga Supergroup sequence is made up of the Roan, Nguba and Kundelungu groups (Selley *et al*, 2018). Mineralisation in the majority of the Katangan Copperbelt orebodies such as at Kolwezi and Tenke–Fungurume is hosted in the Mines subgroup. The mineralisation at Kamoa-Kakula differs from these deposits in that it is located higher in the stratigraphy in the Grand Conglomerate unit at the base of the Nguba Group.

Local geology

The Kamoa-Kakula deposit occurs on a gently folded shelf, overlying a regionally NNE-trending, generally south-plunging basement block composed of Kibaran (1.3 to 1.1 Ga) metasedimentary rocks.

The gentle folding relates to basin inversion during the Lufilian orogeny, however local complexities in the geometry occur reflecting the underlying rifting controls on sedimentation. The rift geometry also controls the development of thick diamictite units of the Grand Conglomérat (Nguba Group) across the project and localised pyrite-rich siltstone units representing areas of sediment starvation.

Pyrite forms the reductant across the Project, with siltstones units hosting the highest grades where in close proximity to the underlying oxidised Roan aquifer (Figure 1). The diamictites represent debris flows into the evolving basin and reworked the upper portions of the siltstones during emplacement, incorporating reductant into their matrix (Turner *et al*, 2023). This has resulted in very strong lithological controls on grade and a very well-defined vertical grade profile and mineral zonation, with the highest grades and highest copper-sulfide species occurring at the base of the mineralised zone (Schmandt *et al*, 2013).

The mineralisation at Kamoa is broadly stratiform, but zones of elevated copper grades appear to be related to inferred growth faults that were active during sedimentation, underscoring an important link between extensional fault architecture and localisation of orebodies. These are marked by abrupt changes in stratigraphic thickness, steepened bedding and rotated mesoscopic faults (Twite *et al* 2019).

Post-Lufilian, there has been little in the way of structural disturbance, with the West Scarp Fault forming the only major brittle offset of the orebody. The West Scarp Fault is a much younger extensional fault set; it transects the project in an approximate north–south orientation to the west of Kamoa and separating Kakula from Kakula West (Gilchrist, 2019).

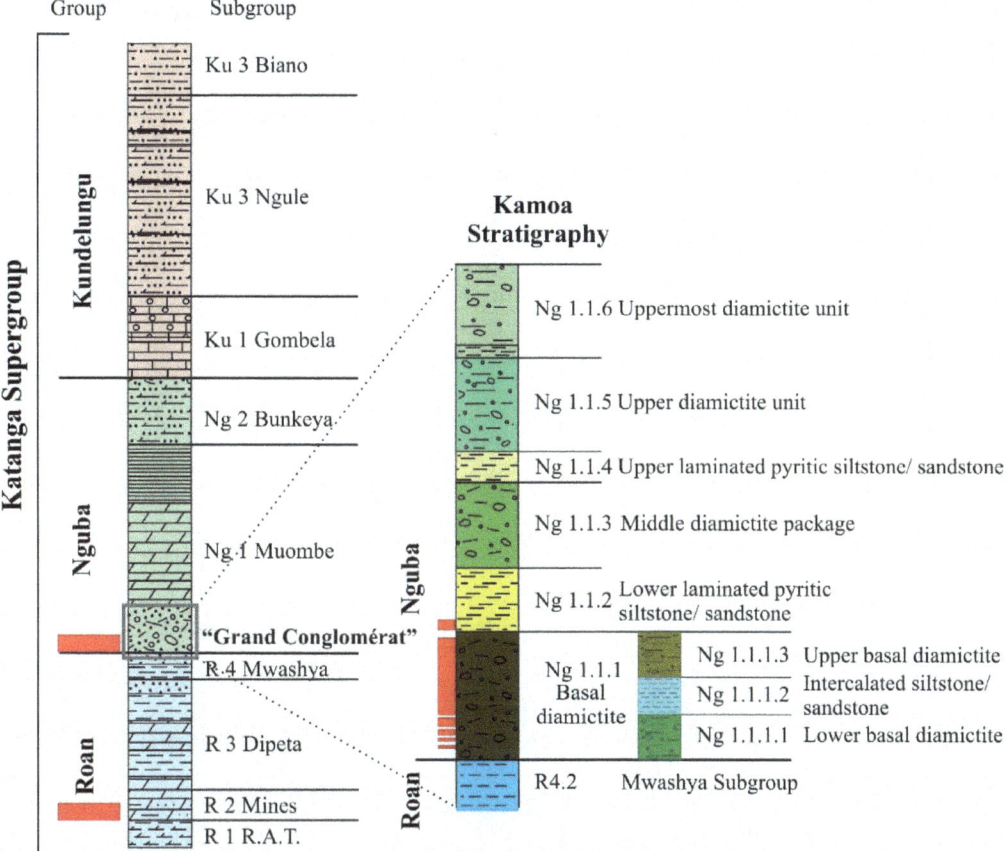

FIG 1 – Generalised stratigraphy of the DRC Copperbelt and detailed stratigraphy column of the Grand Conglomerate within the Kamoa-Kakula deposit. Copper mineralisation (in red) is hosted in the basal diamictite unit (Ng 1.1.1).

Mineralisation

Mineralisation at Kamoa has been defined over an irregularly shaped area of 24 km × 14 km. Mineralisation thicknesses at a 1.0 per cent Cu cut-off grade ranges from 2.3–21.6 m (for Indicated Mineral Resources). The deposit has been tested locally from below surface to depths of more than 1560 m and remains open to the west, east and south.

The vertical position of mineralisation relates to the location of the reductant/s and proximity to the underlying Roan aquifer. Although broadly stratiform, mineralisation does transgress stratigraphy when a lower reductant narrows or pinches out. Mineralisation is strongest and the bottom-loaded profile is best developed when the reductant is in direct, or very close contact, to the Roan aquifer. The mineralisation varies consistently and predictably from one unit to another (Gilchrist, 2019).

Copper sulfide mineralogy at Kamoa-Kakula is vertically zoned with respect to this redox boundary (Figure 2). The sulfide mineral zonation from the bottom to the top is as follows: chalcocite and/or bornite, chalcopyrite and pyrite (Schmandt *et al*, 2013; Hendrickson *et al*, 2015). Near the surface, the diamictites have been leached, resulting in minor localised zones of copper oxides and secondary supergene copper sulfide enrichment.

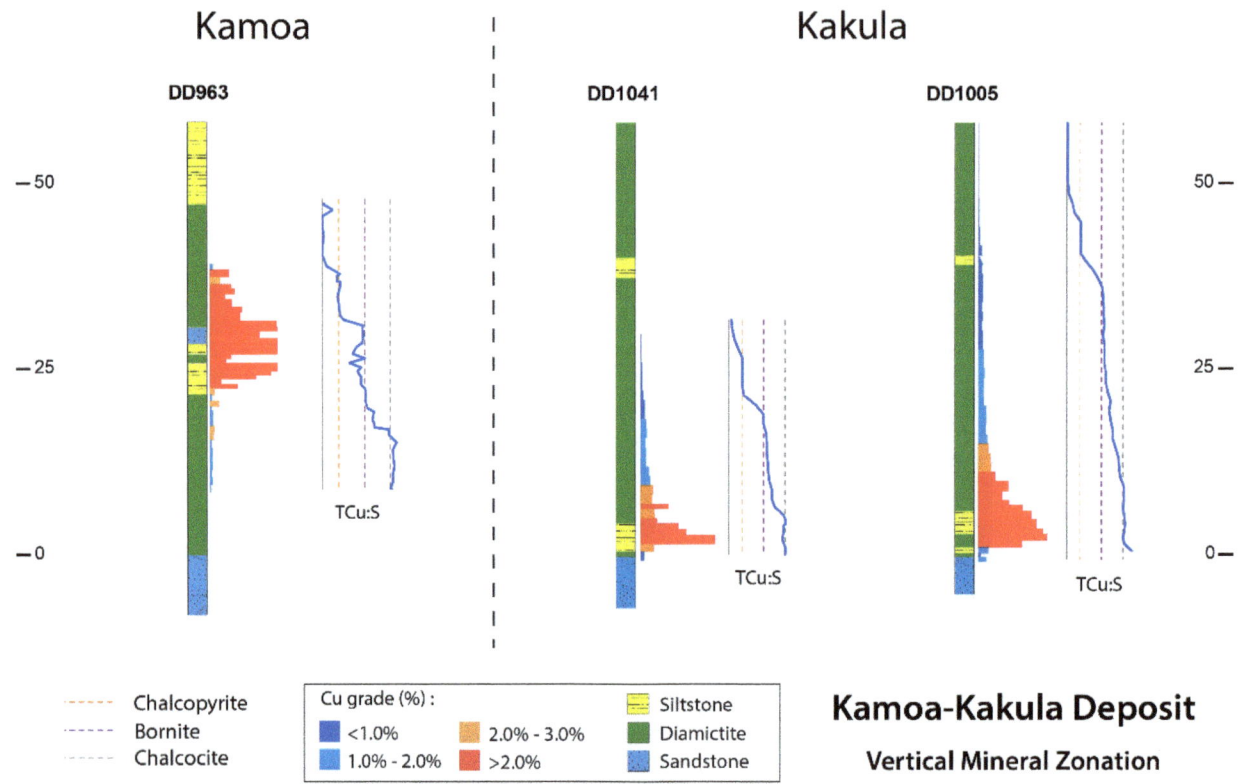

FIG 2 – Drill holes from Kamoa-Kakula illustrating vertical mineral zonation.

The geological characteristics of the Kamoa-Kakula deposits poses challenges to the grade control modelling:

- The orebody is developed over a wide area but individual stratigraphic units can be localised or only a few metres thick.

- Underground information only represents a partial exposure of the deposit. With the strong lithological controls, vertical trends in grade and mineral zonation, it is critical that data is placed in its correct stratigraphic position.

- Local undulations in the orebody geometry require a data coordinate transform prior to estimation so that corresponding parts across grade profiles can be correlated.

OVERCOMING THE COORDINATES CHALLENGE

The grade control model is updated monthly and used to report the mined grade. To adhere to the reporting deadline, the modelling process must deal with the constraints of time and availability of the survey team to collect coordinates for every geology profile and channel sample collected every month. The huge demand for survey services at the mine prevents them from being available to collect the collar position of the data.

To deal with these constraints, Easting and Northing are determined from actual mine outlines and the Elevation is corrected based on monthly survey scans.

Correcting elevation – Datamine process

To correct elevation in Datamine, the required input data are:

- 3D scans of underground development

- 2D outlines of the development

- Geology profile and channel sample data set, including Collar, Survey and Geology to allow manual checking and visual validation of the results.

The operation is performed in two steps.

Step one – Generate points at the base of the development 3D scans

Mining development and stoping are scanned monthly and solids are created in Deswik software. Whilst largely representative of the mined volume, some common challenges are encountered (Figure 3).

- Complexities within the scans (the process of creating scans can generate separate surfaces, empty faces, duplicate faces and duplicate vertices) that can interrupt the Datamine process.
- Underscan due to water or blasted material obscuring the scan.

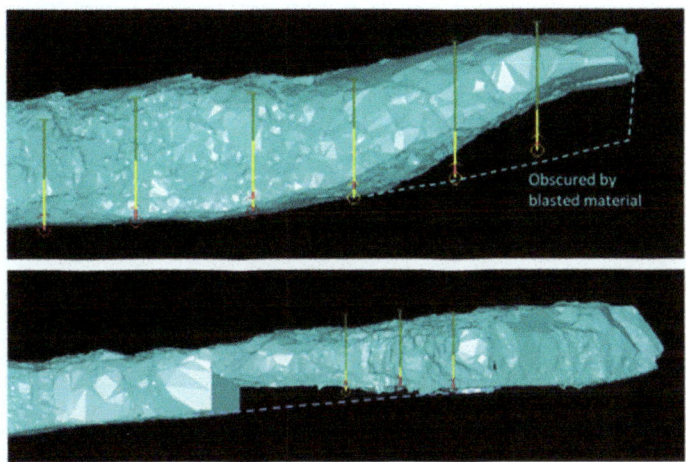

FIG 3 – Examples of scan issues that can constitute source of error.

Points are generated by filling the 3D scans with small blocks (2 × 2 × 1) with subcelling to 20 cm in the vertical to avoid a stepped pattern in the points.

These blocks are then regularised over the full vertical interval in the Datamine software functionality. The coordinate at each block centroid obtained after regularisation is converted into points and the Z component is dropped to the base of the development. This is made possible by calculating the vertical thickness of the drive.

The scanned drives are not always perfectly square in shape and this can create edge effects. A Datamine functionality is used to shrink the development outline and select points within the shrunk area. All other points that do not fairly represent the base of the development especially those at the base of the obscured areas are manually deleted.

The scan complexities are overcome by running the wireframe verify function in Datamine and the errors in the scans are overcome by the surveyor in the validation process.

Step two – Generate wireframe from development points and drape collars to this surface

The points file from the base of development can create artefacts where two drives are at significant elevation differences. To overcome this, an inverse distance power 5 estimate of the Z is used to allow the development elevation to extend beyond the limits of the individual drives. This is done to ensure sidewall profiles and channel samples are draped to the base of the drive and not part way up a slope.

The points generated from the estimate process are then used to create the surface which represents the base of the development of the mine onto which collars are draped to correct the elevation (Figures 4 and 5).

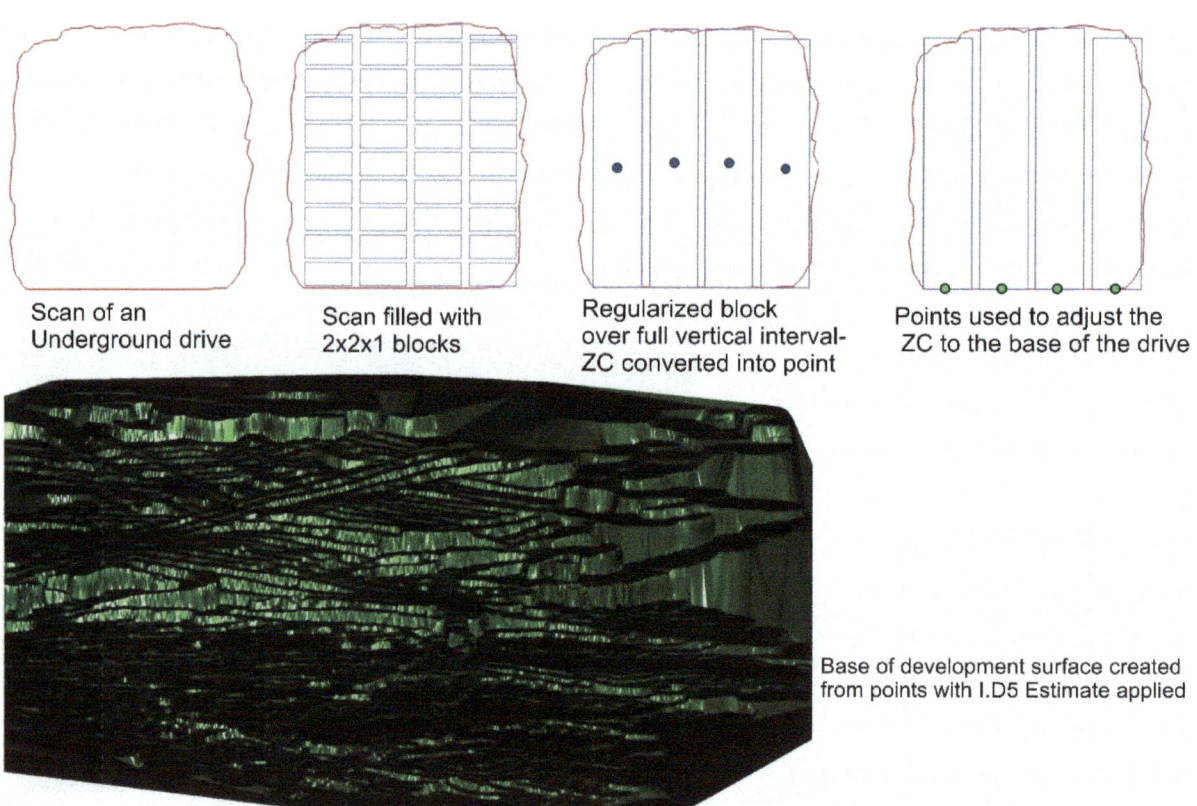

FIG 4 – Illustration of points and surface creation from scans.

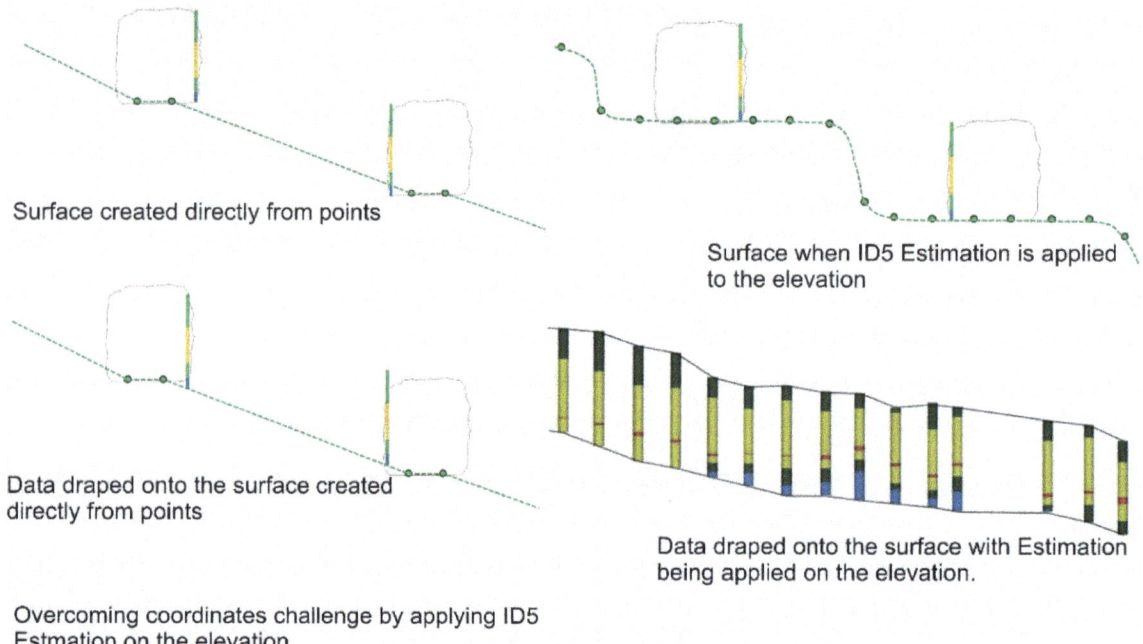

FIG 5 – Image showing challenges in creating the base-of-development surface for Elevation correction.

A visual validation through individual sections is still needed to ensure all the collar data are sitting perfectly at the base of the drive.

OVERCOMING WIREFRAMING CHALLENGES

A key challenge to wireframe modelling at Kamoa-Kakula is to prevent units with variable thickness from pinching out or abnormally thickening simply as a modelling artefact. The problem is compounded as there is insufficient time to carefully model multiple surfaces across the whole mine given production and reporting deadlines.

This challenge is overcome through a combination of Datamine and Leapfrog modelling software functionalities.

Reference surface modelling technique

The wireframe surface modelling process uses all available data, including surface drill holes and underground geology profiles, drilling and channel samples. Each contact of the stratigraphic unit is identified in Datamine, the point file is generated and exported to Leapfrog. A reliable and consistent marker unit is identified and modelled as a reference surface. This first surface created is an initial wireframe to facilitate the generation of structural data (dip and dip direction) to add to the pierce point data. This is achieved in Datamine through the calculation of anisotropy angle to derive the dip and dip direction of every triangle of the wireframe. These dip and dip direction are estimated into a block model and a Datamine functionality is used to write the block model values back to the point data at the corresponding easting and northing. The point file created including coordinates, dip and dip direction is exported to Leapfrog as structural discs. The initial surface is then refined with the structural data (Figure 6).

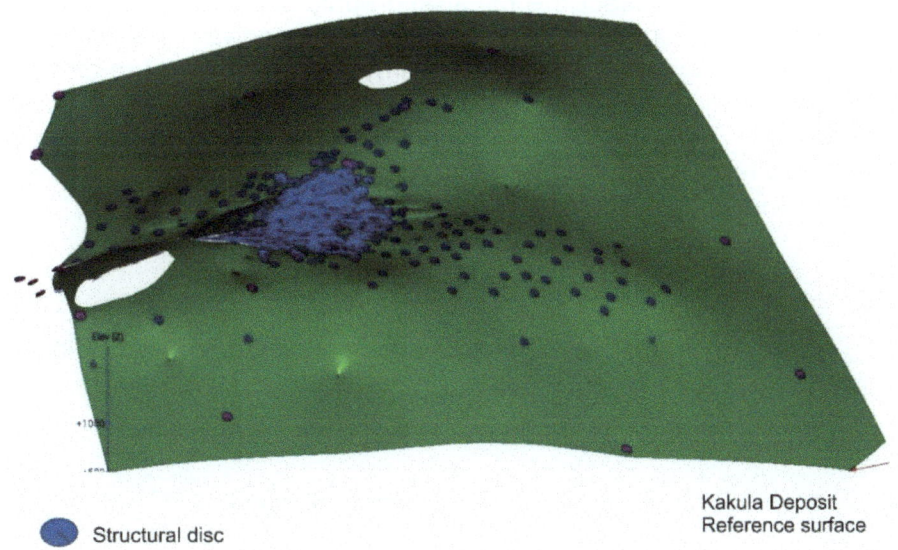

FIG 6 – Reference surface with structural discs used to control its modelling.

Compared with point data, which can lead to a dimpling effect, using structural data where dip and dip direction are defined at every point, has the advantage of smoothing the surface and honouring the trends.

This new surface is smooth, requires minor adjustments along the fault surface and is realistic enough to serve as a reference. A vertical thickness estimate for all units above or below the reference is used to adjust all other surfaces.

Vertical thickness estimation

This process generates vertical thickness for each stratigraphic unit across the orebody. This is made possible through two-pass vertical thickness estimation.

Pass one

The grade control modelling process uses underground data (geological profiles and channel samples) that don't necessarily expose the full stratigraphic unit. The first pass aims to identify only the data with full exposure where both top and bottom contacts for each specific stratigraphic unit are present.

To simplify the process in Datamine, all the stratigraphic units are coded into domains.

The top and bottom contacts for each domain are identified and the vertical thickness per stratigraphic unit is estimated into a block model using the inverse distance power one (ID1) to ensure smooth changes between data points.

Pass two

The vertical thickness estimate is read back onto the partial thickness data using a Datamine functionality. Partial exposures that are thicker than the estimate are identified and incorporated into the data set.

The vertical thickness is rerun per stratigraphic interval and this second vertical thickness model is used to adjust the reference wireframe points and create the additional surfaces. This thickness estimate also acts as a validation step given that the thickness changes are key indicators of a rift geometry and controls on mineralisation. Observed thickness artefacts are investigated and fixed if they prove to be errors in the original data capture.

Overcoming stratigraphic position

Placing data in their correct stratigraphic position is key for grade estimation. The underground data rarely intersect full stratigraphic units. The mineral resource estimate workflow relies solely on surface drill holes that always intersect the full stratigraphic sequence. To merge into the mineral resource estimate workflow, underground data need to be expanded to resemble surface drill holes. The expansion is applied through the addition of dummy intervals at the top and bottom contacts of the geology profiles and channel samples to allow domain selection on each wireframe created in the previous step (step two). The expanded portions are composite to 1 m intervals prior to selection. All underground data now have every modelled stratigraphic unit (domain) coded and can be aligned correctly even when an incomplete profile is present (Figure 7).

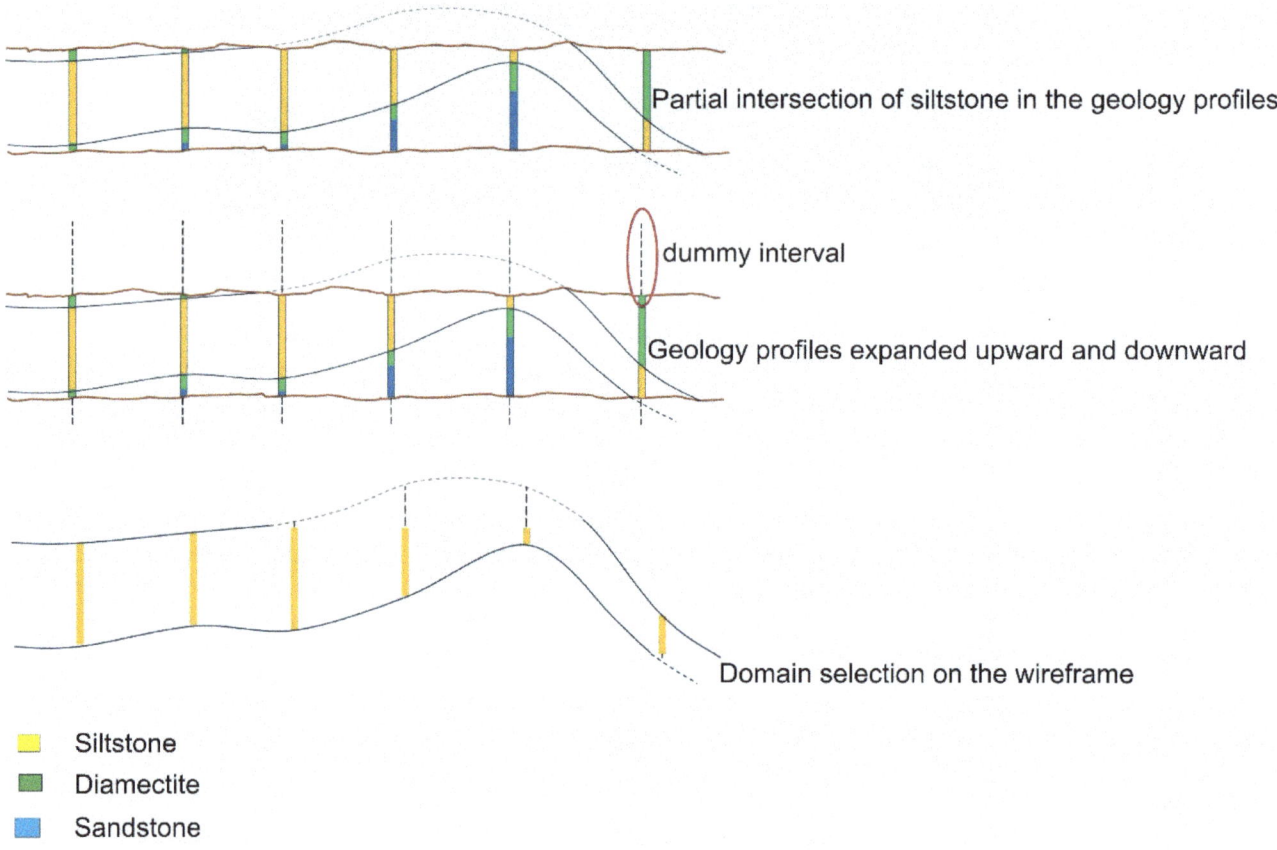

FIG 7 – Illustration of interval expanded on partial stratigraphic unit exposition underground data.

Three-dimensional modelling and grade estimation

The modelled surfaces of stratigraphic and mineralised horizons constitute key elements in setting up a 3D block model prior to grade estimation. Wireframes are merged to allow for filling between surfaces with 5 m × 5 m × 1 m blocks. At Kamoa-Kakula, the stratigraphic units are combined with the mineralised zone to form vertically stacked estimation domains (Figure 8) following compositing sample length from drill holes at 1 m.

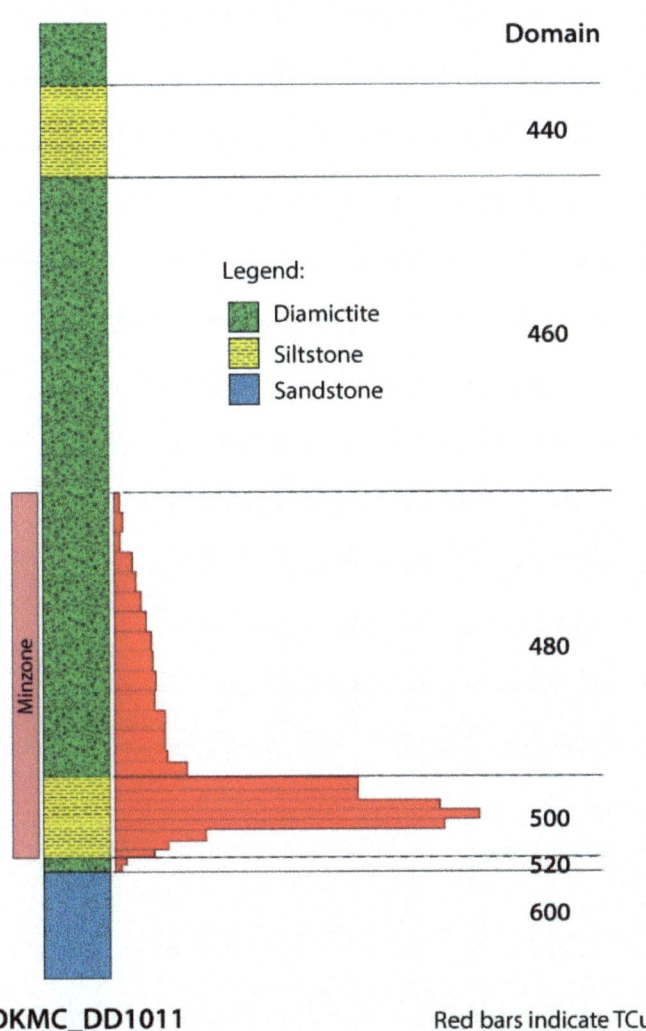

FIG 8 – Illustration of domain coding at Kakula.

In the Kamoa-Kakula deposit, the copper grade profile is bottom-loaded, with the higher grades occurring at the bottom and decreasing upward. To maintain the bottom-loaded grade profile in the grade estimation while dealing with folding and thickness variation in the mineralised zone, the Z-coordinate is transformed ('dilated') on both drill holes and block model to ensure that the vertical grade profiles match between drill holes (Gilchrist, 2019; Latifi and Boisvert, 2022).

The transformation is performed by adjusting the Z-coordinate of the data to 'dilate' the drill hole composites and blocks to the maximum vertical thickness for each domain. This helps to ensure that the lower, middle and upper portions of the grade profile align correctly between drill holes (Figure 9). The 1 m composite and transformed drill holes are now used to perform variography and grade estimation. The block model generated in transformed space is then back-transformed to their original vertical location by setting the centroid of each block back to its original Z-coordinate (Gilchrist, 2019).

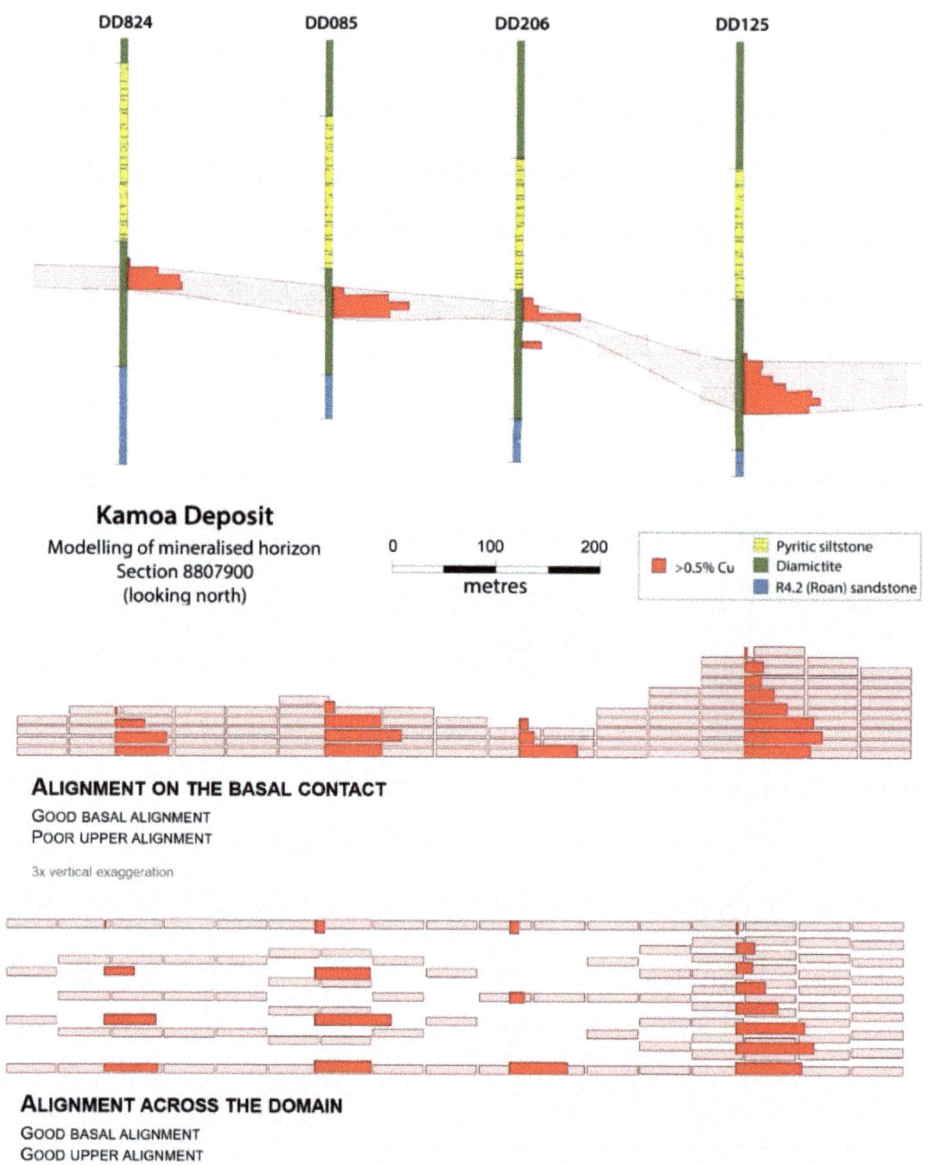

FIG 9 – Dilation on drill holes and block model to align grade distribution.

The dilation generally works well and produces accurate grade estimate (Figure 10).

The entire process is coded in a number of Datamine macros to keep the process constant, efficient and auditable and two to three days are required to perform this process (Gilchrist, 2020).

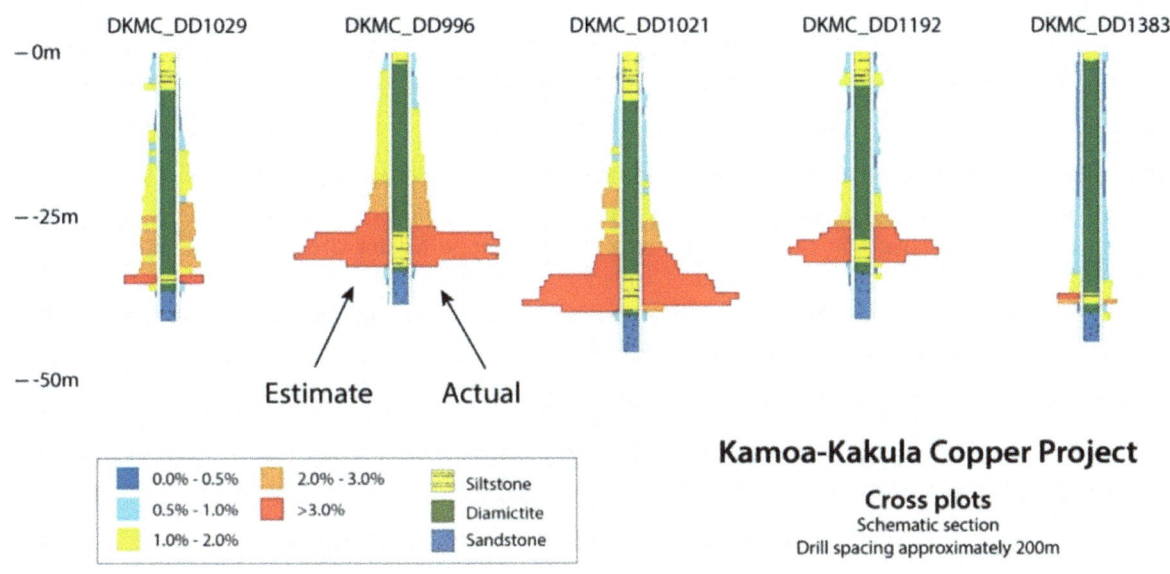

FIG 10 – Grade estimate validation cross-plot.

CONCLUSION

The grade control modelling routine, implemented to facilitate short-term planning, grade control and monthly production reporting, has proven effective despite facing several practical challenges. By innovatively addressing these challenges, the routine ensures the accuracy of grade estimation and compliance with production reporting deadlines.

Key achievements include the development of automated processes to correct elevation data for underground profiles and channel samples using monthly survey scans. This approach mitigates the limitations posed by the high demand on survey resources and enables the incorporation of accurate data into the modelling workflow.

Additionally, wireframe modelling challenges, such as preventing unit pinching out or abnormal thickening, are effectively addressed through a reference surface modelling technique. By utilising structural data to refine wireframes and ensure consistency, the modelling process maintains accuracy and reliability.

The paper underscores the importance of these advancements in supporting efficient mine planning, grade control and production reporting at the Kamoa-Kakula Deposit. Notably, the routine's automation and efficiency allow for timely completion, enhancing overall operational effectiveness.

The grade control workflow benefits from the previously developed Mineral Resource estimation process. Thus, the underground data can be easily incorporated into the resource modelling for its update.

ACKNOWLEDGEMENTS

The authors wish to express appreciation to the management representatives, Mr Franck Twite, Mr Mike Freer and Mr Daan van der Berg of Kamoa Copper SA for their roles in granting approval for the use of company data, for facilitating staff development through opportunities like this one and for providing invaluable general assistance towards this work.

Acknowledgement is also expressed towards Mrs Mbali Nkwali and the entire Kamoa Copper Transformation department for their unwavering commitment to nurturing young talents within the company. Acknowledgements are extended to management and contributors for their support and assistance throughout the project.

Finally, heartfelt thanks to Mr Innocent Mushobekwa for his significant contributions during our discussions.

REFERENCES

Gilchrist, G, 2019. Evolving estimation techniques for an evolving world-class stratiform copper deposit at Kamoa-Kakula, Democratic Republic of the Congo, in *Mining goes Digital: Proceedings of the 39th International Symposium Application of Computers and Operations Research in the Mineral Industry (APCOM 2019)*, pp 192–200 (CRC Press).

Gilchrist, G, 2020. Datamine Macro Writing Guide, Ivanhoe Mines, Johannesburg.

Hendrickson, M D, Hitzman, M W, Wood, D, Humphrey, J D and Wendlandt, R F, 2015. Geology of the Fishtie deposit, Central Province, Zambia: iron oxide and copper mineralization in Nguba Group metasedimentary rocks, *Miner Deposita*, 50:717–737.

Kampunzu, A B and Cailteux, J, 1999. Tectonic Evolution of the Lufilian Arc (Central Africa Copper Belt) During Neoproterozoic Pan-African Orogenesis, *Gondwana Research*, 2:401–421.

Latifi, A M and Boisvert, J B, 2022. Stratigraphic Coordinate Transformation, in *Geostatistics Lessons* (ed: J L Deutsch). Available from: <http://www.geostatisticslessons.com/lessons/stratcoords>

Schmandt, D, Broughton, D, Hitzman, M W, Plink-Bjorklund, P, Edwards, D and Humphrey, J, 2013. The Kamoa copper deposit, Democratic Republic of Congo: stratigraphy, diagenetic and hydrothermal alteration, and mineralization, *Econ Geol*, 108:1301–1324.

Selley, D, Scott, R, Emsbo, P, Koziy, L, Hitzman, M W, Bull, S W, Duffet, M, Sebagenzi, S, Halpin, J and Broughton, D, 2018. Structural Configuration of the Central African Copperbelt: Roles of Evaporites in Structural Evolution, Basin Hydrology, and Ore Location, *SEG Special Publications*, no 21, pp 115–156.

Turner, E, Dabros, Q, Broughton, D and Kontak, D, 2023. Textural and geochemical evidence for two-stage mineralisation at the Kamoa-Kakula Cu deposits, Central African Copperbelt, *Mineralium Deposita*, 58:825–832.

Twite, F, Broughton, D, Nex, P, Kinnaird, J and Gilchrist, G, 2019. Lithostratigraphic and structural controls on sulphide mineralisation at the Kamoa copper deposit, Democratic Republic of Congo, *J Afr Earth Sci*, 151:212–224.

Remote sensing LiDAR for enhanced underground geology mapping – a Nova Mine case study

G Boyce[1]

1. Superintendent – Geology, IGO Limited, Nova Nickel Operation, Fraser Range WA 6443. Email: glenn.boyce@igo.com.au

ABSTRACT

This study examines the integration of LiDAR technology into geology mapping underground, focusing on the Nova mine's early stages (2016) using a Riegl VZ-400 laser scanner. The motivation behind initial research was the hypothesis that reflectance data could be utilised to differentiate between disseminated sulfides and gangue minerals, thus driving more accurate grade control estimations.

While promising, the Riegl VZ-400's cumbersome manual handling and extensive data sets posed challenges in the underground environment. Nonetheless, the Nova geology department successfully employed a method to combine point clouds with the high-resolution DSLR photo data to create precise 3D geology mapping records. Primarily employed to validate grade control mineral resource estimate (MRE) domain boundaries and structural model contacts, it also notably reduced production costs by eliminating the need for mine surveyors to pick up ore contact paint markings. Value extended to other applications, including infrastructure modelling and stockpile volume calculations.

The adoption of remote sensing and photogrammetry in underground geology mapping varies across the mining industry due to instrument costs, training and workflow integration concerns. Logistical issues like instrument weight and man-made reflective surfaces add complexity in an underground environment.

Over eight years, LiDAR technology has rapidly evolved. Nova mine geologists recently developed a method of applying LiDAR data exported from a lightweight iPad Pro, yielding high-quality, aligned, cost-effective colour-textured meshes. This approach, utilising free apps and mining software, offers a feasible alternative for teams seeking digitised mapping workflows without high-end LiDAR expenses, that yield accurate polygon modelling of underground geology exposures.

The Nova Mine case study illustrates both challenges and opportunities in adopting advanced LiDAR techniques in the mining industry. As technology continues to advance, the prospect of more accessible, cost-effective mobile alternatives suggests a promising future for junior mining companies looking to modernise their mapping strategies.

INTRODUCTION

The Nova-Bollinger mine is accessible by road, 160 km north-east of Norseman and 380 km north-east of Esperance in Western Australia (Figure 1).

The mine is within the 425 by 50 km wide Mesoproterozoic-age Fraser Zone of the Albany-Fraser Orogen. A chonolith gabbroic intrusion is interpreted to be the source of the sulfide mineralisation (Barnes *et al*, 2020).

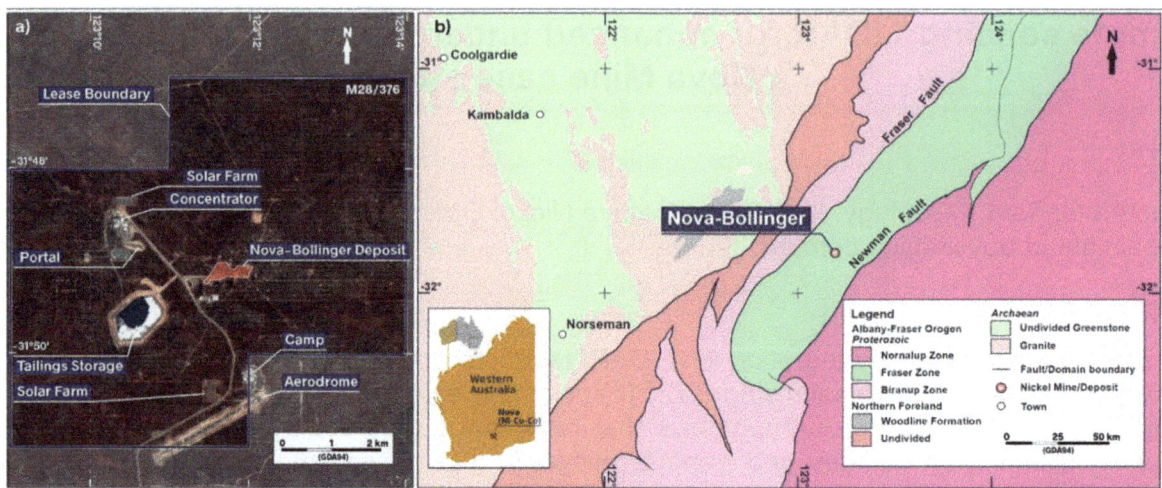

FIG 1 – Nova-Bollinger deposit mine infrastructure, geographical location and simplified regional geology.

GEOLOGY

Nova-Bollinger is in the southern part of the 425 km by 50 km wide Mesoproterozoic age Fraser Zone of the Albany-Fraser Orogen. The Fraser Zone is fault bounded by the Biranup Zone to the north-east and the Nornalup Zone to the south-east. The Arid Basin forms the basement to the Fraser Zone and the Snowys Dam formation of the Arid Basin is the basement package in the Nova-Bollinger area (Spaggiari *et al*, 2014).

The Nova-Bollinger complex that hosts Nova-Bollinger's Ni-Cu-Co sulfide mineralisation has been interpreted as a doubly plunging synform, where a magnetite-bearing footwall gneiss has been identified as the cause of 'The Eye' magnetic feature. The sill complex is a dish-shaped package 2.4 by 1.2 km in plan and up to 450 m in thickness. The rocks of the complex range in mineralogy from peridotite to pyroxenite, to gabbronorite and norite, with both sharp and gradational contacts between different intrusive phases. The mine area is covered by up to 3 m thick regolith and/or transported cover, with oxidation of sulfides in fresh rock down to depths of 20 m in the western end of the Nova area.

Nova-Bollinger's massive sulfide mineralogy is dominated by pyrrhotite (80 to 85 per cent), minor pentlandite (10 to 15 per cent) with lesser chalcopyrite (5 to 10 per cent). Concentrations of up to 5 per cent magnetite also occur locally within more massive sulfides zones. Cobalt is strongly correlated with nickel, as both elements are found concentrated in pentlandite and in minor concentrations with pyrrhotite.

Over the life-of-mine 94-month period to 29 February 2024, Nova has mined and processed 10.54 Mt of ore grading 1.90 per cent Ni, 0.76 per cent Cu and 0.06 per cent Co. At the end of June 2023, the Nova-Bollinger JORC Code (JORC, 2012) reportable *in situ* Mineral Resource estimate (MRE) was classified as ~93 per cent Measured, with ~96 per cent of the Ore Reserve estimate (ORE), classified as Proved Ore Reserve (IGO, 2023). The high JORC Code classification confidences assigned to nearly all the Nova-Bollinger MRE and ORE is supported by the 386 km of high quality, close spaced (nominal 12.5 by 12.5 m drill hole pierce-point spacing through the mineralisation's limits) surface and underground diamond drill testing of the deposit. Additionally, the precision of the MRE geological model has been attributed to thorough confirmation of high-grade ore domains by Nova's routine high-quality 3D laser scan mapping of mine development headings (Figure 2).

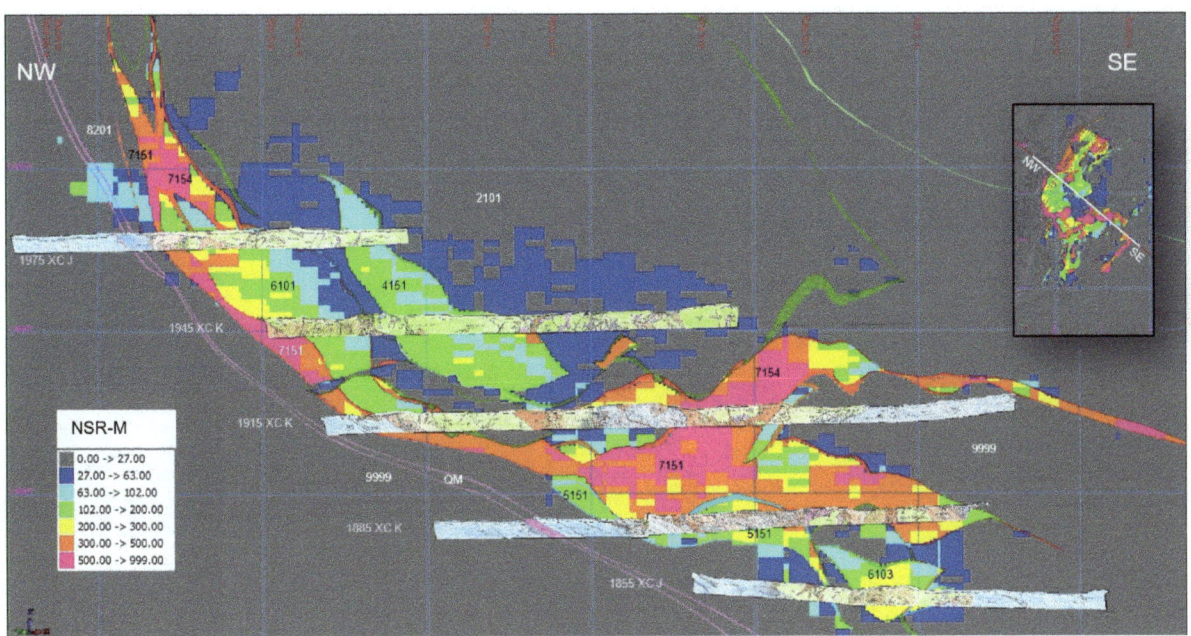

FIG 2 – Confirmation of Mineral Resource estimate (MRE) domains achieved using geology maps at 1:250 scale derived from LiDAR collated in 3D software (Surpac). Plan view sliced at 1915 mRL. NW-SE section of the Nova 1855–1975 mRL levels (looking NE). MRE block model section showing *in situ* net smelter return values (NSR-M in $AUD/ton).

LIGHT DETECTION AND RANGING (LIDAR) USE IN GEOLOGY MAPPING AT NOVA

The Nova geology team successfully implemented a Riegl VZ-400 LiDAR scanning instrument into the underground geology mapping workflow in 2017, which has since formed an intrinsic part of data recording of geology contacts and structures of the mine workings. Replacement of the Nova underground mine's aging Riegl VZ-400 laser scanner accounts for one of the highest costs to the mine geology department, both in production time to analyse the data and fiscally to the 2024 budget. With a new instrument costing over $200 000, a thorough review was conducted to assess the past and future scanning requirements for the Nova geology department. The investigation spawned a case study of the integration of LiDAR technology into geology mapping underground and the discovery that the techniques used during the Nova mine's early stages (2016) of underground mapping using a Riegl VZ-400 LiDAR laser scanner, could now be applied using low budget, light weight, short range mobile LiDAR devices.

Sirius Vision

The original Riegl project scope for Sirius Resources in 2013, stated that with the planned rates of development at Nova-Bollinger, the use of conventional face-mapping would always be extremely limited. It was recognised that implementing a high speed scanning system could be advantageous. The goal was not to replace face mapping, but to transition the mapping process from the underground environment to the surface. Apart from safety and time-saving benefits, it was predicted there would also be a significant increase in mapping accuracy that would ultimately feed directly into the production geological model.

Several systems were evaluated, the Riegl VZ-400 chosen due to the on-board waveform processing capabilities, which showed to be the most applicable to the requirements of the geological department. Most 3D scanners would be able to conduct similar scan density and speeds, however the addition of on-board waveform processing and automatic high-resolution digital photography was thought to provide the geology department the opportunity to extract lithological information quickly and efficiently without any significant interference to development scheduling.

With only one supplier in Perth, Australia, Sirius acquired and conducted the first scan demonstration test work at the Nova core yard in 2015. Various pieces of core were scanned during trails, along with a wall, that was converted to a textured mesh. The scans and mesh were rudimentary volume scans, not georeferenced to any world grids and showed reflectance bands that offered no additional information. However, the demonstration did offer a proof of concept and a look at how geology drill core (rock types) would present in the first 3D scans with colour data derived from the external DSLR camera seen in Figure 3.

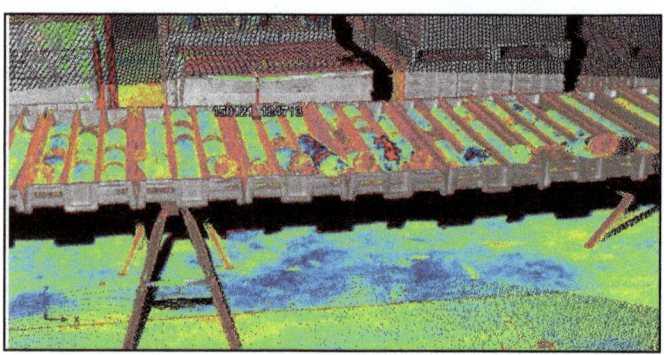
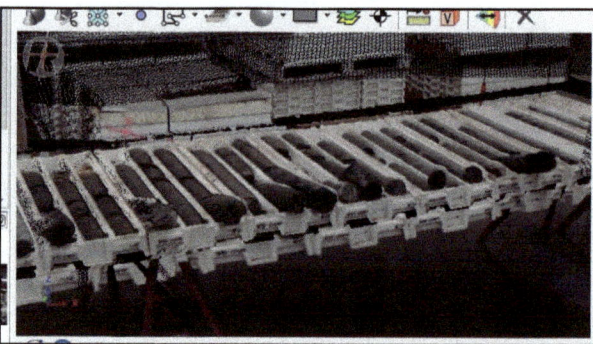

FIG 3 – The first scan data acquired at Nova from September 2015: (left) 'Core Sample reflectance colour scan .jpeg' from initial Riegl VZ-400 scan test work; (right) 'Core Samples natural colour scan .jpeg' showing proof of concept of DSLR imagery from external camera aligned to point cloud.

A small handful of scan projects exist for the Nova box cut and decline during the scanner commissioning in July and November 2016. The Nova Project was not in ore production at this time, with the main milling and processing capital infrastructure construction incomplete. During the handover of the Nova Project from Sirius Resources to IGO and as the project progressed from resource development to active mining, the Riegl VZ-400 was not utilised for face mapping as it had been intended. With the exception of box cut ground monitoring and decline development and a few survey pickups of some faults modelled through resource drilling, the scanner mostly sat idle due to its very large data sets and specialist training required to operate. At this stage the asset was the responsibility of the survey department.

Application of LiDAR for resource estimation and grade control

The mine development tunnelling at Nova was engineer designed to a grid at early stages of the Nova project, based on resource definition drilling. Due to the high variability of the breccia ore domains, it was decided that the resource model should be drilled on a close 12.5 × 12.5 m grade control drill spacing program, leaving the geology department with a time advantage to rely on the mineral resource estimate rather than 'geo control' of mine development drives. Scale ruler, tape and distometer paper template mapping of development faces has always been part of the Nova geology mapping workflow, however much of the time in the early stages of mining was spent analysing grade control drill core data to build the Leapfrog implicit model.

The geology team were processing an average of 500 m of grade control drilling per day. The peak of grade control drill out of the Nova-Bollinger resource was completed in 2021, with over 300 000 m of resource drilling completed. The scan data and geology mapping was therefore used mainly for resource estimation validation during this time, rather than mine design being dependent on geology control or tunnel direction, such may be the case with narrow vein gold style mine development at other projects.

A study early in the mine development, assessed the ability for the Riegl scanner to differentiate between minerals, specifically focusing on representative Nova rock types including metamorphic, magmatic and sulfide rich hand samples and wall scans of the underground workings. The wall scans underground proved problematic for mineral discrimination through reflectance data – since there were so many other highly reflective background noise objects underground like reflective signs, tape, paint and most notably, steel mesh and bolts used for ground support. It effectively made any data captured that had been supported (which was any accessible area of the mine) impossible to run automated grade estimation work on using the Riegl laser, where with the aid of the resource estimate any manual estimation required was conducted by a mine geologist visually.

Scanning every new unsupported development cut without mesh and bolts was attempted but deemed unfeasible due to the fast production rate. The value add was deemed minimal in lieu of conventional methods using a tape measure/distometer, scale ruler and record keeping on paper templates that could later be draped into the mine model at an approximate location accurate to within a few metres.

On the surface in a mining office environment, various samples were assessed twice from the same set-up position with the Riegl scanner. The scanner was set-up to scan directly in front of the samples and laser scan data captured. Without moving the scanner, the next test conducted with the samples turned ~30° on the shelf oblique to the laser incidence. The study found that despite various minerals and rock types having different reflective properties visible by the different range of reflectance intensity displayed in the scan data, the ability to differentiate between minerals was hampered by the angle of incidence and therefore reflection (Figure 4).

FIG 4 – A variety of rocks from the Nova decline (photo on top) scanned with the Riegl scanner at a 0.015° resolution (middle) and scanned with the same conditions but rocks were turned for around 30° (bottom). Rocks are approximately 2.7 m away from scanner; field of view is 1 m. Red colours represent high reflectivity and blue low reflectivity of rocks. Note, the black biotite has a high reflective intensity (red) when sheets are perpendicular to scanner, but low (blue) when turned. Conversely, muskovite has a very high reflectivity perpendicular and still reasonably high when turned. Feldspar-pegmatites with white feldspar have a very high reflectivity, green feldspar has a medium to high reflectivity depending on the angle.

Sulfides exhibited a high reflectance when perpendicular to the laser beam and lower reflectance when positioned on an angle. The reflectivity of high reflective gangue minerals such as silicates, were similar to highly reflective sulfides on a perpendicular angle, and therefore not useful in mine grade estimation of disseminated sulfides in magmatic host rocks that contain an abundance of fine to coarse grained chain silicates (pyroxenes) and olivine.

The study concluded that the geology rock types at Nova underground were easily distinguished by eye without the aid of the laser and geology contacts that fell significantly outside of the resource modelling boundaries could be added to the digital model by standard mining practice of spray-painting significant contacts and picked up by the survey department. It was hoped that with the advancement of machine learning, an algorithm could be written to automatically remove background noise and differentiate minerals in the RiScan software. At this stage of the mine development, no major ore contacts had been exposed, and the study concluded with a recommendation to re-evaluate the value of the Riegl scanner when the main orebody was intersected.

After the conclusion of the reflective intensity tests failed to find value in mineral discrimination for MRE grade estimation in hand sample and underground outcrops, a research and optimisation project for the Riegl scanner was outlined in early 2017.

The aim of the project was to:

- Optimise the use of the Reigl scanner in the underground development.
- Continue with the discussions that had already commenced with the University of Western Australia (UWA) and Mapability on further uses of the Scanner on-site at Nova.
- Investigate other opportunities for the use of the scanner with external sources.
- Develop a standard process and procedure with documentation for new users.

2017 Riegl scanner optimisation project

Application of LiDAR for geology mapping

The Riegl scanner was not being utilised by the mine surveyors, who relied on traditional total station lasers to pick up the mine development and cavity monitoring equipment for stope voids.

The optimisation project was delegated to two geology graduates, who were successful in identifying several alternative uses for the Riegl scan data. A method was derived to accurately locate the geology contacts mapped in the point cloud scan data and export as polylines. Structures and faults in the underground workings could be modelled as planes from regions of points that had like orientation, and then export the millimetre accurate data to the Leapfrog resource model without the requirement for the survey department to pick up any high-grade ore contacts. The geology mapping derived from the scanner provided a millimetre accurate validation of the high-grade domains in the resource estimation model. The laser scans were georeferenced using two reflectors mounted to existing underground control points. The resource model wireframes could easily be snapped to the RiScan polyline data in the Leapfrog software. The data also provided a permanent record of the mine workings long after the development drives had been coated with spray on fibrecrete for ground support.

The in-house development and training of geologists in use of the Riegl equipment and point cloud data processing continues to be challenging – given most geologists do not have any background in survey instruments and software used by laser and remote sensing equipment. This practice, however, saves significant amount of time for the Nova Survey department – since there is no requirement for paint markups of geology or structural contacts to ever be picked up by the Survey team. This allowed the survey department to focus attention on other production tasks.

The digital map data could only be generated as digital polylines in Riegl software (RiScan Pro), exported and plotted onto paper, traced and coloured into simplified IGO geology lithological codes and scanned back in on a printer. The .jpeg imagery could be draped back over mine development solids at 1:250 scale (Table 1) in 3D. This method of geology mapping is time consuming and problematic in that the mapping data is turned into a raster image rather than a digital line. However, the map textured 3D solid data allowed for more easily comparable grade estimations of the block model over the various high-grade ore domains on a mine scale. Each mine ore drive walls and backs was analysed in long section, with commentary on geology and estimation accuracy provided for each specific section of every mapped drive. To date, 18 700 m of development backs and walls have been mapped in the Nova underground development workings using this method.

Areas where the block model did not correlate well with mapping observations were flagged in a list of recommendations for review, and wireframes were adjusted using scanner polyline pickups in time for the next model update. This improved the accuracy of the estimation and therefore mineral resource to processing reconciliation.

TABLE 1

Method for adding digital lines to georeferenced LiDAR point cloud of underground mine workings, to create detailed 1:250 scale 3D geology maps.

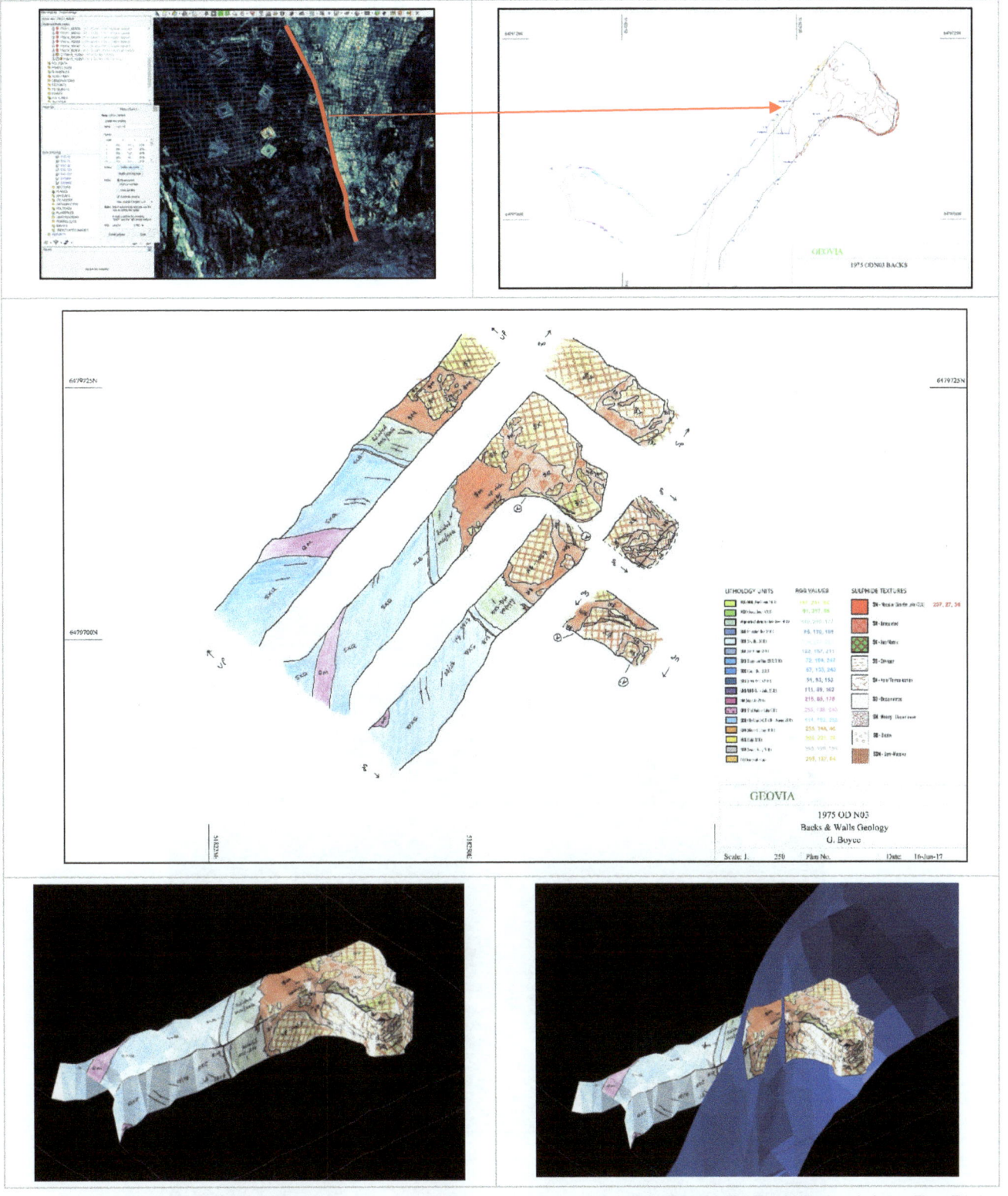

Notes: Top Left: Digital polyline with millimetre accuracy created by snapping to Riegl LiDAR point cloud data. Top Right: Digital polylines are exported from RiScan and plotted with lithology codes added. Centre: The 1:250 scale map is coloured by hand and scanned on a printer to produce a digital .jpeg raster. Bottom Left: The .jpeg raster image is draped over a simplified wireframe mine solid of the mine tunnel. Bottom Right: The ore domain wireframes (blue) used for the mineral resource estimate are analysed for conformity and updated as required.

The ability to take survey accurate structural measurements on planes, joints and fault strikes using the point cloud data, cannot be reproduced from paint markup using conventional laser-disto, tape and compass/clinometer geology mapping tools – particularly in areas inaccessible to physically hold a compass, such as the tunnel roof (backs) and from unsupported ground prior to fibrecreting. Moreover, the LiDAR instrument data remains unaffected by magnetic interference, a notable advantage given that a significant portion of Nova rock types exhibit ferromagnetic properties, thereby mitigating the potential impact on the precision of structural measurements recorded with a conventional magnetic compass.

This has served as a valuable contribution to the Nova structural model, and permanent 3D record for assessing ground support around planned stope brow positions and predicting the likelihood of poor ground conditions in adjacent development drives after opening large voids.

During mining of the orebody, the reflectance data provided some information on timing relationships between the various rock types that had never been observed in mine scale (ca. 5 m) outcrops before. The analysis showed that some areas of the mine had interesting timing relationships that didn't conform well to the existing genesis theory (Figure 5). Use of the Riegl VZ-400 was then fully implemented as the main geology mapping workflow in late 2017 with superior results compared to 'paint markup' mapping.

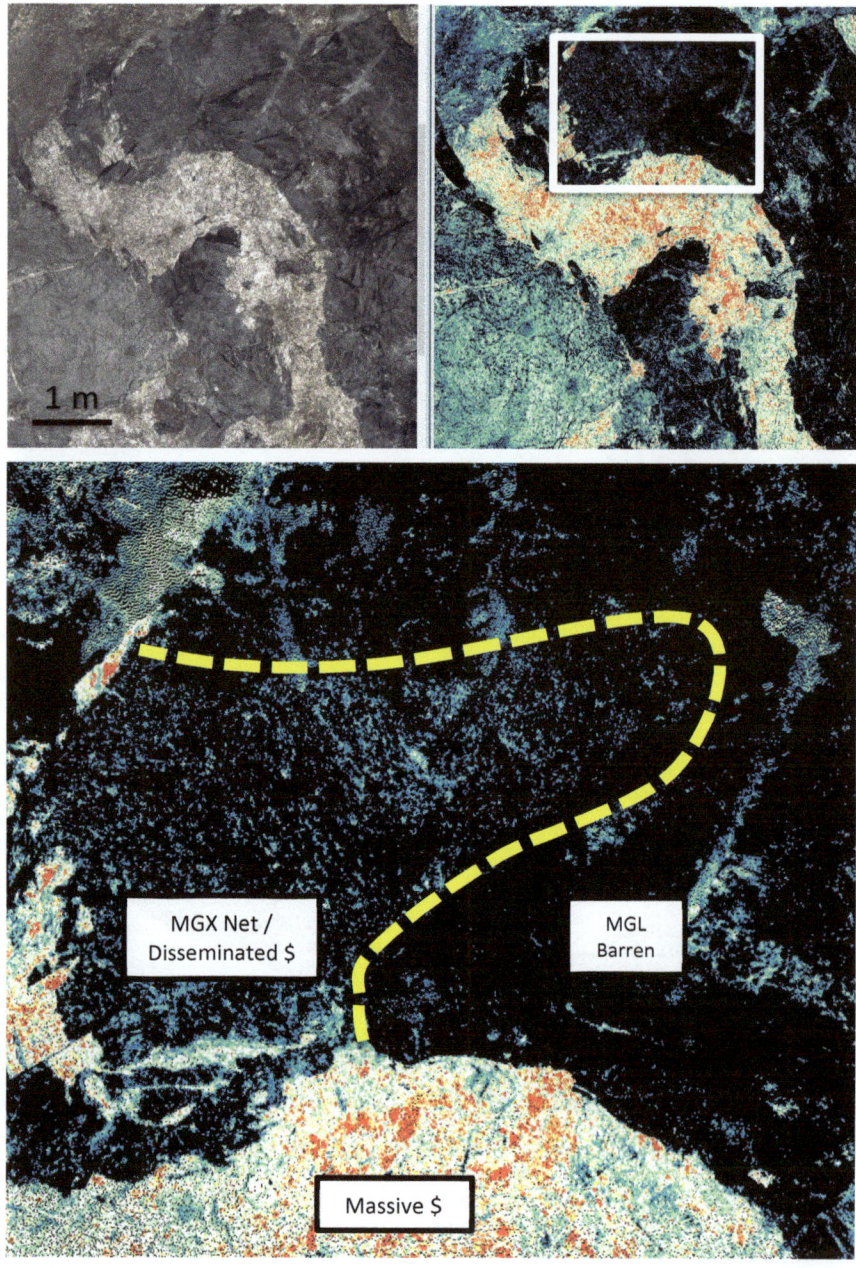

FIG 5 – Riegl VZ-400 data Left: DSLR image of massive sulfide outcrop in mine development face; Right: Magnified LiDAR reflectance intensity data, shows a difference in the reflectance patterns of disseminated sulfides between a leucogabbro on the right (MGL), and a mineralised two pyroxene

gabbronorite (MGX) on the left. Although the reflectance is showing gangue minerals, the different sulfide textures are more apparent between low and heavily disseminated and massive sulfides. The relationship with the massive sulfides shows the two gabbroic rock types were *in situ* before the massive sulfides intruded in a separate pulse. This exposure is contrary to main modelling theory that interprets the massive sulfides percolating out of the MGX host rocks, forming massive sulfide lenses.

2018–2019 alternative applications of mobile LiDAR for the Nova Project

Swell factor study

During the Riegl optimisation project, other values were discovered from experimenting with LiDAR data. A project was launched with an aim to quantify the loose rock (dry) density of broken ore, in order to increase the accuracy of volume measurements on broken rock on run-of-mine (ROM) stockpiles. The scanner was used in lieu of survey total station due to the high accuracy of the point cloud data and the Riegl laser scanner became an intrinsic tool in the study. The project took over six months to complete due to the complexity of gaining access to representative development cuts without disruption to the mine production schedule. A total of 32 development cuts were scanned throughout the mine before and after mining. The ore was loaded directly to trucks, which were weighed over a weigh bridge and the tipped stockpiles were scanned with an aerial drone on the ROM. In 2019, the results presented an average 'swell factor' for broken rock, to be applied in mining operations, along with a density plot comparing nickel grade, which facilitated improved measuring of end of month ROM stockpiles.

Concentrate shed stockpile surveys

The laser point cloud data proved far more accurate for building digital solids for volumetric calculations than survey total station instruments used in the survey department. The Riegl scanner was therefore employed to measure the concentrate stockpiles within the Nova concentrate shed, used for end of month processing reconciliation. As mobile LiDAR devices became more lightweight in the coming years, the large area of the concentrate shed was replaced by a Hovermap drone. The drone can easily cover the back of stockpiles without the need for operators to climb the concentrate piles.

Long range applications

Point cloud data was captured for measuring and monitoring of the water and spoils level at the tailings dam facility at Nova. It was also useful for monitoring potential ground movement in the box cut, where a large area requires more accuracy than photogrammetry from drone flight can offer.

Drill hole pickups, validation and site set-ups

The Riegl scan data was useful for some drill platforms where multiple holes had been drilled. For mine surveyors to pick up drill collars with a total station, it usually requires a two technicians. In an underground environment, it is difficult to see the exact collar location or collar borehole identification (usually written on a pink pin flag) from where the total station is set-up, so it would require either the surveyor to walk back and forth from the collar location to the total station which can be time consuming if there are 10–20 holes drilled from the same drill platform. The geology department could collect collar pickups with a single scan of the drill platform and pick the collars within the 3D environment from the comfort of the office. It was also possible to pick up grade control drill intercepts later in mine life, when the tunnel development eventually intersected grade control drilling and exposed the hole in the tunnel walls or backs. This added validation to some of the drill hole surveys, which averaged around a metre from where the hole survey had projected the drill string.

RIEGL SCANNING NOVA – PROJECT CHALLENGES

Physical limitations

The purpose of a terrestrial scanner such as the Riegl VZ-400, is for larger scale terrestrial land surveying such as open pit mining and large scale civil engineering projects. Hence, it has inherent limitations when conducting operations in an underground tunnel. It cannot capture data less than a 1.5 m distance. There also is no usual requirement to capture the sky in land surveying, it therefore does not scan directly above itself, leaving a 'hole' in the point cloud data on the tunnel roof underground. In new areas of the mine, this requires two scans to infill gaps in the data, which increases the data capture time and captures data on oblique angles.

Lack of peer operation

During the period when the Nova geologists reached the pinnacle of their efforts in integrating the Riegl scan data into the geological workflow, the methods were presented at a geology symposium in Kambalda in late 2017. At that stage, the Nova team were positioned as a leader in the field, since Nova stood as the sole underground mine in Australia employing a VZ-400 for mapping underground geology tunnels. In contrast, other sites relied on photogrammetry or different LiDAR capturing devices that necessitated additional work from the survey department for pickups for each scan position. This left Nova with no peer underground mines in Australia for comparing data sets or workflows.

Operational access

Both traditional 'paper and scale ruler' mapping and digitised strings from scan data necessitate access to mine services and a comprehensive wash-down of the drive underground prior to data collection. This task is time-consuming and challenging due to access constraints during jumbo drill operations, service crew works and bogger operations, which can take several hours for a single heading or lead to overnight shifts to complete. Fibrecreteing on night shift in the mine schedule led to data loss.

Software updates and training

The Nova resource geologists changed the mining software package from Surpac to Datamine Studio RM in 2020. The legacy draped .rgf map files were not compatible and required new draping and plotting techniques. This update took a vacation student around two months to update the entire data set.

Manual handling

Training of new starters in scanner operation takes several days on-site then weeks to master depending on the skill level of the operator. Unfamiliar equipment handling led to back strains/slips, trip hazards climbing stockpiles and traversing underground locations. Equipment technical difficulties also occurred. The scanner is cumbersome to handle when mounted on a tripod, and weighing ~13 kg poses a manual handling risk if incorrect techniques are employed while traversing underground. Additional equipment of a laptop computer and DSLR camera with external flash mount is required, all of which can suffer malfunction.

2019 new model scanner – Z&F Imager5016; Hovermap

Staff required re-training with new procedures and workflow updates which was time consuming.

Research and development can be very time consuming and sometimes doesn't eventuate into value add with adoption of conclusions. The Imager5016 was chosen for its light weight and onboard camera and lighting set-up, but the post processing software was very problematic, with little support and the unit itself has intrinsic design flaws. It's light weight value add is far outweighed by the time lost addressing other issues. Some test work was conducted underground during commissioning of the survey department's drone, where a GoPro camera was mounted to record imagery. The point data was deemed more suited for large volumes; sparser data in the point cloud deemed unsuitable for fine detail mapping of geology.

Data management

New staff found the time-consuming laser scan data collection and post scan processing often lead to falling behind in the actual analysis of geological information and the large data sets demanded high end computer processors and storage.

To date, the Nova geology team have captured over 2800 scan positions; representing more than 3.5 TB (terabytes) of processed scan data (each scan is almost 1 GB file size) that has all been processed in-house in the geology department and represents a significant body of survey accurate mapping data.

Training and licensing

Being a precision survey instrument, the Riegl scanner and data processing software carries significant user training to operate and maintain, that is often very unfamiliar for geologists. Extensive Nova-specific user manuals and procedures needed to be written. Maintenance includes service hours that are programmed into the machine to meet survey calibration certificates – the instrument can only be serviced at the factory in Austria, which usually has at least a two month turnaround due to shipping from Perth. A hire unit is usually available from the supplier while the scanner is off-site and the cost for service is typically around A$20 000.

Future scope of works/research and development/decommission options

As the Nova mine moves towards closure, the continuation of processes need to be constantly evaluated to assess fitness for purpose and time versus value of validating a fairly well defined MRE model.

Now that the Nova orebody is relatively well understood and the mine development has moved towards a more optimised phase, there is an option to phase out development scanning all together in lieu of traditional mapping techniques and pursuit of new and cheaper LiDAR technologies.

The approaching end of mine life has been a contributing factor driving the need for innovation into a low cost alternative.

2024 optimisation test results

Scaniverse (iOS)

Scaniverse is a free software app for Apple iOS, that allows users to easily capture LiDAR point data at 5 m range. The data can either be exported as coloured points, or a textured mesh.

Scaniverse, uses visual information as well as accelerometer/gyro data for tracking. The reconstruction algorithm is based on structure from motion (SfM) and multi-view stereo (MVS) algorithms.

For processing with Detail mode, SfM and MVS is used, plus photogrammetry to build depth maps and estimate scale separately using data obtained from ARKit.

LiDAR device process scans in Area mode with LiDAR data and non-LiDAR devices use a combination of photogrammetry and ManyDepth.

An underground trial of alternative mobile LiDAR devices was undertaken in early 2024, where data was acquired using the Scaniverse app on an iPad Pro tablet, using the iPad's Lidar/photogrammetry sensors – a tool worth ~$2000 and weighing only 1 kg. It took about five minutes to walk the drive to the face and pick up two survey station points – which were the only two points used to georeference the mesh to mine grid. The output rendered meshes were exported/imported into Datamine Studio RM for 3D viewing. Considering there were only two control points – that did not include the prism offset (typically only 0.2 m) – the process delivered an impressively accurately georeferenced result. The accuracy of this method has since been improved by using laser control points in conjunction with prism offsets.

An iPhone 15 Pro has since been acquired at Nova, that uses the same sensors as the iPad used in initial trials. However, it weighs just 220 g, and when mounted to the end of a telescopic 'selfie stick' tripod, alongside an LED light cube, is extremely light weight and simple to operate. Scan times are typically only five mins to scan an entire 5 × 5 × 4 m development tunnel face, walls and backs.

The final test data was captured using iPad Pro LiDAR sensors before implementing it into the current mapping workflow at Nova, was collected underground mid-January 2024 of the Bollinger 1815 XC M mining face. The data was successfully saved as a georeferenced native Datamine triangle file (with textured object mesh) for use in mine design and resource validation. The grade control drilling database was loaded and compared in the 3D project and section of MRE blocks estimation accuracy verified. The georeferencing of the LiDAR wireframe to mine grid was validated in the mine engineering project (Deswik), which showed excellent conformity to surveyed floor strings and control station locations (Figure 6).

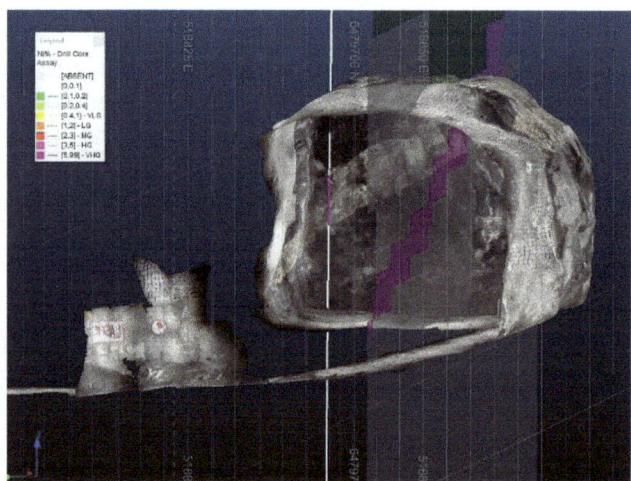
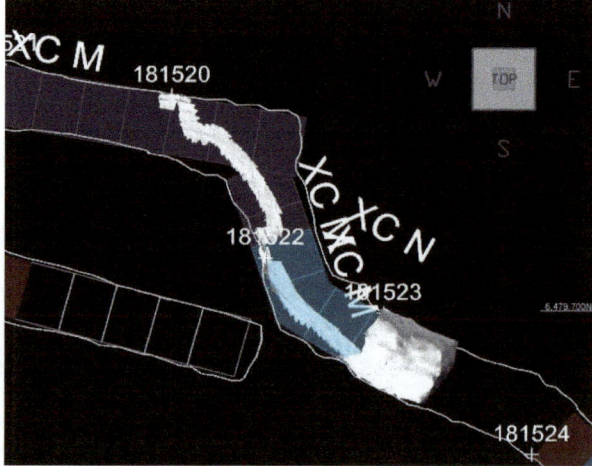

FIG 6 – (left) iPad LiDAR textured wireframe, showing proximal grade control drilling intercept string and section of mineral resource estimate blocks (purple). (right) Wireframe solid georeferencing verified in the mine engineering project.

The drill hole pickup method using the Riegl scanner has been replicated using mobile LiDAR data derived from iPhone sensor (Figure 7). When viewed on a mining scale, there was no noticeable difference between data sets of Survey total station pickup, diamond drill downhole gryo survey (REFLEX) and the iPhone LiDAR scan data.

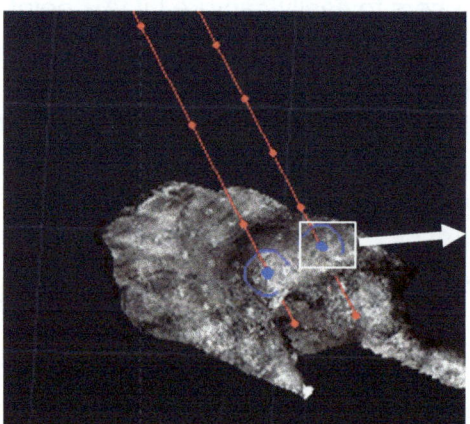

FIG 7 – Images showing three comparative surveys of three different tools. Textured wireframe 3D model of mine development tunnel (Datamine Studio RM), with drill string gyro survey (red line) and total station survey pickup of breakthrough point (blue).

The exercise validated three different data sets, using three different data collection methods:

- The gryo survey from the drillers is deemed accurate (typically if drill hole intercepts development and the survey pickup is within 0.5 m over 100 m hole depth, the gyro survey is considered accurate).

- The textured wireframe model from the iPhone LiDAR sensor is accurate (this scan took about ten minutes to conduct the survey and used only two survey prism control points to resection to mine grid in Datamine).

- The gyro survey (string), LiDAR (wireframe/solid) and survey pickup (point) all correlate well to within <0.3 m of each other. For a mining infrastructure hole, this is a very good result – especially considering these holes were pre-drilled before the development tunnel was advanced to the predicted drill hole breakthrough point.

Some limitations of the iPhone LiDAR sensor is that it is only short range – at 5 m it is sometimes difficult to get reflectance from the tunnel face from supported ground. This is resolved by mounting the mobile device to an extendible pole. Reflectance from water puddles can also cause data errors and ghosting that software filters can't rectify, which can affect the accuracy of georeferencing. There is no ability to re-enter a scan project once it has saved and there is no function to pause data collection during a survey.

GEOREFERENCING

One of the key challenges with 3D scanning of underground mine tunnels, is orienting the scan data to mine grid, otherwise known as georeferencing. The underground environment does not have the luxury of GPS coordinated referencing. The Riegl scanner makes use of the control station prisms set-up by the survey department, needing at least two stations to do a three-point re-section to mine grid. Textured wireframe data derived from Apple LiDAR sensors do not have a central scan position. In order to triangulate the data, it is therefore necessary to collect three or more reference points. Since all the mine workings require survey control stations and lasers, it is no additional work to utilise the same control points when collecting LiDAR data with an iPhone. Markings on walls can also be utilised, so developing tunnels can simply be reference object from the previous excavation.

Due to the bulk mining style of the Nova deposit, it is not essential that geology mapping from Riegl scans be accurate to millimetre scale. The backs and walls map records are at a 1:250 scale. However, accurate ore contacts improve recovery and more accurate reconciliation. Additionally, fault strikes mapped and modelled as accurately as possible can reduce production downtime from unexpected ground movement, or production time lost in rehabilitation or additional ground support plans where stope brows interact with unexpected variances in the structural model when mined. The mobile LiDAR data from an iPhone has already proven the accuracy at 1:250 mine scale is suitable for grade control purposes. Due to poor drilling angles, the high-grade ore lens predicted by the mineral resource estimate domain wireframe (pale blue) is offset by up to 2 m as a stratabound narrow massive sulfide lens (red outline). The ore domain wireframes can be updated using this data on a daily basis as the tunnel progresses, allowing for speedy updates to the resource model and optimised stope design (Figure 8).

FIG 8 – (left) Plan view of a Nova mine tunnel (UPPR 2180 OD) derived from iPhone15 Pro LiDAR sensor in March 2024, showing the high-grade ore lens (red outline) is offset by up to 2 m as predicted by the MRE domain wireframe (pale blue).

CONCLUSIONS

Mobile LiDAR devices have advanced to such a degree that they now comfortably fit in the pocket of a miner's overalls. The Riegl scanner is heavy at 13 kg, and when mounted on a tripod can be cumbersome for operators to move around in an underground environment. The need for lighter more mobile digital mapping device was driven by workplace injuries and large amounts of lost production time processing and managing the large data sets.

Smaller data sets derived from mobile LiDAR to produce detailed textured wireframes for geology mapping are easier to manipulate, store and share than large colour point clouds produced by high end laser scanners.

Quantifying the cost savings to junior companies is difficult, since mine production overheads vary from site to site with the variable time costs for delays to mine schedule waiting for technicians to collect their data and make decisions on mine design. If cost in additional wages are reviewed without other variables, the average mine surveyor wage is ~A$70/hr (2024). Using Nova as a case study, it could be estimated that the turnaround time from surface to a single pickup job at the mining face, including time spent post-processing data, represents an average of three hours per shift. In a 24-hour operation this represents a saving of A$77 000 per annum in contract wages, if the geology department do all geology scan pickups in-house without relying on paint lines and survey pickups.

Utilising Apple's LiDAR technology represents savings of over A$200 000 in capital when compared to high end laser scanning instruments, with the possibility of no ongoing fees for software or hardware maintenance, representing further savings of around A$20 000 per annum.

Advantages of using lightweight, mobile LiDAR devices for geology mapping underground are:

- Using detailed scan models in the geology mapping workflow brings mapping to the comfort of the office, spending less time in the hazardous underground environment.
- Removes some subjectivity in geology rock type interpretations, allowing for remote validation.
- Provides permanent 3D record of geology and structural outcrops after mining.
- Saves time for the survey department by eliminating the requirement for ore contact pickups.
- Provides accurate ore domain boundaries for mineral resource estimate block model reconciliation.
- Potential to bring cost-effective face mapping into the digital 3D space and eliminate paper mapping/interpretation altogether.

A cheap but competitive 3D mapping alternative can be achieved by using iPadPro or iPhone, coupled with free software like Scaniverse and Cloud Compare, and exported to any mining software package such as Datamine Studio RM, Deswik, Micromine, Vulcan, or Surpac, as a textured wireframe object.

As of March 2024, the Nova Geology team has intergrated an iPhone15 Pro's LiDAR sensor and photogrammetry as a key tool for production face mapping. The workflow utilises free Scaniverse software to capture the 3D data, in unison with the current mine resource modelling software (Datamine Studio RM) to georeference the textured wireframe objects to mine grid. The mapped ore contacts remain as digital line work, where wireframes can be mapped directly into the Studio RM projects, and compared to drilling and mineral resource estimate block model without the requirement for printing and draping of raster images.

One of the key problems for mapping in 3D for the Nova team has, and continues to be, the ability to easily share the information from these large data sets. The challenge of how to share 3D mapping data in 2D reporting is being overcome as software and hardware for PCs and mobile devices rapidly advance. The model from Figure 9 can be rotated in 3D in Microsoft Word. The author has made it available to download as a separate file, please note that the file size is 68 MB.

The file can be downloaded from:
https://www.dropbox.com/scl/fi/bjz4dgj0whdhmppo3gkro/240429_Glenn-Boyce_Remote-Sensing_A-Nova-Mine-Case-Study_Appendix1.docx

FIG 9 – Nova underground workings, shown as embedded 3D model. The data was captured using LiDAR sensor on an iPhone 13 Pro, mounted to a telescopic monopod pole and light cube. 3D textured wireframe is an.OBJ file exported from Scaniverse app. The model can be georeferenced using existing mine survey control points of the laser and prisms which can be observed joined to the model via the floor traverse. Textures to note in the model are the ~1 m thick massive sulfide vein cross-cutting the tunnel, which can be digitally mapped and compared to mineral resource estimate in the current production mine software (Datamine Studio RM) after georeferencing. Microsoft Word allows rotation of this embedded 3D object by clicking the rotation widget in the centre of the image.

ACKNOWLEDGEMENTS

The author would like to acknowledge Markus Staubmann for his initial vision of using a laser scanner for geology mapping underground workings in 2013. With the help of Joel Woodage (CR Kennedy) throughout the project, many alternative LiDAR scanning devices and techniques were researched. Nova mine geologist John Chapman developed the method for using the Riegl scanner data to adjust Leapfrog wireframes, which formed the pillar for the development of using scan data to map the Nova underground workings. Nova mine geologist Callum Laming championed the Nova Swell factor study. Nova mine geologist Dr. Sebastian Staude provided initial test work on mineral reflectance intensity using the Riegl scanner. Thank you to Stacy from tech support at Scaniverse for details on the function of the software and iPhone sensors. Thank you to the developers at iFunBox for providing a free software tool for easily transferring iOS data to a Windows OS. The Author would like to thank the Nova Survey Department for their help assessing volumetric data for comparison. Thanks to Paul Hetherington who sparked initial feasibility for iPhone LiDAR as an alternative method for mine mapping, and his detailed contributions to Nova's comprehensive Mineral Resource and Ore Reserve Estimate reports. Thanks to Fletcher Pym for assistance in editing the paper. The author would like to thank IGO Ltd for their vision and support for research and development into LiDAR technologies at the Nova Operation, and their willingness to share the information amongst the wider mining and exploration community.

REFERENCES

Barnes, S, Taranovic, V, Miller, J, Boyce, G M and Beresford, S W, 2020. Sulfide Emplacement and Migration in the Nova-Bollinger Ni-Cu-Co Deposit, Albany-Fraser Orogen, Western Australia, *Economic Geology*, 115(8):1749–1776. doi:10.5382/ECONGEO.4758.

IGO Limited, 2023. FY23 Annual report of Mineral Resources and Ore Reserves, 30 June 2023. Available from: <https://www.igo.com.au/site/pdf/0ad43283-515b-4047-a042-56af27755480/FY23-Mineral-Resources-and-Ore-Reserves.pdf>

JORC, 2012. Australasian Code for Reporting of Exploration Results, Mineral Resources and Ore Reserves (The JORC Code) [online]. Available from: <http://www.jorc.org> (The Joint Ore Reserves Committee of The Australasian Institute of Mining and Metallurgy, Australian Institute of Geoscientists and Minerals Council of Australia).

Spaggiari, C, Kirkland, C, Smithies, R H and Wingate, M T D, 2014. Tectonic links between Proterozoic sedimentary cycles, basin formation and magmatism in the Albany-Fraser Orogen, Western Australia, Geological Survey of Western Australia Report 133. Available from: <https://www.researchgate.net/publication/265907986_Tectonic_links_between_Proterozoic_sedimentary_cycles_basin_formation_and_magmatism_in_the_Albany-Fraser_Orogen_Western_Australia>

Homogeneity and heterogeneity of crushed gold certified reference materials

J Carter[1] and B J Armstrong[2]

1. General Manager, Independent Mineral Standards, Bayswater WA 6053.
   Email: john@imstandards.com.au
2. Operations Manager, Independent Mineral Standards, Bayswater WA 6053.
   Email: bruce@imstandards.com.au

ABSTRACT

The use of control standards, or certified reference materials (CRM), are an important implement when assessing the quality of results from an on-site or commercial contract laboratory. While laboratories operate their own internal quality control, the submission of a blind CRM by the customer provides independent data and additional assurance.

CRMs on the market are pulverised to a fine particle size rendering them analysis-ready. These pulverised CRMs bypass the sample preparation step and are queued at the analytical stage of the process reducing the efficacy of a blind CRM. This is problematic. Sample preparation is a critical part of the laboratory process and a potential source of systematic error and bias.

In addition to pulverised CRMs, crushed iron ore and bauxite reference materials, typically less than 5 mm in particle size, have been included in quality control programs and submitted blind to laboratories for many years. For other commodities however, particularly gold, the manufacture of a crushed reference material has been elusive largely due to the disseminative mineralogical nature of the ores, the gold nugget effect and ineffective manufacturing techniques. For a crushed CRM to be an effective quality control tool, however, it needs to behave like a sample from the field which is, by nature, heterogeneous while maintaining effective between unit homogeneity.

In order to establish the potential for a crushed gold CRM to be used as part of routine quality control in a reverse circulation (RC) drilling program, a number of reference materials with specific gold concentrations were engineered then subjected to an inter-laboratory round robin exercise. The degree to which the materials are homogeneous or heterogenous is discussed in the paper, along with a comparison to a natural ore reference material crushed to 3 mm and subsampled. While providing some insights into the potential for systematic bias to occur in sample preparation for methods such as fire assay, the study also has implications for blind quality control of methods that analyse crushed samples, such as photon assay and cyanide leaches (PAL).

INTRODUCTION

The production of high-quality data models for grade control programs, block models, mine plans, blending and stockpile management relies on accurate and effective sampling and analytical protocols. The mine geologist largely takes responsibility for setting and monitoring protocols from the initial sampling at the drill rig or in-pit and face sampling, through to establishing the on-site or near-site laboratory methods. It is important these protocols maintain a high level of integrity so the mine geologist can confidently interpret the data. Sampling and analytical quality control is therefore fundamental to mining activities and are a critical activity for establishing data reliability and confidence. Incorrect sampling and sample handling procedures may have a significant impact on costs and decision made (Minnitt, 2007).

For the vast majority of cases, the laboratory contract includes two components agreed between the mine and the laboratory – sample preparation, and analysis. It is the responsibility of the mine geologist to include into the program, appropriate quality control tools to independently assess the laboratory results for trueness, precision, accuracy or bias. More so, it is generally not possible to achieve meaningful results without well controlled sampling and sample handling protocols. (Carswell, 2017; Dominy, 2016).

The primary purposes for the sample preparation process in the laboratory are, firstly, to ensure the samples are in a form ready for the analysis to proceed, and secondly the prepared sample is representative of the original submitted sample. In recent years there have been developments in automation and robotisation that enable laboratories to prepare samples with less manual intervention (Knudsen, 2007). Either way, from a sample and quality control perspective, the two fundamental requirements of sample preparation to ready the sample for analysis are the same, particle size reduction (crushing and pulverisation), and mass reduction (splitting and sub-sampling). These two requirements typically involve a primary and/or secondary crusher to reduce the particle size of the sample from the as received size to a nominal top size of 2–3 mm, followed by a sub-sampling step to reduce the mass. Sub-sampling may involve a riffle splitter, manual or automated rotary splitters or linear divider with waste material discarded. The target mass depends on the size of the pulverising bowl, typically 1 kg for LM2 bowls and 2–3 kg for LM5 bowls. Lastly, pulverisation, or grinding, occurs in a sealed vessel containing a puck and/or ring, and rotated with a counterweight to cause the puck and rings grind at high speed against the bowl wall. Following sample size reduction to the powder form, a sub-sample is extracted from the bowl, typically 100 g or 200 g and stored in a labelled pulp packet ready for analysis. The remainder of the pulverised material is often discarded, or returned to the original bag and the pulveriser cleaned in readiness for the next sample.

Recent analytical developments have interrupted historic sample preparation requirements. Photon Assay utilises a 400–500 g sample, and while largely incognisant of the particle size of the sample, bulk density and packing stability of the sample presented to the instrument are important variables that are monitored as part of the analysis. For photon assay, samples are crushed to a nominal top size of 2–3 mm, followed by splitting to the volume of the analytical jar. The pulverisation step is not required and not preferred (Tickner, Preston and Treasure, 2018).

Sample preparation is a critical part of the laboratory process. Samples damaged in sample preparation are difficult to recover, and errors carry through to analysis and compromise the results and therefore the trueness of the resource estimation. Sampling theory (TOS) studies (Gy, 1982; Pitard, 1993; Dominy *et al*, 2018; Minnitt, Rice and Spangenberg, 2007) have demonstrated that at each step in the field to data process a sampling error can occur and an understanding and minimisation of these errors should be a key objective in any quality control program. Dominy, Glass and Purevgerel (2022), in more detail summarise the application of the theory of sampling (TOS) to the sampling value chain including sample preparation and analysis. Within sample preparation, the potential for errors that contribute to analytical integrity to occur are significant. It is therefore critical that quality control strategies are implemented to monitor and assess all aspects of sample handling.

QUALITY CONTROL STRATEGIES

Quality control approaches currently used across the entire industry are fairly common and have been standard practice for many years (Thompson and Howarth, 1976; Abzalov, 2016; Dominy, Purevgerel and Esbensen, 2020) and in summary:

- Blank – the submission of a blank to assess contamination or carry-over, particularly for gold (trace element) and base metal (minor element) is common practice. Contamination can occur at any time within the laboratory, in all stages of sample preparation, and many stages within the analytical process, the severity dependent on the material types, sample preparation practices, laboratory cleanliness, analysis methods and elements of interest. Blanks measure neither precision nor accuracy in the laboratory.

- Field duplicate – a second submitted sample from the same drill interval can be collected submitted as a quality control sample in an attempt to measure laboratory variance and the quality of sample preparation. In practice, however, the splitting error from the field is often significantly greater than the variances within the laboratory, and the field duplicate should not be considered as an identical sample nor used for laboratory quality control. Field duplicates measure the variance in FSE and the sub-sampling technique in addition to compounding errors throughout sample preparation and analysis.

- Sample preparation split – in an attempt to assess quality in sample preparation, some mining geologists request the laboratory to extract a second split at the crushing step. Where a sample weight reduction is required and excess material is available, a second sample for pulverisation can be taken. These splits are taken at laboratory pre-defined sample intervals, and as such may not split a sample of grade or of interest to the exploration company. The sample preparation split may be able to assess precision in the sample preparation process, but will not assess accuracy. If both the primary and split sample are compromised a systematic bias will not be detected.

- Analytical duplicate – a second split from the pulveriser, or a separately weighed aliquot from the sample pulp packet can be requested by the exploration company. The laboratory will label this sample as a duplicate or repeat. The duplicate data received can be used to assess the analytical precision of the laboratory and includes an extraction and weighing sampling error. The laboratory may also conduct their own internal repeats at designated intervals. However, neither a pulverising nor aliquot duplicate determine laboratory bias, but will only be able to assess precision. If the sample was compromised in sample preparation, both the primary and secondary aliquot will be biased.

- Submitted CRM – the primary quality tool utilised by exploration companies for the assessment of accuracy is a submitted CRM (Sterk, 2015). Purchased from a reputable supplier, CRMs are a stable pulverised material, sufficiently homogenised and characterised by a metrologically traceable procedure (ISO 17034:2016). The CRM is accompanied by a certificate containing consensus values, their associated uncertainties and a statement of metrological traceability. Uncertainties can be used to establish control limits that are monitored by the mine geologist. These CRMs are ideally 'blind' to the laboratory, and may have been sourced from the mining company's ore or manufactured from appropriate material. The CRMs are usually packaged in a foil or sachet, and submitted in the calico bags that would normally contain a field sample. The presence of pulverised CRMs in the submitted batch represents a dilemma for the laboratory as these samples need to bypass the sample preparation process, and be returned to the batch at a later step before analysis. The laboratory will exercise a number of sequencing options in order to optimise the re-insertion onto the batch.

- Laboratory QC – the laboratory will insert their own internal CRMs, blanks and repeat assays into the analysis batch at periodic intervals. A CRM will assess accuracy of the analysis, and only with sufficient statistical data collected over time, can they be used to determine method precision. Repeats assay will assess precision, and blanks will assess contamination.

It is the form and use of submitted CRMs that are of particular focus in this study. Despite the warnings in previous sampling studies (Dominy, Glass and Purevgerel, 2022), a quality control tool to adequately assess systematic and random biases in sample preparation is lacking in the industry. CRMs on the market are pulverised to a fine particle size rendering them analysis-ready. Pulverised CRMs bypass the sample preparations step and are queued at the analytical stage of the process reducing the efficacy of a blind CRM. In addition, for the case of analytical methods that test a crushed sample portion (Photon Assay, PAL), the use of a pulverised CRM unlike routine samples is problematic and it is not just from a representative and matrix matching point of view. These methods rely on the crushed nature or rock/particle properties of the sample to maintain method integrity. For best practice QA/QC, the properties and nature of the CRM used should be similar to the routine samples – as prepared and analysed.

Crushed CRMs containing a certified concentration of gold have been elusive, largely due to the disseminative mineralogical nature of the ores, the gold nugget effect, or ineffective manufacturing techniques. To state the conundrum from another perspective, for a crushed CRM to be an effective quality control tool, it needs to behave like a sample which by its nature is heterogeneous, while at the same time maintain fit for purpose between unit homogeneity. The manufacture of such a CRM is not trivial.

For bulk ores (iron ore, bauxite), crushed CRMs with consensus values and their uncertainties have been used for many years and are available. Crushed CRMs have been designed to eliminate the use of a blank that presents a high risk to submitted sample contamination. Furthermore, crushed CRMs of bulk ores allow the quality of the reported results to be assessed throughout the whole laboratory process not just the analytical component. The practice of using a crushed iron ore CRM for sample preparation quality control has been demonstrated through a case study by Carter and Armstrong (2023). The case study demonstrated the effectiveness of a crushed CRM to detect a systematic bias brought about by faulty dust extraction on a primary crusher in an automated laboratory. The equipment fault had a significant bearing on the accuracy of the analytical results for samples that passed through the crusher. The error would not have been detected without the use of a crushed CRM passing through the complete sample preparation process.

CRUSHED REFERENCE MATERIAL FOR GOLD

While the case for crushed reference materials in bulk commodities has been well established, for gold bearing ores the manufacture of a crushed CRM requires some thought. To be similar to a field RC sample, a crushed CRM product would need to have a maximum size of 5 mm and packaged in 500 g up to 2 kg units. For routine submission of a crushed CRM, the product needs to be homogeneous between units, at a variance level at or below the typical level of variance of the laboratory method. Between unit variance is difficult to achieve for gold ore based certified reference materials, due to the nugget effect (Brand, 2015).

To circumvent the problem of manufacturing a crushed certified reference material from gold ore stocks, Independent Mineral Standards engineered and manufactured three crushed gold CRMs with a nominal top size of 3 mm, where the distribution of gold was at the microscopic level. The patented product was manufactured from an engineered rock, with dispersed gold embedded within the mineralogical structure. The material was further prepared by multi-stage homogenisation and sub-sampling. The final product was packed at nominal 500 g and 2 kg units, in labelled heat-sealed bags for individual use in their entirety. The units were then subjected to a variety of within batch (between unit), and within unit tests to understand the level of inherit homogeneity and heterogeneity and therefore their suitability as a routine quality control tool.

Homogeneity and certification study

During the packaging stage, units were selected for homogeneity and characterisation studies. The homogeneity of each crushed gold CRM was performed by sending 15 samples selected throughout the batch to a single laboratory and conducting analysis of each sample in triplicate. The results are shown in Table 1. The variances determined are applicable only when the entire contents of each unit are analysed or prepared. In the case of pulverisation, subsequent sub-sampling of the prepared units were taken, then analysed.

TABLE 1

Crushed gold CRM homogeneity study results, grade and between sample mean variances.

Au CRM	Sample preparation and Pb fire assay (2 kg unit)			Photon assay (500 g unit)		
	Mean (g/t)	SD (g/t)	Relative SD (%)	Mean (g/t)	SD (g/t)	Relative SD (%)
IMS-235	0.24	0.004	1.72	0.23	0.019	8.12
IMS-236	0.74	0.012	1.57	0.74	0.022	3.05
IMS-237	2.14	0.035	1.65	2.10	0.050	2.37

Following these studies, the batches were found to be sufficiently homogeneous to proceed to the characterisation study. Relative standard deviations for sample preparation and fire assay were all below 2 per cent which is exceptionally low, and similar to the performance of pulverised CRMs that only measure the variance in the analytical component. The homogeneity results did not identify any nugget effects in the material. The higher standard deviations for photon assay are typical for the method for gold grades at these levels.

Certified values were obtained for 2 kg bagged material using traditional sample preparation and fire assay. A total of ten laboratories received five 2 kg samples and were requested to conduct routine sample preparation, followed by fire assay for the determination of gold concentration. Australian based laboratories tended to crush and split the sample to a mass for pulverisation in an LM5, nominally 2.5 kg mass. Canadian based laboratories tended to crush then split the sample to a mass for pulverisation in an LM2, nominally 1 kg mass. In both regional cases, fire assay was conducted between 25 g and 50 g masses with either an AAS or ICP finish.

Values for Photon Assay were certified using 500 g bagged materials with no further sample preparation performed. Samples were transferred to the photon assay jars in their entirety. A total of seven photon assay machines were used to analyse five samples. The property values and associated uncertainties from the characterisation studies are shown in Table 2.

TABLE 2

Crushed gold CRM characterisation study results.

Au CRM	Sample preparation and Pb fire assay (2 kg unit)			Photon assay (500 g unit)		
	Certified (g/t)	Within lab SD (g/t)	Relative SD (%)	Certified (g/t)	Within lab SD (g/t)	Relative SD (%)
IMS-235	0.23	0.006	2.6	0.22	0.025	11.4
IMS-236	0.72	0.024	3.3	0.72	0.028	3.9
IMS-237	2.08	0.063	3.0	2.07	0.072	3.5

The characterisation results in Table 2 show good agreement between fire assay and photon assay for the multi-laboratory exercise. The results demonstrate the materials are fit for purpose as quality control tools, with relative standard deviations below 4 per cent, with the exception of the low-level CRM from photon assay where the variance of the measurement method dominates uncertainty as the grade approaches the detection limit. The relative standard deviation of 3–4 per cent is higher than those obtained during the homogeneity study and higher than typical pulverised CRMs (1–2 per cent). This was attributed to the contribution to variance from the ten laboratories who processed the samples, possibly due to differences in sample preparation and fire assay method.

Comparison to natural ore – homogeneity

To illustrate the homogeneity performance of the engineered material, a crushed CRM was also manufactured from a natural ore. Precured from mill feed material in a greenstone-hosted gold mine in Western Australia, the ore was subjected to the same crushing to a top size of 3 mm followed by the same multi-stage homogenisation and sub-sampling steps as the engineered material. The natural ore reference material was packaged in 2 kg with further sub-sampling to 500 g units. The 2 kg units were submitted to two commercial laboratories for sample preparation and fire assay. The 500 g units were analysed by the PAL method without any further sample preparation or sub-sampling. In this way the homogeneity can be compared to the engineered reference material of a similar grade, with difference attributable to the between unit homogeneity. The results are shown in Figure 1.

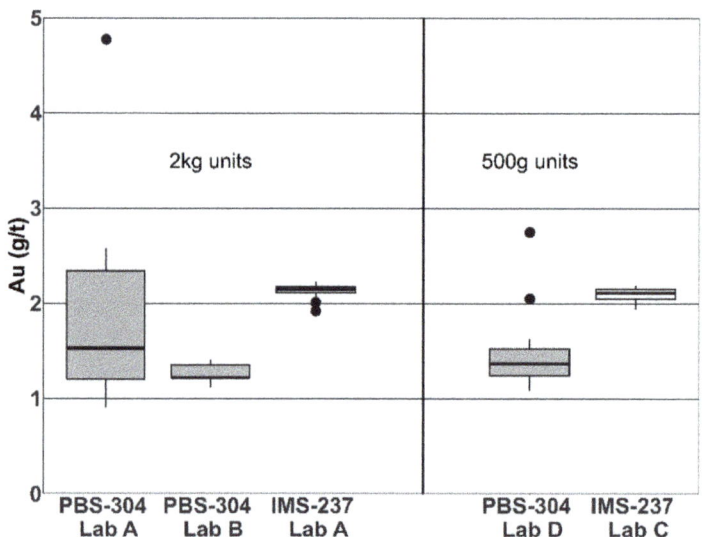

FIG 1 – Comparison of homogeneity study results for a natural ore reference material (PBS-304) and an engineered reference material (IMS-237). Outliers are shown as dots.

The study demonstrates the between unit homogeneity of the engineered product is superior to the natural ore. The variances observed in the data for the natural ore CRM render the product unsuitable as a candidate reference material. Seemingly random spikes in grade are indicative of the material containing poorly disseminated gold throughout the material. The natural ore material was not subjected to a characterisation study.

Heterogeneity study

In order to understand the potential for contribution to uncertainty from sample preparation, the engineered reference material was subjected to a within bag variance test. This test was designed to understand the level of homogeneity-heterogeneity within each unit. A random selection of 2 kg units for IMS-236 were selected from the batch and split into 4 × 500 g samples followed by analyses in their entirety by photon assay. The first study involved careful splitting by rotary splitter divider into eight segments, with two segments from opposite sides of the carousel combined into each 500 g sample. The second study involved taking a 'grab' sample from the 2 kg unit by pouring the sample into each of the four 500 g photon assay jars for analysis. This latter technique would be considered poor sampling practices, failing Gy's test for each particle to have equal opportunity to be included in each sub-sample (Gy, 1982). For each of the 500 g jars, multiple analysis was conducted in order to understand the contribution to uncertainty from the analysis and the sample splitting technique. In this way, the within, and between jar variances could be visualised and assessed. The results are shown in Figure 2. Each figure is shown on the same y-scale.

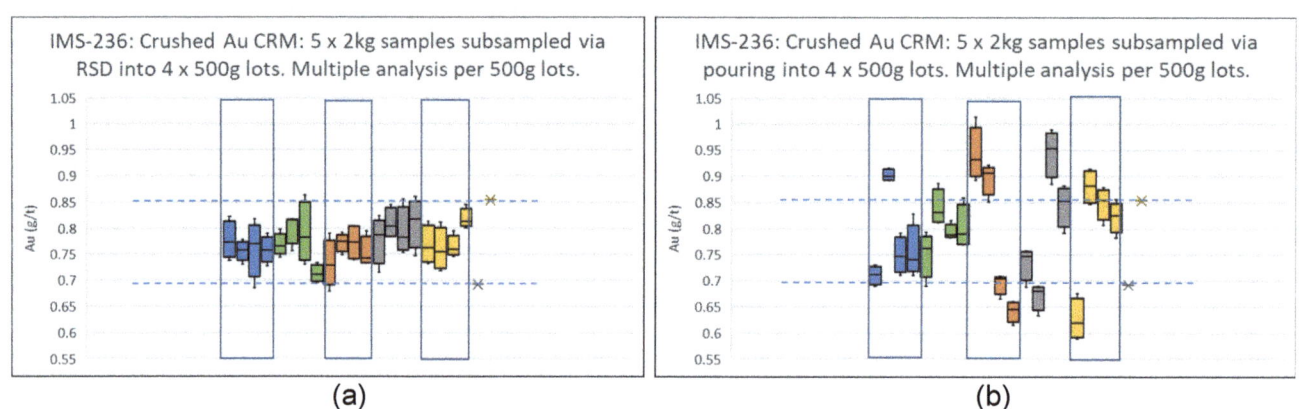

FIG 2 – (a) Photon assay results – 2 kg IMS-236 CRM split to 500 g lots using rotary splitter divider. (b) Photon assay results – 2 kg IMS-236 split to 500 g lots using a grab sample approach.

The test results in Figure 2, is objective evidence of the impact of splitting method, and quality of the sampling practices on the variance between replicate aliquots. The results demonstrate that when the CRM is split correctly, the within jar variance is similar to the between jar variance. The dotted lines in Figure 2a are three times the standard deviation from the mean and would represent a set of control limits that could be applied to these CRMs. It is clear from Figure 2b, when the 2 kg CRM is poorly sub-sampled into the 500 g jars, a significantly higher between aliquot variance occurs, with some of the results falling outside the control limits and therefore failing the quality control test. The results are confirmation the design of the crushed CRM enables sample preparation biases to be detected.

The splitting of a 2 kg unit received by the laboratory, into a 500 g sub-sample for pulverisation is not an uncommon practice. What these results demonstrate is that a bias may occur and be undetected when a single subsample is pulverised and analysed. If the exploration company requests a crushing split duplicate to be analysed and split biases are occurring, it may be possible to detect the poor practices are the cause. Random biases can be somewhat accommodated over a large drilling campaign with sufficient data over time and only if the magnitude of the bias is acceptable. However, more importantly, if the bias includes a systematic component, that is, all samples are biased high or low in sample preparation, the nature of the problem is more serious for mining geology. We were unable to confirm with these results, however, if there was any systematic bias because the analysis order was not tracked against the loading order and a trend from analysis to the split sample was not established. A subsequent investigation was conducted to establish if the sample preparation biases observed are potentially systematic.

It is worth noting a similar trend in Figure 2b could result from a field sample that contains nuggety gold which could mask the detection of splitting quality. In fact, methods of analysis for extremely nuggety gold ores sometimes includes preparation and analysis to extinction, or collection of bulk samples, so that a better measurement of gold grade can be obtained (Roberts, Dominy and Nugus, 2003). With the use of a crushed CRM demonstrating correct splitting practices, a comparison of the CRM duplicate data to the variance between a sample split could be used as a measure of the 'nuggetiness' of the samples and determine if an alternate sample preparation and analytical regime is warranted.

INVESTIGATION INTO SAMPLE PREPARATION BIAS

Characterisation study results

In order to determine if systematic biases can occur in sample preparation, and if crushed CRMs can detect them, an additional investigation was conducted using the samples analysed during the characterisation study. The characterisation study for sample preparation and fire assay involved sending five units to ten laboratories. The certified values and uncertainties for all of the CRMs are shown in Table 2, however, when a comparison between the results from all of the laboratories are visualised in box plots (Figure 3), an interesting trend emerges.

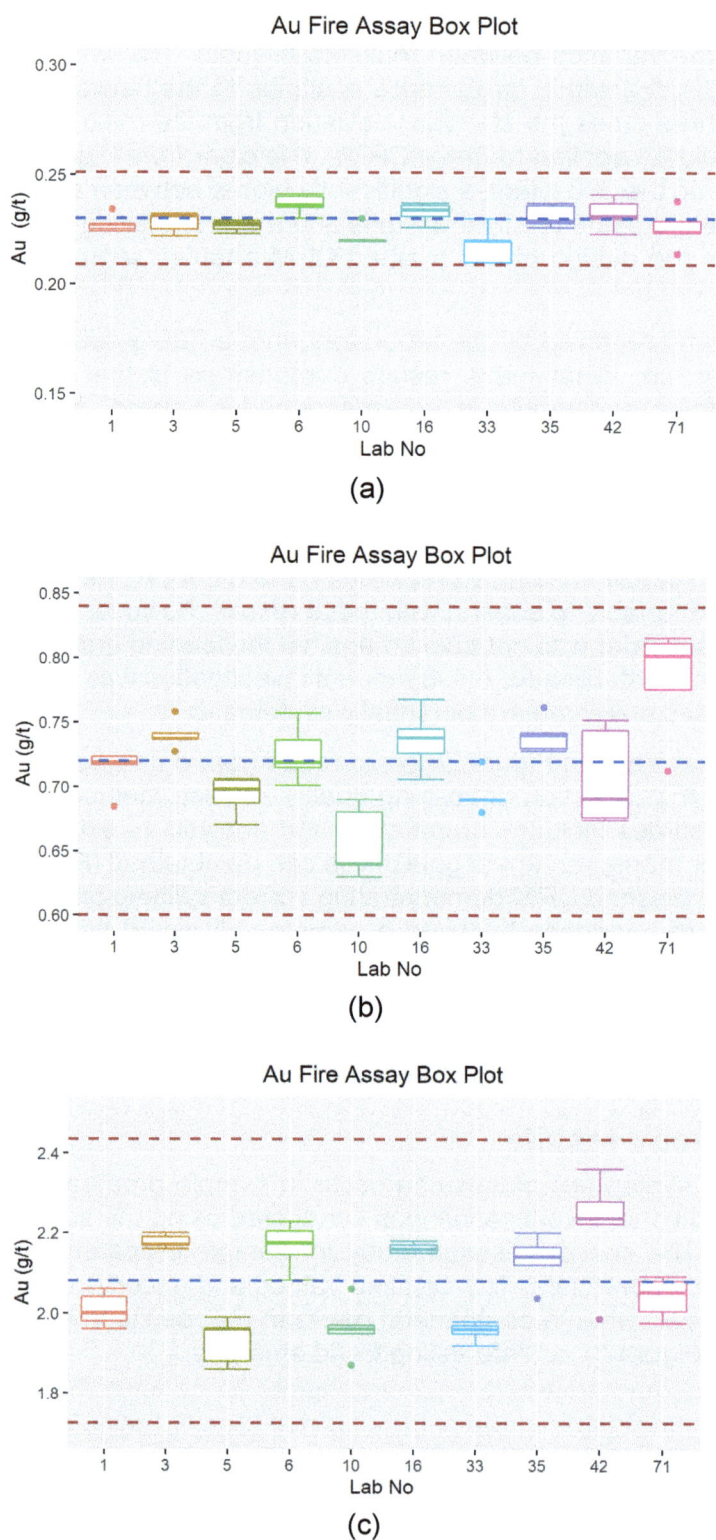

FIG 3 – (a) Low-grade IMS-235 characterisation study box plot. (b) Medium grade IMS-236 characterisation study box plot. (c) High-grade IMS-237 characterisation study box plot.

While not as evident in the lower grade, as the grade of the CRM increases, there appears to be two populations of data. The spread of the data explains why the standard deviations of the characterisation studies are significantly higher (2.6–3.3 per cent) than the homogeneity studies (<2 per cent). Some laboratories have reported results above the mean and others below. The trend appears to be similar for all three grades, but more pronounced at the highest grade. This is an interesting development. Table 2 shows the mean or certified values between fire assay, and photon assay are in very close agreement. Both of these methods are expected to determine the total gold concentration of the sample. The hypothesis is the bias occurs in sample preparation with the loss, or concentration of gold-bearing material. Alternatively, there could be an analytical bias between each of the laboratories.

In the case of IMS-237 (Figure 3c), the z-scores for sample preparation fire assay method of the individual laboratory means were calculated using the global mean and standard deviation from the photon assay results. The results suggest for two of the laboratories, the bias is significant (z-score > 2).

Umpire analysis

To test if the bias is occurring in sample preparation, or during subsequent analysis, pulverised samples were returned from four of the laboratories and subsequently labelled as A, B, C and D respectively. The selection was made from four laboratories belonging to different global groups and where biases were both positive and negative.

The returned pulverised samples prepared at laboratories A, B, C and D, were then sent to an alternate single laboratory for analysis. The results are shown in Figure 4. The samples were re-numbered and submitted blindly and routinely to the alternate re-analysis laboratory without reference to the nature of the investigation.

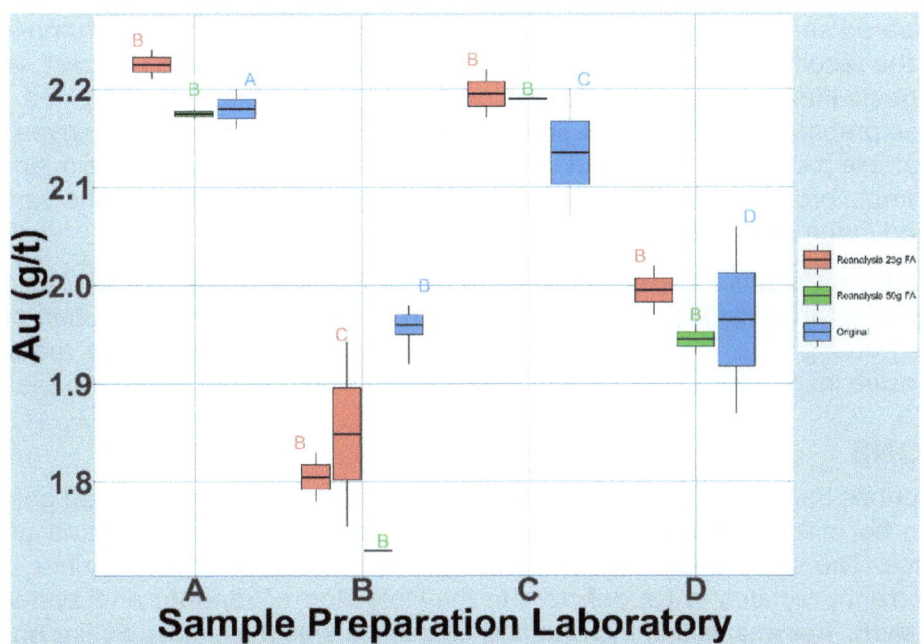

FIG 4 – Fire assay results for pulverised samples prepared at laboratories A, B, C and D (x-axis), then resubmitted and analysed at laboratory B and C as indicated adjacent to each box. Re-analysis was performed at both 25 g and 50 g masses.

In Figure 4 the original results are shown in blue (the furthest box plot to the right of each laboratory indicated on the x-axis), and represents the original analysis for the submitted samples. In these cases, sample preparation and fire assay analyses were conducted at the same laboratory in the same submitted batch. The original results from laboratories A and C biased high compared to the mean, and laboratories B and D where biasing low. All of the pulverised samples returned were then submitted to laboratory B for repeat analysis. To complete the re-analysis exercise, pulverised samples prepared at laboratory B were also sent to laboratory C for analysis. Repeat analysis results are shown in orange for 25 g fire assay (first box from the left for each preparing laboratory), and green for 50 g fire assay (second box from the left for each preparing laboratory).

The first observation to make is the bias does not trend significantly with the mass used to conduct the analysis. Both 25 g and 50 g fire assay results are aligned in Figure 4. The instruction to conduct 25 g assay, or half weight, included a request to maintain the same flux quantities so that if there were any difficulties with chemistry of the material and extraction during fire assay, the results should bias low for 50 g fire assay. While there is a slight low bias for the larger mass, the bias is not noteworthy.

The more significant trend in the box plots from Figure 4 is that the repeat analysis at laboratory B and C matches the bias that occurred in the analysis from where the original samples were prepared. In addition, as a quality control check, the repeat analyses at laboratory B, for the resubmitted samples are consistent in bias with the original results prepared at laboratory B. This informs us that the analysis performed at laboratory B is providing consistent results, and there are no biases occurring from analytical batch to batch.

To determine if the overall high/low bias was occurring in sample preparation, or in the analysis, the alignment between the repeat analysis (orange and green boxes), and the identification of the preparing laboratory on the x-axis should be noted. If the bias was occurring in the analysis, one would not expect the re-submitted sample results to closely align with the original results from where the samples were first prepared. However, the repeat analyses in the case of laboratory B and C closely align with the origin of the prepared samples in both cases, therefore indicating a systematic bias has occurred in sample preparation.

Umpire analysis implications

One of the common quality control strategies exploration companies employ is to conduct duplicate analysis of prepared samples from the main contract laboratory at a repeat or umpire analysis. The comparison of the results is given as confirmation of trueness of the main contract laboratory assay. This study suggests there could be a confirmation bias at the umpire laboratory. If a systematic bias occurs in sample preparation from the original submitted sample, the umpire analysis will not be able to detect this sample preparation bias and will confirm the original analysis. The submission of a field duplicate for sample preparation at an umpire laboratory may not be ideal as the variances from the field are included in the umpire result.

While evidence for a potential systematic bias to occur in sample preparation has been obtained, further investigations are required to understand the root cause. Closer inspection of the laboratory techniques used during crushing, splitting, pulverisation and sub-sampling are required, in addition to a comprehensive inspection of equipment conditions including dust collection systems.

CONCLUSIONS

This study describes the homogeneity and heterogeneity properties of a crushed gold CRM that has the potential to be utilised as a quality control tool during the preparation and analysis of mine geology samples. The study describes typical protocols utilised for quality control, and identified a potential for current practices to be deficient in the detection of random and systematic biases in sample preparation. The analysis of the crushed gold CRM was performed by fire and photon assay methods, with good agreement in the total gold for each of the materials produced.

Sub-sampling, splitting and sample preparation trials have indicated that there is a potential for both random and systematic biases to occur in sample preparation, that may be difficult to detect with current quality control practices. The systematic bias from a crushed gold CRM with certified value of 2.1 g/t could be as high as ± 0.2 g/t, representing a total relative bias of 10 per cent which is significant for mine planning and delineation of ore and waste.

While this study has identified the potential for a systematic bias to occur, the root cause of the bias has not been identified. Evidence suggests crushing and splitting processes may be a contributor, however pulverisation and sub-sampling processes cannot be ruled out. Further investigations are required to determine the parameters in the laboratory that are contributing to the bias and find evidence to confirm or refute the observations.

ACKNOWLEDGEMENTS

Independent Mineral Standards acknowledges the assistance provided by Chrysos Corporation Limited for performing photon assays.

REFERENCES

Abzalov, M, 2016. Quality control and assurance (QAQC), *Applied Mining Geology*, pp 135–159.

Brand, N W, 2015. Gold homogeneity in certified reference materials; a comparison of five manufacturers, *Explore*, 169:1–24.

Carswell, J T, 2017. Best practice sampling QA/QC for gold and base metal mining – including how to assess the applicability of underground reverse circulation grade control sampling, in *Proceedings of the Tenth International Mining Geology Conference*, pp 297–304 (The Australasian Institute of Mining and Metallurgy: Melbourne and Australian Institute of Geoscientists: Crows Nest).

Carter, J and Armstrong, B J, 2023. Homogeneity assessment of crushed and pulverised iron ore certified reference materials – implications for laboratory quality control, in *Proceedings of the Iron Ore Conference 2023*, pp 2–11 (The Australasian Institute of Mining and Metallurgy: Melbourne).

Dominy, S C, 2016. Importance of good sampling practice throughout the gold mine value chain, *Mining Technology*, 125(3):129–141.

Dominy, S C, Glass, H J and Purevgerel, S, 2022. Sampling for resource evaluation and grade control in an underground gold operation: A case of compromise, in *Proceedings of the WCSB10 Conference*. Table 1.

Dominy, S C, O'Connor, L, Glass, H J, Purevgerel, S and Xie, Y, 2018. Towards representative metallurgical sampling and gold recovery testwork programmes, *Minerals*, 8(5):193.

Dominy, S C, Purevgerel, S and Esbensen, K H, 2020. Quality and sampling error quantification for gold mineral resource estimation, *Spectroscopy Europe*, 32:21–27.

Gy, P, 1982. *Sampling of particulate materials theory and practice* (Elsevier: Amsterdam).

International Organization for Standardization (ISO), 2016. ISO 17034:2016(en) – General requirements for the competence of reference material producers.

Knudsen, T, 2007. Fully automated sample preparation and analysis in the mining industry, *IFAC Proceedings Volumes*, 40(11):153–158.

Minnitt, R C A, 2007. Sampling: the impact on costs and decision making, *Journal of the Southern African Institute of Mining and Metallurgy*, 107(7):451–462.

Minnitt, R C A, Rice, P M and Spangenberg, C, 2007. Part 1: Understanding the components of the fundamental sampling error: a key to good sampling practice, *Journal of the Southern African Institute of Mining and Metallurgy*, 107(8):505–511.

Pitard, F F, 1993. *Pierre Gy's sampling theory and sampling practice: heterogeneity, sampling correctness, and statistical process control* (CRC Press).

Roberts, L S, Dominy, S C and Nugus, M J, 2003. Problems of sampling and assaying in mesothermal lode-gold deposits: case studies from Australia and North America, in *Proceedings of the Fifth International Mining Geology Conference*, pp 387–400 (The Australasian Institute of Mining and Metallurgy: Melbourne).

Sterk, R, 2015. Quality control on assays: addressing some issues, paper presented to 2015 AusIMM New Zealand Branch Annual Conference, Dunedin.

Thompson, M and Howarth, R J, 1976. Duplicate analysis in geochemical practice, part I, Theoretical approach and estimation of analytical reproducibility, *Analyst*, 101(1206):690–698.

Tickner, J, Preston, R and Treasure, D, 2018. *Development and Operation of Photon Assay System for Rapid Analysis of Gold in Mineral Ores* (ALTA: Australia).

Interrogating the space between surveys – an assessment of deep diamond drill hole survey intervals and the impact on resource confidence at Olympic Dam

A Chapman[1] and D Clarke[2]

1. Mine Geologist, BHP Olympic Dam, Adelaide SA 5000. Email: andrew.p.chapman@bhp.com
2. Principal Resource Geologist, BHP Olympic Dam, Adelaide SA 5000. Email: david.clarke@bhp.com

ABSTRACT

Deep diamond drilling in hard rock is becoming more common across the mining industry as the need for defining additional mineral resources increases. Exploring for deep geological targets greater than 1500 m below the surface is regularly achieved using intentionally curved drill holes, often via navigational drilling. Curved drill holes provide an optimal angle to intersect mineralisation as well as increase drilling efficiency but require close monitoring of survey measurements to identify and track the hole position underground. Understanding the resolution for survey intervals is imperative to ensuring that the downhole data is reliable. This data is used to inform accurate spatial locations of drill holes, which is integral to orebody definition, target generation and resource classification.

The orientation of a drill hole can be measured using a north seeking gyro from inside the drill casing at any desired interval utilising the overshot assembly. These measurements, called single shots, are recorded, and visualised in 3D software to determine accuracy to the original planned hole trace or desired target. At the Olympic Dam mine, mineralisation data and knowledge are sparse below the current extents of the active mining operations (1000 m below surface). Acquiring knowledge beyond this requires drilling holes to lengths of up to 3000 m from surface. Drill holes of this length will generally have over 100 single shot survey measurements recorded during drilling. This presents a problem for drilling productivity versus data quality due to compounding error downhole traded off against increasing survey times with depth.

This study provides a comparison of drill hole surveys at different intervals along deep directional drilling at Olympic Dam. A baseline downhole spacing at 6 m was established to allow subsampling at coarser intervals for comparison. This allowed for the absolute error seen in the coarser spaced surveys to be assessed against the baseline, which forms a proxy for suitability of the subsampled survey intervals. This assessment showed that deviation from the baseline increased with coarser spaced intervals, but was variable downhole, highlighting the importance of understanding where a survey plan can be optimised to maximise drilling efficiency while balancing data quality.

INTRODUCTION

Olympic Dam is Australia's largest underground sublevel open stoping mine, producing around 10 Mt of ore per annum (Badenhorst, O'Connell and Rossi, 2016). The resource footprint typically sits above 1000 m from surface, however, following the discovery of deeper mineralisation in 2006, recent drilling combining surface and underground diamond drill rigs has defined ore to greater than 2000 m below surface (BHP, 2024). The deeper mineralisation target has also been imaged with seismic velocity profiles, which have complemented the drilling data and highlight the opportunity of further understanding the deeper part of the Olympic Dam Breccia Complex (ODBC) (Ehrig *et al* 2023).

Targeting mineralisation at depth is often achieved through curved drill holes and navigational drilling. Curved drill holes allow the target mineralisation to be intersected on an optimal angle for orebody characterisation as well as improving drilling efficiency. During drilling, understanding the precise location of these curved drill hole traces underground relative to the intended target is important. Modelling of any geological data gathered from the drill hole relies upon the accuracy of this information. Furthermore, understanding if a drill hole will interact with development, infrastructure, or other underground hazards is a critical control to managing potential risks.

The orientation of a drill hole can be measured using a north seeking gyro from inside the drill casing at any desired interval utilising the overshot assembly. These measurements, called single shots, are recorded, and visualised in 3D software to determine accuracy to the original planned hole trace or desired target. The industry standard for calculating the drill hole path between consecutive single shots is the Minimum Curvature method (Jamieson and Knight, 2017). Most mining software has this method built into de-surveying functions. The frequency as to which survey single shots are obtained will impact the quality of the drill hole path. Kinks or bends (commonly known as doglegs in the mining industry) in a drill hole trajectory will occur during drilling due to geological factors or can be influenced by the set-up of the drill string and drilling technique. If survey frequency is too coarse, incremental deviations from plan may not be observed, introducing error, which as the hole progresses can incrementally increase. Conversely, if survey frequency is too fine, there will be a penalty in lost productivity due to the additional time required to take these measurements.

An analysis of single shot frequency was undertaken from two deep surface drill holes drilled at Olympic Dam. Six metre single shots were measured to allow a calculation of absolute error from coarser survey frequencies including 18 m, 30 m and 48 m. The resulting error was interrogated to determine which spacing provided the most reliable drill hole path, whilst optimising drilling productivity. The comparison was assessed statistically as well as visually to identify trends along the hole path.

Mineral Resources are classified based on a combination of geological confidence, assumed continuity, drill hole spacing, drill hole orientation with respect to mineralisation, quality assurance of assays, reconciliation studies as well as the accuracy and quality of the point data. Within Table 1 from JORC (2012) it is noted that, it is the responsibility of the Competent Person to consider all the criteria that should apply to the study of a particular project. In the location of data points, it is specified that the accuracy and quality of surveys used to locate drill holes (ie collar and downhole surveys), trenches, mine workings and other locations used in Mineral Resource estimation are to be assessed and commented on. With this in mind accurate location of data points is essential, and proper survey procedures and control systems must be established to ensure a high degree of confidence in the downhole positions of drill samples when assessing their impact to Mineral Resource classification.

DEPOSIT GEOLOGY

The ODBC (Figure 1) is the host for the world-class Iron Oxide Copper Gold (IOCG) deposit Olympic Dam. The ODBC is hosted within the Middle Proterozoic Roxby Downs Granite (RDG) and is concealed by ~350 m of flat-lying unaltered and unmineralised Neoproterozoic to Cambrian sedimentary rock units. Within the ODBC, there is a mineralised rock volume approximately 6 km long, 4 km wide and 0.8 km deep. However, pre- to syn-mineralisation structural events record a normal offset of up to 3 km in the southern area of the deposit (Clark and Ehrig, 2019). This normal offset is coincident with a density anomaly target known as the OD Deeps, which sits below the current resource outline. Recent drilling of the OD Deeps has outlined an area more than 2 km along strike and more than 1 km in-depth (BHP, 2024) where IOCG mineralisation style appears analogous to the main orebody at Olympic Dam. For further description of the deposit geology there are many published papers including Reeve *et al* (1990), Haynes (1995), Ehrig, McPhie and Kamenetsky (2012).

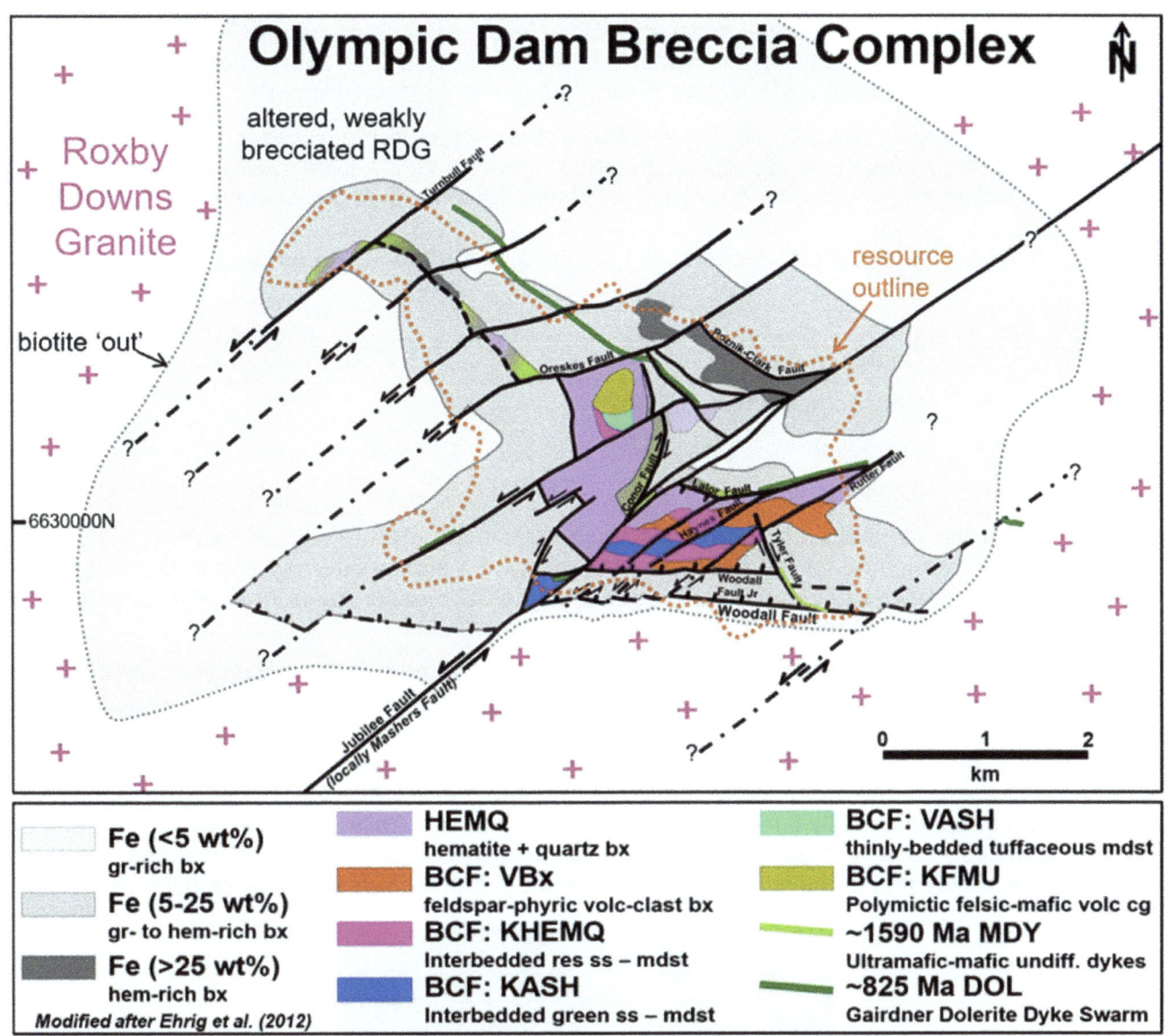

FIG 1 – Simplified geological plan of Olympic Dam at approximately 350 m below surface. The resource outline is indicated by the orange dotted line.

METHOD

To effectively determine the introduced error through reduced survey intervals, two drill rigs executing diamond drilling from surface to downhole depths of greater than 2100 m were selected to conduct single shot survey measurements at increments of 6 m. The holes selected, RD4582 and RD4597, were curved drill holes with inclination ranges of -79 to -51° and -74 to -51° respectively. RD4597 contained minor zones of navigation drilling, where single shot surveys are taken every 3 m. Navigational drilling surveys were omitted from the comparison in this study due to standard industry practice of shortening survey interval spacing while navigational drilling because of the high rates of azimuth and/or dip changes within these zones.

For each hole, the 6 m data was sub-sampled to generate a new data set that replicated a spacing of 18 m. For example, the 6 m and 12 m surveys were removed leaving only the 18 m survey. Then the 24 m and 30 m were removed leaving only the 36 m survey and so on. This filtering was then repeated so that a 30 m spaced data set and a 48 m spaced data set were generated. The 6 m data set was considered the 'true' path of the drill hole given that it was the highest density of survey information and is believed to be typically a high enough resolution to capture the geometry of the drill hole path.

The resultant four data sets; 6 m, 18 m, 30 m and 48 m for each drill hole were de-surveyed in Deswik CAD 2023 software using the minimum curvature method. This process produced an X, Y,

Z coordinate for each of the survey interval depths. Error for each coordinate was determined by calculating the difference between the subset data and the baseline data. The three errors were summed together to produce the Absolute Error at each survey measurement.

Additionally, the dogleg severity (DLS) at each survey measurement was calculated using the formula below as a means to identify sections of the drill hole where navigational drilling had occurred, or where erroneous data may exist. DLS was calculated as degrees per 30 m.

$$DLS = \left(\frac{30}{\Delta MD}\right) \times \cos^{-1}\{\cos(I_2 - I_1) - \sin I_1 \times \sin I_2 \times (1 - \cos(A_2 - A_1))\}$$

where:

DLS = Dogleg Severity

ΔMD = Change in Measured Depth

I = Inclination

A = Azimuth

Visualisation of the resultant drill hole paths were performed in Leapfrog Geo software using the minimum curvature method for de-surveying. This allowed a simple interrogation of the differences between drill hole traces of each data set, as well as the 3D measurement of an absolute variation from the 6 m 'true' case to each of the filtered sample data points.

Figure 2 provides an example where hole trajectories (azimuth and dip) start and end with the same values and are both the same length. However, there is uncertainty if the drill hole followed path one or path two. Without appropriate survey frequency, the location of where the drill hole bend has occurred could be missed. In this example, the length between survey measurements for both lines is 120 m measured downhole depth (simulated in Deswik CAD 2023), which results in a displacement of approximately 8 m at survey measurement two. While survey frequency in diamond drilling will unlikely be as coarse as 120 m, the cumulation of many small uncertainties due to an inappropriate survey frequency could still be significant over the course of a deep drill hole.

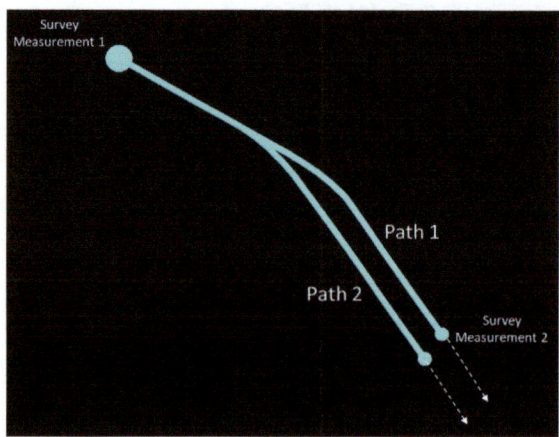

FIG 2 – Simulated drill hole path (blue) showing two different traces are possible even though the survey measurement data is the same at the start and end of the interval.

A simple sensitivity test to assess the effect of varying downhole survey intervals on Mineral Resource classification was conducted. The two drill holes used in the study were from the OD Deeps drilling program and support the inferred resource category below 1000 m from surface. The resource classification models were assessed using a combination of drill hole density (drill hole count per panel in a block model) and geological continuity (polygon strings) in Vulcan and Leapfrog. The panel size used for inferred is 100 m × 100 m × 100 m and at a notional 70 m × 200 m drill spacing.

The initial step involved defining the range of existing survey intervals and establishing a set of scenarios to assess for the varying survey intervals. A simple baseline resource category model was generated using the expected 6 m surveys. The downhole survey data was then updated to reflect the modified intervals, and potential resource category models were re-created using the same methodology as in the baseline model. The results were compared visually with the original model to identify changes between and within the potential resource classification categories with the emphasis on the impact to inferred category. The interpretation of results considered geological context, outlining the implications for mineral resource classification.

Assumptions and limitations

The 6 m baseline for this study is assumed to be the 'true' path. Six metres was chosen for practical reasons as it is the same as the drill pipe length and coincides with the frequency the driller will retrieve core. Olympic Dam mineralisation typically contains disseminated sulfides within breccia matrix and clasts. Abrupt changes in the brecciation are usually related to sub-vertical faults, while alteration in the host rock is gradational from the outside of the RDG inwards. Drilling is typically designed to cross faults or intersect mineralisation at a perpendicular to sub-perpendicular angle where a 6 m survey frequency should capture any geological change.

ANALYSIS

Absolute error versus depth and 3D visualisation

All measured survey frequencies showed an increase in absolute error with depth (Figure 3). Another common feature between RD4597 and RD4582 was a linear increase in absolute error between 150–750 m downhole (Figures 3a, 3d). This section of the hole was designed to drill with minimal curve, achieved by drilling with a wider core diameter, which holds the drill string straighter. To see a linear increase in error during this straight part of the hole is unexpected and suggests the 'true' hole path and subsampled data were diverging from one another at a constant rate.

To interrogate the cause of the increasing error from 150–750 m, the dogleg severity was assessed (Figure 4). DLS is a measure of the change in dip and/or azimuth of a drill hole between survey measurements and is expressed in degrees per length. As such, where DLS is measured as high (>3°/30 m), the drill path is turning at a high rate, which could be a result of natural bending, or a kink in the path. A kink will represent the greatest chance that an interpolated hole trace will differ from the 'true' trace creating a high absolute error. This is because a survey frequency that is too coarse to detect the likely location of the kink along the path, will introduce uncertainty in the subsequent drill hole path (Figure 2).

DLS increases with depth, observed by the regression lines in Figure 4. This is expected given that the deeper part of the hole is designed to curve, however, there is elevated DLS at the start of the hole in the 6 m data seen in Figure 4. This feature is much more subtle in the 18 m data, suggesting a kink was not identified. Figure 5 shows 0–150 m downhole in RD4597 and 0–300 m downhole in RD4582 have lower and erratic error trends before the linear trend is established, suggesting that the location for a kink is in this depth range.

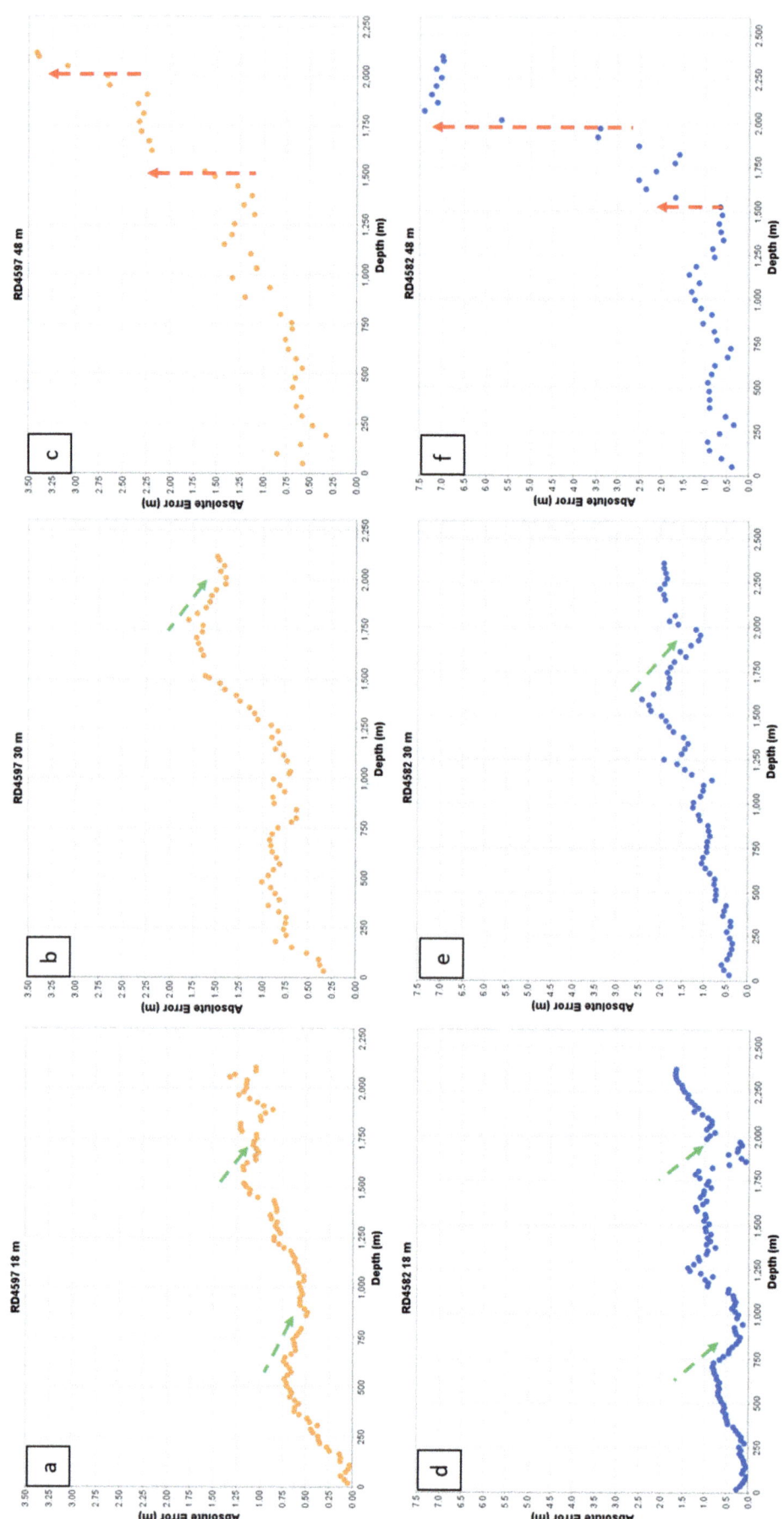

FIG 3 – Absolute error versus depth. Green arrows indicate where error is trending towards zero. Red dashed arrows indicate steps in the data. (a) RD4597 18 m data. (b) RD4597 30 m data. (c) RD4597 48 m data. (d) RD4582 18 m data. (e) RD4582 30 m data. (f) RD4582 48 m data.

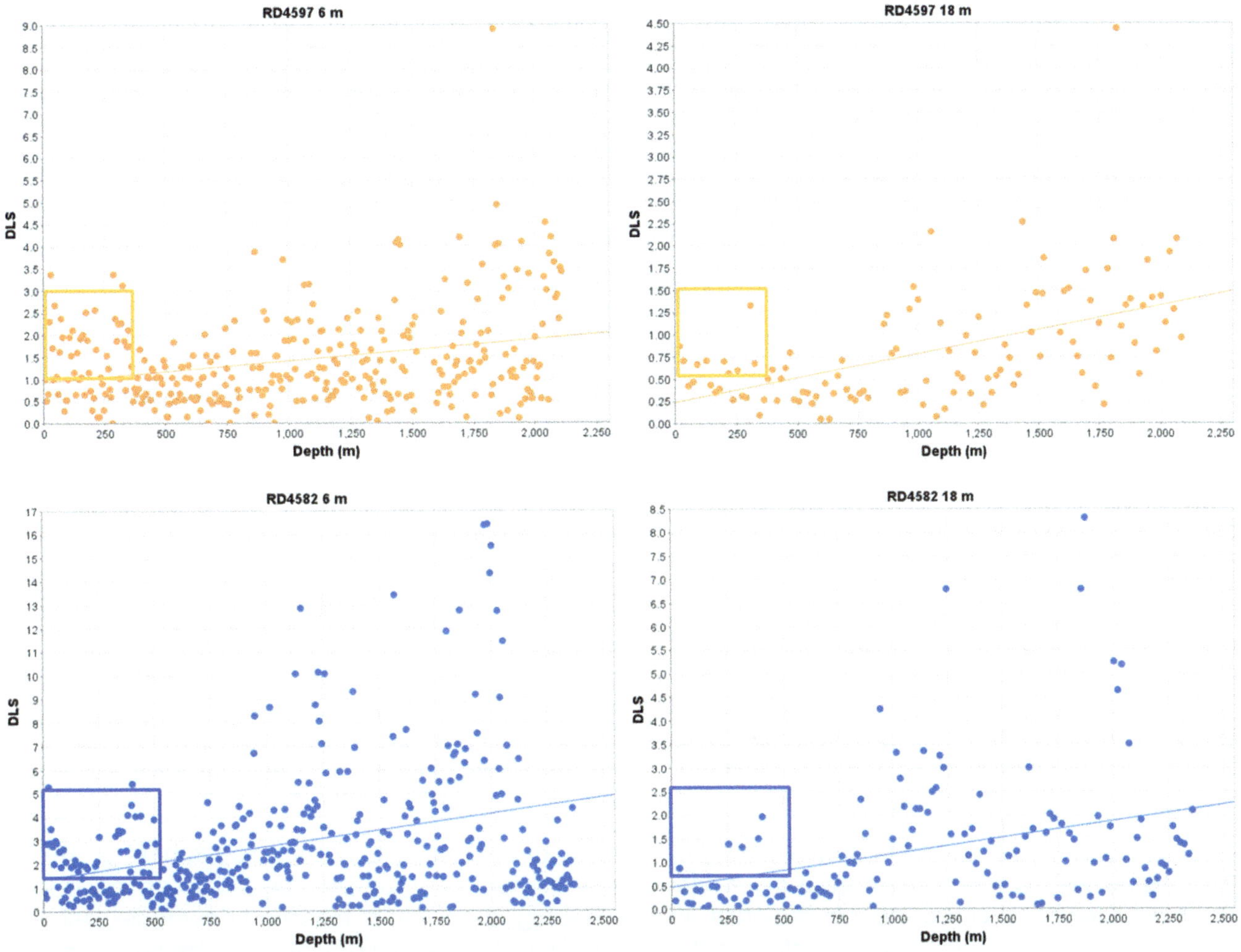

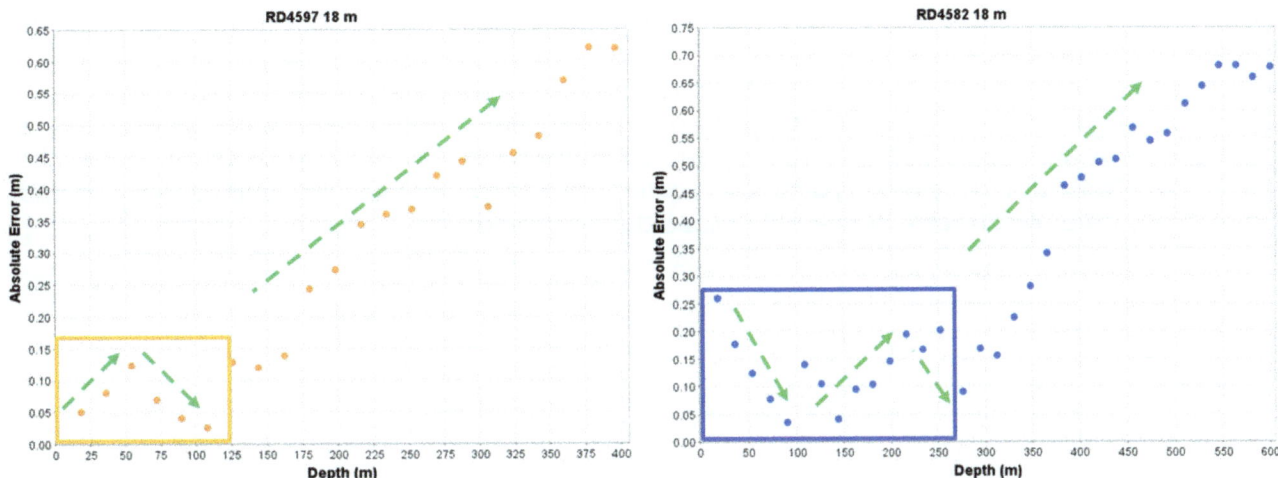

FIG 4 – DLS for the 6 m baseline survey and 18 m survey data sets show the same trend. Solid line is simple regression line. Boxes in 6 m data (left) show high DLS in 6 m data, which is not present in 18 m data (right).

FIG 5 – Absolute error versus depth for top of hole. Orange and blue boxes show a low error at start of hole, before a linear increase is established (long green arrows).

It is likely a kink was present in the top 300 m of the hole due to the below reasons, described as *Environmental Factors*:

- At steep drill hole inclinations, there can be greater sway in the tool compared to shallower inclinations where the tool can rest on the drill hole wall.

- The cover sequence at Olympic Dam extends to approximately 350 m below surface, which is drilled with PQ diameter core. The survey tool is fitted with centralisers/stabilisers which are designed to hold the survey camera still, however it is anecdotally easier for the tool to move within this diameter hole.

- Movement in the wireline from weather conditions has a greater impact on survey camera stability at this depth. As the depth of the survey tool increases, the weight on the wireline increases stability and will see less movement from weather conditions.

- The presence of aquifers and voids within the cover sequence, which will potentially allow increased water turbidity at the survey tool position while it is taking a measurement causing instability.

Given the length of the drill hole, error in the survey data at the start of hole can incrementally build as drilling continues, decreasing confidence in the drill hole position downhole. It is recommended that 6 m survey measurements are obtained for 0–200 m, at which point environmental factors will have less impact on the survey data confidence. This depth coincides with a lithology change in the cover sequence at Olympic Dam, so is not likely ubiquitous across other deposits.

Curvature of the drill hole path increases at depths greater than 750 m due to hole design. At these depths, the 18 m data set had trends where absolute error reduced, highlighted by green arrows in Figures 3a, 3d. This feature was less clear in the 30 m data set (Figures 3b, 3e) and was absent in the 48 m data set (Figures 3c, 3f). The absence of this feature suggests the 48 m survey frequencies are smoothing the drill hole trace and are unable to resolve complexity observed in the finer survey traces. Furthermore, there are apparent sharp steps in the 48 m data, highlighted by red arrows in Figures 3c, 3f. These steps may be the result of the drill hole trace quickly deviating from the 'true' value and are unlikely to be accurate.

To interrogate the stepped changes seen in the 48 m spaced data, a visual assessment of the drill hole trace was performed (Figure 6). A visual assessment is important because absolute error removes the vector component of the error. Where the highest error is observed downhole, data has the lowest accuracy. However, this does not necessarily correlate to a deviation of the drill hole path. For example, error in the X plane may be +0.1 per cent in one survey, while in the next consecutive survey it may be -0.1 per cent, which can balance out the deviation. The visual assessment confirmed a correlation between stepped data and a deviation from the 6 m 'true' path. This was seen clearly in the RD4582 data at 1800 m downhole and is also present in RD4597 at 1500 m downhole. This suggests the 48 m survey spacings introduce uncertainty to the drill hole trace, particularly in the deeper parts of the holes (>1500 m downhole) and their use should be excluded when drilling deep diamond holes. It is worth noting that at depths greater than 2000 m, that the survey tools began to experience temperature related faults. Only surveys passing internal quality assurance checks are utilised within this data set, however, still have potential to impact results at this depth. Temperature related issues also limited the integrity of multishot and continuous end of hole survey data, which were hence not utilised during drilling.

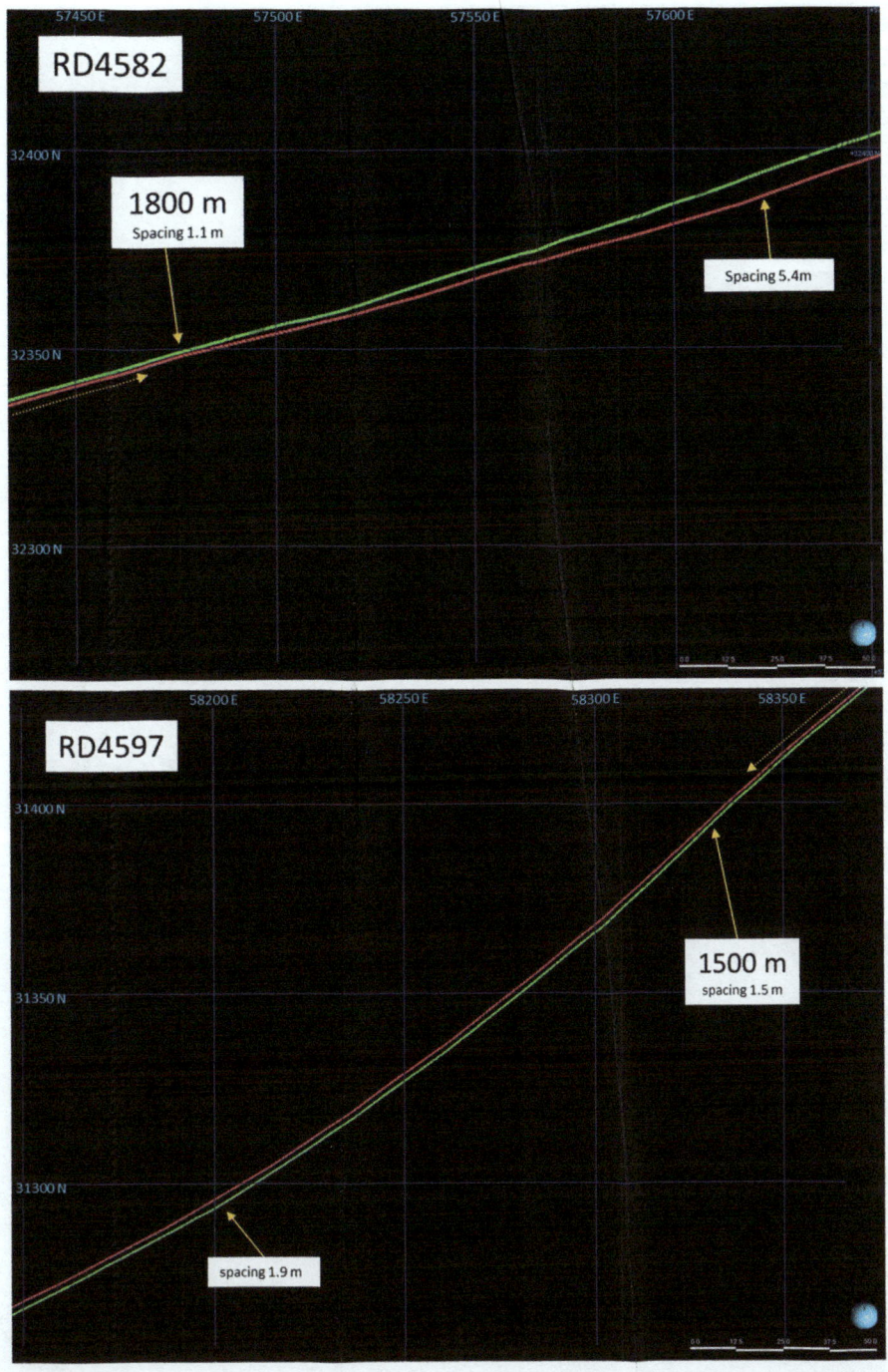

FIG 6 – Top: RD4582 stepped change at ~1800 m. Red line is 48 m spaced data, green line is 6 m spaced 'true' data. Orange dashed arrow is direction of drill hole. Bottom: RD4597 stepped change at ~1500 m. Red line is 48 m spaced data, green line is 6 m spaced 'true' data. Orange dashed arrow is direction of drill hole.

Even with the presence of kinks introduced from environmental or other factors, Figure 7 shows that the drill hole traces for all subsampled data sets have a good correlation to the 6 m baseline when looking in plan view. Table 1 shows that the 18 m drill hole trace was closest to the 'true' 6 m trace at EOH, followed by the 30 m, while the 48 m was furthest away. A maximum discrepancy of 5.5 m from the 'true' path at end of hole is considered very close given the ~2300 m length of the drill hole.

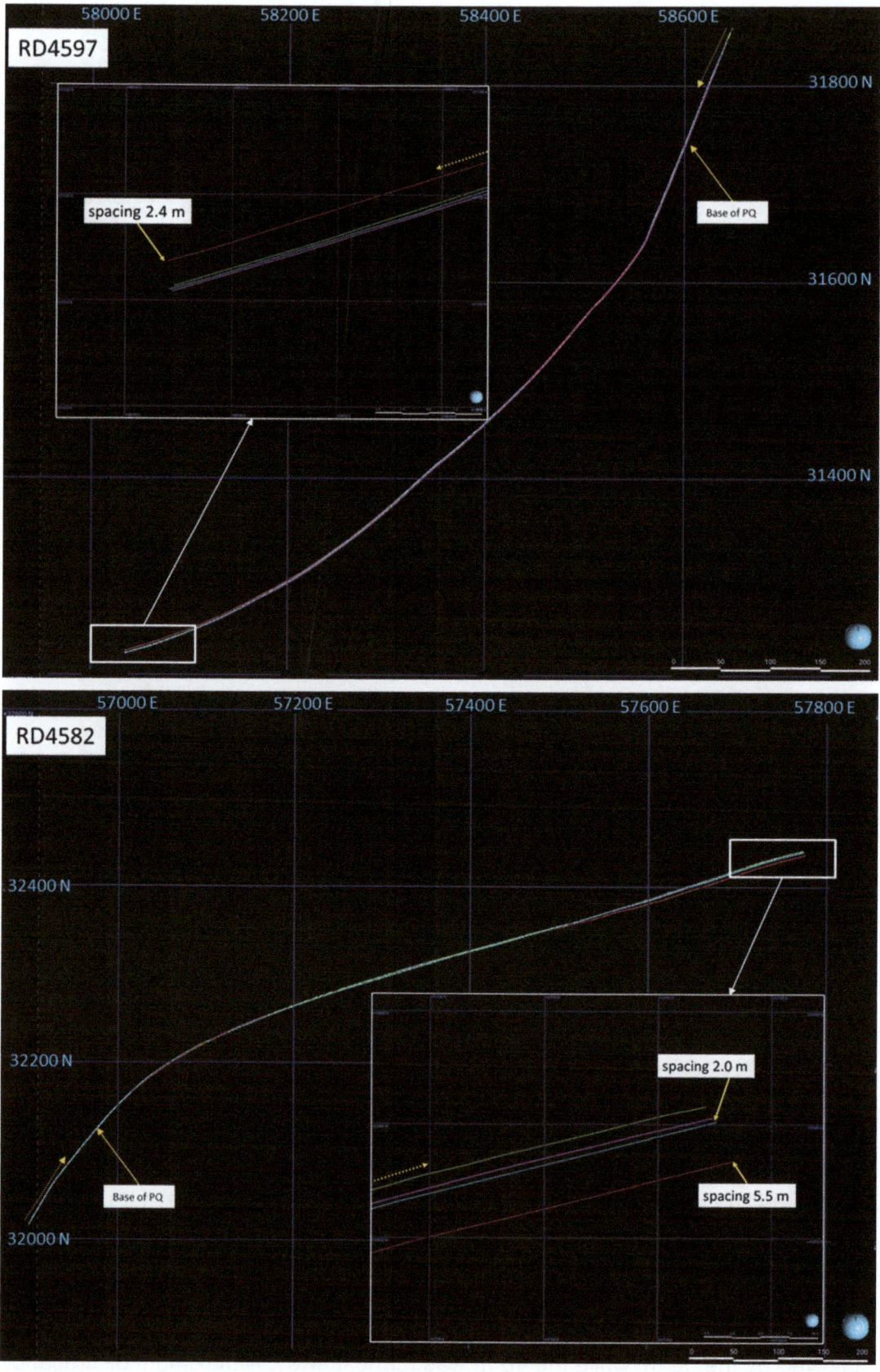

FIG 7 – Plan view of 6 m 'true' baseline data (green) 18 m data (pink), 30 m data (blue), 48 m data (red). Inset images show the deviation at end of hole. Orange dashed arrow is direction of drill hole.

TABLE 1

Measured distance between 6 m baseline and final survey for each survey spacing.

Survey frequency	Distance from 'true' (m) RD4582	RD4597
18 m	1.6	0.8
30 m	2	1.3
48 m	5.5	2.4

The intended use of the survey data as well as the style of mineralisation of the orebody needs to be considered. If these drill holes are planned to define a resource to a specific spacing; 100 m × 100 m for example, the error in the 48 m data set should be considered unacceptable as it is >5 per cent of the target spacing. The 30 m spaced data set for RD4582, however was only 2.0 m away from the 'true' data, which is deemed acceptable at 2 per cent of the target spacing. Olympic Dam mineralisation consists of bulk disseminated sulfides, which further supports the use of a 30 m spaced data set given the continuity of mineralisation. However, future drilling would likely need to target closer spacing such as 20 m × 20 m. Keeping this in mind, while a 30 m survey interval would be suitable for the current scope of work, an 18 m baseline survey frequency is recommended.

Sensitivity test

The results from simple sensitivity test showed very little difference between the choice of drill hole survey intervals (6 m, 18 m, 30 m and 48 m). With the greatest difference between the baseline inferred model from the widest survey interval (48 m). The choice of survey intervals impacts the precision and reliability of resource categories, with a trade-off between detail and cost efficiency. Balancing these factors becomes crucial in maintaining the reliability of inferred resources. Ensuring that the chosen drill hole survey intervals align with the geological complexity of the deposit while minimising the risk of underestimation or misinterpretation as overly frequent surveys may result in data redundancy and increased costs.

CONCLUSIONS

Drill hole survey monitoring is an integral part of the data collection process during exploration and resource definition. Understanding the most suitable frequency for survey intervals for a drill program impacts a range of downstream products in modelling and estimation. This study has demonstrated that a simple baseline comparison undertaken early in a drill program can be used to effectively inform the optimal survey frequency intervals.

The study concluded that drill hole path uncertainty exists in the cover sequence at Olympic Dam and is attributed mainly to environmental factors including difficulties centralising the survey tool in the hole, presence of aquifers and weather. Environmental factors are unlikely to be isolated to Olympic Dam and should be considered during similar studies within future drill programs. Analysis concluded that drilling at Olympic Dam within the cover sequence should implement survey intervals of 6 m between 0–200 m due as this section is likely to be most impacted by environmental factors, followed by 18 m surveys.

Given the close spatial correlation between 30 m survey frequency and the 'true' drill hole trace, this study found 30 m to be a suitable survey frequency for the base of the cover sequence to end of hole. However, with the future use of this data in mind, an 18 m survey frequency is recommended to minimise the risk to future data products including resource estimation. With any chosen survey frequency, there is still the ability to conduct higher resolution surveys at times where high DLS variability is observed or is expected in the drill hole. For example, when drilling with a more flexible barrel, or around geological complexities. It is expected that with the above recommendations, a dynamic survey interval plan is developed by the responsible geologist where appropriate.

Due to the combination of the highest absolute error as well as smoothing of kinks observed in the 48 m data set, this study concludes that this survey interval spacing is avoided. Furthermore, 6 m interval spacing is deemed unnecessary given the little difference observed between survey frequencies in the sensitivity test.

Survey readings are a streamlined process and do not have a large impact on rig productivity. However, over a 2300 m drill hole the change from 18 m to 30 m surveys would reduce the need for approximately 40 shots, equating to half a shift of rig productivity per hole drilled. This time saving can quickly add up over a large drill program and should be considered in the planning phase. Additionally, it is recommended that the future use of any geological data should be considered when making any decision that may introduce uncertainty. After all, the challenge of dealing with uncertainty in historic geological data is not unique to Olympic Dam and should not be overlooked.

ACKNOWLEDGEMENTS

The authors would like to thank the team of geologists, drillers and drill coordinators who are integral to executing a drill program effectively and safely. Many geologists from the Olympic Dam Geoscience team have contributed to material used in this paper, but specifically Kathy Ehrig, Superintendent Geometallurgy, for her support and encouragement to produce and distribute this piece of work as well as Lachlan Macdonald, James Taylor, Matthew Goldman and Lauren Brown for reviewing the content.

REFERENCES

Badenhorst, C, O'Connell, S and Rossi, M, 2016. New Approach to Recoverable Resource Modelling: The multivariate case at Olympic Dam, *Geostatistics Valencia QGAG*, ch 19, pp 131–149.

BHP, 2024. Operational Review for the half year ended 31 December 2023, 2024. Available from: <https://www.bhp.com/news/media-centre/releases/2024/01/bhp-operational-review-for-the-half-year-ended-31-december-2023>

Clark, J and Ehrig, K, 2019. What controls high-grade copper mineralisation at Olympic Dam?, in *Proceedings Mining Geology 2019*, pp 222–236 (The Australasian Institute of Mining and Metallurgy: Melbourne).

Ehrig, K, McPhie, J and Kamenetsky, V, 2012. Geology and mineralogical zonation of the Olympic Dam iron oxide Cu-U-Au-Ag deposit, South Australia, *Geology and Genesis of Major Copper Deposits and Districts of the World: A Tribute to Richard H Sillitoe* (eds: J W Hedenquist, M Harris and F Camus), Society of Economic Geologists, Special Publication, 16:237–268.

Ehrig, K, Schijns, H, Townsend, J, Haddow, D and Shawcross, M, 2023. Seismic velocity (Vp) vs alteration mineralogy in the Olympic Dam IOCG deposit, in *Proceedings of the Fourth AEGC: Geoscience – Breaking New Ground 2023*, pp 1–10.

Haynes, D W, 1995. Olympic Dam ore genesis: a fluid mixing model, *Econ Geol*, 90:281–307.

Jamieson, A and Knight, S (ed), 2017. Introduction to wellbore positioning, ver V09.10.17, University of the Highlands and Islands. Available from: <https://pure.uhi.ac.uk/en/publications/introduction-to-wellbore-positioning> [Accessed: 11 August 2023].

JORC, 2012. Australasian Code for Reporting of Exploration Results, Mineral Resources and Ore Reserves (The JORC Code) [online]. Available from: <http://www.jorc.org> (The Joint Ore Reserves Committee of The Australasian Institute of Mining and Metallurgy, Australian Institute of Geoscientists and Minerals Council of Australia).

Reeve, J S, Cross, K C, Smith, R N and Oreskes, N, 1990. Olympic Dam copper-uranium-gold-silver deposit, *Geology of the Mineral Deposits of Australia and Papua New Guinea* (ed: F E Hughes), pp 1009–1035 (The Australasian Institute of Mining and Metallurgy: Melbourne).

Enhancing geometallurgical ore domaining through improved orebody knowledge – a case study from the Sotiel Cu-Pb-Zn-Ag Mine, Spain

R Dale[1], M Biven[2], A Claflin[3], J Gonzalez[4] and M Puerto[5]

1. Project Geologist, Sandfire Resources, West Perth WA 6005.
   Email: roseanna.dale@sandfire.com.au
2. Plant Metallurgist, Sandfire MATSA, Almonaster la Real, Huelva 21130, Spain.
   Email: michael.biven@matsamining.com
3. Plant Metallurgist, Sandfire MATSA, Almonaster la Real, Huelva 21330, Spain.
   Email: aaron.claflin@matsamining.com
4. Processing and Metallurgy Supervisor, Sandfire MATSA, Almonaster la Real, Huelva 21330, Spain. Email: joseandres.gonzalez@matsamining.com
5. Project Geologist, Sandfire MATSA, Almonaster la Real, Huelva 21330, Spain.
   Email: miguel.puerto@matsamining.com

ABSTRACT

Geometallurgy is a critical discipline for optimising mineral processing, combining geology and metallurgy. At Sandfire MATSA's Sotiel Cu-Pb-Zn-Ag mine in Spain, the presence of complex pyrite poses significant challenges to achieve consistent metal recoveries. This study illustrates how the collection of spatially distributed mineralogical and metallurgical data can enable the mapping of distinct ore domains, identifying the geological controls of metallurgical performance and improving operational performance.

A comprehensive data collection campaign was undertaken, encompassing detailed geological logging, mineralogical and metallurgical analysis from across the Sotiel ore domains. These data sets were used to create a geometallurgical ore domain model, revealing the spatial relationship between ore textures and metallurgical behaviours.

Results demonstrate that an understanding of pyrite content plays a pivotal role in defining distinct ore domains. These have varying target mineral liberation, with pyrite texture and composition significantly influencing metallurgical performance. By incorporating this newly refined knowledge into a geometallurgical block model, the Sotiel Mine can improve predictability of metal recoveries across the deposit resulting in the development of more precise ore blending strategies, enhancing the overall process efficiency and profitability.

Insights gained from the Sotiel case study serve as a valuable guidance for orebody characterisation and domain mapping in similar complex mineralogical settings, ultimately enabling more efficient mineral resource utilisation in the mining industry. Finally, it also highlights the important role played by geoscience in adding value to the mining operational value chain.

INTRODUCTION

Geometallurgy stands at the intersection of geology, mineralogy, geochemistry, geostatistics, extractive metallurgy, engineering and economics. It is dependent on a quantitative comprehension of fundamental resource attributes, including mineralogical composition and texture and the spatial distribution and variability of these characteristics. At the fundamental level, mining involves extracting minerals, rather than isolated elements and mineral processing deals with complex particles and not pure mineral grains. Geometallurgy explores how these factors interact with the mining and beneficiation processes, to develop comprehensive, geologically and spatially based 2D and 3D predictive models to optimise all facets of the mining process.

The depletion and scarcity of mining resources have encouraged the industry to deal with increasingly diverse orebodies, emphasising requirements for geometallurgical investigations due to the consequential impact of natural variability on processing recoveries. Geological uncertainty presents challenges in both mining and ore processing phases. Mineral processing plants are typically optimised for stable conditions, unforeseen fluctuations in feed characteristics can, unfortunately, result in suboptimal processing performance. A major aspect of all geometallurgical studies is the quantification of uncertainties and if possible, their reduction (Frenzel *et al*, 2023).

BACKGROUND

Sandfire MATSA (MATSA) produces copper, zinc and lead-silver concentrates from the Aguas Teñidas, Magdalena and Sotiel Mines in Huelva province, southern Spain (Figure 1). The deposits are located in the Iberian Pyrite belt (IPB), a large zone of VMS mineralisation that extends over 230 km in length and 40 km in width between Seville (Spain) and Lisbon (Portugal). The Aguas Teñidas and Magdalena mines are located in the northern zone of the IPB, characterised by massive sulfides hosted in pumice-rich volcaniclastic rocks, in areas marginal to volcanic domes, whilst the Sotiel Mine is in the southern IPB and is characterised by pyrite-rich, black slate hosted deposits. Sulfide mineralisation at all the three mines is typically strata bound, structurally controlled massive sulfide lenses comprised primarily of pyrite, chalcopyrite, sphalerite and galena.

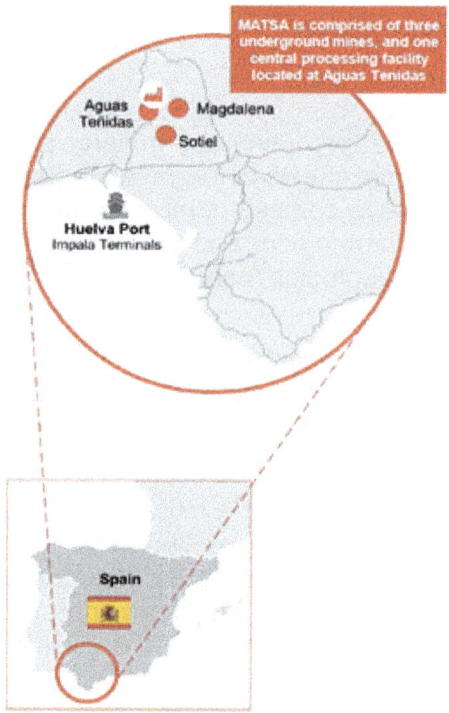

FIG 1 – Location of MATSA's Aguas Teñidas, Magdalena and Sotiel Mines in Huelva, Spain.

The Sotiel Mine currently consists of five deposits spanning over 3 km: Calabazar, Sotiel, Sotiel East, Migollas and Elvira (Figure 2). Each deposit is made up of multiple lenticular bodies of massive sulfides, with areas characterised as being polymetallic (Cu, Zn and Pb) or more copper enriched (Cu, minor Zn, Pb). A Cu/Zn ratio is currently used to classify the ore for mining, stockpiling, and processing, designating ores as cupriferous or polymetallic and treating them in different circuits depending on that ratio.

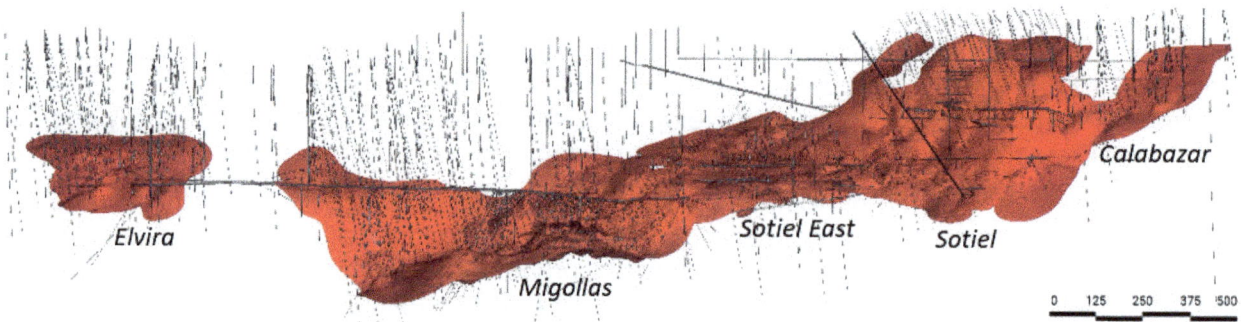

FIG 2 – View south of Sotiel underground mine. Red wireframes indicate modelled massive sulfides, with drilling and development shown.

The Sotiel underground mine is approximately 400 m deep with mining being conducted through bottom-up longhole open stoping with unconsolidated and cemented rock fill. Ore and waste material is hauled to the surface via 40 t trucks that are loaded underground by load-haul-dump machines. Once on the surface, ore is delivered and segregated in stockpiles based on the polymetallic or cupriferous classification, before being trucked again 38 km to the run-of-mine (ROM) next to the processing plant located at the Aguas Teñidas Mine.

The Central Processing Facility located at Aguas Teñidas consists of three processing lines. The facility can treat both polymetallic and cupriferous ore with an overall capacity of 4.7 Mtpa. Line 1 and 2 have been in operation since 2009 and feature two-stage crushing, a SAG and ball mill grinding circuit and three tertiary mills. The product P_{80} for Line 1 is 50–60 µm and 35–40 µm for Line 2. In both lines the copper rougher flotation reports to a three-stage cleaning circuit. Line 1 treats solely cupriferous ore both from Aguas Teñidas, Magdalena and Sotiel. Sotiel polymetallic ore can create a zinc concentrate high in mercury so it is sent mostly to Line 2 which can run in both cupriferous and bulk polymetallic configurations. Both lines handle 25 per cent each of total ore treatment. Line 3 has been in operation since 2015 and features three-stage crushing, a Primary and Secondary ball mill and four tertiary mills with a product P_{80} of 35 µm. Line 3 treats ore from both Aguas Teñidas and Magdalena and runs in a bulk polymetallic configuration producing a copper and lead concentrate, with the tails stream going to a zinc circuit. Line 3 handles 50 per cent of the overall ore treatment.

Previous campaigns indicate that Sotiel ore has slower flotation kinetics compared to ore from the other mines at MATSA. Sotiel material is often shown to be highly reactive during treatment. It tends to have a reducing Eh (oxidation/reduction potential) and very low dissolved oxygen concentrations. Sotiel ore can benefit from a fine grind in terms of improved liberation though this tends to exacerbate its reactive nature by increasing the mineral surface area. As a result, the oxygen demand is assumed to be quite high after fine grinding through the tertiary mills. Previous test work has shown that when this oxygen demand is met, the flotation kinetics improve.

ROM oxidation simulation test work has also shown that if Sotiel ore is allowed to oxidise on the ROM it will lead to higher levels of soluble lead compared to other ore types at MATSA. High levels of lead in solution are detrimental to flotation performance as seen in previous Sotiel campaigns. It also causes sphalerite and pyrite to be activated for recovery in the copper concentrates.

Pyrite variations

Pyrite, the most prevalent sulfide mineral found in both base and precious metals deposits, holds substantial economic and environmental significance due to its abundance and close association with other minerals. Its processing presents many challenges, particularly in base metal sulfide flotation where it can become a problematic waste sulfide. With the depletion of conventional high-grade metal orebodies, there is a growing need to process complex orebodies rich in pyrite to fulfill the rising demand for metals. Unfortunately, these complex orebodies tend to be non-responsive to traditional methods for pyrite depression. The necessity for higher pyrite rejection causes projects and operations to resort to complicated processing flow sheets (Jefferson *et al*, 2023).

Depending on ore genesis and later recrystallisation, pyrite can display varied textures, chemical compositions and electrochemical properties. These variations have been reported to affect pyrite's flotation response at different levels, though the mechanisms are not well understood. Different pyrite textures may influence flotation by altering the electrochemical properties of pyrite and the ways it interacts with the pulp (Jefferson *et al*, 2023).

Pyrite may be present in many crystal structure forms and textures. Pyrite textures are classified by grain size, spatial relationships with other minerals, and crystallisation level; disseminated, agglomerated, euhedral, subhedral, anhedral and framboidal/spongy-melnikovite, typically in the fine class of between 5–40 micron (Figure 3) (Jefferson *et al*, 2023). The preservation of framboids and fine crystalline pyrite is evidence of low-temperature formation and the absence of any significant metamorphism (Craig, Vokes and Solberg, 1998). Pyrite responds to increasing metamorphic temperatures primarily through recrystallisation, evident in the growth of euhedral crystals from existing anhedral grains, the development of annealed textures and increases in grain size (Craig, Vokes and Solberg, 1998).

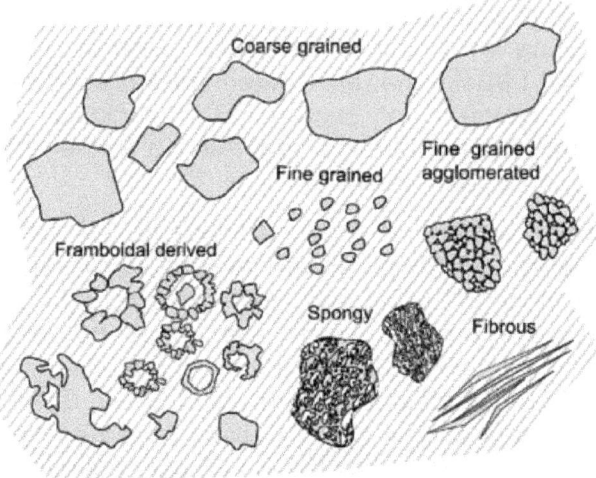

FIG 3 – Pyrite texture schematic examples (Jefferson *et al*, 2023).

Minor and trace element content of sulfide minerals can vary. Pyrite can incorporate impurities during formation, commonly As, Co and Ni, which are likely to either occur as small inclusions of other minerals or as ion substitution in the crystal lattice (Craig, Vokes and Solberg, 1998). Coarse euhedral pyrite has lower concentrations of minor and trace elements than other fine-grained, subhedral, or anhedral textures. In general, complex pyrite textures seem to present a higher content of impurities such as As, Au, Ag, Zn, Pb, Sb and Cu (Jefferson *et al*, 2023). The presence of impurities in pyrite crystal structure influences its electrochemical properties and flotation behaviour in various ways (Jefferson *et al*, 2023). Lehner *et al* (2007) studied the electrochemical reactivity of synthetic pyrite doped with As, Ni and Co, finding that the presence of impurities (mainly As), increased the reactivity of the pyrite. A decrease in pyrite rest potential with increased Cu and As content can be observed across studies, seemingly confirming that pyrite impurity content increases its reactivity (Jefferson *et al*, 2023).

Poorly crystallised, fine and framboidal pyrite are expected to exhibit higher reactivity than coarse-grained pyrite due to their formation and characteristics. Small particle sizes and large surface area increase the reactivity of pyrite, and maximum oxidation of the pyrite surface occurs along pits, cracks, pores, and solid and liquid inclusions (Lottermoser, 2010). In framboidal pyrite, grains are exceptionally small, forming dispersed particles or agglomerating into small spheroidal masses. Such pyrite is more reactive than other pyrite morphologies because of the greater surface area and porosity (Lottermoser, 2010). Barker, Gerson and Menuge (2014) suggest that the increased surface area increases galvanic interaction with other minerals and their floatability. Can, Özçelik and Ekmekçi (2021) found that in copper sulfide ore samples containing framboidal pyrite lower copper recoveries were obtained, and Fe was more difficult to depress than in samples with no reported framboidal pyrite.

The use of electrochemically active grinding media causes galvanic interaction between pyrite and the grinding media, which results in an effect on the electrochemical behaviour of pyrite (Moslemi and Gharabaghi, 2017). Selective flotation of base metal sulfide minerals from pyrite is strongly influenced by the type and concentration of metal oxidation species produced during grinding (Peng *et al*, 2003). These species may affect the floatability of minerals, rendering the surfaces hydrophilic or leading to inadvertent activation (Moslemi and Gharabaghi, 2017). The pyrite content of the flotation feed plays an important role in increasing the oxidation of minerals as it consumes dissolved oxygen in the flotation pulp (Owusu *et al*, 2014).

OBJECTIVES

The objective of this study is to undertake geometallurgical ore characterisation through the collection of primary (inherent properties of a rock or ore) and secondary (behaviour of an ore during processing) ore property data. Key primary ore properties include mineralogy, texture, and geochemical composition, whilst secondary ore properties include flotation response, reagent consumption and metal recovery. To support this, a sampling scheme is designed to collect representative samples covering the geological variability within the deposit.

Utilising metallurgical test data at laboratory scale, we aim to establish links between the primary and secondary characteristics for each ore type, enabling effective mine-to-plant optimisation. The aim is to use the existing drill hole databases and grade control models to domain ore types and textures and create a 3D geomet model, aiming to reduce technical risks in the operations by identifying areas of positive and negative material impact, thus increasing prediction confidence.

METHODOLOGY

Logging and ore domain characterisation

A logging campaign was undertaken to characterise the orebodies by their specific geological, mineralogical and textural characteristics throughout the deposit. This was done by prioritising the current ore reserve and future mining areas of the mineral resource. Logging was completed in a stepwise progression of cross-sections through each orebody at the Sotiel Mine. At each location holes were logged gathering information on any variability in ore mineralogy and texture visible at the macro scale. The existing multi-element assay data from these logged drill holes was used to interrogate geochemical trends occurring throughout the different orebodies, lithologies, mineral types and textures.

The Sotiel orebodies were observed to predominantly consist of three main ore textural types: massive, banded, fractured (Figure 4). These textural variations are present throughout both the cupriferous and polymetallic zones, therefore samples of six ore types (massive-copper, massive-polymetallic, fractured-copper, fractured polymetallic, banded-copper and banded-polymetallic) needed to be collected.

FIG 4 – Macro scale textural characteristics used to define first pass ore types. (a) massive sulfide texture. (b) banded sulfide texture. (c) fractured sulfide texture.

To be representative of these contrasting ore domains and the wide range of grades and textures, 20 samples of drill core from specific 1–2 m intervals were collected (Table 1). For each individual orebody, if this ore type was present, a sample was collected for representivity. These variability samples were tested separately and not composited. Using these first pass ore types for mineralogical and metallurgical analysis would enable the creation of geometallurgical ore types, based on the similarities or differences seen in the analysis.

TABLE 1

Variability samples of 20 ore types collected from the Sotiel deposit.

Sample ID	Sotiel Orebody	Ore Type
GM0042	Calabazar	Massive Copper
GM0043	Calabazar	Fractured Copper
GM0044	Calabazar	Fractured Polymetallic
GM0045	Sotiel	Banded Polymetallic
GM0046	Sotiel	Fractured Copper
GM0047	Sotiel	Massive Polymetallic
GM0048	Sotiel East	Massive Polymetallic
GM0049	Sotiel East	Fractured Polymetallic
GM0050	Sotiel East	Fractured Copper
GM0051	Sotiel East	Massive Copper
GM0052	Migollas	Halo Mineralisation
GM0053	Migollas	Fractured Polymetallic
GM0054	Migollas	Massive Polymetallic
GM0055	Migollas	Massive Copper
GM0056	Migollas	Fractured Copper
GM0057	Migollas	Banded Polymetallic
GM0058	Elvira	Massive Polymetallic
GM0059	Elvira	Fractured Copper
GM0060	Elvira	Massive Copper
GM0061	Elvira	Massive Polymetallic

Primary ore properties

The samples were sent for quantitative optical mineralogical assessment with MODA Microscopy (Burnie, TAS), and consisted of -3.35+1.17 mm chips prepared from diamond core samples. Polished thin section grain mounts were prepared by Adelaide Petrographic Laboratories (Adelaide, SA). A subset of each sample was submitted to McKnight Mineralogy (Ballarat, VIC) for quantitative X-ray diffraction (QXRD), to ALS for multi-element geochemistry (Burnie, TAS) and a split underwent the standard grade control assaying at the MATSA on-site laboratory.

Post optical analysis, select samples containing the best representations of pyrite and arsenopyrite variations were composited and polished mounts sent to CODES University of Tasmania for Laser Ablation Induced Coupled Plasma Mass Spectroscopy (LA-ICPMS) elemental analysis. Thin sections containing the best representations of all other sulfide mineral variability were sent to the Central Science Laboratory University of Tasmania for elemental analysis with Electron Microprobe Analysis (EPMA).

Secondary ore properties

The grain mounts were logged on an unbiased regular grid (~1.8 mm square grid covering 55 × 25 mm) using a 53 µm circular mask to simulate a nominal 53 µm grind. Sotiel ore is milled finer, however this analysis provides a robust first pass analysis highlighting trends in the data before physical test work. Over the whole 55 × 25 mm area of the mount, 400 masked areas of 53 µm diameter were logged microscopically by estimating the area% of each mineral present. The logging results were tallied, and the area per grain percentage (APG%), mineral association, and liberation (free, binary, ternary, and quaternary composites) parameters computed for all minerals.

At ALS, splits from each ore type sample underwent an ethylene diamine-tetra acetic acid disodium salt (EDTA) leach routine and a natural pH test. The EDTA leach is used to determine the extent of metal sulfide (copper, lead, zinc, iron) surface oxidation generated by the exposure of ground ore to surface oxidation. The natural pH of each pulverised sample is tested at 15-minute and 60-minute intervals.

RESULTS

Primary ore properties

Quantitative optical mineralogy assessments identified the predominant sulfide minerals in all ore type samples as pyrite, chalcopyrite, sphalerite, galena with minor occurrences of arsenopyrite. Additionally, trace amounts of tetrahedrite, pyrrhotite, stannite, bournonite and marcasite were observed. Non-sulfide gangue minerals include interstitial siderite, chlorite, dolomite, quartz and muscovite.

Optical microscopy confirmed pyrite to be the most abundant mineral. Three types of pyrite are observed:

1. Spongy, fine-grained spheroidal grains (py1).
2. Fine to medium-grained subhedral to euhedral grains (py2).
3. Coarse-grained subhedral to euhedral grains (py3).

Py1, also known as melnikovite has a complex texture that varies from colloform bands and consistently sized spheroids/framboids typically in the range of <10–30 microns (Figure 5a) to extensive zones of porous aggregates (Figure 5b). The uniform size of the spheroidal shapes and their spongy-porous nature suggest a biogenic origin. As discussed above, py1 is more reactive due to its greater surface area and porosity. Py2 ranges in size from 10–40 micron and forms aggregates of subhedral to euhedral grains (Figure 5c). Py3 typically ranges from 40–125 micron and are usually well-formed euhedral grains (Figure 5d). Commonly, py2 and py3 are observed with cores and relics of py1 (Figure 6) and it is hypothesised that py1 is the primary pyrite, with recrystallisation over time refining to coarser grained py2 and py3. Each pyrite type is present across all ore types.

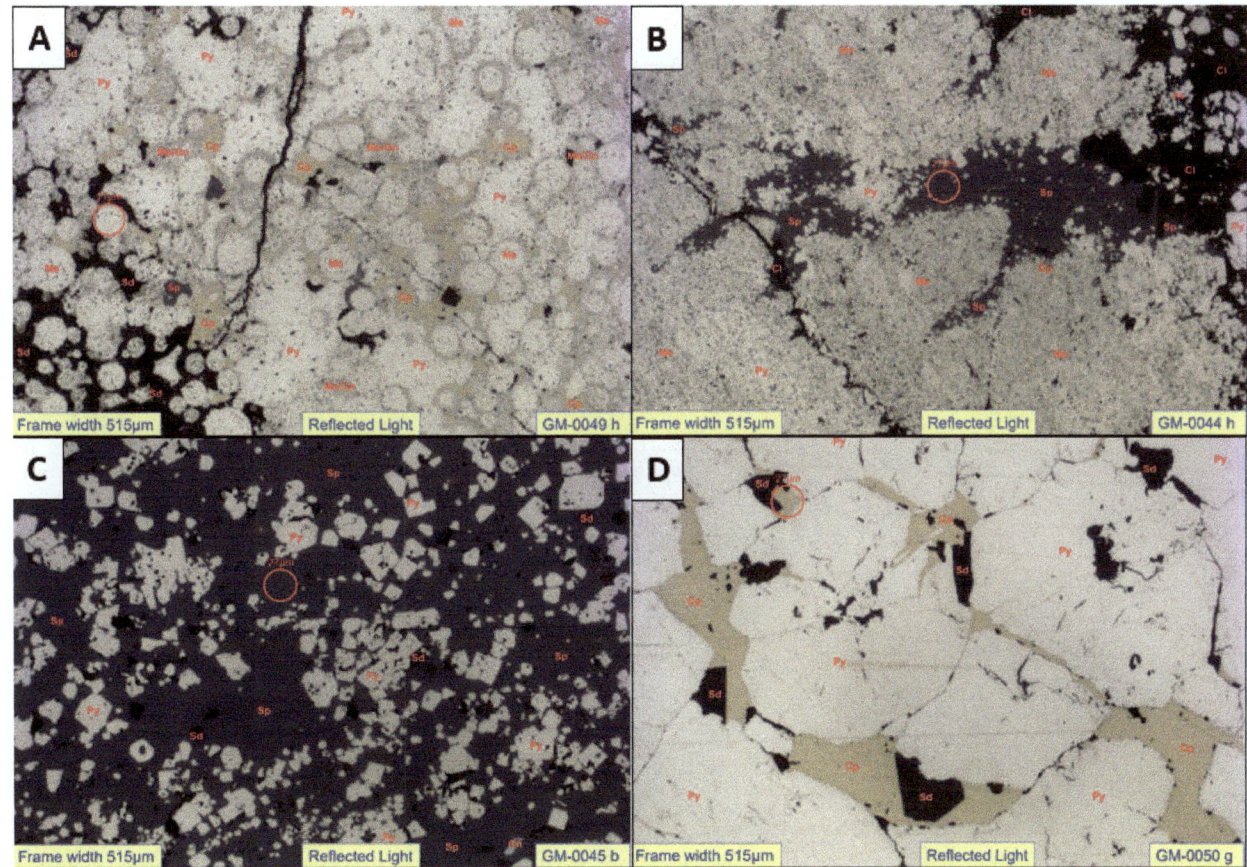

FIG 5 – Pyrite types present at Sotiel: (a) py1, spheroidal-framboidal melnikovite with interstitial chalcopyrite. (b) py1, massive area of fine-grained porous melnikovite within interstitial sphalerite. (c) py2 fine-grained sub-euhedral grains within massive sphalerite. (d) py3, coarse-grained sub-euhedral grains with interstitial chalcopyrite. Abbreviations: Py = pyrite, Me = melnikovite, Cp = chalcopyrite, Sp = sphalerite, Gn = galena, Sd = siderite, Cl = chlorite.

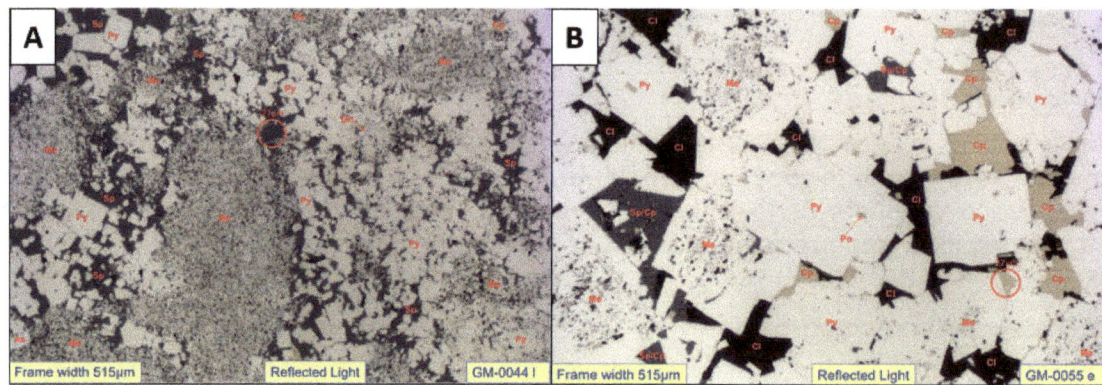

FIG 6 – (a) gradational boundaries py1 and py2 with ultra-fine interstitial sphalerite. (b) porous cores of py1 within large euhedral py3 with interstitial chalcopyrite and sphalerite. Abbreviations: Py = pyrite, Me = melnikovite, Cp = chalcopyrite, Sp = sphalerite, Cl = chlorite.

Chalcopyrite occurs in association with pyrite and varies from ultra-fine to coarse-grained. It is present either interstitially to or within fractures of py2 and py3 (Figure 7a), occupying the porous spaces of py1 where it can be less than 20 µm (Figure 5a) and can also be observed replacing pyrite. Sphalerite is associated with pyrite and galena (Figures 7b, 7c), is irregular in shape and ranges from ultra-fine to large, massive zones. It occurs interstitially or within fractures of py2 and py3 and in the porous spaces of py1, where it can be less than 20 µm (Figure 6a). Chalcopyrite disease is common in sphalerite and ultra-fine blebs of chalcopyrite can be observed throughout (Figure 7b). Galena is associated with pyrite and sphalerite, is irregular in shape and is often appearing as fine-medium intergrowths throughout larger massive sphalerite zones (Figure 7c). It is present interstitially or within fractures of py2 and py3, or in the porous spaces of py1 where it can be less than 20 µm.

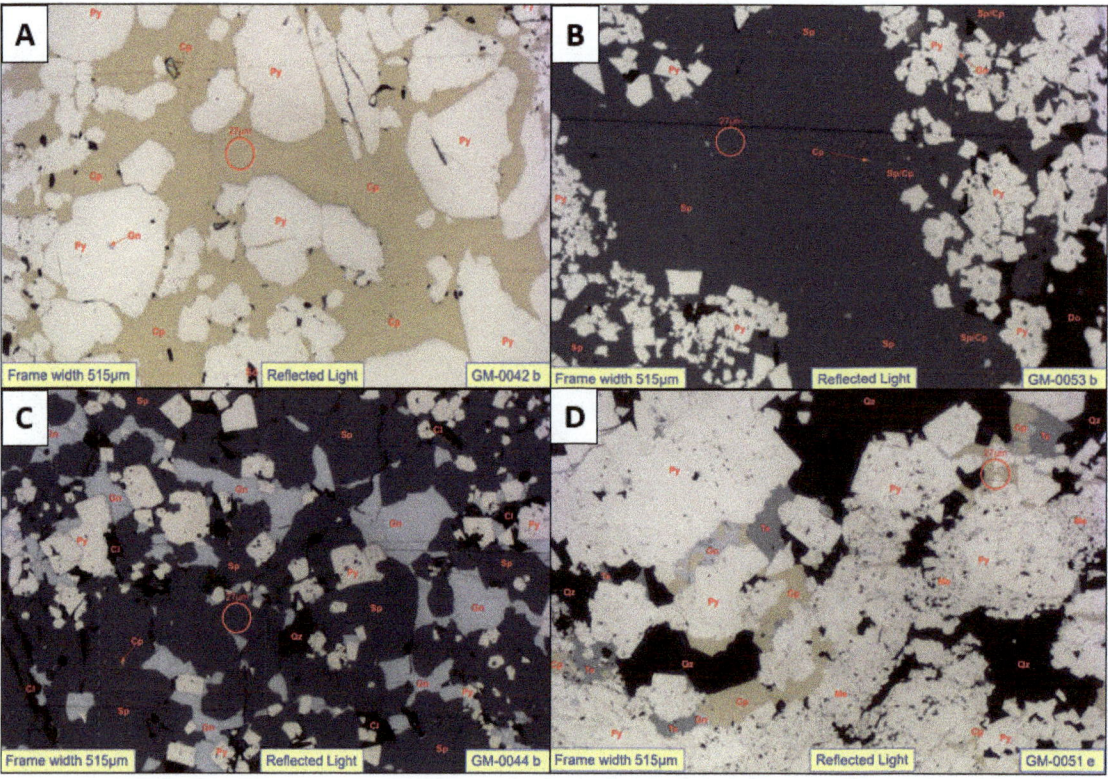

FIG 7 – (a) chalcopyrite filling interstitial space and partially replacing pyrite aggregates. (b) coarse-grained chalcopyrite diseased sphalerite with fine pyrite aggregates. (c) annealed sphalerite and galena with fine-grained pyrite aggregates. (d) fine-coarse grained pyrite with interstitial chalcopyrite and tetrahedrite-galena intergrowths. Abbreviations: Py = pyrite, Me = melnikovite, Cp = chalcopyrite, Sp = sphalerite, Gn = galena, Te = tetrahedrite, Sd = siderite, Cl = chlorite, Qz = quartz, Do = dolomite.

Tetrahedrite is often associated with galena in the form of intergrowths, suggesting simultaneous precipitation (Figure 7d). Arsenopyrite is seen as fine to coarse-grained euhedral crystals. Rare fine-grained pyrrhotite, bournonite and stannite were observed within the massive sulfides.

The averaged QXRD data in Figure 8a demonstrates that in all massive ore types, pyrite content exceeds 70 per cent, in contrast to fractured and banded ore types where pyrite content is, on average, below 70 per cent. As expected, the copper ore types exhibit lower-than-average levels of sphalerite and galena, whilst being enriched in chalcopyrite (Figures 8b, 8c, 8d). Typically, the massive ore types are observed to contain significantly lower amounts of target minerals chalcopyrite, sphalerite and galena than the fractured and banded counterparts (Figure 8).

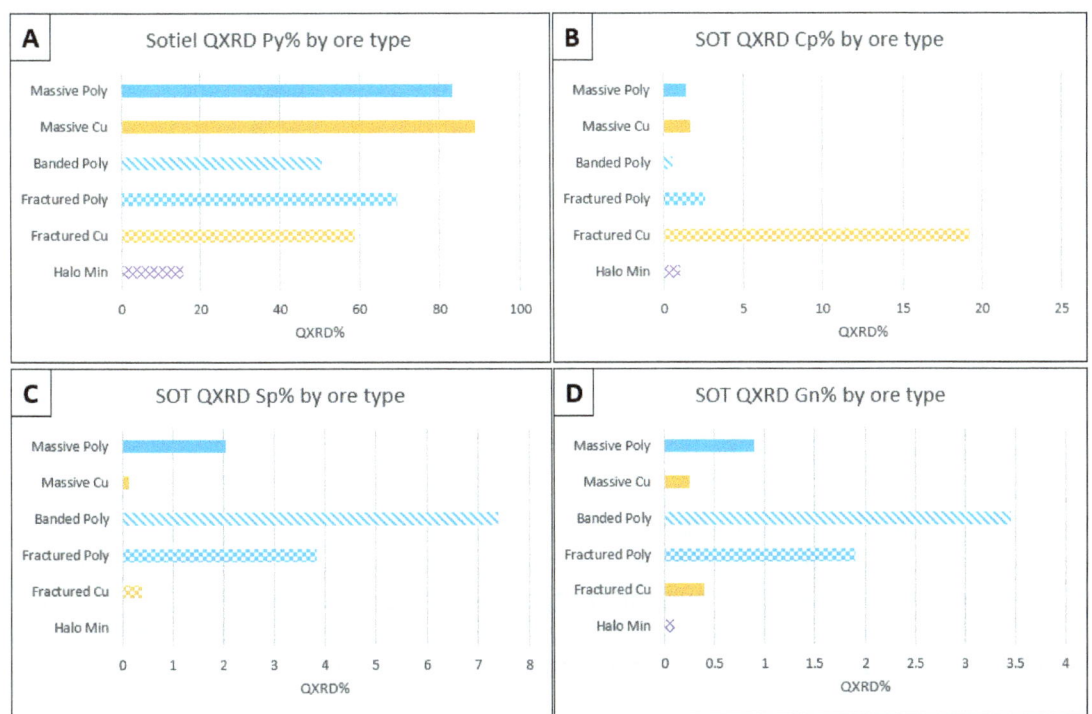

FIG 8 – Sotiel QXRD for each ore type: (a) average pyrite%. (b) average chalcopyrite%. (c) average sphalerite%. (d) average galena%.

The LA-ICPMS elemental data was utilised to measure trace element concentrations in different variations of pyrite identified through optical microscopy. By examining the ablated pits in the polished mounts, the data could be categorised as py1, py2, or py3. As illustrated in Figure 9, from the 105 ablated pyrite grains, it is evident that py1 generally exhibits higher trace element concentrations than py2 and py3 for most elements measured except Cu, Zn, Cd and Te. An elevation in trace element impurities within pyrite correlates with an increase in its reactivity. It is important to note that the LA-ICPMS laser spot size is 19 µm, as depicted in Figures 5–7, py1 and py2 can exhibit a fineness of less than 10 µm. Consequently, it is plausible that some of the point data would encompass ultra-fine minerals residing within the porous interstices of py1 and py2, as opposed to constituting solid substitution in the mineral lattice.

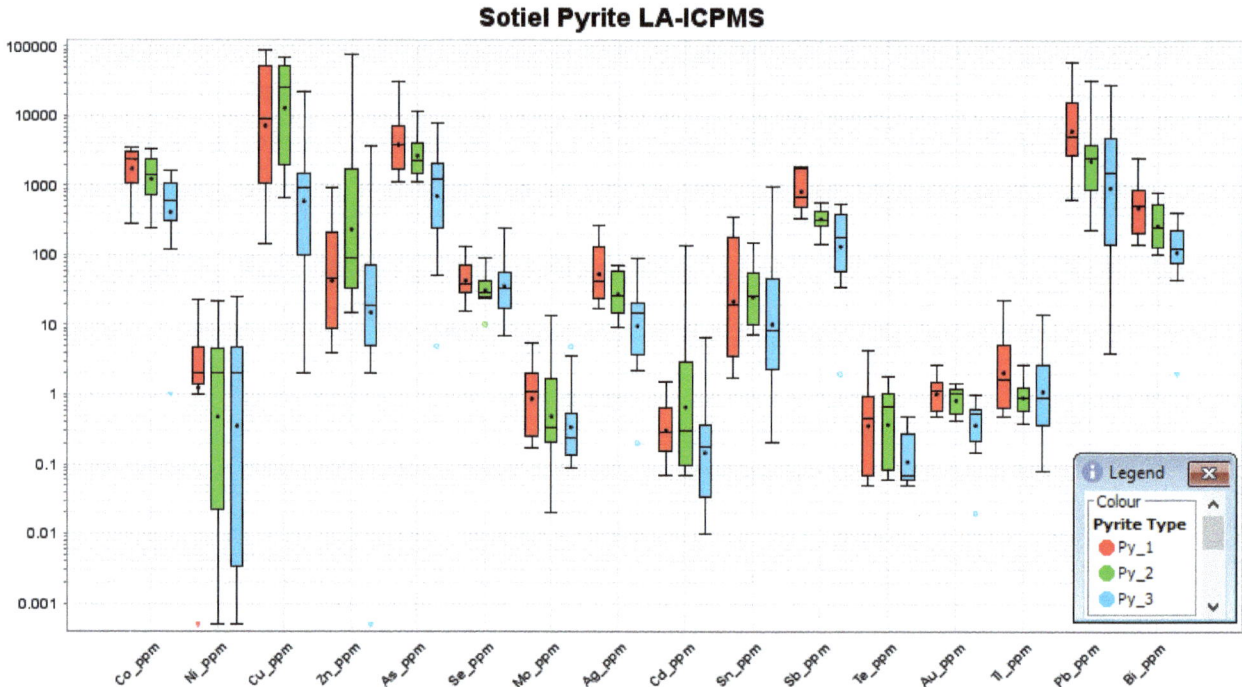

FIG 9 – LA-ICPMS elemental data for py1, py2 and py3.

Secondary ore properties

The liberation analysis of the simulated 53 µm grind shows interesting results for the examined ore types. In samples of the massive copper ore type, none of the target minerals achieve liberation at a 53 µm grind, displaying high concentrations of complex ternary and quaternary composites (Figure 10). In contrast, the fractured copper ore types all have liberated chalcopyrite, a positive sign when the grind size is much finer. Sphalerite and galena have no liberated particles, which is not entirely unexpected due to their low content in the cupriferous ore.

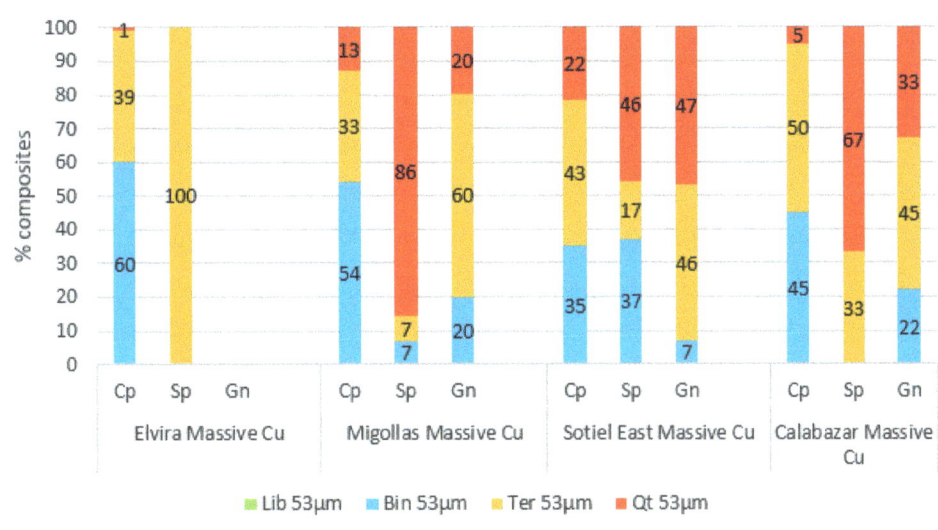
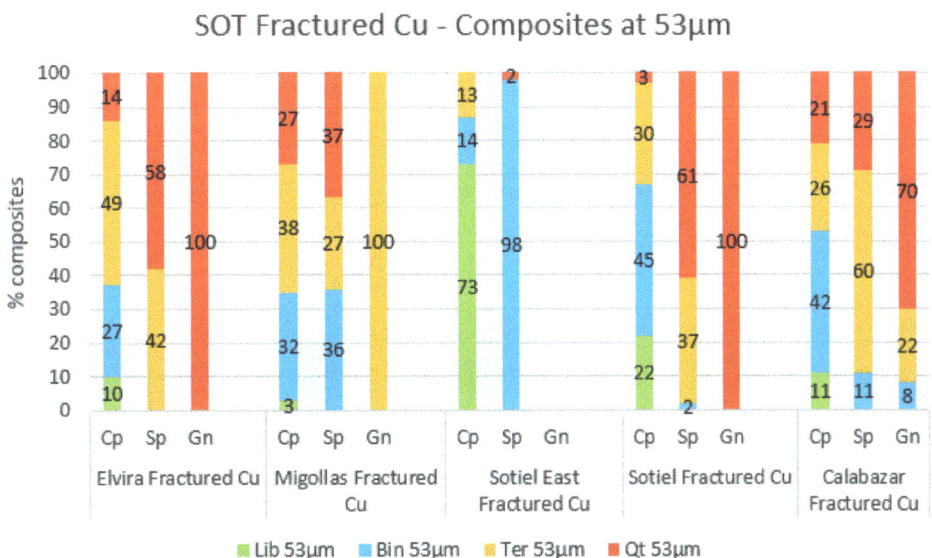

FIG 10 – Composites at 53 µm: Massive copper (top) versus fractured copper (bottom). Abbreviations: Cp = chalcopyrite, Sp = sphalerite, Gn = galena, Lib = liberated, Bin = binary, Ter = ternary, Qt = quaternary.

Within the samples of the massive polymetallic ore type, none of the target minerals achieve liberation at a 53 µm grind, displaying high amounts of complex ternary and quaternary composites (Figure 11). Similarly, in certain fractured (and banded) samples, chalcopyrite grains achieve liberation. However, there is no significant difference in the liberation of sphalerite and galena, both of which are present in substantial amounts of complex ternary and quaternary composites.

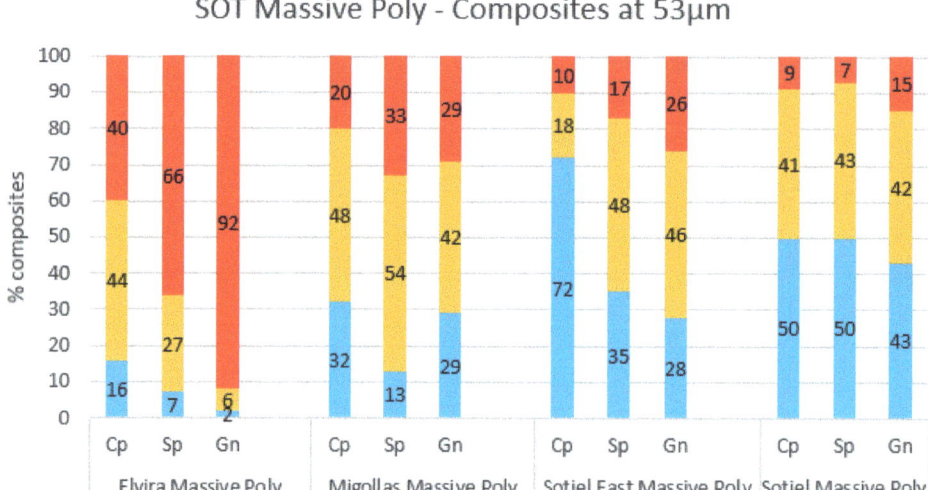

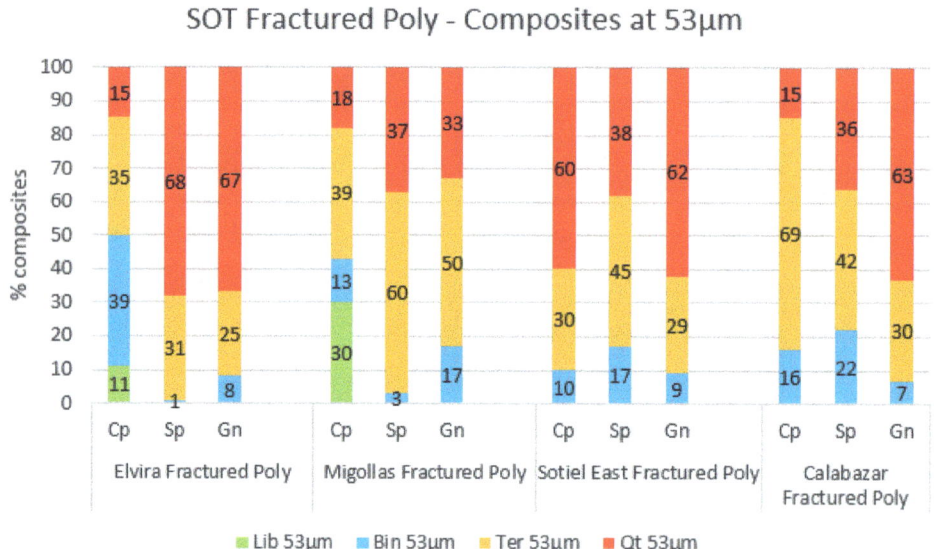

FIG 11 – Composites at 53 µm: Massive polymetallic (top) versus fractured polymetallic (bottom). Abbreviations: Cp = chalcopyrite, Sp = sphalerite, Gn = galena, Lib = liberated, Bin = binary, Ter = ternary, Qt = quaternary.

Figure 12 illustrates the mineral associations for each simulated 53 µm grain. Approximately 90 per cent of sphalerite and galena are found in composites with pyrite, whilst approximately 70 per cent of chalcopyrite occurs in composites with pyrite. Non-sulfide gangue is present in 40–50 per cent of the 53 µm composites involving all target minerals. The qualitative estimation of fine-grained melnikovite (py1) was conducted during the optical microscopy assessment. However, due to the gradational nature of the relationship between py1 and py2–py3 (Figure 6), these figures are subject to uncertainty.

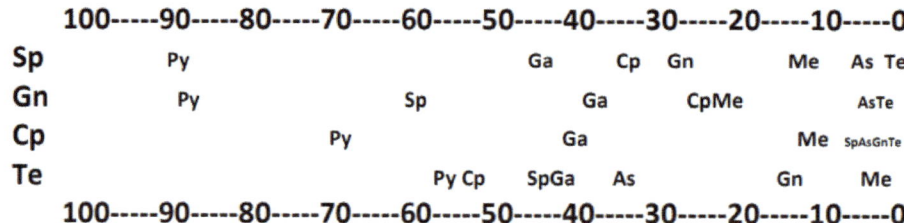

FIG 12 – Sotiel 53 µm grain mineral associations. Abbreviations: Py = pyrite, Me = melnikovite, Cp = chalcopyrite, Sp = sphalerite, Gn = galena, Te = tetrahedrite, Ga = non-sulfide gangue (from Gary McArthur, MODA Microscopy)

APG%, interpreted as an indicator of grain size, is a metric calculated as the total area percentage for each mineral in a sample divided by the number of times the mineral occurs. Figure 13 demonstrates that for chalcopyrite, sphalerite, and (less so) galena as the mineral content increases, so does the APG% (ie the associated grain size). It must be noted that for sphalerite and galena the overall APG% is significantly lower, potentially explaining the overall low liberation of these minerals presented in Figures 10 and 11. As pyrite content increases, the chalcopyrite APG% noticeably decreases. Hence, it can be interpreted that pyrite content, and the interstitial space chalcopyrite has available to fill, either in the ultra-fine porous voids of py1 or in between py2–py3 are inversely proportional. Interestingly, this relationship is not noticeable between pyrite content and sphalerite and galena (Figure 13). Arguably this might be a consequence of the general fine-grained nature of these minerals, irrespective of the ore type, explaining the poorer liberation in Figures 10–11.

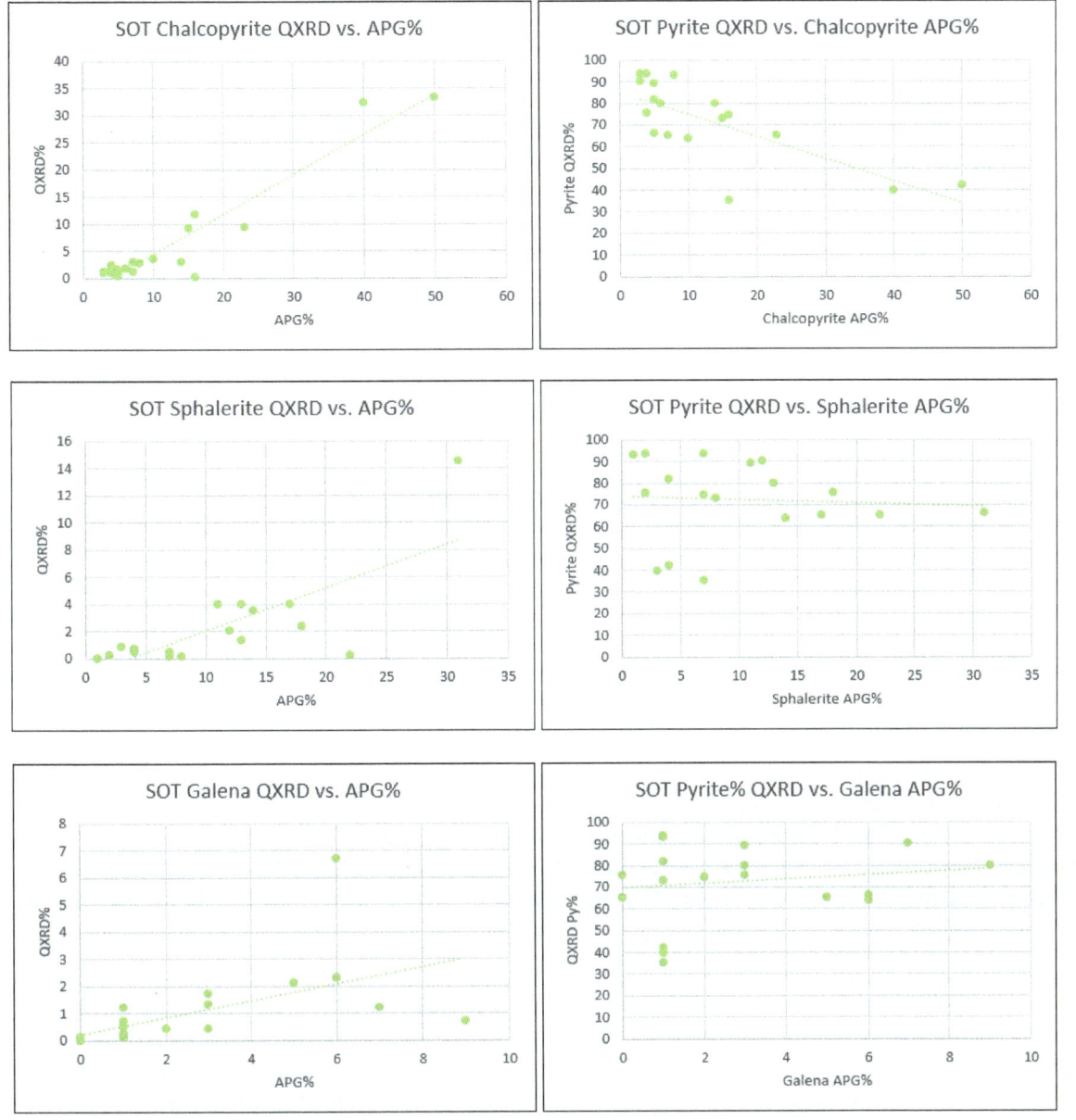

FIG 13 – Sotiel target mineral QXRD content versus APG% and target mineral APG% relationship to QXRD pyrite content.

APG% grain size proxies for chalcopyrite, sphalerite, and galena have positive correlations with their respective 53 µm liberated + binary composite data. Figure 14 presents the chalcopyrite and sphalerite examples of the relationship between the grain size proxy and the simulated 53 µm composite data.

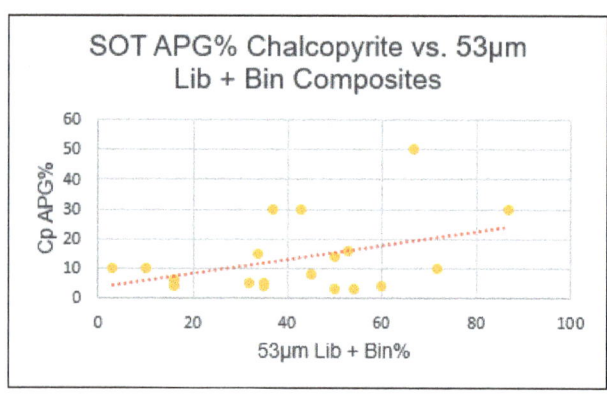
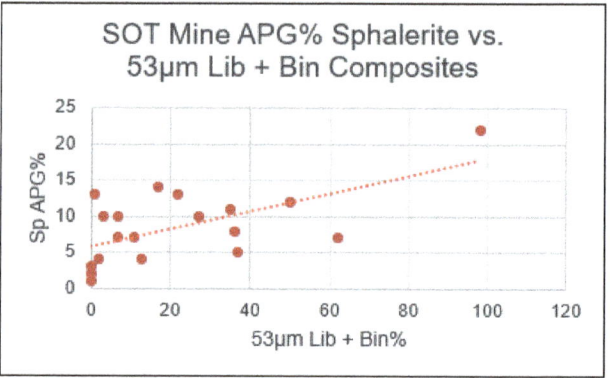

FIG 14 – Sotiel chalcopyrite APG% (left) and sphalerite APG% (right) versus their 53 µm liberated + binary composite data.

The results of the EDTA leach and pH indicate that pyrite content can potentially be used as a proxy for domaining the presence of reactive pyrite. Figure 15 shows a correlation between QXRD pyrite content and the number of lead ions in solution, as well as pH increasing as the amount of pyrite decreases. This can be interpreted as massive-high pyrite ore types with a high content of reactive py1, producing galvanic reactions with surrounding sulfide mineral surfaces and creating a strongly reducing environment. These are the conditions seen at times in the MATSA processing plant.

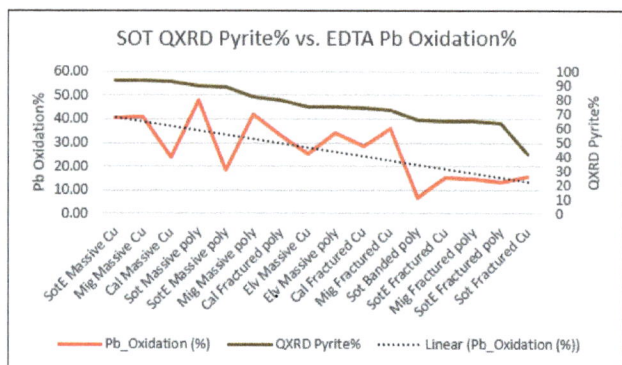
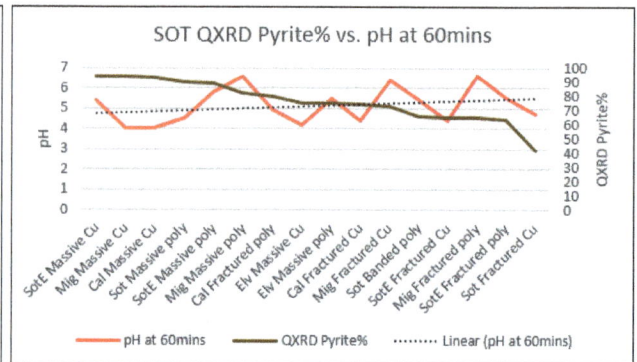

FIG 15 – Sotiel Pyrite QXRD% versus Pb Oxidation from EDTA leaching (left) and pH (right).

GEOMETALLURGICAL ORE DOMAINING

This study shows that pyrite content can be used as a robust proxy for the presence of the complex, reactive and fine-grained py1. This particular form of pyrite is considered a contributing factor to the occurrence of ultra-fine chalcopyrite grain sizes leading to low target mineral liberation. This highly reactive pyrite species is responsible for causing low dissolved oxygen concentrations, producing conditions that hinder flotation of target minerals and promote favourable flotation conditions for pyrite and gangue.

The QXRD data for each ore type revealed that in massive ore types (either copper or polymetallic), the average pyrite content exceeds 70 per cent. Conversely, in fractured-banded ore types (either copper or polymetallic), pyrite content is less than 70 per cent. Analysis suggests that banded and fractured ore types have similar mineralogical and metallurgical results. These two ore types have similar pyrite contents and cannot be geochemically distinguished, therefore they have been combined. This clear pyrite content distinction can be a driver in the definition of geometallurgical ore domains.

Conversion of element data to minerals is required, prior to commencing ore domaining. This limitation is a consequence of the absence of modal mineralogy data in the drill hole database. The sulfide mineralogy must be calculated from the available assays which have all been completed at the MATSA on-site laboratory, initially using theoretical mineral compositions (to be refined later based on LA-ICPMS and EMPA data). The calculation sequence involves determining chalcopyrite, sphalerite and galena first. Subsequently, by utilising the remaining sulfur after accounting for these minerals, the calculation for pyrite can be performed. To validate these mineral calculations the QXRD data from the 20 ore type samples was compared to the mineral grade calculated from their MATSA lab assays. Figure 16 illustrates a significant correlation between the mineral calculations and QXRD values, with R2 values exceeding 0.98. This provides high confidence in the ability to use the existing drill hole database assays for domaining.

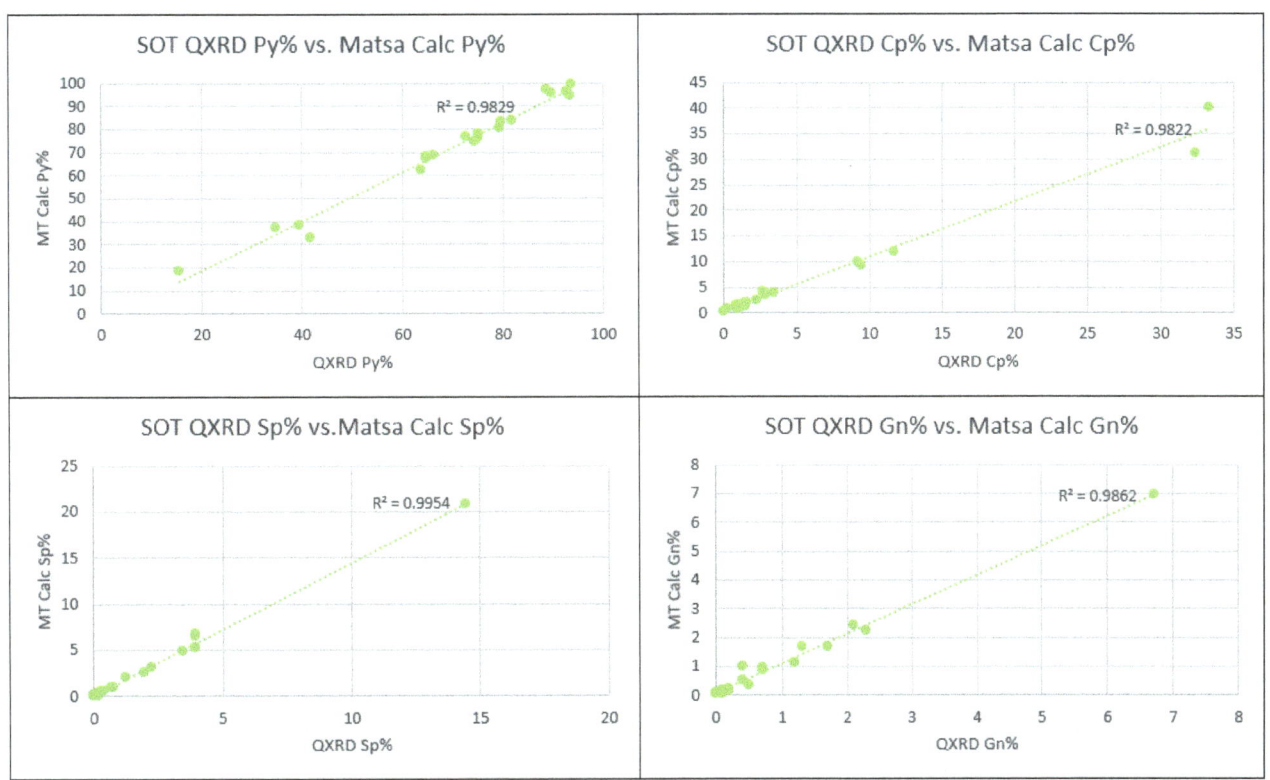

FIG 16 – QXRD mineral grades versus calculated mineral grades.

In addition to adding these calculations to the drill hole assay database, which will allow further geochemical investigation into the geometallurgical domains, the mineral calculations were applied to all blocks in the current grade control block model. This process resulted in the development of a sulfide mineral domain map (Figure 17), representing the initial iteration of the geomet block model.

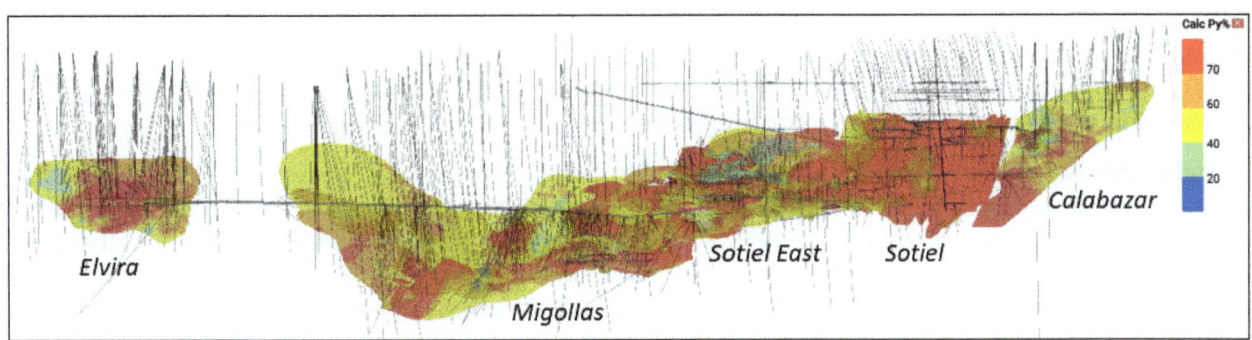

FIG 17 – View south. Sotiel sulfide mineral domain map, showing areas of high and low pyrite.

By utilising the correlations between target mineral APG% with pyrite content and the metal grades (Figure 13), regression calculations have recently been incorporated into the geomet block model to serve as proxies for chalcopyrite, sphalerite and galena grain size. For chalcopyrite, APG% can be predicted using the variables Cu ppm and Py per cent, with the model explaining 89 per cent of the variation (Figure 18). Figure 19 illustrates the prediction of sphalerite APG% using only the variable Zn ppm, with 80 per cent of the variation explained by the model. As previously demonstrated (Figure 13), sphalerite grain size was found to be unaffected by ore type or pyrite content. It is important to recognise that assuming the processing behaviour for an entire deposit based solely on the primary properties of each block can lead to oversimplification. Therefore, further analysis is necessary.

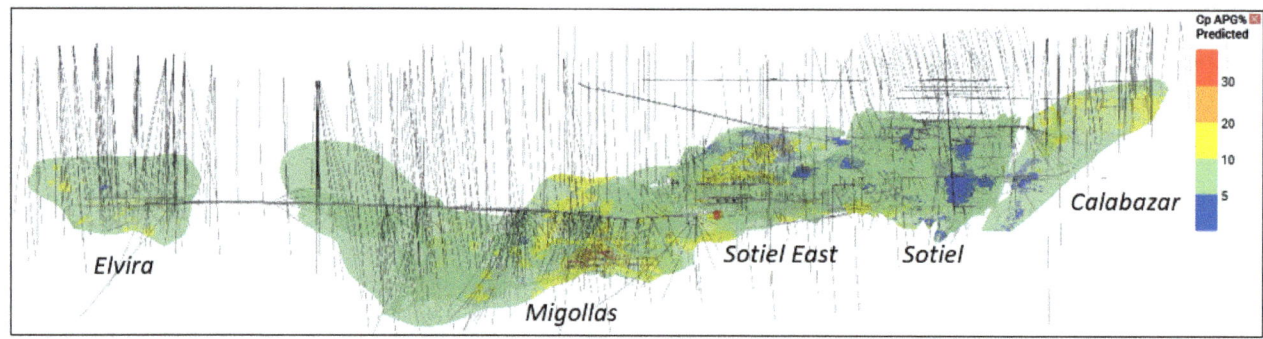

FIG 18 – Sotiel view south. Showing chalcopyrite APG% prediction as indication of grain size variability.

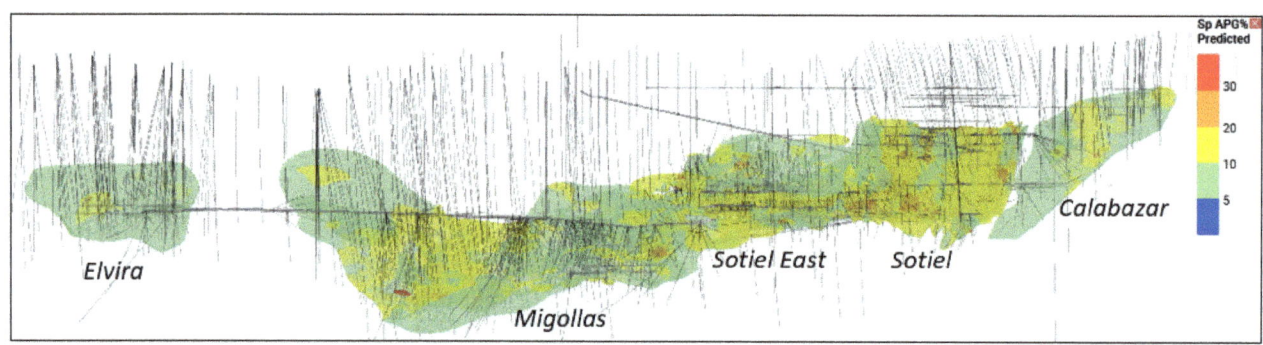

FIG 19 – Sotiel view south. Showing sphalerite APG% prediction as indication of grain size variability.

APPLICATION AND IMPACT

The first iteration of the geomet model is currently in use at MATSA, enabling the geology team to report the pyrite content for each stope firing before mining. The geology, engineering and processing teams can now engage in discussions about mineralogy during planning and stope meetings, being well-informed about the pyrite content and potential target mineral grain size for each stope. It is understood that a stope with elevated pyrite content requires expedited mining due to the economic implications associated with the speed of oxidation and decreased metal recovery. The mine control management system has been configured to include pyrite content for each stope and the 'number of days since firing' data, ensuring constant availability of this information. This set-up promotes the seamless flow of information from mining to processing.

Optimisation during flotation requires dynamic circuit control to accommodate variations in mineralogy. Consistency in mineralogy naturally improves control and stability of the flotation circuit. Given limitations in ore blending capacity and the acknowledged variability in reactivity of the ore fed into Line 2, there can be inconsistent chemistry within the mineral slurry. Identifying and modelling contributors to this high reactivity aids in developing a pulp chemistry environment that is resilient to these fluctuations, as well as assisting in characteristic prioritisation for ore blending.

Predicting the recovery and concentrate grade expected from Sotiel also aids in achieving production targets. Having this data is important in deciding when to campaign Sotiel material and how it should be integrated into production schedules. Increased knowledge of this ore type provides information on how to better process this material, the optimum grind size required, expected reagent usage and whether it can be blended with other ores.

CONCLUSION

The literature demonstrates that the texture and composition of pyrite plays a critical role in determining its surface characteristics and flotation behaviour in complex sulfide ores. Pyrite particles exhibiting a complex texture are recognised for their substantial impact on pulp chemistry, leading to adverse effects on flotation performance.

Six different ore types from the Sotiel deposit (massive-copper, massive-polymetallic, fractured-copper, fractured polymetallic, banded-copper, and banded-polymetallic), were used to investigate the effects of ore texture and mineralogy on the processing performance and enable spatial domaining. As a result, the Sotiel ore can be simplified into four geomet ore types:

1. Massive copper with high pyrite content <70 per cent.
2. Massive polymetallic with high pyrite content <70 per cent.
3. Fractured-banded copper with low pyrite content >70 per cent.
4. Fractured-banded polymetallic with low pyrite content >70 per cent.

Optical mineralogy assessments revealed the presence of all pyrite types (py1, py2 and py3) in each ore domain, while QXRD indicated that massive ore types exhibit the highest pyrite concentration (>70 per cent). According to EDTA and pH data, these high-pyrite ore types demonstrate heightened reactivity, indicating an increased abundance of the fine-grained, reactive py1. The significance of py1 lies in its fine-grained texture, which diminishes target grain size and liberation, particularly for chalcopyrite. Additionally, its extensive surface area and electrochemical properties escalate the galvanic interaction rate with other metal sulfide minerals and its oxygen demand adversely affects plant pulp chemistry. In contrast, fractured-banded ore types exhibit a lower pyrite content (<70 per cent), coarser grain size with improved liberation for chalcopyrite and lower oxidation rates, suggesting a considerably reduced presence of py1.

By employing straightforward mineralogy calculations validated through QXRD and multi-element assays, it is feasible to leverage the existing drill hole database and grade control models for the creation of block models that spatially domain crucial high-pyrite and low-pyrite zones. This has allowed for the dissemination of this information into all forecasting activities on a weekly to quarterly basis. Recently, the APG% prediction calculation for chalcopyrite, sphalerite and galena was incorporated into the geomet block model. Once additional metallurgical test work is completed, this information will be utilised to establish tangible liberation and recovery parameters that will contribute to life-of-mine planning.

This method has successfully been applied to MATSA's Aguas Teñidas and Magdalena deposits, highlighting its usefulness for other pyrite rich deposits in the IPB and similar deposit types elsewhere. It emphasises the importance of orebody characterisation and domain mapping and how linking new mineralogical and metallurgical analysis with existing spatial geological data sets can add value to the mining operational value chain.

FUTURE DIRECTIONS

The existing Sotiel geometallurgy domains must undergo flotation tests and further analysis. This analysis aims to identify proxies for estimating target mineral recoveries and other parameters that can be observed and defined within these domains, the current reactive pyrite and grain size proxies are just the beginning. The insights gained from this study offer the potential to create a multivariable model encompassing geological characteristics, target mineral recoveries and future reagent or collector usage.

Validating predictions from geometallurgical modelling and adjusting relevant models occur through the use of true operational outcomes. If the predictions prove inaccurate, adjustments are made to the models based on reconciliations. This is crucial to decrease the uncertainties about the attributes of each mining block.

Pyrite textures are recognised and frequently attributed as the root cause of challenges in flotation circuits (Jefferson *et al*, 2023); however, their abundance often remains unknown. The absence of information regarding the relative abundance and distribution of complex pyrite textures in an ore limits our capacity to forecast flotation performance and formulate suitable recovery or depression strategies.

ACKNOWLEDGEMENTS

The authors wish to thank Sandfire and Sandfire MATSA for their support and approval to present this work. Thank you to Gary McArthur at MODA Ore Microscopy and Katie Barnes and Peter Munro at Mineralis for your superb analysis, wealth of knowledge and willingness to guide and mentor a geologist wanting to become a geometallurgist.

REFERENCES

Barker, G J, Gerson, A R and Menuge, J F, 2014. The impact of iron sulfide on lead recovery at the giant Navan Zn-Pb orebody, Ireland, *International Journal of Mineral Processing*, 128:16–24. doi:10.1016/j.minpro.2014.02.001.

Can, I B, Özçelik, S and Ekmekçi, Z, 2021. Effects of pyrite texture on flotation performance of copper sulfide ores, *Minerals*, 11(11):1218. doi:10.3390/min11111218.

Craig, J R, Vokes, F M and Solberg, T N, 1998. Pyrite: physical and chemical textures, *Mineralium Deposita*, 34:82–101. doi:10.1007/s001260050187.

Frenzel, M, Baumgartner, R, Tolosana-Delgado, R and Gutzmer, J, 2023. Geometallurgy: Present and Future, *Elements Magazine*, 19(6):345–351. doi:10.2138/gselements.19.6.345.

Jefferson, M, Yenial-Arslan, U, Evans, C, Curtis-Morar, C, O'Donnell, R, Parbhakar-Fox, A and Forbes, E, 2023. Effect of pyrite textures and composition on flotation performance: A review, *Minerals Engineering*, 201:108234. doi:10.1016/j.mineng.2023.108234.

Lehner, S W, Savage, K S, Ciobanu, M and Cliffel, D E, 2007. The effect of As, Co and Ni impurities on pyrite oxidation kinetics: an electrochemical study of synthetic pyrite, *Geochimica et Cosmochimica Acta*, 71(2007):2491-2509. doi:10.1016/J.GCA.2007.03.005.

Lottermoser, B, 2010. *Mine Wastes Characterisation, Treatment and Environmental Impacts*, (third edition) (Springer: Berlin).

Moslemi, H and Gharabaghi, M, 2017. A review on electrochemical behaviour of pyrite in the froth flotation process, *Journal of Industrial and Engineering Chemistry*, 47:1–18. doi:10.1016/j.jiec.2016.12.012.

Owusu, C, Abreu, S B, Skinner, W, Addai-Mensah, J and Zanin, M, 2014. The influence of pyrite content on the flotation of chalcopyrite/pyrite mixtures, *Minerals Engineering*, 55:87–95. doi:10.1016/j.mineng.2013.09.018.

Peng, Y, Grano, S, Fornasiero, D and Ralston, J, 2003. Control of grinding conditions in the flotation of galena and it's separation from pyrite, *International Journal of Mineral Processing*, 70:67–82. doi:10.1016/S0301-7516(02)00153-9.

Dynamic production grade cut-off – integrating financial understanding into daily mine geology

M Darvall[1], G Sternadt[2] and M Van Ryt[3]

1. Mine Project Geologist, Aeris Resources, Cracow Qld 4719.
   Email: mdarvall@aerisresources.com.au
2. MAusIMM, Principal Mine Geologist, Aeris Resources, Cracow Qld 4719.
   Email: gsternadt@aerisresources.com.au
3. AAusIMM, Senior Geologist, Aeris Resources, Cracow Qld 4719.
   Email: mvanryt@aerisresources.com.au

ABSTRACT

Cracow Gold Mine is a narrow-vein, quartz-hosted, epithermal orebody, mined using predominantly longhole stoping and filling methods. Drive position is critical to the potential success or failure of stopes and mine geologists are key decision-makers in determining drive placement and the immediate and ongoing assessment of ore viability.

Cracow Gold Field has been mined intermittently since the early 1930s, with the current underground workings active since 2003. Previous mining efforts have extracted the core of the immediately accessible lodes, with current mining efforts targeting the remnants and splays off these larger systems. These margin lodes are areas of high variability and geological complexity.

Mining occurs at the limits of economic feasibility and often under fluctuating constraints. As a result, standardisation of cut-off grade, which is the minimum grade of ore that is considered economically viable to mine, is not appropriate. Instead, mine geologists must carefully consider a variety of factors, including the depth of the material from the surface, the likelihood of grade continuity, the potential for geotechnical problems and other risks. This generates multiple possible outcomes of ore viability.

Cracow Gold Mine has successfully implemented multi-factor assessment of ore/waste cut-offs using a wide range of inputs, fundamentally driven by informed geological knowledge. Close communication and liaison with engineering, mining and mill operations yields real-time feed-back on the viability of material extraction and gold recovery and the consequent cut-off grade adjustments required to ensure profitable ore production. The resulting decision flow produces a dynamic approach to production grade cut-offs that maximises the value of the orebody.

INTRODUCTION

Cracow Gold Mine is situated approximately 350 km north-west of Brisbane and 50 km south-south-east of Theodore. It is a narrow-vein, low-sulfide epithermal gold deposit (Braund, 2006) mined using predominantly open stoping and cut-and-fill stoping. The gold is contained in structurally controlled quartz veins hosted in sub-aerially emplaced volcanics and intermediate intrusives of the Camboon Volcanics (Jones, 2002).

Gold was first noted at Cracow in 1875. Mining commenced in 1933 (Johnston, 1993) and the area has been intermittently mined across the Golden Plateau deposit and immediate surrounds ever since. The Western Vein Fields were targeted in the 1990s in response to changing geological thinking that brought low sulfide epithermal deposits to prominence. Mining focus shifted from the Golden Plateau to the Western Vein Fields which have been mined near-continuously since 2003.

Aeris acquired Cracow Mine in 2020 and have progressed mining significantly, including delivering a record year of production throughput in FY2023. Cracow operations treated 667 kt of ore and produced over 48 000 ounces of gold for the year at an all in sustaining cost of $2326 per ounce. During this time the mine has concurrently replaced depletion with in-mine discovery and resource extension.

The orebody in the Western Vein Fields comprises multiple anastomosing lodes which vary in grade, width and extent (Figure 1). The major veins are commonly subvertical with minor subsidiary splays and linking structures dipping from vertical to less than 45°. Grades are typically restricted to the quartz veins and terminate abruptly at the vein margins. Rare examples of silicious alteration carrying economic mineralisation are thought to include fine veins of quartz.

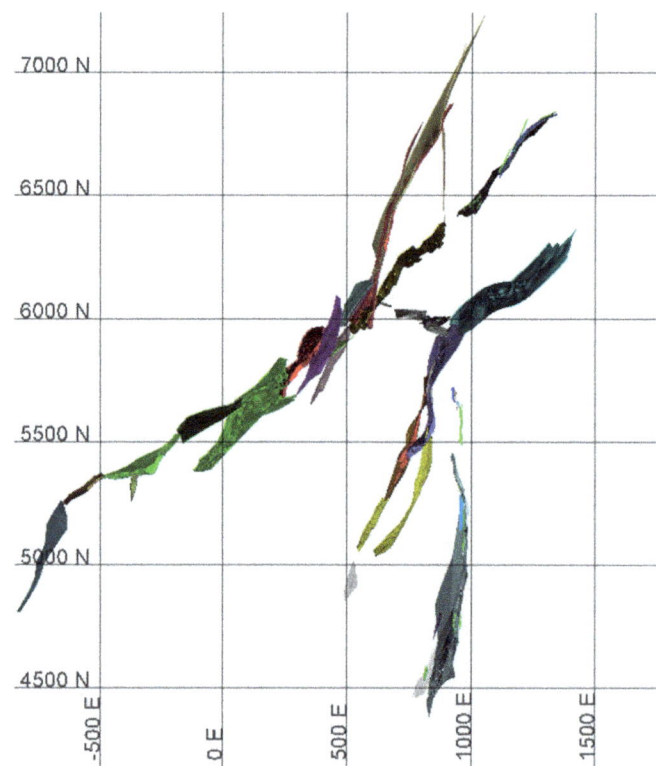

FIG 1 – Overview of the Western Vein Field lodes in plan view. Local grid.

Regional scale structures with generally north–south strike, have provided the foundation for the multiple subsidiary faults that host economic mineralisation around critical dilation zones created by changes in dip, strike and fault reactivation (Micklethwaite, 2009). The cores of these subsidiary faults were the continuous, high-grade orebodies that were preferentially mined as discrete lodes. Around the margins of the high-grade cores are stockwork breccia zones and lower order structures that are intermittent in nature with increased variability.

Ongoing infill drilling and geological reinterpretation is determining the interplay between each of the lodes. A degree of interconnectivity has become apparent that was previously not fully acknowledged. The contiguous nature of the lodes has yielded opportunities to identify additional ore zones through careful reinterpretation, supported by appropriately spaced infill drilling, around pre-existing mining areas. This includes significant extension of lodes both up and down dip and along strike and into the hanging wall and footwall zones.

MINING CONSTRAINTS

The core high-grade zones in the Western Vein Field have been targeted consistently since commencement of mining in the area. Preferential extraction of the core zone has left a halo of lower grade breccias and splay veins that have a high degree of complexity, as well as lode margins that are commonly less continuous in grade.

Mining these zones requires assessment of geological features and potential hazards from remnant voids. Areas must be individually risk assessed and monitored for grade viability as well as mining risks. In most instances hazards caused by remnant mining issues are identified well in advance of mining and mine planning alleviates the risks. Ongoing monitoring directly ahead of mining activities will occasionally encounter unexpected hazards and at times pockets of opportune ore.

The response of halting, re-evaluating and re-planning the mining advance requires flexibility and rapid decision-making as minimising disruption to the production cycle is critical. Where mineralisation is encountered unexpectedly – in short strike splays and breccia zones – ore/waste discrimination must occur in almost real time. A framework for ore cut-offs and a set of criteria for assessment are in place to support decision-making, but it remains a judgment call by the mine geologist at the time.

Mining geologists at Cracow are integrated members of the production team, focused on providing relevant geological input. The level of variability encountered during mining places the mine geologist as a key decision-maker on day-to-day ore/waste discrimination calls. These occur at a range of scales, from marking up oversize on the ROM through to judgment calls on stope material recovery in response to geological risks.

COST FACTORS

Segmentation of cut-off grades is determined according to incremental costs. Basic guiding costs have been determined by considering the cost per stage of production with respect to the cash cost and overheads applicable to each stage. The extraction process is divided into seven practical stages:

1. Full cost recovery including overheads: The full cost of producing gold with allocations for non-cash cost elements such as depreciation, amortisation and company administration. This is used to assess the viability of a new area requiring capital development.

2. Operating cost recovery: The site level costs directly applicable to gold production including all operational development, stoping and backfilling and extraction. This applies to areas where capital development is already in place for assessment of proposed operational development.

3. Stoping incorporating cemented rock fill (CRF): Operational costs where development drives are complete and stoping requires backfilling with CRF for geotechnical stability. The CRF component adds a degree of additional cost and risk assessment around geotechnically challenging areas in addition to standard stoping costs.

4. Stoping with rock backfill: Operational costs where development drives are complete and stoping with uncompacted waste backfill is sufficient for geotechnical stability. This is part of the stope optimisation effort and is used to determine critical drive lengths and stope design.

5. Already blasted material: Any material that is either blasted as part of necessary development or as an adjunct to production ore and is still underground. This includes overbreak in stopes, production and capital development and historic material such as old backfill or rehandle material. Real-time decision-making is often focused on these areas.

6. Processing including General and Administration: Any material that is already transported to the ROM, adjacent to the crusher and mill complex. The inclusion of administrative overheads is critical in strategic assessment of historic sources proximal to the crusher.

7. Processing cash cost only: Any material that is immediately available to the crusher or is expected to be transported to the ROM area, for example recovered historic dumps.

These base-level cost cut-offs are applied as appropriate by the mine geologist and adapted taking into consideration the practical impact of the specific conditions. This includes responding to variable conditions and acknowledging the likelihood of variance in grade from the expected grade and recovery costs in a practical manner. As development progresses the change in the incremental cost recovery curve impacts the day-to-day decisions while maintaining the discipline of full cost recovery; changes to any plan or design are vetted against this metric.

Full cost recovery is based on given assumptions around stope performance, operational constraints and gold recovery. The expected viability of the material is impacted by geological, engineering and production factors – including metallurgy – and the interactions between all three. As a mining geologist, a basic understanding of the engineering and production constraints is critical in determining the key geological aspects to observe and communicate through to the operators and downstream customers. Direct liaison with production personnel has been central to decision-making around ore/waste discrimination as the production environment is highly fluid.

Longhole stoping in narrow vein orebodies is predicated on correct and accurate positioning of the development drives. Stope design is heavily influenced by the absolute position of the drives and their position relative to the orebody and each other. Geological knowledge of grade and cost cut-offs factor into the adjustment of drive designs in real time with the end goal of optimising stoping. This incorporates aspects of understanding the engineer's role and methods of work in stope design and how that is influenced by geological factors, particularly structural controls.

Drives design is based on expected ore position from diamond drilling and any available development above or below, and adjusted based on detailed geological assessment of each face. Face observations and sampling provide a much greater density of data than standard grade control or resource definition drilling do and so are critical in refining drive position. There are multiple interacting and, at times, conflicting parameters in drive optimisation that are considered when directing the drives:

- Maximising ore available for stoping to facilitate ore recovery.
- Minimising overbreak.
- Longhole production drill rig accommodation.
- Ore drive trafficability.

All of these factors impact the likely cost of development, stoping and ore recovery and so impact the real cut-off grade.

Typically, straighter drives are considered preferable to multiple changes in direction. Straight drives are easier to navigate for operations personnel in mining equipment, they simplify stope design and reduce the likelihood of weakly-supported rock-spires projecting into the stope. However orebody geometry and structural controls frequently compel changes in drive direction at the scale of the individual development cut, approximately four metres. The drive is consistently realigned to minimise exposure to the major structures adjacent to the lode and to optimise lode position for operational activities (Figure 2).

FIG 2 – Drive positioned with ore proximal to left wall to minimise undercut of bounding fault zone. Red paint mark-up lines delineate ore zones, with zonal differentiation highlighted to support refinement of lode interpretation. Survey pick-up of the lines ensures accuracy for modelling applications.

The impact of ground conditions on stope performance is significant and may either reduce or increase the costs of stoping. Typically, controlling and bounding structures in the Cracow orebody are wide zones that have been intensely sheared and altered. Drive positioning and the interaction of bounding and cross-cutting structures does increase potential for overbreak dilution and underbreak ore-loss during stoping, but can also create a very clean surface for the stope to break to. The effective cut-off grade is considered in the context of the expected stope performance (Figure 3).

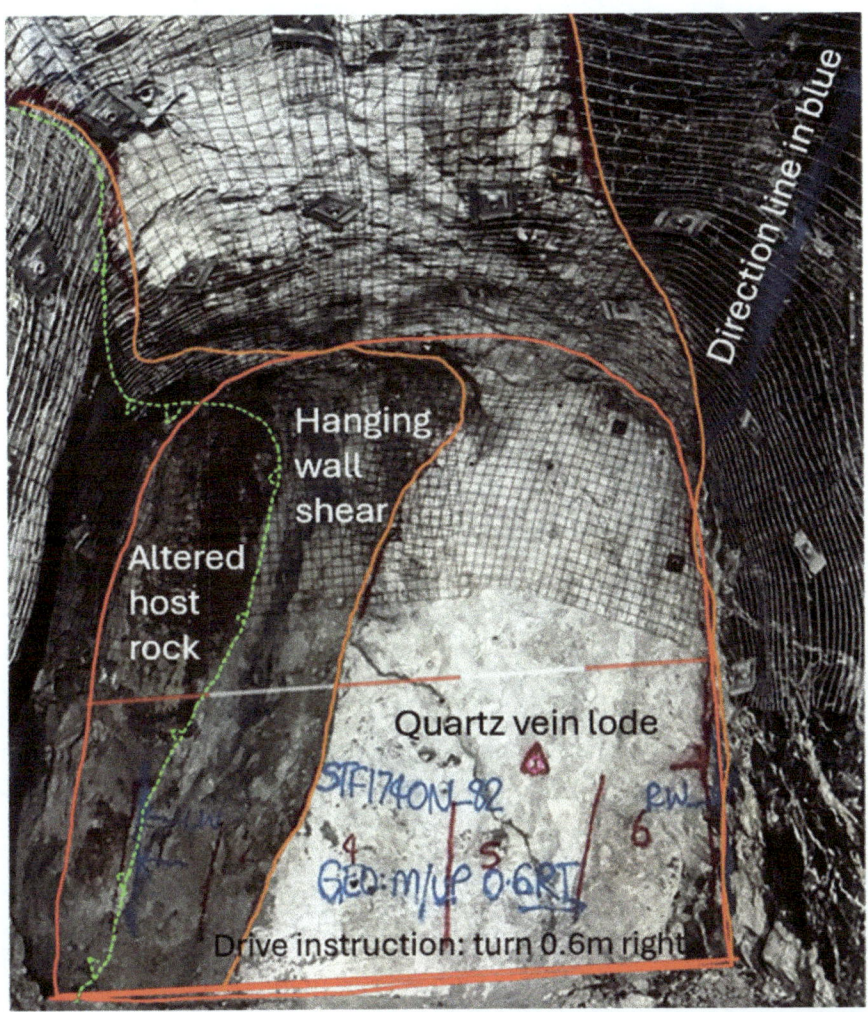

FIG 3 – Quartz lode with intensely sheared bounding structure veering hard right. Drive design has been adjusted by the mine geologist to suit, as marked in the backs in blue and written on the face. The purpose of this directive is to realign the drive so the left hand wall is almost at the hanging wall contact of the lode.

The drive profile can have a critical impact on the success of the development drive so drive design takes into consideration the interactions with geotechnical and geological features. Local responses to different conditions include basic amendment to drives such as implementing a 'shanty' profile (Figure 4). These changes require appropriate liaison with both operations and technical personnel. The mine geologist instigates and coordinates the change process where appropriate.

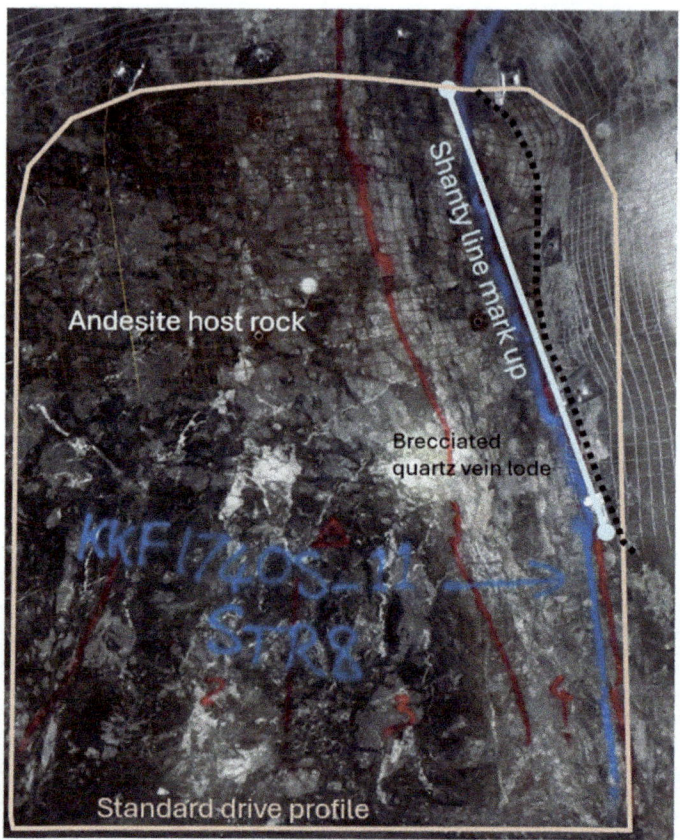

FIG 4 – Shanty profile on sub-vertical lode, bounded by heavily sheared hanging wall fault (black dotted line) with a standard profile overlaid. Note the area in the top right corner, dominated by the hanging wall fault, that would have been included in a standard profile and is excluded from the shanty profile. The preceding cut was bored to a shanty profile and minor break-out to the fault has rounded out the corner slightly, but notably the fault surface remains intact. The shanty can reduce the likelihood of overbreak into the hanging wall during both development and stoping, reducing the cost of mining and improving the viability of the stope.

Where overbreak is more likely, the cost profile for recovering the contained ounces shifts but not in direct correlation with the basic recovery cost calculations. Additional factors are considered including the viability of overbreak separation – typically as oversize, the risk to equipment and the likelihood of overbreak containing additional low-grade ounces.

The understanding of variable grade cut-offs also informs drive termination or extension. Incremental costs determine the viability of progressing a drive. Predicted full cost recovery includes the cost of capital and operational development, as overheads, in the grade cut-off of planned stopes. Development drive designs provided by engineers assume that full cost recovery will be achieved from the stopes accessed along the designed drive length; the planned drive will be paid for by the planned stopes. The incremental cost of extending drive development does not carry the same overheads so extending the drive can be developed at lower cut-off grades, assuming stoping will also be extended.

There is potential also to extend drives that will not be stoped. The rapid change in dip and strike of the orebody can produce plunging chutes of ore that terminate abruptly above the backs position, and sub-horizontal offsetting faults can locally displace ore laterally. This results in areas that may not provide a viable stope despite having mineralisation in the face. The extension or termination of a drive in this situation is directly linked to the cost of development with little or no overheads to consider; it's almost a cash cost decision only. However development without stoping is an expensive mining method and has the highest cut-off grade.

INPUTS TO DECISION-MAKING

The key information available during decision-making includes the raw data from drilling, face grades in levels above and below the current position and the estimated grade from block modelling. These are the base-level information available, they are enhanced significantly through direct observation by the mine geologist, including potentially face sample grade or stockpile grabs of the material under consideration. Critically, the face sample grade or stockpile grabs grade is considered in conjunction with the other data; over-reliance on individual grab samples is not accepted due to the accuracy limits of grab and chip samples (Dominy, 2016).

Additional considerations are the visual assessment of the material, evidence of faults and changes in ore strike that indicate that grade may either be diluted or changing in response to geological conditions, and variations in proximal indicators such as alteration and multiple phases of brecciation. Understanding of the local variables impacting grade help determine if the apparent grade is suitable for assessing whether the ore is above cut-off. The anastomosing, fault-controlled nature of the lode commonly results in rapid changes of lode direction and width. Given development cuts are four metres or greater the grade at the face may be significantly different from the majority of the cut. Likewise lode-scale faults can offset or terminate the vein, impacting grade within individual cuts.

Balancing the geological and grade information are the realities of ore production. Fundamental production constraints include trucking and tramming distance comparing the default alternative for both ore and waste. Where waste movement is likely to equal or excede costs of ore movement it potentially changes the incremental value profile. Conversely, where depth and resultant trucking distances are particularly long, cost is driven up relative to backfill movements which increases the effective grade cut-off.

Ongoing assessment of material viability incorporates real-time feedback on grade response throughout the process. Mill feed grades are monitored at six hourly timescale and often respond quickly to changes in source material; at most the lag time through the process is usually 24 hours from feed to mill grade response. The feedback loop can be directly applied to assessment of underground feed stocks.

Due to the appropriate scaling of milling infrastructure to the orebody's size, processing rates match mining rates extremely well; it is rare to have either a shortfall in production or an excess of ROM stocks. There is near real-time grade response to underground calls ensuring that mine geologists' calls are constantly evaluated. Importantly, this is done in an environment that is genuinely supportive of improvement; accountability is in the form of situational review across the decisions and systems in place rather than the purported failings of a given individual. This enhances future decision-making and trust in the review cycle and in the supporting systems.

Strong links between the mine and mill allow for direct communication between metallurgists and geologists. Issues around recovery and potential problematic materials – clays and unusual sulfides for example – are rapidly escalated at the operational level. Cut-off grade adjustments are then applied in real time to ensure profitable material is prioritised to the ROM and lower recovery mineralisation is classified appropriately. While formal meetings occur between mine and mill personnel on a weekly basis, there is almost daily communication between the metallurgists and mine geologists.

VISUAL DECISION-MAKING

The categorisation of cut-off grades is displayed prominently to ensure direct association between cost impacts and day-to-day ore/waste discrimination. This helps to ensure the cut-off grades are incorporated into relevant decisions. The information is not tucked away in a lengthy work instruction or obscure file or folder on the shared drives, but is pinned on the door of the mine geology office.

Face photographs and, more recently, LiDAR drive records are frequently reviewed both formally and informally to assess decision-making. Orebody outlines and faults are painted up for survey pick-up and to enhance visual distinction in the face photos and for survey pick-up. Drive directions are painted on the face and marked up on the backs. These items support the review process. The emphasis during review is on having a defensible reason for each decision rather than a hind-sight driven, judgemental 'right' or 'wrong' categorisation and this approach encourages rational decision-making.

CONCLUSION

The challenges of mining the margins of a narrow vein orebody relate to the discontinuous and rapidly changing nature of the ore and the presentation of potential opportune ore. Critical decisions that impact real cash flows occur contemporaneously with operational activities. Mine geologists must have the flexibility and confidence to make these crucial decisions in a timely manner to optimise operational activities; there is rarely the luxury of time.

The framework of variable cut-off grades provides a basis for ore-waste discrimination that the mine geologist can apply. Local contextual information and application of geological understanding of the orebody, combined with an awareness of production and engineering constraints, are used to adapt the framework appropriately for individual situations.

Decision-making is expected to be rational and defensible with clear review of the key decisions in a non-judgemental environment; accountability, not blame. The systemic and transparent nature of the key metrics, and the review process, promote confidence in the system and encourage proactive and timely decision-making.

Implementing the variable cut-off framework has yielded significant value gains for Cracow. While total value is difficult to ascertain due to uncertainty regarding what the decision otherwise would have been in each individual circumstance, the development of a consistent approach has produced a noticeable improvement. Estimates of value gain are in the order of three to six percent improvement in ore-waste discrimination. This does not translate directly to ounces as it is the lower grade materials that are most impacted and in some cases results in classification of material as waste that would previously have been classed as ore.

REFERENCES

Braund, K, 2006. Geology, geochemistry and paragenesis of the Royal, Crown and Roses Pride low sulphidation epithermal quartz vein structures, Cracow, south-east Queensland, thesis, School of Physical Sciences, The University of Queensland.

Dominy, S C, 2016. Importance of good sampling practice throughout the gold mine value chain, *Mining Technology*, 125(3):129–141. doi:10.1179/1743286315y.0000000028.

Johnston, M D, 1993. The Cracow Goldfield – a review of geological thinking, Cracow Gold Mine.

Jones, J A, 2002. Geology of the Camboon Volcanics in the Cracow area, Queensland: Implications for the permo-carboniferous tectonic evolution of the New England Fold Belt, thesis, Department of Earth Sciences, The University of Queensland.

Micklethwaite, S, 2009. Mechanisms of faulting and permeability enhancement during epithermal mineralisation: Cracow Goldfield, Australia, *Journal of Structural Geology*, 31(3):288–300. doi:10.1016/j.jsg.2008.11.016.

Rapid drill hole planning for underground delineation drilling at the Platreef PGE-Cu-Ni deposit, South Africa

G Gilchrist[1] and J Chitambala[2]

1. Vice President, Resources, Ivanhoe Mines, Johannesburg, South Africa.
   Email: george.gilchrist@ivanplats.com
2. Resource Manager, Ivanhoe Mines, Johannesburg, South Africa.
   Email: joshuac@ivanhoemines.com

INTRODUCTION

Platreef is a world-class, high-grade PGE-Cu-Ni deposit discovered by Ivanhoe Mines on the northern limb of the Bushveld Complex in South Africa (Figure 1).

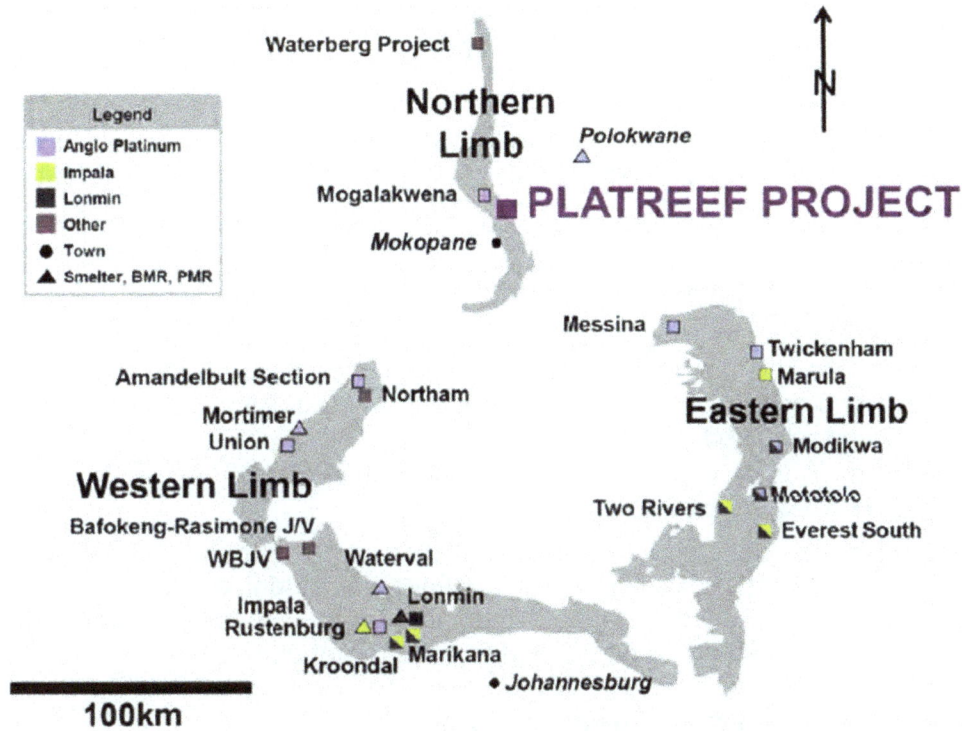

FIG 1 – Locality map of the Platreef Project.

In contrast to the narrow (approximately 1–2 m thick) Merensky and UG2 reefs that have historically dominated PGE production from the Bushveld, the zone of preferential mineralisation within the Platreef (locally termed 'T2') consists of mineralised pyroxenites and harzburgites approximately 20 m thick. Interaction of the intruding magma with highly reactive sediments has likely contributed to the anomalous thickness but also added complexity regarding short-scale changes in geometry and thickness, requiring an infill delineation drill program prior to mining. At depths of 700–1200 m, it is impractical to do this from surface.

Determining the scope and cost of this delineation program during the studies phase, required planning over 20 years of drilling across the entire mining footprint, whilst retaining the flexibility to adjust to changes in the mine plan. A largely automated technique was developed to achieve this.

METHODOLOGY

Figure 2 provides a snapshot of the process described.

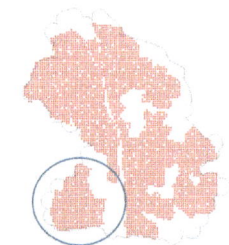
1. Set up a 25 m grid nodes; circled area is typical mining block.

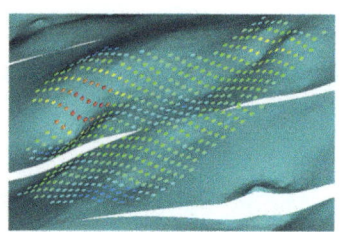
2. Project nodes to top-of-T2 and calculate dip and dip-directions.

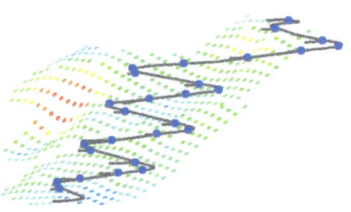
3. Identify possible drill sites from planned footwall development

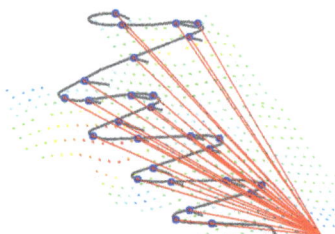
4. Create drillholes linking each target point to every possible drill site

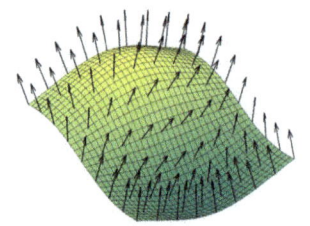
5. Calculate drillhole length and intersection angle – for ranking

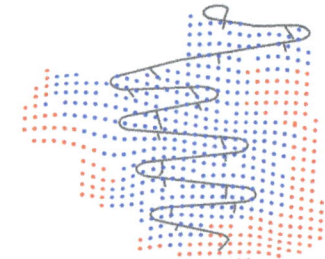
6. Identify best single drillhole for each target or identify if target is unobtainable

FIG 2 – Summary diagrammatic outline of the rapid drilling procedure.

Setting up target drilling grid

A 25 m × 25 m grid in 2D is set-up for individual mining areas; outlines are wider than the planned mining to accommodate the possibility of the tighter grid defining wider high-grade zones. The distance from each grid node to an existing hole is determined. All targets closer than 12.5 m are deemed redundant and removed.

Projecting grid nodes to the top of T2 surface

The target grid nodes are projected to the modelled surface defining the top contact of the T2. The wireframe's dip and dip-direction values, at all locations coincident with projected grid nodes, are extracted. This allows for the 3D orientation of the plane of the orebody (vector data), required for determining the intersection angle, to be defined at each target point.

Obtain coordinates for potential drill sites

The only manual step in the process requires identifying coordinates for suitable drill sites well above or below the target mining block from the mine plan. This could be sites specifically provisioned for delineation drilling, cover drilling (for water and gases) or other logistical use. It is critical that the mining schedule allows for the drill sites to be available well ahead of planned mining of the block to be tested.

Construct drill holes from target to each drill site

Each target point (grid node) is connected to every identified drill site in the mining block. Each connective is assigned drill hole properties such as drill hole identifier, dip, azimuth, start depth (and coordinates), end depth (and coordinates) and length. The process is repeated for all target locations in the mining block.

Determining intersection angles with the plane

Calculation of intersection angles between drill holes and the 3D planes of the orebody involves use of vectors methodology, using the drill hole azimuth and dip, as well as the dip and dip direction data captured for each target location.

Instead of assuming general strike and dip for the plane, and dip and azimuth for the drill hole, undulations inherent in the orebody are honoured by reading vector data off the constructed wireframe at each grid node.

In geological terms, planes are described by their dip and strike. In mathematical terms however, a plane is represented by the combination of a point (pole) on the plane and a vector orthogonal to it.

Therefore, dip and dip-direction of the plane at the target location constitutes one vector set, while the dip and azimuth of the planned drill hole constitutes the complimentary vector set.

To be used as vectors, the dip and dip-direction information has to be converted to a triplet (X, Y and Z) of *direction cosines* for the normal (pole) vector to the plane.

The **cosine** between the drill hole and the **pole to the plane** at a given target is calculated using Equation 1:

$$\cos(\Theta) = l_1 l_2 + m_1 m_2 + n_1 n_2 \qquad (1)$$

where,

Θ is the angle between two lines determined by the scalar or dot product of two unit vectors

l, **m** and **n** refer to the XYZ direction cosines of the vectors

The **intersection angle** (Θ) between the drill hole and the **plane** at a target is calculated using Equation 2, with the terms as explained for Equation 1:

$$\Theta = |a\cos(l_1 l_2 + m_1 m_2 + n_1 n_2) - 90°| \qquad (2)$$

Identify best single drill hole per target location

For each grid node, all possible drill holes are ranked by their intersection angle (the closest to 90° the better) and then by length (the shorter the better). A weighted ranking based on these two variables allows the selection of the 'best' drill hole per target grid node.

A minimum intersection angle (30°) and maximum drill hole length (250 m) are used to filter drill holes.

For some target locations, none of the drill holes meet these criteria. In such a case the target location is then deemed to be unobtainable from the current plan. Such dill sites are highlighted to motivate for designing of dedicated drill platforms as alternatives.

CASE STUDY

Shown in Figure 3 are results of initial implementation and feedback from the process. Note the high number of unobtainable locations in red.

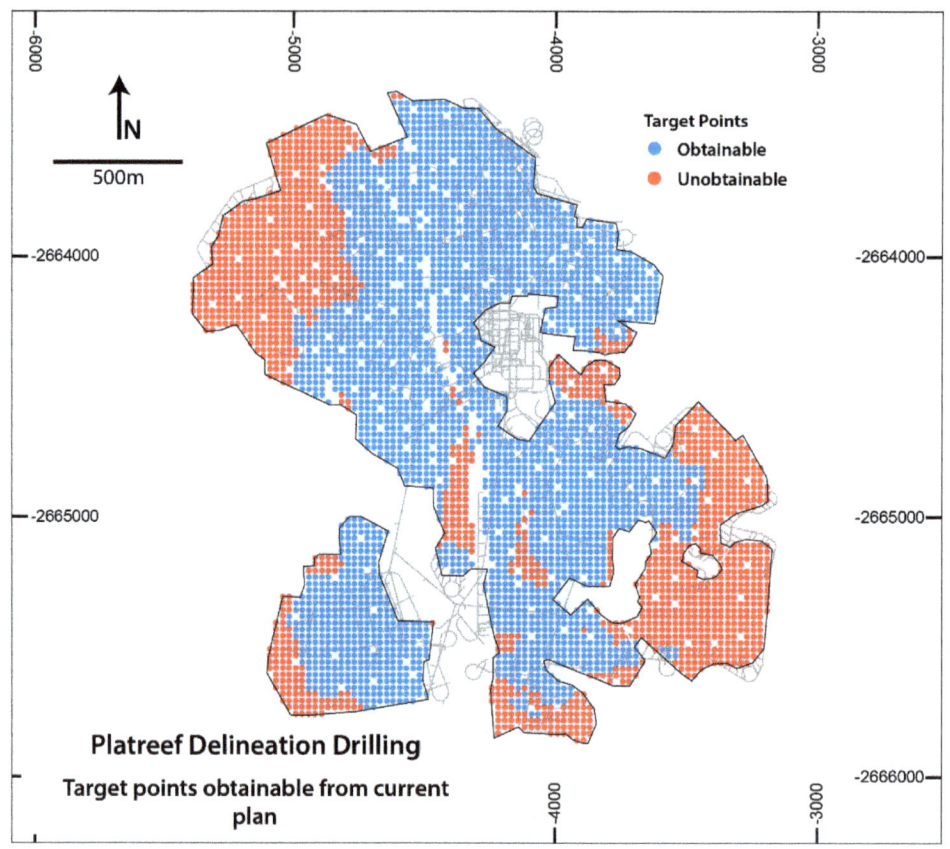

FIG 3 – Initial implementation outcome.

Following the high number of unobtainable locations, additional mitigative drilling locations were proposed, as shown in Figure 4. The whole exercise was repeated, leading to the improvements observed in Figure 5.

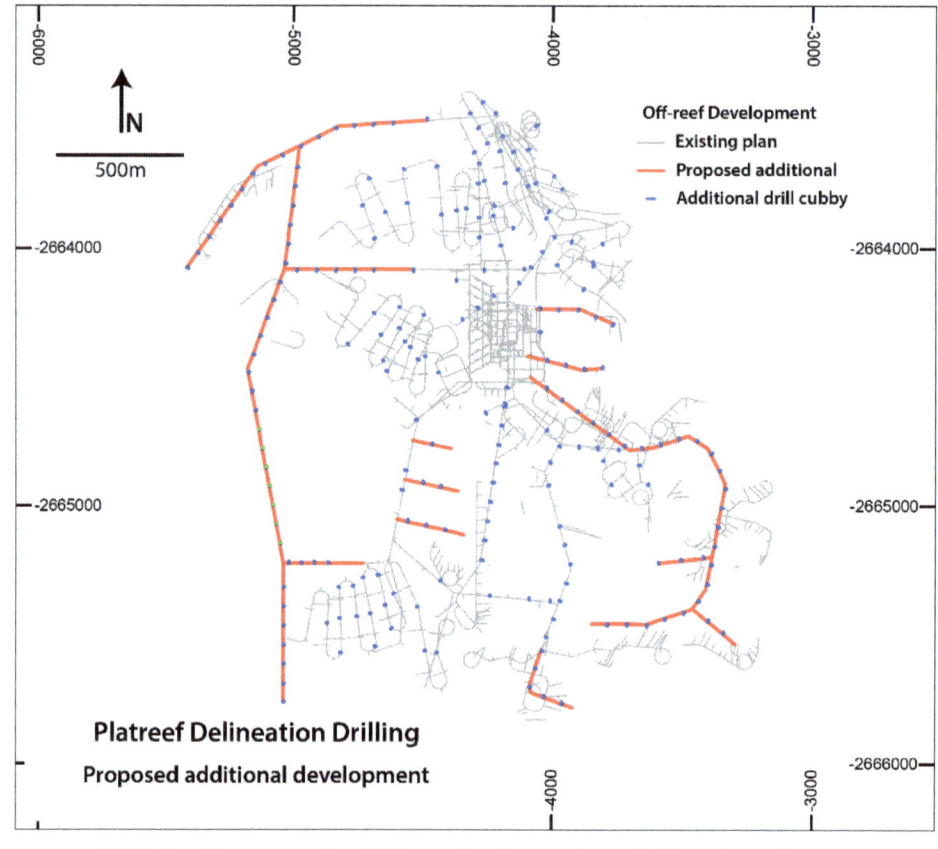

FIG 4 – Proposed additional, alternative drilling locations.

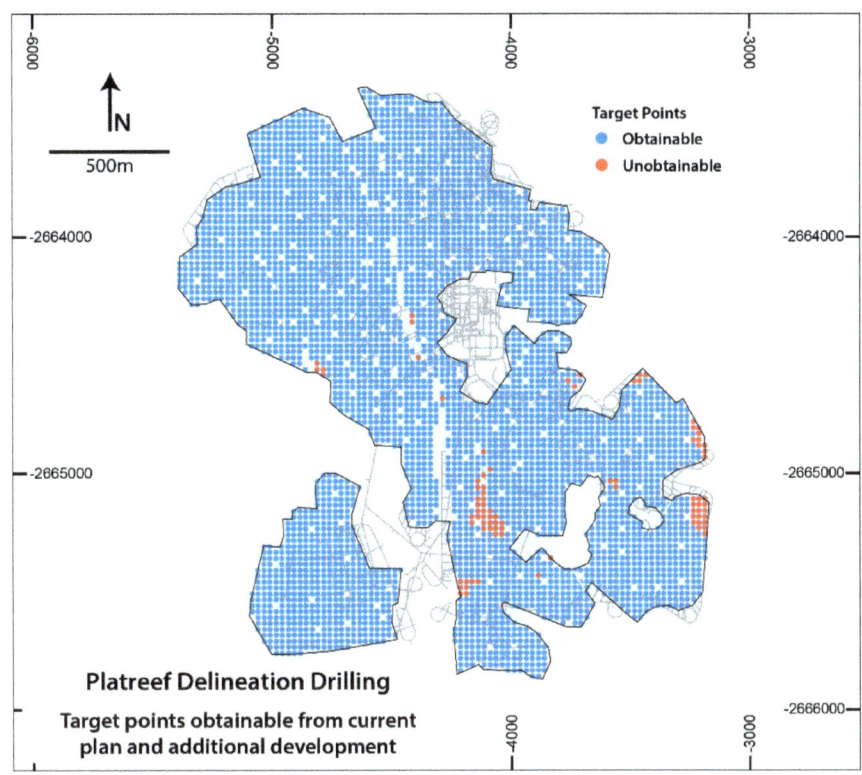

FIG 5 – Outcome of revised drilling platform layout.

DISCUSSION AND CONCLUSION

All the above steps are consolidated into Datamine Software macro. The steps are easily implementable in any scripting platform such as Python. The process is largely automated. The only manual input required is the identification of distinct mining areas and determination of drill sites coordinates. The parameters (grid size, ranking criteria etc) can easily be adjusted to meet unique situations.

The macro output is considered provisional. Checks need to be conducted for any impractical drill hole designs. The macros are intended to make the process more repeatable and save time on the planning process so that more time can be spent on analysing outcomes.

Outputs include a fully designed drill program, drill metres, number of individual drill holes and the number of drill holes per drill site. This allows for budgeting the number of rigs, monthly or quarterly drill metres and staff necessary to support the program. With the mine now in development, the geologists can use the process to rapidly update the drill program to accommodate changes in design to the short-term plan.

The process is also flexible, enabling different scenarios to be tested and provide justifications of any adopted drilling approach. And, as demonstrated in the case study, the process can be used to propose design layouts that are likely to yield optimal drill locations.

Use of near mine exploration resource estimates in assessing strategic growth at the Super Pit Kalgoorlie

J Ireland[1]

1. FAusIMM, Geology Manager Discovery, JIreland Group, Northern Star Resources, Kalgoorlie WA 6443. Email: jireland@nsrltd.com

ABSTRACT

The Kalgoorlie Super Pit, operated by Northern Star Resources Ltd, is located adjacent to the city of Kalgoorlie-Boulder, Western Australia. The operation spans the renowned Golden Mile, a site that has yielded an impressive 60 million ounces of gold extracted from the Super Pit. In June 2014, the project had an estimated five-year mine life, with mining activity predicted to end in 2019.

In 2015, a study was conducted to assess the impact of incorporating all available drilling data into a resource estimate, with the aim of understanding and evaluating the potential growth of any future cutbacks. The deposit boasts a storied history, with a mining legacy spanning more than 130 years. With a long and varied ownership history, this has resulted in numerous discrepancies and inconsistent documentation of drilling and assay methods within the drilling database. Underground mining activities prior to 1988 often relied on partially sampled data, as the highly visible nature of mineralisation led previous companies to exclusively assay for gold within observable ore zones. Integrating this irregular sampling data into modern resource estimates has proved complex and the data has been omitted for Resource and Reserve reporting. This resulted in the omission over of 9886 holes for use in ore domaining. The study used Leapfrog software implicit modelling to encapsulate mineralisation greater than 0.9 g/t Au in partially sampled holes located outside mineralised domains. Three resource block models were created to assist in evaluating the full potential of the project.

Open pit optimisations conducted on the resource block models played a pivotal role in identifying the potential of the Fimiston South cutback. Positive results from the optimisations provided justification for exploration drilling to further evaluate the southern extents of the deposit. Over 232 728 m of drilling was completed between January 2017 and December 2022, resulting in the addition of over 8.8 Moz of Indicated and Inferred Mineral Resource declared in March 2021.

This work unlocked significant value and extended the mine life of the Fimiston open pit operation by more than ten years. It is an example of using all available data to create conceptual mineralisation models and guide critical exploration decisions within a well understood, world-class gold deposit.

INTRODUCTION

The Kalgoorlie Super Pit, operated by Northern Star Resources Ltd, is located adjacent to the city of Kalgoorlie-Boulder, Western Australia. The operation spans the renowned Golden Mile, a site that has yielded an impressive 60 million ounces of gold extracted mainly from the Fimiston-style mineralised lodes (Figure 1). The deposit boasts a storied history, with a mining legacy spanning more than 130 years. There is a wealth of available drilling data, however, a long and varied ownership history has resulted in numerous discrepancies and inconsistent documentation of drilling and assay methods within the drilling database. Underground mining activities prior to 1988 typically relied on partially sampled drill holes, as the highly visible nature of mineralisation led previous companies to exclusively assay for gold within observable ore zones.

The mining processes have evolved over time, ranging from narrow and highly selective underground mining to bulk mining methods such as longhole stoping and open pit mining using large mechanised equipment. Before the initiation of the Fimiston open pit in 1989, mining practices were selective, reflecting the higher head grade requirements prevalent at that time. The peak year of production was in 1903 when 1.27 million ounces of gold was produced at an average grade of 41 g/t Au (from WA Department of Mines 1903 Annual Report, in KCGM, 2022). Underground mining of the Golden Mile ceased in 1990 as the project transitioned to a large scale open pit mining operation, complemented by continued underground mining of the Mt-Charlotte deposits to the north. The principal ore source within the Fimiston open pit are the remnant low-grade haloes of the high-grade lodes mined historically from underground. The presence of numerous voids and the long history of mining present ongoing challenges for resource estimation.

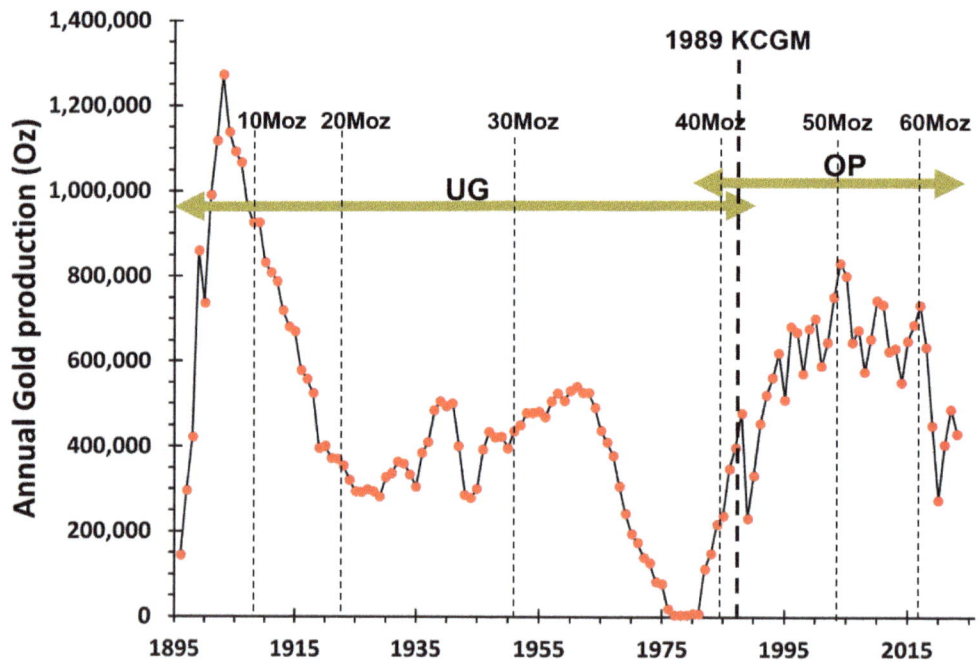

FIG 1 – Historical gold production from the Golden Mile (data from KCGM, 2022).

History

In June 1893, Irish prospectors Paddy Hannan, Thomas Flanagan, and Dan Shea discovered gold in Kalgoorlie, sparking a gold rush in what became known as 'Hannan's Find'. Sam Pearce and William Brookman named and staked the Golden Mile a few weeks later. Within five years, the Golden Mile had yielded half a million ounces of gold, leading to the establishment of the twin towns of Kalgoorlie and Boulder. Over the next decade, 49 mines, 42 mining companies, 100 headframes and over 3.5 km of underground workings were established across the bustling goldfield.

By 1950, four major companies controlled Golden Mile mining leases, dwindling to two by 1970, with only the Mt Charlotte underground mine remaining active at the northern end. The late 1980s witnessed the revival of the Golden Mile when Alan Bond aimed to consolidate leases with the vision of building a cost-effective, large-scale open pit mining operation. Bond's efforts led to an eventual Joint Venture of Gold Mines of Kalgoorlie (a Normandy Mining subsidiary) with Homestake Gold and the formation of the Kalgoorlie Consolidated Gold Mines (KCGM) management company in 1989.

In 2001, Barrick Gold Corporation acquired Homestake Gold, and Normandy Mining merged with Newmont Mining Corporation. Both entities retained the KCGM Joint Venture until late 2019. In November 2019, Saracen Mineral Holdings acquired Barrick's 50 per cent interest in the KCGM joint venture, followed by Northern Star Resources purchasing Newmont's 50 per cent interest in early December 2019. In February 2021, the merger of Saracen Mineral Holdings and Northern Star Resources resulted in the Golden Mile coming under single ownership for the first time in its history.

Regional and local geology

The Fimiston gold deposit is located in the Kalgoorlie district of the Eastern Goldfields Province, the most prolific gold producing region of the Archean Yilgarn Craton of Western Australia (Figure 2). The geology of the area has been well documented over the years by various authors.

The Kalgoorlie stratigraphic succession occurs in the lower part of the Archean Norseman-Wiluna belt and comprises a 3–4 km-thick basal sequence of ultramafic and mafic rocks, including the Golden Mile Dolerite, overlain by a 1 km-thick volcano-sedimentary sequence.

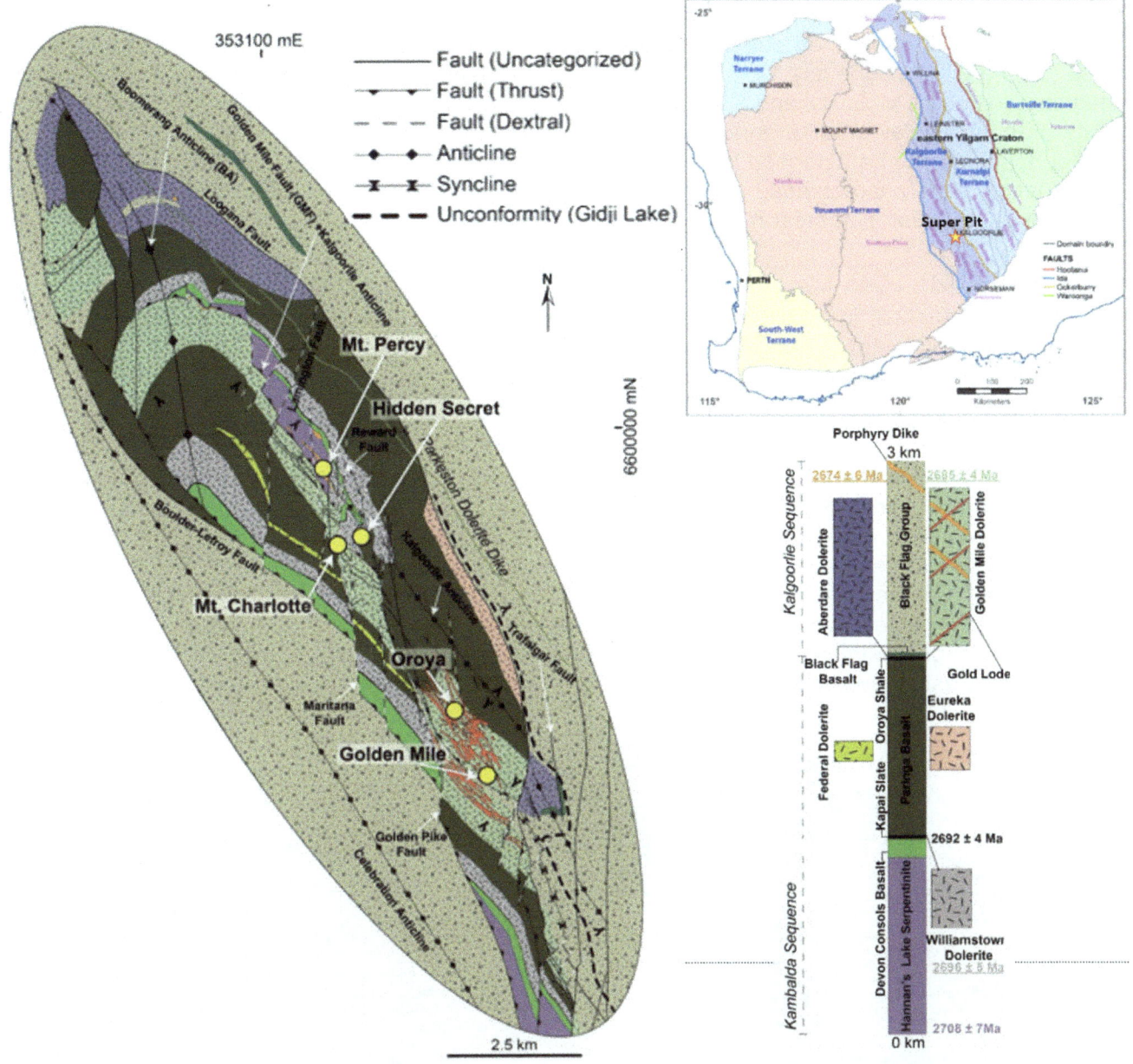

FIG 2 – Plan view of local the geology of Golden Mile (McDivitt *et al*, 2020) and South-western Australia Regional Geology (Czarnota, Blewett and Goscombe, 2010).

The regional deformation history includes D1 folding and thrusting, D2 upright folding about NNW-trending axes, D3 sinistral transcurrent faulting, and continued ENE-WSW regional shortening. Mafic volcanism occurred around 2.71 to 2.69 Ga, and the main regional deformation occurred between 2.68 and 2.61 Ga. The greenstone belt has been deformed and regionally metamorphosed to grades ranging from lower greenschist to amphibolite facies.

The Golden Mile gold deposit is situated on the limbs of an upright F1 syncline-anticline pair that was later folded into the sub-regional scale, north-west plunging F2 Boomerang Anticline (Tripp, 2013). The Golden Mile Dolerite is the principal gold host and comprises a 700 m thick layered tholeiitic sill emplaced at the contact between the Paringa Basalt and overlying sedimentary rocks of the Black Flag Group (Figure 2). The Paringa Basalt is a 400 to 750 m thick mafic volcanic sequence, grading from high magnesium basalt at the base to more evolved tholeiitic basalt towards the stratigraphic top. A 50 to 200 m thick high-iron tholeiite unit occurs at the top of the sequence and forms a consistent stratigraphic marker across the deposit (Gauthier, Hagemann and Robert, 2007).

The Golden Mile Dolerite has been subdivided in ten units based on petrographic and geochemical characteristics (Travis, Woodall and Bartram, 1971). Units 1 and 10 form the basal and upper chill margins respectively. Units 2 and 3, the basal pyroxene cumulate units, are characterised by very high Cr, Ni and MgO concentrations, with a decreasing trend upward into the intrusion. Units 4 and 5 consist of medium grained sub-ophitic textured gabbro. Units 6, 7 and 8 are the magnetite-rich units (10–15 wt% magnetite). Unit 6 is enriched in elements with a strong partition coefficient into magnetite, such as V, Cr, Ni and Cu. These elements are in turn strongly depleted in Units 7 and 8. Units 6 and 7 display gradual enrichment trends in Zr, TiO_2 and Fe_2O_3 towards Unit 8. Unit 8, the granophyric unit, is characterised by high SiO_2, Zr, TiO_2, P_2O_5 and Fe_2O_3. Unit 9 displays a gradual fractionation trend characterised by higher Zr, V, TiO_2 and lower Cr, Ni, MgO towards the contact with Unit 8 (Gauthier, Hagemann and Robert, 2007).

Mineralisation

Gold mineralisation at the Golden Mile is distributed along and mainly east of the NNW-trending Golden Mile fault system. Two predominant mineralisation styles are recognised: Fimiston Lodes and Mt Charlotte-type stockwork quartz-carbonate vein systems.

The Fimiston Lodes, which account for the bulk of economic mineralisation, are generally 1–2 m wide but locally up to 20 m wide and can extend over 1000 m along strike and up to 1200 m vertically (Figure 3). The lodes comprise gold and telluride-bearing quartz-carbonate veins that are enveloped by zones of disseminated carbonate-quartz-pyrite alteration. The Fimiston Lodes occur in three principal orientations: the dominant Main N140°/80°W and Caunter N115°/55° to 80°W Lodes, and the lesser Cross Lodes N050°/80° to 90° N-S (Finucane, 1948; Figure 3). Lodes to the west of the Golden Mile Fault display good lateral and vertical continuity whereas those in the Eastern Lode System are segmented by a series of steeply east-dipping reverse faults (Figure 3). Mineralisation style and lode geometry are controlled by a complex structural history, and by rheological and chemical contrasts between host lithologies.

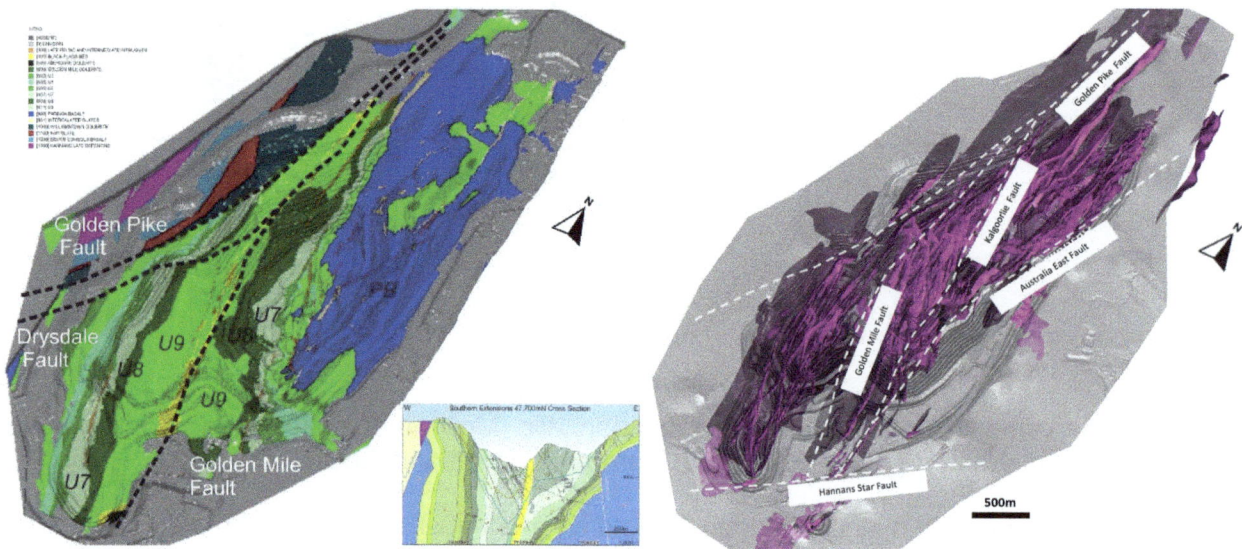

FIG 3 – Golden Mile lode model and geology.

Life-of-mine

In June 2014, KCGM annual mining production was approximately 70 million tonnes (Mt) with a milling capacity of 12 Mt, producing some 630 000 ounces of gold. Mining was carried out from the Fimiston open pit, with a small contribution from the Mt Charlotte underground mine. The Fimiston pit measured 3.4 km in length by 1.5 km in width and extended to a depth of approximately 400 m, to make it the largest open pit gold mining operation in Australia at the time. The Fimiston open pit was made up of several cutbacks and phases, namely Brownhill, Morrison, Chaffers, Trafalgar and Golden Pike. The Golden Pike and Morrison cutbacks represented the final pit expansions at the mine and extended the mine plan from year 2014 to 2019. In June 2014, KCGM had a total mineral resource of 4.6 Moz (KCGM, 2014).

Drilling

The history of drilling at the Golden Mile is intricately woven into the narrative of its rich mining heritage. Since the initial discovery of gold in 1893, drilling has played a pivotal role in delineating and extracting the valuable resource. Diamond drilling was used at Fimiston from the early days (Figure 4). In the 1970s two major innovations meant that diamond drilling became much more intensive in nature: the advent of the carbide-tungsten impregnated diamond bits and core tubes retrieved by wirelines. Historically, intervals of quartz-carbonate veining and visible wall rock alteration were selectively sampled. These partially sampled holes were documented in physical logbooks and later transcribed into databases in the 1980s. Sampling was done using manual or hydraulic core splitters until the introduction of the modern core saw in 1993.

FIG 4 – Diamond drill rig, underground at Fimiston (taken from WA Department of Mines, 1904).

Drilling database

Validation of KCGM drill hole data for use in resource estimation includes checking the accuracy of drill hole collar locations, depth, azimuth and dip, together with a QA/QC analysis of assay results. Drill holes are subsequently assigned a confidence code that reflects the reliability of these checks, as listed in Table 1.

TABLE 1
Drill hole confidence codes (Valliant, 2014).

Pre-1988 drill holes

	Confidence	Description
0	Unchecked or unreliable	Collar coordinates, assay and or downhole survey information has not been checked and/or is considered inaccurate
1	Semi-accurate	Significance (<2 m), discrepancy between collar location or survey data recorded in the original log and the drafted plan. Hole drilled from stope (ie RL uncertain). Collar coordinates and/or survey not recorded in original log
2	Reliable	Imperial coordinates correctly converted. Data matches plans. Assay intervals and conversion checked.
3	Accurate	Surveyed Oroya East Grid. Downhole surveyed. Fire assayed in g/t.

Post-1988 drill holes

	Confidence	Description
0	Unreliable	Collar coordinates unreliable or drill hole terminated prematurely and not assayed or QA/QC failed and not redeemable.
1	Semi-accurate	Azimuth data not collected from hole or no reliable downhole survey data or sample intervals uncertain. Sample QA/QC passed
2	Reliable	Collar coordinates known but azimuth data assumed from set out or limited reliable downhole survey data. Sample intervals certain. Sample QA/QC passed
3	Accurate	Reliable collar surveyed including azimuth reliable downhole. Surveys at least every 50 m downhole (ie unaffected by magnetic dolerite). Sample intervals certain. Sample QA/QC passed

Fully or partially sampled drill holes at a confidence level of two or three were used in the 2015 Fimiston SEP resource block models. This differed from the 2014 resource modelling process where only fully sampled holes above a confidence level of two were used. Grade control data was excluded from both data sets.

Partially sampled holes were mainly drilled from underground positions prior to the 1980s (Figures 5 and 6).

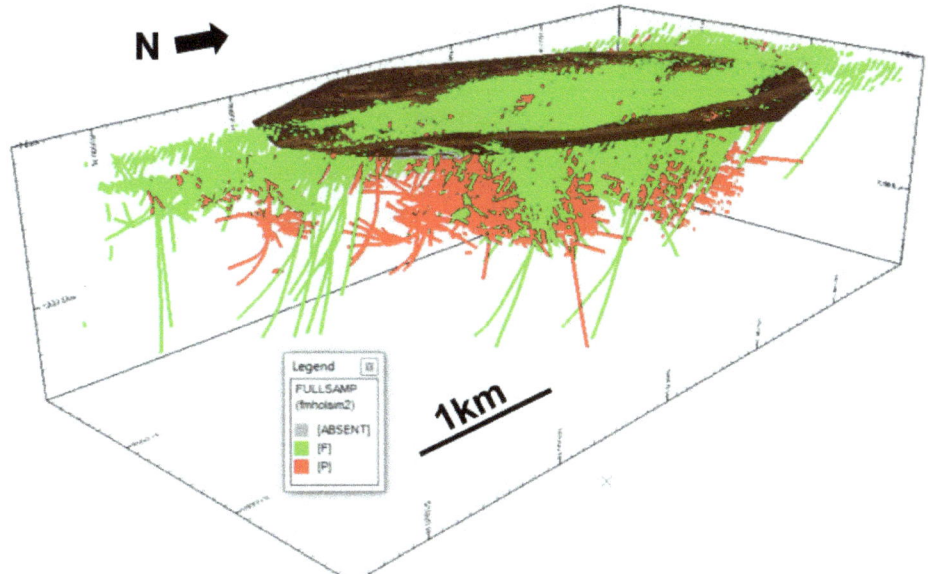

FIG 5 – View looking north-west showing location of fully sampled (green) and partially sampled (Red) holes relative to the 2014 Reserve pit shell.

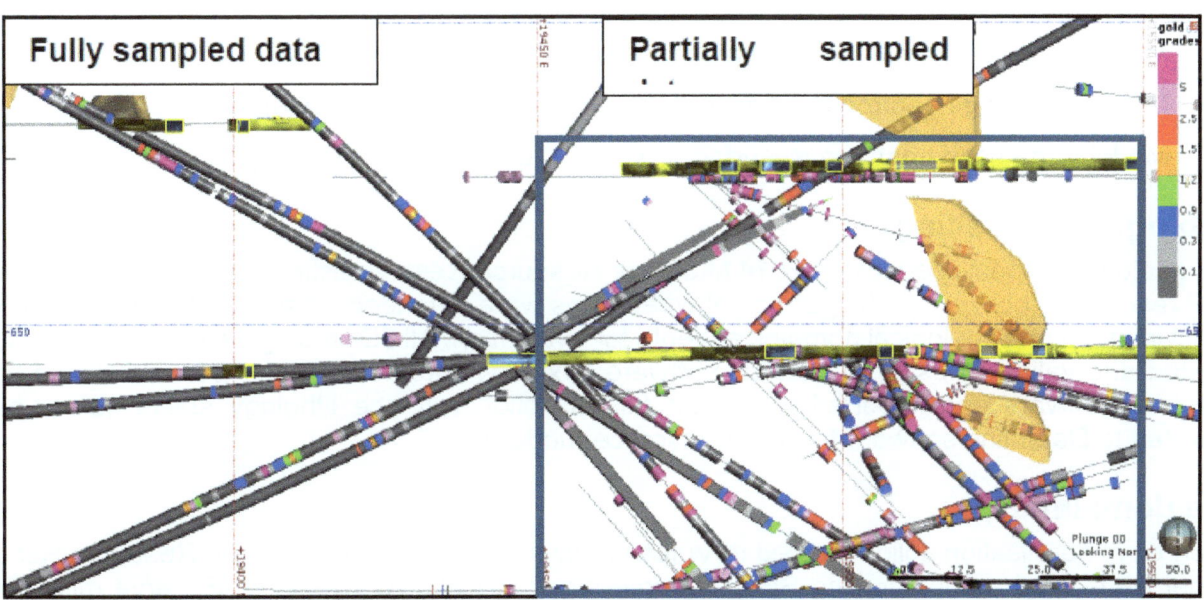

FIG 6 – Section along 19450E showing typical zone of fully sampled and partially sampled drill holes.

Geostatistical work was carried out in 1997 by Mining and Resource Technology to assign gold grades to the un-sampled intervals of partially sampled holes (Horton, 1997). This involved the use of Monte Carlo simulation to statistically analyse grade decay from the Fimiston main lodes. Grade generally decayed by 2 m and the predicted grades were generally low tenor (0.0–0.8 g/t Au). A default grade of 0.14 g/t was universally applied to all unsampled intervals of Mt Charlotte-style stockwork mineralisation, based on the assumption that gold in these systems is primarily hosted within quartz veins with no grade decay from the veins. The simulated grades were then incorporated into the KCGM drill hole database as a separate column titled Au SIM and used to create conceptual mineralisation models.

The inclusion of the partially sampled drilling resulted in the addition of 9886 holes and 483 996 samples for use in ore domaining (Table 2).

TABLE 2

Summary of the final drilling database used for the SEP resource block models.

Hole type	Number of holes	Number of samples
Partially sampled confidence 2	9872	482 252
Partially sampled confidence 3	14	1417
Fully sampled confidence 2	32 503	731 433
Fully sampled confidence 3	75 894	851 698
Total	118 283	2 066 800

Modelling

In March 2015, a resource block modelling process was initiated to address a diminishing mine life and uncertainties surrounding the future of the Fimiston operation. Three Strategic Exploration Planning (SEP) resource block models were developed as integral components of the November 2015 Mid-Year Fimiston All Data Inventory Model. These models combined actual assay data with the simulated grade data generated for the partially sampled drill holes. The three models each considered a 'what if' scenario for mineralised drill hole intercepts located outside modelled mineralisation domains:

- Model one used the 2014 Mid-year Resource model (Base case).
- Model two used Leapfrog isosurfaces with an 80 m search (Mid case).
- Model three used Leapfrog isosurfaces with a 120 m search (Blue sky).

Geological modelling

The geological model used for the 2014 Mid-Year Resource/Reserve estimation was adapted for the SEP resource block model. All geological wireframes were extended from -900RI to the -1600RI to assist in projecting geological domains at depth for evaluation. All 3D lithology wireframe solids were created and validated using Vulcan 3D software. Oxide wireframe surfaces were produced using Leapfrog software and imported to Vulcan for validation with the lithology solids before being exported to Datamine software for use in resource modelling.

Ore domaining

Open pit mineralisation solids created in Vulcan software for the 2014 Mid-Year Resource/Reserve model were used as built for the SEP resource block models. All Resource/Reserve mineralised zones were interpreted using a nominal cut-off grade of 0.3 g/t Au. The section intervals for interpretation are typically 20 m for well-drilled areas, and out to 100 m for peripheral areas with sparse drilling. Bench ore block plans at 10 m intervals and underground level plans at approximately 30 m intervals were also used to guide the interpretation as were historically mined out shapes.

Over 32 700 samples grading above 0.9 g/t Au were identified outside of the modelled mineralisation wireframes. Utilising numerical modelling in Leapfrog software, grade intercepts external to the existing Resource/Reserve domains were captured in wireframes using Leapfrog numeric interpolants. The lodes were captured using a series of search parameters where grades were captured and projected 80 m and 120 m along trend from the isolated intercepts.

Leapfrog mineralisation domains was constrained by stratigraphy based on host rock and spatial location relative to the Golden Mile Fault, Golden Pike Fault and Drysdale Fault. Six geological domains were defined:

1. Western Lodes are hosted by Golden Mile Dolerite and are located on the western side of the Golden Mile Fault.

2. Eastern Lodes are hosted by Golden Mile Dolerite and are situated on the eastern side of the Golden Mile Fault.

3. Northern Lodes are hosted by Paringa Basalt.
4. North-eastern Lodes are hosted by Golden Mile Dolerite to the north-east of the Drysdale Fault.
5. Stockwork domains along the Drysdale Fault.
6. Stockwork domains along the Golden Pike Fault.

This approach aimed to provide geological constraints for currently unclassified and unconstrained material in the waste domain without the immediate need for manual wireframing.

The assignment of mineralisation domains for the Leapfrog derived domains was based on host rock and spatial location relative to the Golden Mile Fault, Golden Pike Fault and Drysdale Fault. Seven domains were defined based on their structural position as outlined in Figure 7.

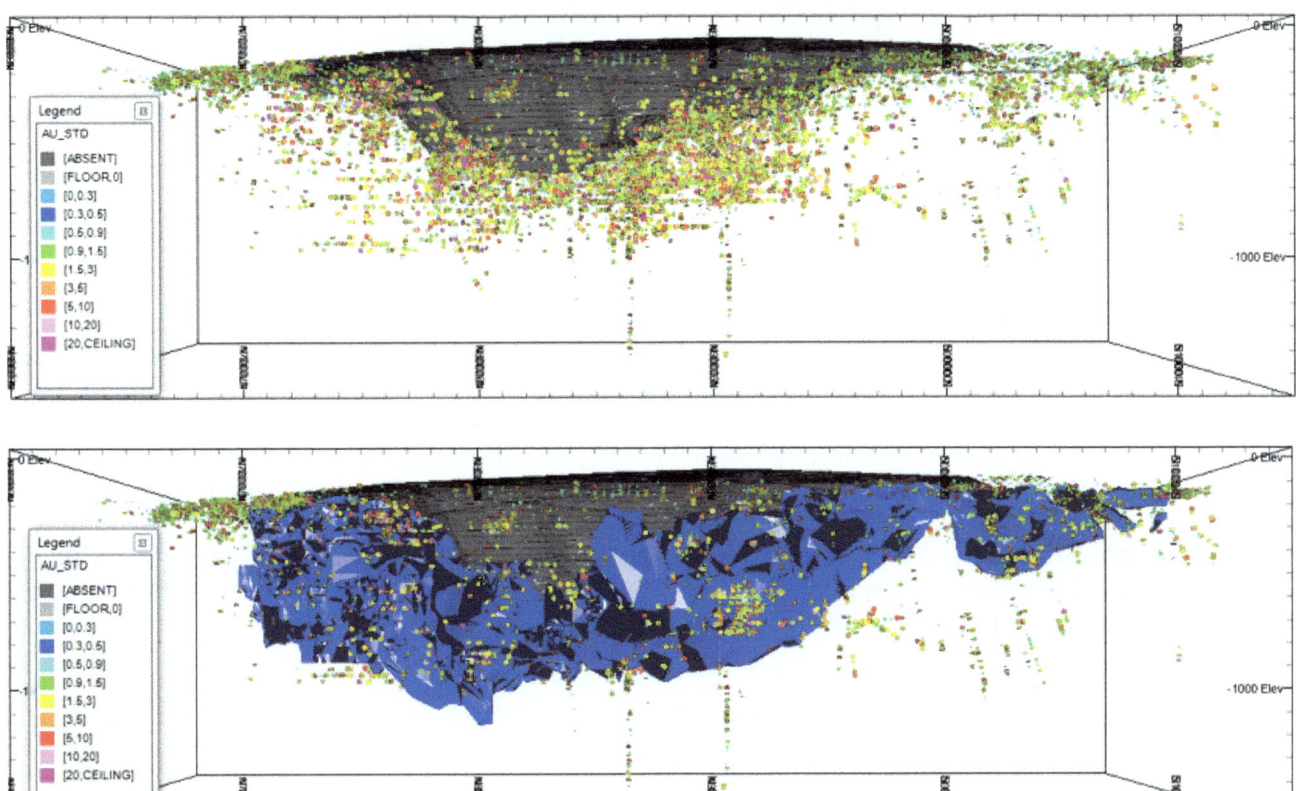

FIG 7 – Distribution of 32 700 samples of grade >0.9 g/t Au outside mineralisation envelopes.

Midpoints along the resource lode wireframe interpretations allowed for the generation of wireframes that were then used to guide the numerical searches. Each geologically domained area had two grouped 0.9 g/t Au Leapfrog isosurfaces: Leapfrog scenario one, with isolated grade extrapolated 80 m and Leapfrog scenario two, with isolated grade extrapolated 120 m (Figure 8).

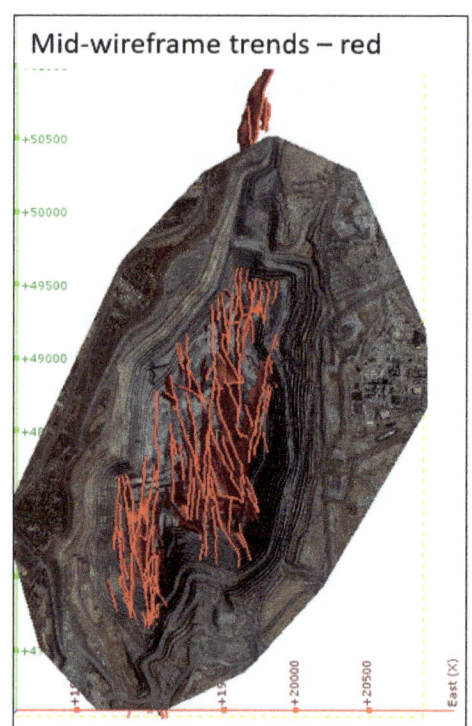
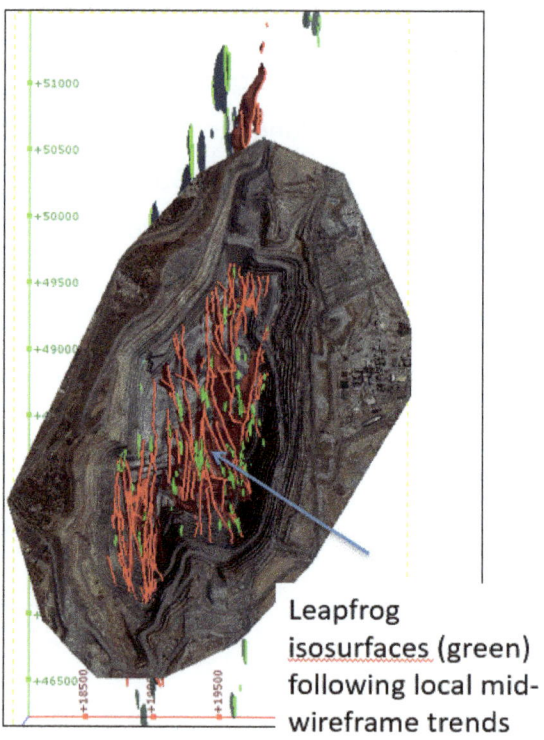

FIG 8 – Structural trends were generated from the midpoints of the resource wireframes and used to guide the Leapfrog isosurfaces.

Not every isosurfaces generated with the partially sampled assays was used due the sheer volume of isosurfaces required to encapsulate the 32 700 samples unmodelled. Isosurfaces with a volume of greater than 40 000 m^3 were used for block modelling (approximately 300 new domains), Figure 9. It is crucial to emphasise that the resulting Leapfrog-derived mineralisation domains were not intended as wireframes of resource model quality but rather as evaluation tool for sensitivity studies, leveraging all available data.

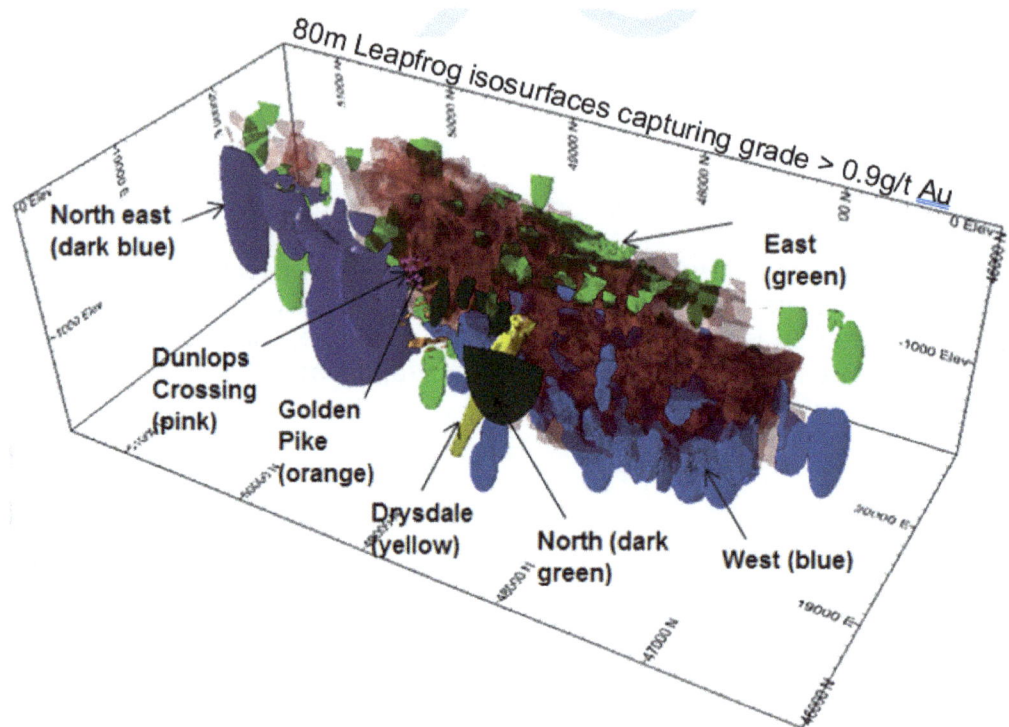

FIG 9 – Rotated view looking east with grades >0.9 g/t Au captured by Leapfrog relating to their geological positions projected 80 m.

Estimation

The estimation method used for the 2014 Mid-Year Resource/Reserve model was maintained for the SEP resource block models. The model included estimating the additional Leapfrog isosurfaces and assigned the average grade of the lode to wireframes, if it did not have sufficient samples. Prior to the study the lodes were assigned a grade of 0.14 g/t Au if insufficient samples were available. The block model extents were also extended from -900RI to -1600RI.

Ordinary Kriging was not attempted for the Leapfrog derived domains due to the large number of leapfrog wireframes. All estimates were completed using Inverse Distance squared. This method was selected due to the time required to estimate each new Leapfrog wireframe.

Optimisations

Prior to the SEP resource block models of 2015, KCGM optimisations indicated no possibility of extending the open pit mine at depth. However, pit boundaries were restricted by the limits of the drilling, the limits of the block model and lode wireframes, and the parameters used in the resource and reserve block model for pit optimisations.

The SEP resource block models differed from previous models by the inclusion of all inferred and unclassified data for optimisations. For previous optimisations these were all assigned a grade of 0.14 g/t Au. A gold price of A$1515 was used for the evaluation based on the gold price in Nov 2015. Underground evaluations were conducted as a separate study.

All three SEP resource block models indicated that the area just south of the pit had not been adequately tested by drilling and that it held a potential of between 6 and 11 million ounces below the 2015 mined pit shell. It was concluded that if estimated resources under the 2015 pit were increased by up to 10 per cent then the life-of-mine could be increased by over ten years. Following numerous reviews and reiterations of the models, the optimisations led to a large-scale drilling campaign at Fimiston South, which was initiated in January 2017 (Figures 10 and 11).

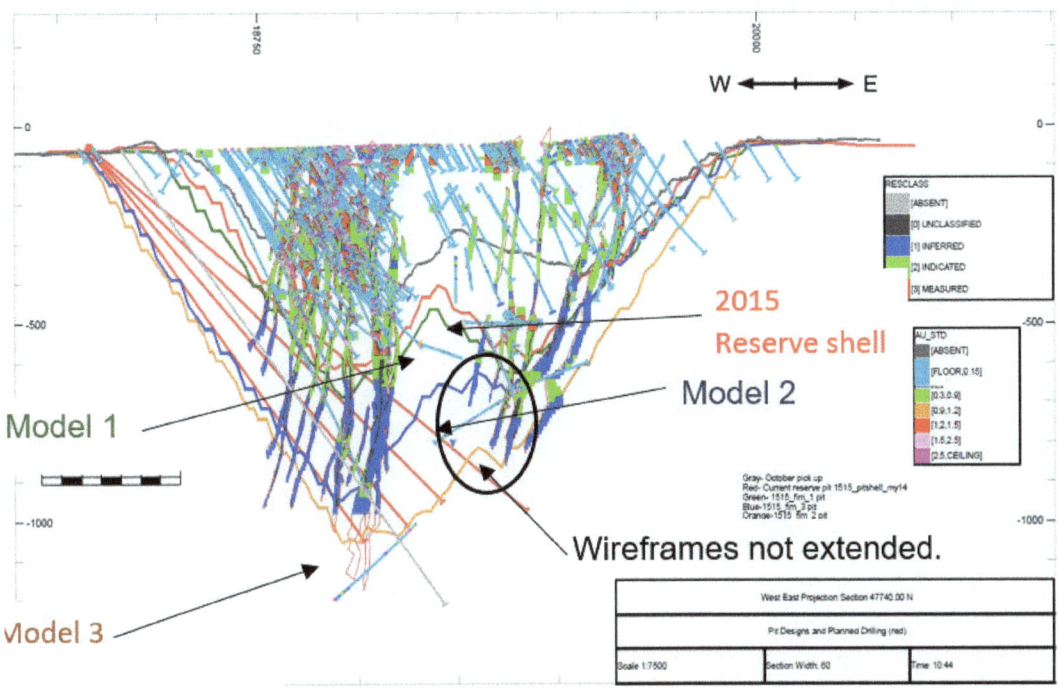

FIG 10 – Cross-section from 2015 (477400 N) showing pit optimisations at A$1515 grade shell optimisations with planned drilling (red) targeting inferred material (blue) to be upgraded. With lodes not extended at depth.

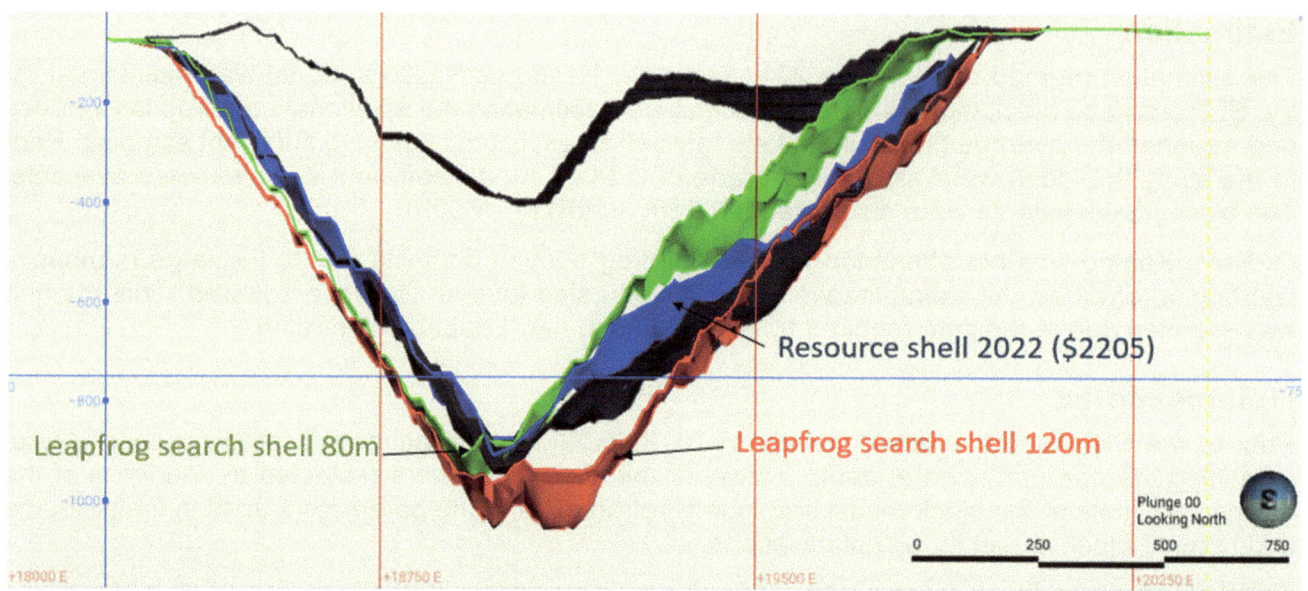

FIG 11 – Cross-section (47600 N) showing pit optimisations in comparison to old optimisations.

Drill planning

To define the potential of the Fimiston South cutback, a drilling program comprising a series of parent and daughter holes was designed to further define lodes at depth within the SEP Blue Sky Model (120 m continuity). Drilling through numerous historical stopes had never been attempted on such a large scale at the Fimiston deposit and it was unclear if the program would be successful.

Reaching target areas was challenging with diamond drilling intersecting multiple open stopes at depth (up to 13 open stopes in a single drill hole) and historical underground infrastructure such as wooden beams and drilling carts. Intersecting planned targets was also a challenge due to a lack of optimal drilling platforms, particularly for the Eastern Lode System. Navigational drilling techniques were implemented to direct drill holes towards targets.

Multi-element and spectral data was collected at 5 m intervals in every hole to allow for the discrimination of local geological units and for the characterisation of hydrothermal alteration assemblages (Figure 12). This in turn helped improve the understanding of the geology and mineralisation.

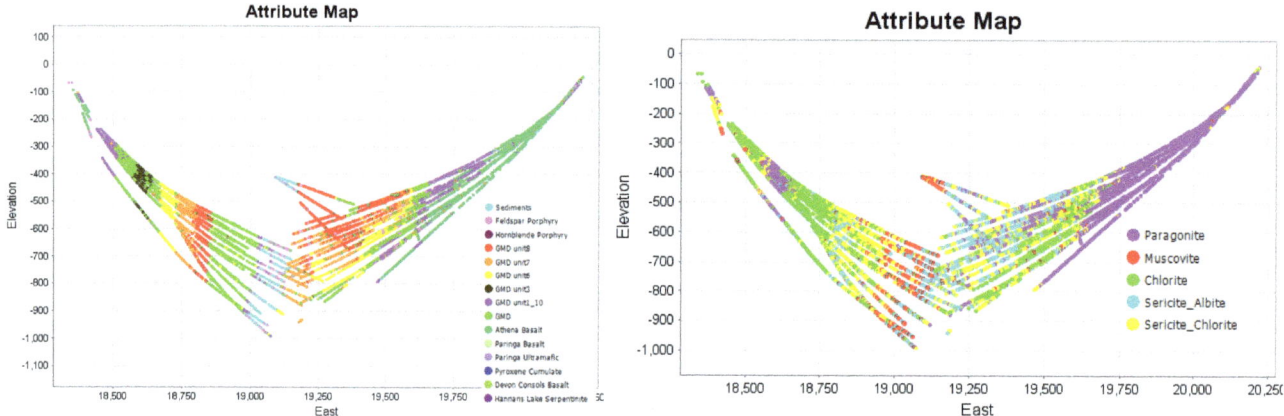

FIG 12 – Drilling from 2017 to 2021 though Fimiston South section 47700N with one in five multi-element and spectral sampling.

Results

Drilling confirmed the presence of the interpreted lodes and assay results were generally in line with the statistical simulations. Assay results confirmed the presence of mineralisation at depth and drilling continued until beyond Dec 2021 (Figure 13).

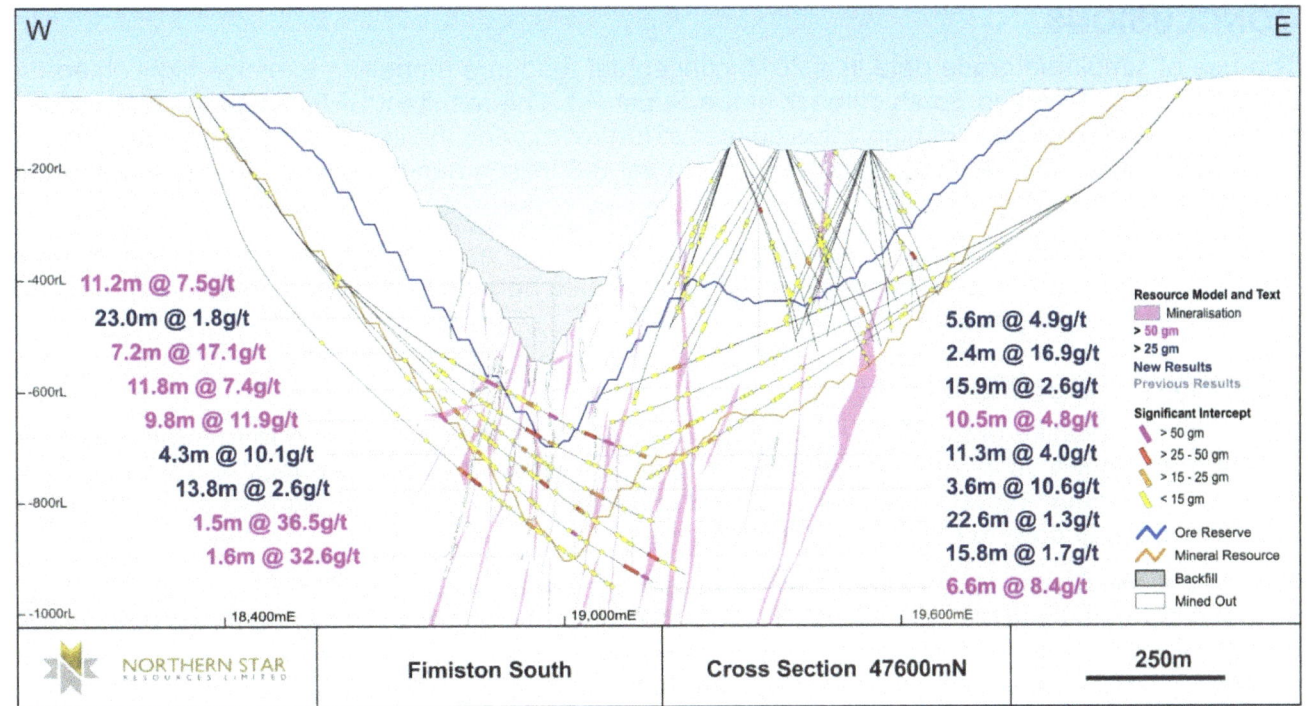

FIG 13 – Cross-section (47600 N) showing drilling results (Northern Star Resources Limited, 2021).

In December 2018, a maiden resource was released for the Fimiston South for a total of 3 Moz Au (KCGM, 2018). Drilling over five years has increased the resource of the Fimiston South to a total of 8.8 Moz (Table 3). Results were released in May 2021 (Northern Star Resources Limited, 2021).

TABLE 3

Resource development drilling metres over six years at Fimiston South.

Year	Diamond holes	Diamond (m)	RC holes	RC metres	Total (m)	Oz increase per annum
2017	54	45 623	1	270	45 893	0
2018	80	40 073	14	3736	43 809	3 014 433
2019	102	55 495	0	0	55 495	1 293 367
2020	43	23 221	71	23 181	46 402	1 792 200
2021	57	24 101	63	17 028	41 129	2 711 000
Total	336	188 513	149	44 215	232 728	8 811 000

Results released by Northern Star Resources Limited on 31 March 2021 for Fimiston South was a total Reserve of over 5.4 Moz and a total of 8.8 Moz in Mineral Resources extending the Super Pit Life-of-mine to 2034 (Northern Star Resources Limited, 2021).

Exploration drilling programs arising from the SEP resource block modelling exercise had the following benefits:

- Areas with the highest economic potential were targeted.
- Planned drill holes fell within realistically mineable envelopes.
- Open cut versus underground mining boundaries were defined.

CONCLUSIONS

The use of simulated grade data in a 2015 conceptual resource modelling exercise highlighted the potential for the Fimiston South cutback at the Super Pit. The resultant SEP resource block models assisted in targeting areas of high value ounces to further expand the life-of-mine. Exploration drilling programs justified by the study added a total of over 8.8 million ounces over a five-year period and extended the mine life until 2034.

SEP resource block models in well-known deposits offer an effective way to assess various mineralisation continuity scenarios and upside potential to help drive mine exploration and resource development drill programs.

ACKNOWLEDGEMENTS

The author would like to thank Northern Star Resources for their permission to use the data from KCGM in this publication. And would also like to acknowledge the many geologists, field staff, core yard staff, environmentalists, metallurgists, and engineers whom have worked over the Golden Mile and helped the deposit to grow in all directions.

The author would like to acknowledge the numerous teams who worked on the Fimiston South drilling programme and their dedication. Without the continued efforts of all the teams the programme could not have occurred. A special thanks to Ibrahim Ormai and Josh Walley who worked closely in the initial stages of the study, with much collaboration and enthusiasm. Thank DDH1, AusDrill and Topdrill drilling companies' for their persistence in pushing drilling though large stopes during challenging times.

REFERENCES

Czarnota, K, Blewett, R S and Goscombe, B, 2010. Predictive mineral discovery in the eastern Yilgarn Craton, Western Australia: an example of district scale targeting of an orogenic gold mineral system, *Precambrian Research*, 183:356–377.

Finucane, K J, 1948. Ore distribution and lode structures in the Kalgoorlie Goldfield, in *Proceedings of the Australasian Institute of Mining and Metallurgy*, pp 111–129.

Gauthier, L, Hagemann, S and Robert, F, 2007. The geological setting of the Kalgoorlie Golden Mile Gold Deposit, Kalgoorlie, WA.

Horton, J, 1997. Fimiston Default Grade Simulation, Internal report.

KCGM, 2014. KCGM Processing Gold to 2029, media release, 5 February 2014, Kalgoorlie Consolidated Gold Mines. Available from: <https://s24.q4cdn.com/382246808/files/doc_news/archive/CR_MER_140205_KCGMProcessingGoldTo2029.pdf>

KCGM, 2018. Competent Person Report, unpublished internal report, Kalgoorlie Consolidated Gold Mines.

KCGM, 2022. Data from WA Department of Mines from 1896 to 1982, and KCGM internal data from 1982, unpublished internal report, Kalgoorlie Consolidated Gold Mines.

McDivitt, J A, Hagemann, S G, Baggott, M S and Perazzo, S, 2020. Geologic Setting and Gold Mineralization of the Kalgoorlie Gold Camp, Yilgarn Craton, Western Australia, *SEG Special Publications*, 23:251–274.

Northern Star Resources Limited, 2021. The Australian Gold Miner - For Global Investors, KCGM Site Visit, May 2021. Available from: <https://www.nsrltd.com/investor-and-media/asx-announcements/2021/may/investor-presentation-kcgm-site-visit>

Travis, G A, Woodall, R and Bartram, G D, 1971. The geology of the Kalgoorlie Goldfield, in *Symposium on Archaean Rocks* (ed: J E Glover), Geol Soc Australia, Pub 3, pp 175–190.

Tripp, G, 2013. Stratigraphy and structure in the Neoarchaean of the Kalgoorlie district, Australia: critical controls on greenstone-hosted gold deposits.

Valliant, W, 2014. Technical report on the Kalgoorlie Consolidated Gold Mines, Western Australia, unpublished, 14 p.

WA Department of Mines, 1904. Annual report, J J Dwyer collection, 5816B/275.

Mine-scale reflection of a multi-stage regional structural history in the orebody geometry of Cutters Ridge Open Pit at Mungari, Western Australia

R Lumaad Paras[1], R Gordon[2], R Gulay[3] and J Strang[4]

1. Project Mine Geologist, Mungari Operations WA 6430. Email: rogie.lumaadparas@evolutionmining.com
2. Geologist, Xirlatem, Kalgoorlie WA 6430. Email: rick@xirlatem.com.au
3. Mine Geologist, Mungari Operations WA 6430. Email: rosanna.gulay@evolutionmining.com
4. Formerly Senior Mine Geologist, Mungari Operations WA 6430.

ABSTRACT

The regional structural history of the Kalgoorlie Terrane is now generally accepted and fairly uncontroversial (even if not always consistently named and described), but the practical application of this deformation regime is usually limited to interpreting and ranking exploration projects. The Cutters Ridge deposit at Mungari, provides a rare picture of most of the regional structural history in a single deposit that is mappable at the mine scale. Detailed mapping in the relative freedom of the immediate post-production environment was able to identify these important structural relationships and explain why Cutters Ridge was able to form at an economic scale and continuity at this location. Understanding such relationships can be applied to the exploration and development of extensions, repeats and analogues. Most importantly, however, this exercise has demonstrated that there is an enduring value to diligent mapping through increasing the production team's confidence and ability in recognising the importance of key geological relationships.

The granophyric zone of the Powder Sill Gabbro is a favourable site for gold mineralisation due to the concurrence of favourable iron-rich chemistry and rheological tendency for catastrophic brittle failure. Mapping at Cutters Ridge shows that an early faulting event propagated a fault along the sill margin placing the reactive granophyric zone against a substantial fault zone. The main mineralising event at Kundana, has later reactivated that architecture with high fluid pressure along the fault, driving fracturing and a loosely aligned vein network (ie a poorly developed stockwork) within the reactive granophyric zone. The low angle of incidence of the reactivated mineralising fault has resulted in 400 m of strike extent within a zone that was otherwise too thin (~20 m thick) to host significant mineralisation. A post mineralisation D_4 fault is mapped at the southern end of the orebody offsetting the main lode.

The practicalities of mapping in an active production environment mean that restricted access and superseding production priorities result in discontinuous mapping that can ultimately bring together a spatially correct map but may miss identifying the geological concepts relevant to mineralisation. Rather than dismissing the pit as mined, the Mungari Open Pits team built on the real-time mapping with a detailed post-production mapping campaign that focused on identifying key geological relationships that are very difficult to discern with real-time mapping. Most importantly, having the frontline geologists mapping in a pressure-free environment and concentrating on geological relationships that control mineralisation is a valuable development opportunity. The Mungari Open Pits team is now applying these skills in a real-time setting and under the day-to-day production pressures to better understand the mineralisation context of the next pit project and ultimately refine their modelling and ore-waste discrimination.

INTRODUCTION

Cutters Ridge is 28 km west of Kalgoorlie, located between the Kunanalling and Zulieka Shear Zones of the Yilgarn Craton (Figure 1). The 450 m long open pit, with a 180 m long satellite pit operated from April 2020 to February 2023 producing 2.6 Mt @ ~1 g/t for 86 koz.

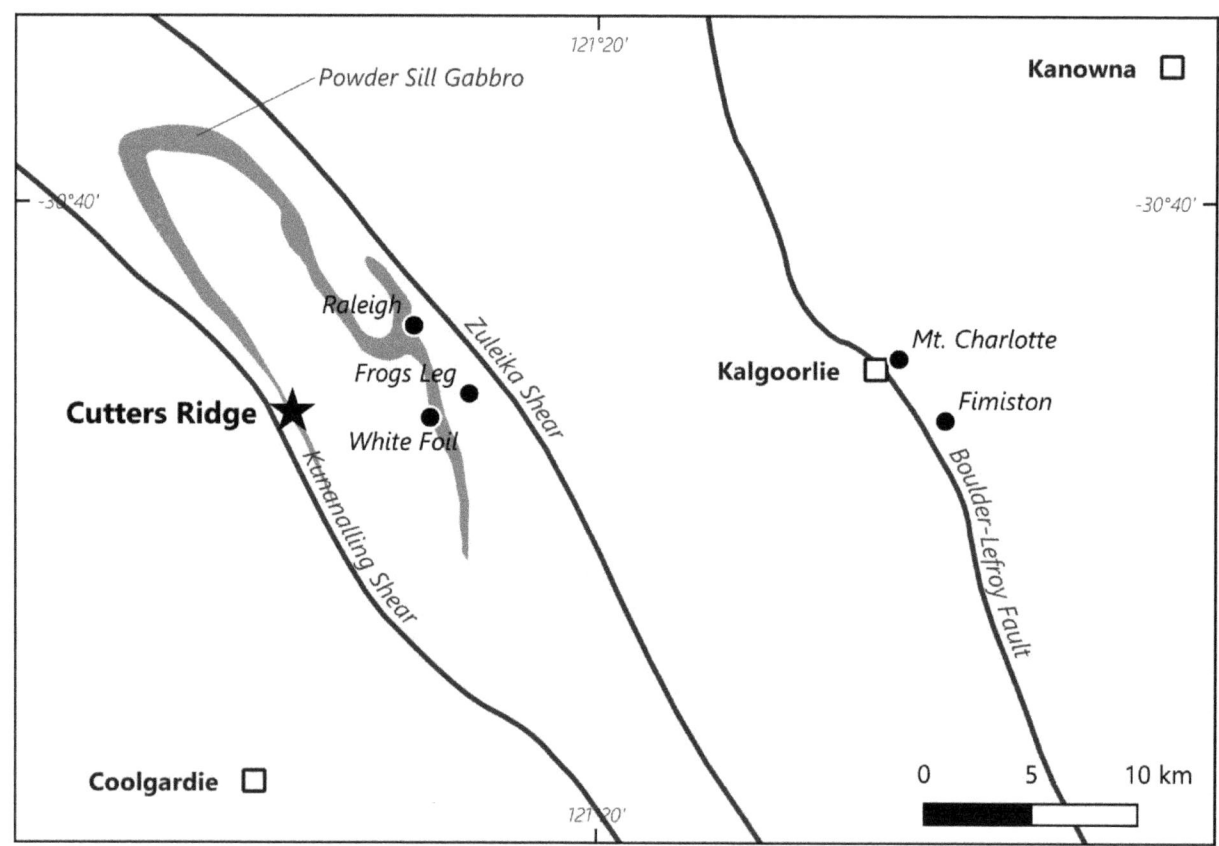

FIG 1 – Cutters Ridge Open Pit is located on the eastern edge of the Powder Sill Gabbro, 25 km west of Kalgoorlie.

Cutters Ridge mineralisation is a network of semi-planar quartz-pyrrhotite-pyrite (±scheelite, galena and sphalerite) veins hosted in a strongly differentiated mafic intrusive sill, the Powder Sill Gabbro. This mineralised zone is around 25 m wide extending 400 m along strike on, or close to, the eastern margin of the mafic intrusive. The network veins commonly have strong albite haloes extending up to 20 cm into the host rock in addition to a ten-metre-scale pervasive, but less intense, albite alteration zone that gives a bleached appearance to the gabbro and roughly correlates with gold mineralisation.

The mining was supported by a mine geology team of nominally six personnel across two rosters who conducted mapping on an ad hoc basis after general grade control and reconciliation duties. Whilst mineralisation is visual at Cutters Ridge through abundant veining and albite bleaching, the ore-waste cut-off grade is inside the mineralisation envelope and not visually apparent. As such, grade control relied on reverse circulation (RC) grade control drilling and pre-defined dig blocks from the grade control block model rather than mapping or ore spotting. The grade and tonnage of dig blocks defined from the grade control block model then formed the basis for Mine to Mill reconciliation. Under this process, geological pit mapping was always part of the mine geologist's duties with the aim of better informing the modelling, but it was never a time-sensitive or priority part of that process.

The tight geometry of both mineralisation and the pit design made mapping access during production particularly problematic for Cutters Ridge. The resultant map coverage was consequently quite disjointed which prompted a decision to launch a dedicated mapping campaign immediately post-production. The dedicated campaign proved to be a very valuable exercise and revealed geological relationships that demonstrate three of the four main regional deformation events of the Kalgoorlie Terrane exposed at a mesoscale. This exposure of the regional structural history in a single pit and consequential understanding of the structural setting of Cutters Ridge in turn allows for a deeper understanding of why mineralisation was able to form at an economic scale and continuity at this location.

METHODOLOGY AND RATIONALE

Whilst mapping was completed during the mining of the Cutters Ridge open pit, it was as a series of individual maps of exposures available at the time of mapping. It would have been possible to pull these individual maps together with the assistance of drill data to produce a spatially correct map, but that would provide little coherent insight into the geological relationships to further explain and understand the mineralisation at Cutters Ridge to assist future development of extensions and analogues. A dedicated mapping campaign guided by an external structural geology specialist aimed to map the entire pit in a consistent manner to achieve this deeper level insight.

As the pit was nearing completion, the opportunity was seized to map immediately after mining equipment was pulled out to achieve unhindered access free from production pressures and thereby achieve the best possible map and maximise the training and development aspects of the exercise.

GEOLOGICAL FINDINGS

With ad hoc mapping already available, the post-production mapping campaign focused on identifying important geological relationships that explain the location and geometry of the Cutters Ridge mineralisation (Figure 2). Important observations are:

- The eastern contact of the sill is not an intact intrusive contact but rather an array of fault slivers as evidenced by siltstone-sandstone rocks occurring on either side of lenses of gabbro with strong subhorizontal ductile fabrics near perpendicular to the fabric in the sedimentary rocks on either side (Figure 3).

- The faults creating those slivers run approximately along the eastern margin of the differentiated gabbro but cut into it at a very low angle in a way that has sliced off the outer differentiated zones of the gabbro. This has resulted in the granophyric zone, the most conducive to hosting mineralisation, having over ~200 m of strike exposure against these faults despite having a true width of only 15–25 m.

- The granophyric zone of the gabbro that hosts most of the mineralisation at Cutters Ridge is a 100 m-scale lensoidal geometry resulting from primary differentiation of the intrusive body. This differentiation is reflected in similar small scale differentiation processes that resulted in 10 cm-scale similarly shaped lenses of leucocratic (feldspar-rich) gabbro within melanocratic (feldspar-poor) zones (Figure 4).

- Mineralised veins occur as an array of thin planar veins (Figure 5) that are variably oriented when viewed in the horizontal plane but are all steeply dipping in the third dimension. This vein geometry is a stockwork in 2D only (referred to by site geologists as a 'pseudo-stockwork') and means that veins are generally oriented about a single subvertical axis that likely approximated the σ_2 stress direction during faulting, fracturing of the surrounding rock and vein infill. This 2D (rather than 3D) variability in vein orientation implies syn-mineralisation strike slip displacement along the controlling fault system, potentially with a Riedel type geometry to fracture initiation in the σ_1–σ_3 plane (Robert and Poulsen, 2001; Jacques, Muchez and Sintubin, 2022).

- Lower density veining, reflected in lower gold grades, correlates with areas of the granophyric gabbro lens that become increasingly distal from the fault planes as the two features are not exactly parallel and diverge from one another toward the north.

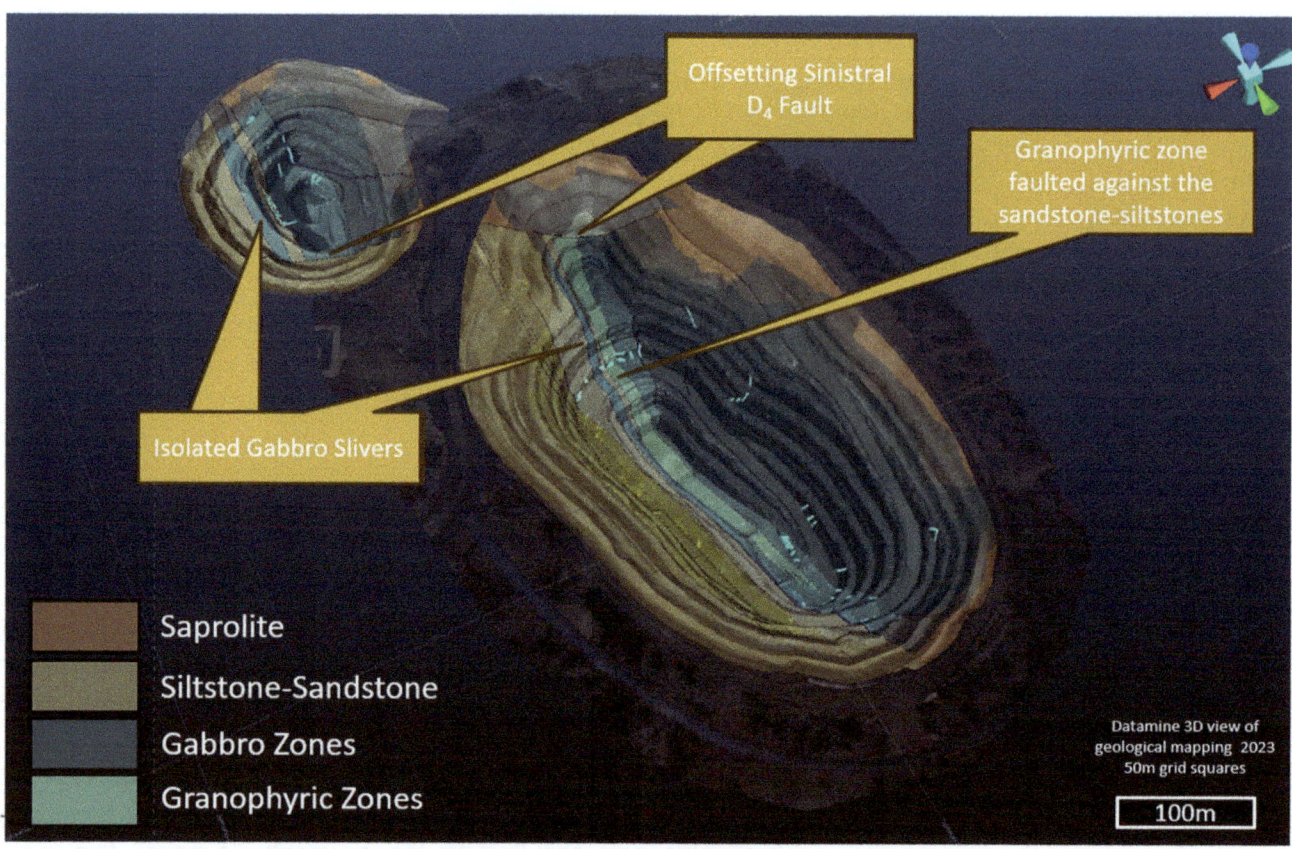

FIG 2 – The final product of the Cutters Ridge Mapping Campaign with some of the key relationships highlighted.

FIG 3 – A fault sliver hosts strongly S_2 foliated gabbro at a high angle to the faults positively identifying those faults as D_3 timing.

FIG 4 – Centimetre scale leucocratic lenses in the melanocratic gabbro exposed within the pit provide a useful small-scale analogue with which to visualise the hundreds-of-metre-scale granophyric lens that extends across and beyond the entire Cutters Ridge pit.

FIG 5 – The network veining Cutters Ridge mineralisation consists of an array of planar centimetre-scale veins, the vast majority of which are steeply dipping but variably oriented in the horizontal plane. A pit exposure of this veining in a low-grade mineralised zone is shown (a); beside a stylised 3D depiction with the same viewing aspect (b). Approximately 20 cm albite alteration halos are common around many of the gold mineralised veins (a).

The regional structural history of the Kalgoorlie Terrane is well understood in a regional context although the terminology varies significantly (Swager, 1997; Vielreicher, Groves and McNaughton, 2016). An early north–south shortening event penecontemporaneous with granite doming is not apparent at Cutters Ridge, but all other major deformation episodes are:

- The regional D_2 ENE–WSW shortening event is responsible for the earliest phase of faulting seen in the pit. A major fault at a very low angle to the eastern gabbro intrusive margin has sliced across the sill displacing the outer differentiated zones and juxtaposing the reactive granophyric zone against the sedimentary wall rock and a potential fluid conduit, the fault zone. D_2 is also responsible for the main ductile fabric in the gabbro, S_2.

- During D_3-related east–west shortening, the D_2 fault zone has been reactivated resulting in disoriented slivers of S_2 foliated gabbro positioned between the planes of the pre-existing fault system. High fluid pressures during D_3 have heavily fractured the surrounding rock (both the sedimentary package to the east and gabbro to the west) with a steeply south plunging common axis to these veins likely reflecting the σ_2 direction during fracture and vein formation. This network of veins is the mineralising event, with the zones of higher density of fracturing and veining producing the orebody. The more brittle rheology of the granophyric zone has favoured mineralisation.

- The dominant structural feature of Cutters Ridge, a sharp sinistral fault that offsets the southern tip of the orebody by 250 m, is a characteristic feature of the regional D_4 deformation.

In terms of both timing and structural context, Cutters Ridge is comparable to the nearby Raleigh deposit despite what the distinctly contrasting mineralisation styles might first suggest (Raleigh being narrow vein and Cutters Ridge being stockwork). Both deposits are D_3 mineralisation at ~2640 Ma (Vielreicher *et al*, 2015) in association with fault slivers in a fault array running approximately along the margin of the Powder Sill Gabbro.

Another important understanding gained from this piece of work is that Cutters Ridge has been able to accommodate a sizeable stockwork orebody (400 m long) in a mafic intrusive with a very thin (~20 m) granophyric zone due to the very low angle of incidence of the mineralising fault against that zone. The inherited D_2 fault architecture has placed the most favourable unit for mineralisation, the granophyric zone, against what was to become in D_3, a fluid rich mineralising fault system.

The mapping observations underpinning this deeper understanding of Cutters Ridge are not beyond the skillset of an undergraduate-qualified mine geologist, but are a step beyond spatially correct rock type mapping. Elements such as delineating the rather diffuse boundaries of the granophyric zone, tracking the ductile fabrics either side of faults and collecting a single consistent set of vein orientation data across the pit would have been very difficult in a production environment. It is unlikely that these critical observations, which ultimately pull the story together of why Cutters Ridge is what it is, would have been made in a production setting with time pressures, discontinuous and interrupted access and alternating rosters.

AN INVESTMENT IN HUMAN CAPITAL

Ore-waste discrimination, whether it be by grade control modelling or defining dig blocks, is one of the primary responsibilities of any mine geology team and a deep understanding of the orebody is critical to taking this responsibility beyond the coarse, 20 m-scale (ie drill spacing) lode geometry (Cooke and Fitzgerald, 2011). Actively seeking to understand the geological relationships that influence the fine-scale geometries and distinguishing features of mineralisation is a skill that can be greatly enhanced by comprehensive pit mapping.

Mapping in near ideal conditions (no production pressures and uninterrupted access) enabled the mine geology team to relate the regional structural history to specific observations within the pit and in turn understand how identifiable geological features will affect the geometry of mineralisation. For example, at Cutters Ridge D_4 faults will sharply truncate any mineralisation, but mineralisation will likely bleed across D_3 faults with preferential vein density and grade in the gabbro, but still developed in the adjacent sedimentary rocks. Similarly, vein density and grade will decrease as the granophyric gabbro zone diverges away from the controlling D_3 faults.

Fully understanding these relationships retrospective to mining may seem like an academic exercise, however as mining, and industry in general, moves toward a better appreciation of human capital and developing skillsets (Madgavkar *et al*, 2023), Cutters Ridge demonstrates that maximising the developmental opportunity by mapping post-production, and focusing on geological relationships, adds significant value beyond the incremental gains (ie fine scale ore-waste discrimination) of solely undertaking real-time production pit mapping.

Upskilling and increasing the confidence of the mine geology team allows for the application of higher-level, more-refined geological skills in the production phase of future pit projects. The skills refined by the Cutters Ridge mapping exercise and the confidence gained are being actively employed in the next pit operation at Mungari. Dig face and blasthole map data are actively collected with the understanding of geological relationships used to integrate this somewhat crude data into a geological model beyond a pure spatially correct map. The frontline team are critically questioning some of the modelled lode geometries and dynamically making decisions to refine markups beyond the inherited dig block designs.

CONCLUSIONS

The Cutters Ridge open pit exposes direct manifestations of three of the four major regional deformation events of the Kalgoorlie Terrane at a mappable scale. These key geological relationships help explain the scale and continuity of mineralisation at this location, but only became apparent through a dedicated mapping campaign with uninterrupted access to the entire pit immediately after mining ceased. The ultimate goal of any mapping by the mine geology team is a better understanding of the orebody to achieve better ore-waste discrimination, so the value of a comprehensive mapping campaign after mining has completed might be questioned. Seeking to fully understand an orebody post-production has value for exploration and resource development in the area, but the biggest value is as a continuous improvement exercise that better equips the mine geology team to actively seek out and identify key geological relationships in that achieve a better understanding of the next orebody to be mined and apply dynamic decision-making in real time.

ACKNOWLEDGEMENTS

The authors wish to thank Brad Daddow for instigating and supporting the dedicated mapping project on which this paper is based. The authors note the alacrity with which newer members of the Mungari Open Pits team have sought to join the authors in implementing the continuous improvement initiatives developed from this project.

REFERENCES

Cooke, D and Fitzgerald, M, 2011. The Value of Geological Human Capital in the Improvement and Maximisation of Ore Deposit Value – The Kalgoorlie Consolidated Gold Mines Experience, in *Proceedings Eighth International Mining Geology Conference 2011*, pp 59–74 (The Australasian Institute of Mining and Metallurgy: Melbourne).

Jacques, D, Muchez, P and Sintubin, M, 2022. Conjugate, Riedel-shear vein arrays: A new type of en echelon vein system, *Journal of Structural Geology*, 163:104725.

Madgavkar, A, Schaninger, B, Maor, D, White, O, Smit, S, Samandari, H, Woetzel, J, Carlin, D, Chockalingam, K and Renaud, L, 2023. Managing human capital: Performance through people [online], *McKinsey Global Institute*. Available from: <https://www.mckinsey.com/mgi/our-research/performance-through-people-transforming-human-capital-into-competitive-advantage#/> [Accessed: 02 Feb 2023].

Robert, F and Poulsen, K H, 2001. Vein Formation and Deformation in Greenstone Gold Deposits, in *Structural Controls on Ore Genesis* (eds: J Richards and R Tosdal), pp 111–155 (Society of Economic Geologists: Littleton).

Swager, C P, 1997. Tectono-stratigraphy of late Archaean greenstone terranes in the southern Eastern Goldfields, Western Australia, *Precambrian Research*, 83(1–3):11–42.

Vielreicher, N, Groves, D and McNaughton, N, 2016. The giant Kalgoorlie Gold Field revisited, *Geoscience Frontiers*, 7(3):359–374.

Vielreicher, N, Groves, D, McNaughton, N and Fletcher, I, 2015. The timing of gold mineralization across the eastern Yilgarn craton using U–Pb geochronology of hydrothermal phosphate minerals, *Mineralium Deposita*, 50(4):391–428.

Ore control based on value – experiences in design and execution

C Morley[1]

1. FAusIMM(CP), Owner – Reconciliation & Ore Control Knowledge Services (R&OCKS), Perth WA 6000. Email: craig@rockservices.com

ABSTRACT

Ore control based on value is an execution approach employed in mines to maximise the economic value of extracted ore. It involves integration of geology, geometallurgy, mine planning and operational execution considerations to optimise mining using the value of the material being extracted rather than only grade. This emphasizes identification and exploitation of high value ore zones and material blends, while minimising the delivery of low value material to processing plants. By focusing on the bottleneck (the stage in the mining process that limits production) this approach enhances overall operational efficiency and financial returns such as EBITDA (earnings before interest, tax, depreciation and amortisation).

The application involves several key steps. Ore control models are constructed from drilling and sampling activities that provide geometallurgical and geological data, enabling identification and classification of mineralisation patterns and variability that define the zones for mining. Value based decision-making algorithms and optimisation techniques are then applied to prioritise the extraction. These algorithms include the grade of the material, geological and geometallurgical characteristics, mining and processing costs, market prices and potential blending strategies, to ultimately determine the material to be sent to the processing plant.

By optimising mining activities based on value rather than grade, the delivery of material that lowers value and displaces feed of higher value ore/blends, is prevented thus optimising operating costs. It is value that determines what is sent for processing, stockpiled for blending or discarded as waste. Ultimately this approach enables mining companies to increase productivity, recovery and profitability and thus maximise the value extracted from their mineral resources.

INTRODUCTION

Ore control based on value aims to maximise the value delivered from Mineral Resources during mining. This value-based approach spans data collection, resource and short-term geology (ore control) modelling, reserve, short-term mine planning, ore definition, blasting, loading, hauling, stockpiling and feed to the processing plant as shown in Figure 1. It focuses on how we can effectively and efficiently consider all the minerals we mine, their various characteristics and elements that come with them and how teams can work together with this information to improve mining processes and ultimately increase profitability.

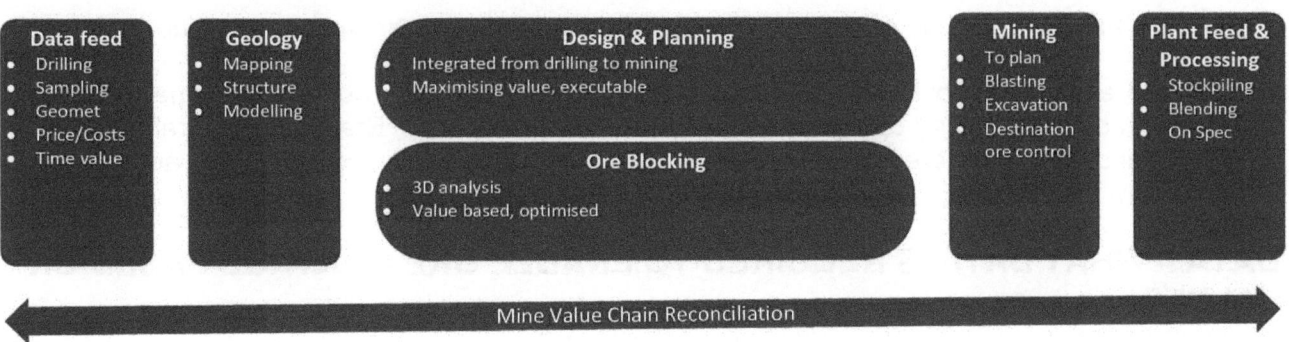

FIG 1 – The ore control based on value framework.

The guiding principles of this approach are:

- We mine rocks, not grades and so all quality and geometallurgical characteristics that impact on the potential value to be extracted must be considered during ore definition.

- Processes must be developed that make the impact on value transparent across the mining value chain such that all technical disciplines understand the role they need to play to maximise value delivery.
- Ore control planning and execution decisions are holistically value based.
- Processes must be objective and produce repeatable results.

During the mining process it is important to focus on the value of all elements, impact of deleterious characteristics of the ore and the cross functional work involving different technical disciplines. All these activities help to optimise the quality of the feed to the plant and therefore contribute to the improvement of both productivity and profitability.

Often, mining operations predict annual revenue based on metal grade, a fixed price for a single metal/commodity and the tonnage of total ore blocks fed to the plant at a name-plate throughput rate and metal recovery, less mining costs. The operational reality is that, even under stable mining and plant conditions, ore throughput and commodity recovery fluctuate because of natural rock characteristics that are present within an orebody, such as variations in hardness and mineralogy, that effects recovery. As a result, not all ore blocks or even truckloads, of a similar grade will provide the same contribution to value as shown in Figure 2.

Which truck has the most value?

1	2
Tonnage: 350	Tonnage: 350
Grade: 1.3% Cu	Grade: 1.3% Cu
Rock type: Porphyry	Rock type: Breccia
Copper minerals: Cu(ox) 0.3% Cu(Sul) 1.0%	Copper minerals: Cu(ox) 0.1% Cu(Sul) 1.2%
Recovery: 80%	Recovery: 85%
Value: US$18,200 per truck	**Value: US$23,205 per truck**
Hardness: 14 (BWI)	Hardness: 13 (BWI)
Throughput per hour: 3,000tph	Throughput per hour: 3,500tph
Total value of feed over 1 hour: US$156,000	**Total value of feed over 1 hour: US$232,050**
Total value of feed over 1 year: US$1,367M	**Total value of feed over 1 year: US$2,033M**

FIG 2 – Example of how rock type and characteristics can impact on value.

Most of the examples provided in this paper are sourced from the authors personal experience working with open pit mining operations in South Africa, Chile, Peru, Brazil and Australia. However, the ideas and principles are also considered applicable to any mining operation and extraction methodology.

DATA – WHAT DATA IS REQUIRED TO ENABLE ORE CONTROL BASED ON VALUE?

Ore control based on value is enabled by knowledge of:

- **Revenue generating elements** including both primary and by-products; for example, South African Platinum mines can recover nickel and copper along with the platinum group elements.
- **Price penalty elements**: for example, in iron ore mines in Australia and Brazil the potassium, phosphorous, silica and alumina content are important to steel makers and if specifications in shipments are not met then price penalties result in lower value being achieved.

- **Cost driving elements** including elements deleterious to mining (eg clays) and mineral processing (eg talc) as well as elements that need specific management processes to eliminate or mitigate safety, health or environmental risks (eg silica minerals in dust and acid generating materials in waste dumps).
- **Geometallurgical elements/parameters** that impact on plant performance such as processing throughput and product recovery.
- **Mining and processing costs** that are subject to production efficiency and variables such as fuel and electricity costs.
- **Time value of information;** for example, having data available well in advance of mining to facilitate planning and optimisation of blending activities to improve throughput and recovery in the processing plant.

Geoscientific data must be *spatially representative* for all mineral content, physical and geometallurgical properties to enable a comprehensive ore control model to be built for medium to short-term planning and mining. Proxies can be useful, if strong correlations exist with primary data. For example, geochemical data can provide a good basis to model lithology if all the lithologies present are well understood and the appropriate geochemical data is available.

Fundamental to sampling is a scientific assessment of the sampling nomogram and selection of safe and appropriate processes for sample acquisition and handling. Any data that is gathered must be accompanied by appropriate quality assurance and quality control (QA/QC) procedures that allow real time responses to sampling or analytical quality issues. Gathering data well ahead of mining allows for comprehensive QA/QC and proper data management systems as well as improved medium/short-term mine planning and blasting processes. Ore control data, which is used to deliver the value from the orebody, should be seen as just as important as exploration or Resource definition data and attract the same amount of effort in QA/QC.

Examples of approaches that can be investigated and used to improve the collection of a full suite of data include:

- Reverse circulation (RC) drilling on statistically appropriate tight spacing, with automated sampling, well in advance of mining.
- X-ray fluorescence (XRF) and X-ray diffraction (XRD) scanning devices for geochemical and mineralogical analysis.
- Fourier Transform Infra-red (FTIR) analysis for mineralogy, geochemistry and density.
- Hyperspectral scanning of drill core and RC chips for texture and mineralogy.
- Measurements of drilling performance (for example the ROCMA System) to derive estimates of hardness and degrees of fracturing.
- Bench scale metallurgical test work to assess/update recovery potential.
- Batching ore through processing plants and reconciling between ore control predictions and plant performance to validate geometallurgical assumptions.

GEOLOGY

Mapping, structure and modelling

Ore geometry is often controlled by lithological and structural geology features that increasingly are not being routinely mapped or modelled as part of the traditional grade control processes. A common excuse is because of safety considerations, however a number of technologies are now available that ensure mine geologists do not need to approach blasted faces and can still safely capture appropriate mapping data quickly and in considerable detail. These include using smart phone or tablet applications and 3D scanning technology with photogrammetry via fixed cameras or drones. This means real-time updates to 3D models of ore controlling features are possible, leading to the limitation of ore loss and dilution that impacts directly on recovered value.

In addition to traditional sampling methods, photogrammetry techniques in underground mines can be used to map mining faces and correlate mined volumes to typical stratigraphic grade profiles, which is then used to estimate the production grade and ore characteristics of mining panels and reconcile back to the Resource model.

The ability to rapidly update 3D geological models using field observations as a direct input has a clear benefit to the accuracy of ore block designs and mining tactics. Implicit modelling software has now reached a level of commercial development that it is considered superior to manual, explicitly drawn interpretations, providing models that are objective and repeatable (Hodkiewicz, 2014). Initial set-up of an implicit modelling process can be time consuming however the benefits accumulate rapidly. This approach is even more powerful when combined with a data acquisition process that deliver input data well ahead of production such that questions about the model can be validated by re-visiting exposures in the mine and running multiple iterations of the model prior to mining.

Modern software and hardware have the capability to load and display large data sets, which can reveal structural and ore trends that might be missed if the focus is only on one blast at a time. Figure 3 shows a large blasthole sample data set that clearly shows a folded structure within grade data using the 'X-ray plunge projection' methodology described by Cowan (2014).

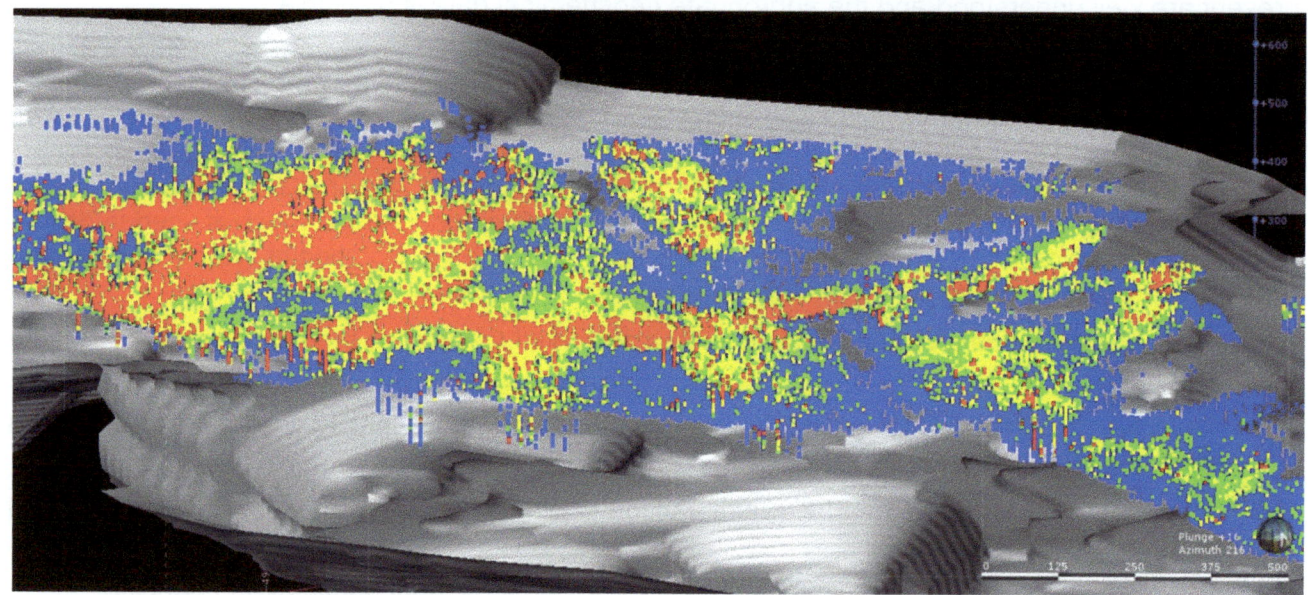

FIG 3 – Folded structures revealed in large blasthole data set (after Cowan, 2014).

Design and planning – integrated planning across the value chain

At many operations mine planning focuses on equipment and material movements only. Including the activities involved in ore control data acquisition into the short-term mine plan and mining schedule is important to ensure that all activities in the mine are coordinated and safely executed to the right level of quality and detail at the right time (Ortiz and Magri, 2014). Active engagement by mine geologists in the operational routines is critical to embed and sustain ore control based on value. Stakeholder alignment and change management required to establish the correct routines are much more effective when a clear value proposition can be put forward.

RC drilling activities involve large mobile equipment operating in the mine in addition to blasthole drills, loading and hauling equipment and other mobile plant needed to operate the mine. Integrating the ore control drilling and sampling activities into the daily operational management processes, by including them in short-term mine planning, ensures safe and timely execution of the ore control data acquisition program, without interfering with production. Just as important to including these activities in the mine plan, is that the results are used to ensure that the value delivered can be maximised. As an example, a key metric for many mines is the Annual Budget – it becomes a target (unfortunately in tonnes) to be achieve at almost any cost or methodology! By including the ore control data collection activities into the mines' 18–24-month planning cycle it is possible to schedule mining during the annual budget planning period on the actual ore control results rather than on a Resource estimate. This value add is explained in more detail in the next section.

Adding value by acquiring data well ahead of production

Significant value can be added by moving from 'just-in-time grade control' in favour of ore control data gathering well ahead of production. This can be achieved with RC drilling (Ortiz, Magri and Libano, 2012; Ortiz and Magri, 2014) 6–18 months ahead of the time when the ore will be mined. When collecting data ahead of production the value comes from:

- Gathering fit for purpose quality samples that provide a reliable and representative view of geological contact locations and rock properties, as opposed to making do with blasthole samples that cannot pinpoint rock type contacts and are very difficult to sample correctly and therefore have very poor quality (Pitard, 2008).

- Removing time and cost from blasthole drilling by removing the sampling activities from the process. At It is common to discover that the dollar per metre manually sampling cost savings in blasthole drilling more than paid for the RC drilling ahead of production, which along with improved sample quality and early availability of data helped motivate for the RC drilling process and costs. Experiences detailing a 54 per cent reduction in sampling and assaying costs at the Mogalakwena platinum mine are documented in Kirk, Muzondo and Harney (2011).

- Having time for quality geometallurgical test work, which typically requires more complex laboratory tests and time compared to grade only assays.

- Acquiring spatially representative geometallurgical data that enables three-dimensional modelling of mineral/metal recovery which is a key input to estimating the value of each mining block, thereby determining the most value adding material destination.

- Understanding the spatial distribution of ore zones, lithologies and structures *at the mining scale* for areas larger than a single blast, such that the continuity of value driving parameters can be assessed with high confidence and used in planning mining tactics. For example:
    o Where to start mining and how to progress the development across a full bench.
    o Design of blasting in order to control ore loss and dilution by separating ore and waste into different blasts and choosing the mining direction that leads to less mixing of materials (eg up-dip as opposed to along strike).
    o Identify where different rock properties present risks and opportunities to optimise blast designs and blending of material feed to the plant (eg hard versus soft areas, clay zones, wet zones etc).
    o Forewarn the processing plant that problematic ores will be delivered, or identify blending opportunities that may mitigate these issues for the plant and optimise recovery.

- Allowing for follow up field investigations and/or iterations of mine plans to identify the most valuable options.

- Providing the ability to re-assess strategies and tactics with high confidence in the event of changes in the mine such as equipment breakdown or geotechnical issues.

ORE BLOCKING

3D analysis and ore block modelling

'Just-in-time grade control' practices can lead to a rushed 2D assessment of data and design of ore blocks on a blast by blast basis, with the risk of misallocating ore and waste, particularly at the boundaries between blasts. Full 3D analysis of all drill hole data (exploration, resource definition and ore control drilling), mapping, structural models and ore block designs for previously mined areas is required to properly model the spatial continuity of the geology and thereby capture the most value.

In many operations the approached to traditional grade control modelling has been to apply the standard industry practice of building a block model and interpolating variables into the blocks via ordinary kriging methods. Typically, the block size is chosen based on the selective mining unit size. Through concept level studies into using computer aided optimisation of ore blocks, it has been shown that a much finer resolution ore control model dramatically improves the ability to demarcate higher value ore blocks for mining. Geostatistical conditional simulation methods have been found to be best suited to producing fine resolution models that reduce misclassification (Vasylchuk and Deutsch, 2016) and facilitate a risk-based view of optimal ore block selection.

Value based ore definition

Defining ore blocks using only the primary economic element (eg total copper grade) can miss opportunities to leverage value from ore that carries recoverable quantities of by-product elements (eg molybdenum), particularly in cases where the mine has complex processing routes (for example flotation and heap leach). In these cases using knowledge of specific mineralogy and the associated differences in metal recovery for different minerals can lead to different ore destination decisions and an increase in value.

Iron ore mines in both Australia and South Africa apply a material classification system that incorporates the geology, grades, density and product yield for both direct ship and processing plants. This material classification allows ore blocks to be designed based on recovered product with tonnes and grades as a strong proxy for value. Daily mine schedulers can access the underlying product yield data and respond quickly if plant feed blends need to change to achieve product specifications.

The author has experience working with mine ore control geologists making ore/waste decisions using a US$ value of polymetallic ore, as well as associated recovery, to calculate profitability. Based on geometallurgical characteristics some ore types with high recoveries secondary metals are more profitable than ore types with the primary metal grades that are just above the low-grade cut-off. This is shown on Figure 4 where the change to using profitability results in some material that was being allocated to long-term stockpiles but has higher value (the large blue $ symbol on Figure 4) than some material being sent to the concentrators (small red $ symbol on Figure 4), is now sent to the concentrators, resulting in increased revenue.

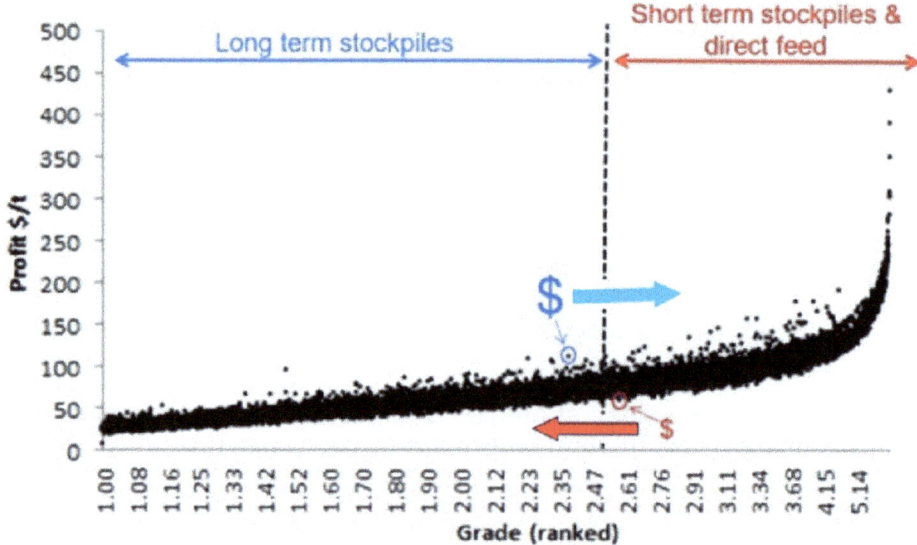

FIG 4 – Grade plotted against profit (inclusive of 4E and base metal value).

Computer aided optimisation of ore blocks

A guiding principle of ore control based on value is to apply methods that generate objective and repeatable results. Despite best efforts and diligence, experience shows that when different people interpret multi-parameter data sets and complex geological settings they produce different ore block designs, with a consequent difference in the value of those designs. Even the same person doing the same ore block design on different days may well see things differently and produce diverse results.

Computer optimisation techniques have been widely used in a broad range of industries and with today's technology the problems of optimising ore block designs are readily overcome. Computer aided ore block optimisation has been studied and used in consulting work and in production on many mines for over 20 years (Shaw and Khosrowshahi, 1997). In the authors experience studies have shown potential to improve value outcomes by 2–5 per cent in most cases, but also by as much as 30 per cent compared to the current practices. In addition, the ability to optimise for either volume or grade facilitates an objective and controlled response to changing economic conditions and provides a tool for assessing alternative ore control approaches. In the case where ore control data has been acquired well in advance of production the mine is well placed to assess likely scenarios and/or respond in a controlled way to changing conditions.

PLANNING AND EXECUTION

Short-term mine planning of plant feed for weekly, monthly and quarterly periods requires higher confidence in local orebody geometry and content than a Mineral Resource at Measured category is designed to provide. The closer spaced ore control drilling data facilitates building a higher resolution model for short-term mine planning thus delivering significant value. 'Just-in-time grade control' is an inadequate approach for this reason.

At many open pit mines the ore control strategy has moved away from 'just-in-time grade control' to RC drilling ahead of production and this change has allowed the construction of higher resolution geological models for short-term planning. Embedding these new models into the planning process, including better design of blast blocks to control ore loss and dilution, facilitates targeted blasting practices that are optimised for the different rock types. Such a change in the planning process can deliver significant cost savings in drill and blast as well as improved recovery of ore and the quality of final plant products. An example from the iron ore sector is preserving the lump fraction of the ore through smarter blasting which can directly add value because of the price premium for lump iron ore products.

MINING

The value of the ore block design can be destroyed if mining of the block is executed poorly. The importance of continuous and direct field monitoring of mining activities by mining geologists cannot be overstated. Only by having a presence (live or virtual) in the mine can the mining geologist see the impact of mining decisions taken on an hourly basis during excavation and ensure the estimated value is delivered.

Several practical approaches that can be taken to controlling ore loss and dilution during mining and to realise the estimated value of the ore block design are discussed.

Train the operator and influence the key performance indicators.

Excavator operators are often interested in what they are doing and value having relevant geological and orebody knowledge that enables them to do a better job. Mining geologists must develop a relationship with the operators and this should include some training for the operator. Relevant topics include:

- Overview of the different rock types and what they look like in the field (colour, fabric).
- The concept of bulk density and how this varies by rock type. Quite often the operator can distinguish between ore and waste on the basis of mass and the number of bucket loads to fill a truck.
- Mining tactics to control ore loss and dilution such as defining the preferred direction of digging and why. For example, approaching a dipping ore contact perpendicular to the strike and mining up-dip so the rill angle has the most acute angle with the ore boundary.

Mining geologists must spend time in the mining machine to understand the capability and constraints specific to that machine and operator. Just as the mining geologists must educate the operator, the operator must also educate the mining geologists. Ore block designs must be practical from a loading point of view in order that the estimated value is delivered and the only way the mining geologist can do this is to become familiar with the mining equipment and how it is being operated.

Operators will strive very hard to deliver on their performance contract key performance indicators and so these must align with the value drivers for the operation. Controlling ore loss and dilution will always be a significant value driver, whether this translates to more or less selective mining practices will depend on the orebody, commodity and type of operation. In all cases if the ore loss and dilution are *controlled* then the expected value is much more likely to be *delivered*.

Blast movement

In operations where blasting is required to facilitate mining it is necessary to ensure the blasting process, the ore block designs, field mark-ups and excavation are coordinated and aligned to maximise value recovery.

Blasted ground will move, therefore the ore block boundaries will move. The movement is not linear in 3D with different degrees of movement occurring in different parts of the blast as a result of the blast design (Thornton, 2009; Isaaks, Barr and Handayani, 2014). If this movement is not accounted for then unplanned ore loss and dilution will occur and the estimated ore block value will be negatively impacted.

Accounting for blast movement can be done in several ways including using visual markers and remote methods such as Blast Movement Monitor® methods described by Cocker and Sellers (2014) and high-speed video capture described by Rogers *et al* (2012). As shown in Figure 5 blasting simulation tools have become available and greatly enhancing the ability for Geologists to use technology to model movement and mixing during blasting (Poupeau, Hunt and La Rosa, 2019). Whichever approach is taken value is maximised when the blasting process produce consistent results and are informed by high resolution spatial understanding of rock properties such as hardness, texture and fracturing.

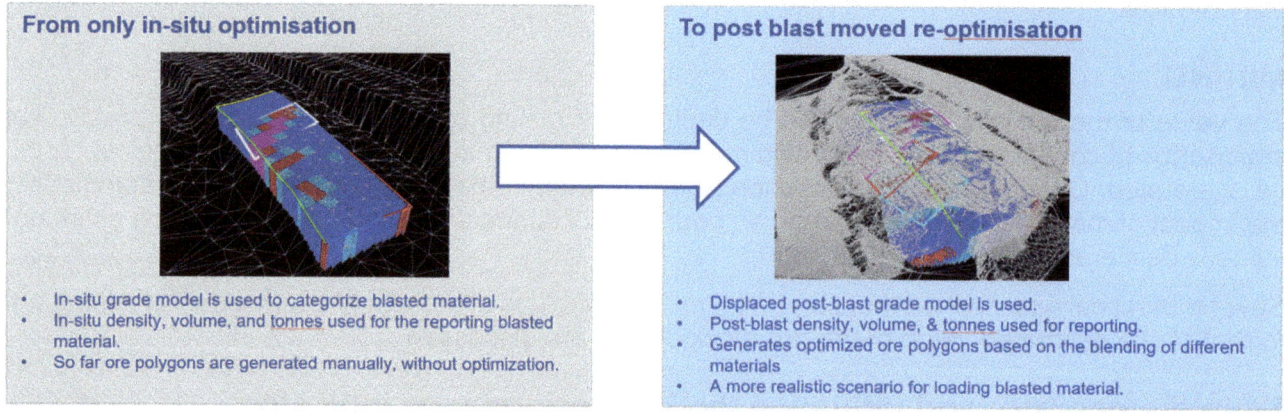

FIG 5 – Use of blasting simulation software (OrePro 3D) to reoptimise ore separations post-blast.

Ore demarcations

Ore demarcations are used to guide and control excavation and must be clearly visible at all times to the operator. Visibility can be achieved by marking lines in the field with wooden stakes, tapes and placards, or it can be achieved digitally with display monitors and global positioning systems presenting real time locations to the operator in the excavator cab and that can be view remotely by mine geologists. Regardless of the method the mining geologist must communicate the ore block design to the mining crew each day and then monitor ore mining during excavation to ensure ore loss and dilution are controlled.

Minimising the frequency of mining across ore block boundaries on night shift is beneficial to controlling ore loss and dilution. Digital systems that provide live data to the operator have an advantage over field mark-ups which can be difficult to see clearly in the dark, even when mobile lighting plants are used.

Material tracking

The old management adage 'you can't control what you don't measure' is particularly relevant to mining. Unless good quality material tracking data is collected and analysed at least daily, but preferably in real time, it is likely that ore loss and dilution will result. This happens when truckloads of ore or waste are delivered to the incorrect destination.

Modern fleet management systems are often under-utilised because the primary purpose of these systems is perceived as being fleet management rather than ore control. However, these systems can be very effective in correcting manual destination errors and both validating destinations and delivery results. In addition, mine value chain reconciliation is also an essential feedback loop that is used to identify process improvements for ore control, that material arrives at the correct destination and for validating Mineral Resource and Ore Reserve estimates and improving value delivery (Morley, 2014).

Mining to design and spatial compliance to plan

It seems obvious, but mining to the mine design is incredibly important! Sterilisation of ore due to not mining to design is an avoidable disaster. It is not uncommon for the modelled boundaries of ore zones to shift when closer spaced data becomes available and this can create risks and opportunities. Continuous monitoring of actual toe and crest positions and floor elevations can add enormous value to an operation by identifying opportunities to recover more ore or avoid sending waste to the plant. Poor control of the bench floor can lead to invalid ore block designs, additional cost due to limiting the speed at which trucks can safely travel and additional fuel burn due to uneven travel surfaces and tyre wear.

Spatial compliance to plan also seems like an obviously good idea but is actually very challenging for most operations. Unplanned events happen in mines for a wide variety of reasons, such as geotechnical slope failures, bad weather, unplanned machine breakdowns etc. Regardless of the excuses the impact of poor spatial compliance to plan on ore control is that as the mine becomes reactive – because mining did not happen where planned, it is necessary to source the ore from a different location. Sometimes it is not possible to access other sources of ore and it is a new challenge to provide ore with the expected properties for processing. Poor spatial compliance to plan destabilises not only the ore control process but can also eat into future ore sources and impact the processing plant, to the degree that recovery of product and hence value, is negatively impacted.

PLANT FEED

Ehrig *et al* (2014) correctly state that 'Geologists should be the technical stewards of mineral Deposits' and are responsible for the value-adding step of transforming mineralogical models into models that quantify mining and metallurgical performance across the entire mineral deposit. Leveraging geometallurgical data the ore control process must be designed to deliver the planned blend of materials to the plant that optimises product recovery and plant performance in order to deliver the expected value. Understanding what the composition of this blend needs to be is often worked out by trial and error, a process that can destroy value. Maximising value requires a deep understanding of how the process circuit responds to the different rock types with all their related geometallurgical and physical characteristics. Parameters such as hardness, density, texture, mineralogy and particle size distribution can all have an impact on the process performance and must therefore be understood and modelled in the ore control process (McKay *et al*, 2016). Even when the plant is operated in a consistent and stable manner the product recovery, throughput and value will vary because of the spatially variable nature of the orebody.

In circumstances where the optimal feed to the plant is not well defined then controlled batching of ore through the plant and reconciliation, can be a very effective method to understand plant responses to specific material types. Pilot plant parcel treatment and full-scale batch trials can be difficult to execute without interrupting production, but good planning and a common understanding of the value to be gained can facilitate the process. Even modest gains (eg 1 per cent) in recovery and/or improvements to plant throughput will generate large value over time.

Ore control processes such as computer aided optimisation of ore blocks and use of stockpiles can be used to reduce the variability of ore feed to the plant to optimise and stabilise plant performance. Stockpiles built in such a way to homogenising reclaimed plant feed carry an additional material handling cost, but often this cost can be more than counter-balanced by reduced costs in running the plant and/or additional revenue from improved recovery.

VALUE RECONCILIATION

Reconciliation has shown that changes to the ore control process and the adoption of ore control based on value can lead to additional EBITDA – for example in the authors experience while working with seven operations value delivery of US$831M was achieved over a period of seven years. This result does not include value added where benefits could be claimed by other changes/improvements – for example where improvements in plant throughput also occurred it was deemed too difficult to separate ore control's contribution from the Plants' improvement value delivery and so no value was included in the ore control results. This additional value contribution significantly strengthened that ability of the mine geologists to execute initiatives, while at the same time strengthening the cross functional interaction at each mine.

CONCLUSIONS

A focus that all mines deal with rocks rather than grades and that geometallurgical characteristics, along with the quantities of metals influence value, can fundamentally alter any operations approach to ore control. Experience has shown objective and repeatable processes in ore control combined with transparency of the impact of each step in the mining value chain has a significant effect on value delivery. Deep cross functional understanding of the value driving characteristics of the ore and waste, throughout the mining value chain, leads to improved predictability and profitability of the ore control, planning, mining and processing activities. Value or profit thinking, rather than a simple tonnes, grade or cost focus, is required to identify the correct work, processes and technologies to build the most value adding ore control and ultimately mining system.

ACKNOWLEDGEMENTS

The author wishes to thank Alastair Cornah, Marius Strydom, Heath Arvidson, Wendy Ware, Georgina Rees, Sergio Spichiger, Nicolas Cruz, Scott Jackson, John Vann and numerous mines operational staff who have contributed to the many discussions and projects that have culminated in the ore control based on value approach described in this paper.

REFERENCES

Cocker, A and Sellers, E J, 2014. Modelling Blast Movement for Grade Control at an Open Cut Gold Mine, in *Proceedings of the Ninth International Mining Geology Conference 2014*, pp 377–386 (The Australasian Institute of Mining and Metallurgy: Melbourne).

Cowan, E J, 2014. 'X-ray plunge projection' – understanding structural geology from grade data, in *Mineral Resource and Ore Reserve Estimation – The AusIMM Guide to Good Practice*, AusIMM Monograph 30, second edition, pp 207–220 (The Australasian Institute of Mining and Metallurgy: Melbourne).

Ehrig, K, Liebezeit, V, Smith, M, Macmillan, E and Lower, C, 2014. Geologists and the Value Chain - How Material Characterisation by Modern Mineralogy can Optimise Design and Operation of Processing Facilities, in *Proceedings of the Ninth International Mining Geology Conference 2014*, pp 5–13 (The Australasian Institute of Mining and Metallurgy: Melbourne).

Hodkiewicz, P F, 2014. Rapid Resource Modelling – A Vision for Faster and Better Mining Decisions, in *Proceedings of the Ninth International Mining Geology Conference 2014*, pp 183–188 (The Australasian Institute of Mining and Metallurgy: Melbourne).

Isaaks, E, Barr, R and Handayani, O, 2014. Modelling Blast Movement for Grade Control, in *Proceedings of the Ninth International Mining Geology Conference 2014*, pp 433–439 (The Australasian Institute of Mining and Metallurgy: Melbourne).

Kirk, G, Muzondo, T and Harney, D, 2011. Improved Grade Control using Reverse Circulation Drilling at Mogalakwena Platinum Mine, South Africa, in *Proceedings of the Eighth International Mining Geology Conference 2011*, pp 329–340 (The Australasian Institute of Mining and Metallurgy: Melbourne).

McKay, N, Vann, J, Ware, W, Morley, C and Hodkiewicz, P, 2016. Strategic and Tactical Geometallurgy – A Systematic Process to Add and Sustain Resource Value, in Proceedings of the Third International Geometallurgy Conference 2016, pp 29–36 (The Australasian Institute of Mining and Metallurgy: Melbourne).

Morley, C, 2014. Guide to Creating a Mine Site Reconciliation Code of Practice, in Mineral Resource and Ore Reserve Estimation – The AusIMM Guide to Good Practice, AusIMM Monograph 30, second edition, pp 755–764 (The Australasian Institute of Mining and Metallurgy: Melbourne).

Ortiz, J M and Magri, E J, 2014. Designing an advanced RC drilling grid for short-term planning in open pit mines: three case studies, The Journal of The Southern African Institute of Mining and Metallurgy, 114:631–637.

Ortiz, J M, Magri, E J and Libano, R, 2012. Improving financial returns from mining through geostatistical simulation and the optimized advance drilling grid at El Tesoro Copper Mine, The Journal of The Southern African Institute of Mining and Metallurgy Transactions, 112:15–22.

Pitard, F, 2008. Blasthole sampling for grade control – the many problems and solutions, in Proceedings of the Sampling Conference 2008, pp 15–21 (The Australasian Institute of Mining and Metallurgy: Melbourne).

Poupeau, B, Hunt, W and La Rosa, D, 2019. Blast induced Ore Movement: The missing step in achieving realistic reconciliations, in Proceedings of the 11th International Mining Geology Conference 2019, pp 309–327 (The Australasian Institute of Mining and Metallurgy: Melbourne).

Rogers, W, Kanchibotla, S, Tordoir, A, Ako, S, Engmann, E and Bisiaux, B, 2012. Solutions to reduce blast-induced ore loss and dilution at Ahafo gold mine in Ghana, from SME Annual Meeting.

Shaw, W J and Khosrowshahi, S, 1997. Grade control sampling and ore blocking: optimisation using conditional simulation, in Proceedings of the Third International Mining Geology Conference 1997, pp 131–134 (The Australasian Institute of Mining and Metallurgy: Melbourne).

Thornton, D, 2009. The implications of blast-induced movement to grade control, in Proceedings of the Seventh International Mining Geology Conference 2009, pp 147–154 (The Australasian Institute of Mining and Metallurgy).

Vasylchuk, Y V and Deutsch, C V, 2016. Non-linear estimation for grade control, CCG Annual Report 18, paper 132, pp 132-1–132-14.

Hidden between the data, small but significant high-grade structural lodes

R Reid[1], D Magautu[2], M Lucien[3], L Gagau[4], M Mondo[5] and E Mudinzwa[6]

1. MAusIMM, FAIG, Group Resource Geologist, Harmony Gold, Brisbane Qld 4069.
   Email: ronald.reid@harmonyseasia.com
2. Geology Superintendent, Morobe Goldfields, Lae Morobe PNG.
   Email: delilah.magautu@harmonyseasia.com
3. Resource Specialist, Morobe Goldfields, Lae Morobe PNG.
   Email: mark.lucien@harmonyseasia.com
4. Senior Mine Geologist, Morobe Goldfields, Lae Morobe PNG.
   Email: lydia.gagau@harmonyseasia.com
5. Mine Geologist, Morobe Goldfields, Lae Morobe PNG.
   Email: michael.mondo@harmonyseasia.com
6. MAusIMM, Geology Superintendent, Morobe Goldfields, Lae Morobe PNG.
   Email: enos.mudinzwa@harmonyseasia.com

ABSTRACT

The Hidden Valley low-sulfidation carbonate-base metal gold-silver mine in Morobe Province, Papua New Guinea, is operated by Morobe Goldfields Consolidated, a Harmony Gold Mining Company Limited subsidiary. The current Mineral Resources stand at 3.1 Moz of gold with 1.2 Moz of gold in Reserves. Some 2.5 Moz of gold has been mined to date.

A historic drill hole in the Hidden Valley-Kaveroi (HVK) data set (HVDD049) contained a long intersection of 56 m at 10 g/t gold and 733 g/t silver, which also included significant missing core comprising 19.3 m (34 per cent) of the intersection. This intersection was located above the hanging wall of the main Kaveroi orebody. Including this intersection in the Resource estimate resulted in a big high-grade ball in the model that was termed 'the Big Red Ball' based on the high-grade colour bin. With little data available to validate the intersection, it was decided to remove the hole from the estimation data set to prevent estimate contamination. The morphology of the pit precluded drilling the area until the pit progressed to allow the establishment of drill platforms. One deep-probing RC hole drilled some years later intersected 19.5 m at 8.9 g/t gold and 212 g/t silver in the same area. This sparked a concerted drill campaign to understand what became known as 'the Big Red Lode', or simply 'Big Red' ahead of time.

Structurally, this lode is now recognised as a conjugate fault striking 20° more westerly than the Kaveroi orebody but dipping west and perpendicular to Kaveroi. The dip of the mineralised lens has led to an investigation of other possible Bid Red lodes that have been missed. The tenure of the grade required a careful blending strategy through the Mill so as not to flood the circuit with metal, especially the very high silver grades. Reconciliation of the material milled so far has shown a positive reconciliation even when using un-cut grades to estimate the ore blocks.

INTRODUCTION

The Hidden Valley-Kaveroi deposit is in the Morobe Province of Papua New Guinea, 210 km NNW of Port Moresby and 90 km SSW of Lae (Figure 1) in the steep, heavily forested, mountainous terrain in the upper headwaters of the Watut River at an altitude of 2500–2600 mRL (Pascoe, 1991). The mine is operated by Morobe Goldfields Consolidated, a 100 per cent owned subsidiary of Harmony Gold Mining Company Limited. Access to the mine is via a maintained gravel road from Bulolo and sealed roads from Bulolo to Lae.

The Hidden Valley deposit was discovered in 1985 by CRA Exploration (CRAE) during a regional stream sediment exploration program (Nelson, Bartram and Christie, 1990; Pascoe, 1991). A follow-up mapping and trenching program discovered mineralised granodiorite in a landslip in the upper reaches of Hidden Valley Creek; sampling returned a horizontal intersection of 100 m at 3.0 g/t gold and 45 g/t silver (Nelson, Bartram and Christie, 1990). Systematic soil sampling and subsequent drilling by CRAE defined the Hidden Valley and, later, the Kaveroi deposits in the early 1990s.

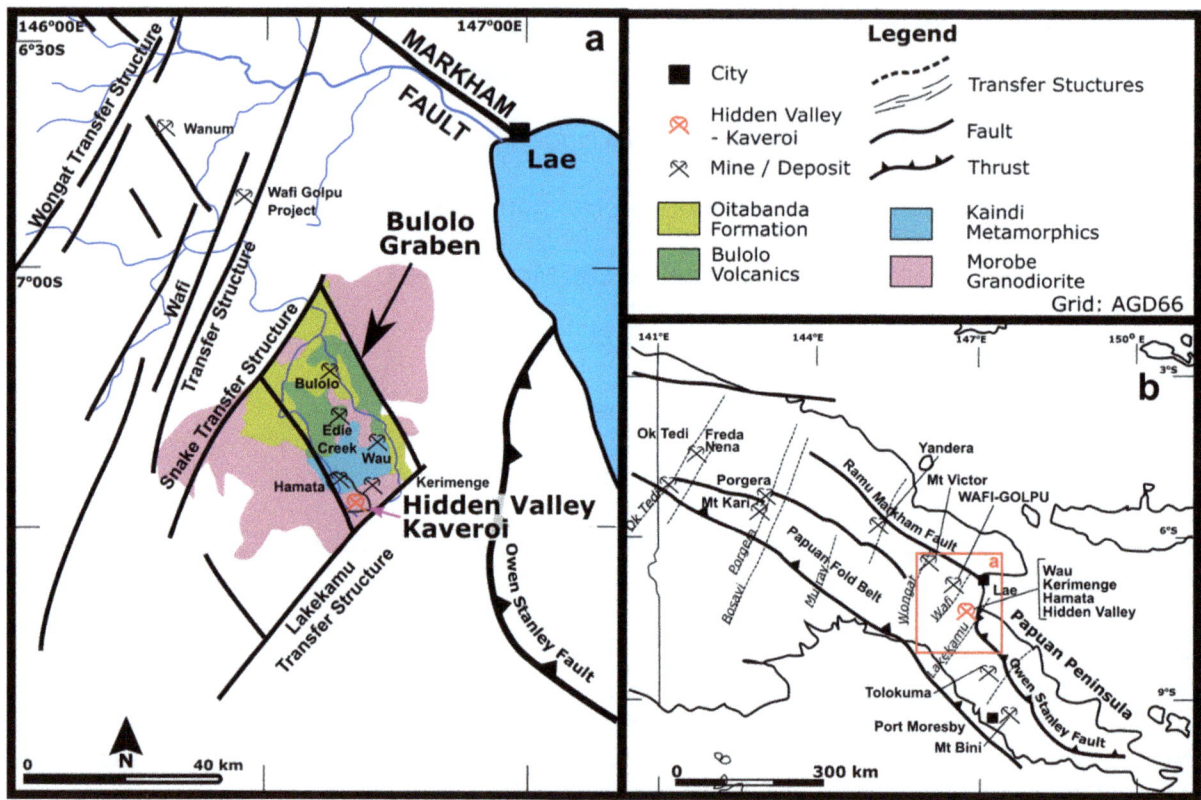

FIG 1 – Location of Hidden Valley-Kaveroi and the Hidden Valley Mine operation after Nelson, Bartram and Christie (1990), Corbett (1994) and Sapiie and Cloos (2004).

The current Hidden Valley Mineral Resources stand at 3.1 Moz of gold and Ore Reserves at 1.2 Moz of gold, with 2.5 Moz of gold mined from the Hidden Valley, Kaveroi and Hamata deposits since mining commenced in 2008 (Harmony Gold, 2023). The current life-of-mine will see mining finish in 2028; however, the HVXX mine life extension is currently in feasibility and seeks convert the balance of the Mineral Resource to extend mine life to 2032.

In addition to Hidden Valley-Kaveroi, other deposits located in the area are the Wau, Edie Creek, Hamata and Kerimenge epithermal deposits and the Bulolo placer deposits (Figure 1). These deposits have been exploited since the 1920s, extracting 115 t of gold (Dow, Smit and Page, 1974; Porter GeoConsultancy Pty Ltd, 2003).

A drill hole in the HVK Resource data set (HVDD049) contained an intersection of 56 m at 10 g/t gold and 733 g/t silver but also included significant missing core comprising 19.3 m (34 per cent) of the intersection (Figure 2). This intersection was located 100 m above the hanging wall of the main mineralised Kaveroi lens, outside the known orebody. Including this intersection when running the Resource estimate resulted in a high-grade blowout called 'the Big Red Ball' based on the high-grade colour bin of +1.8 g/t. The intersection could not be adequately validated due to the missing core component and a lack of drill holes around it. Moreover, poor quality control on the core orientation process meant that there was little confidence in the structural readings of the structure's orientation. While there were veins in the remaining core section, their stockwork nature and a lack of adjacent drill intersections precluded a simple interpretation. Unfortunately, the pit's morphology prevented drill testing of this area until the pit had progressed enough to allow the establishment of accessible drill platforms. Thus, the hole was removed from the data set used for the Resource estimation. The intersection became a footnote in the resource documentation until mining advanced to the point where a deep probing RC grade control drill hole could be drilled and the grade control program intersected the lens' narrow top. This paper describes the geology of the Big Red lens and discusses how the geology department managed the high-grades through the Mine to Mill process.

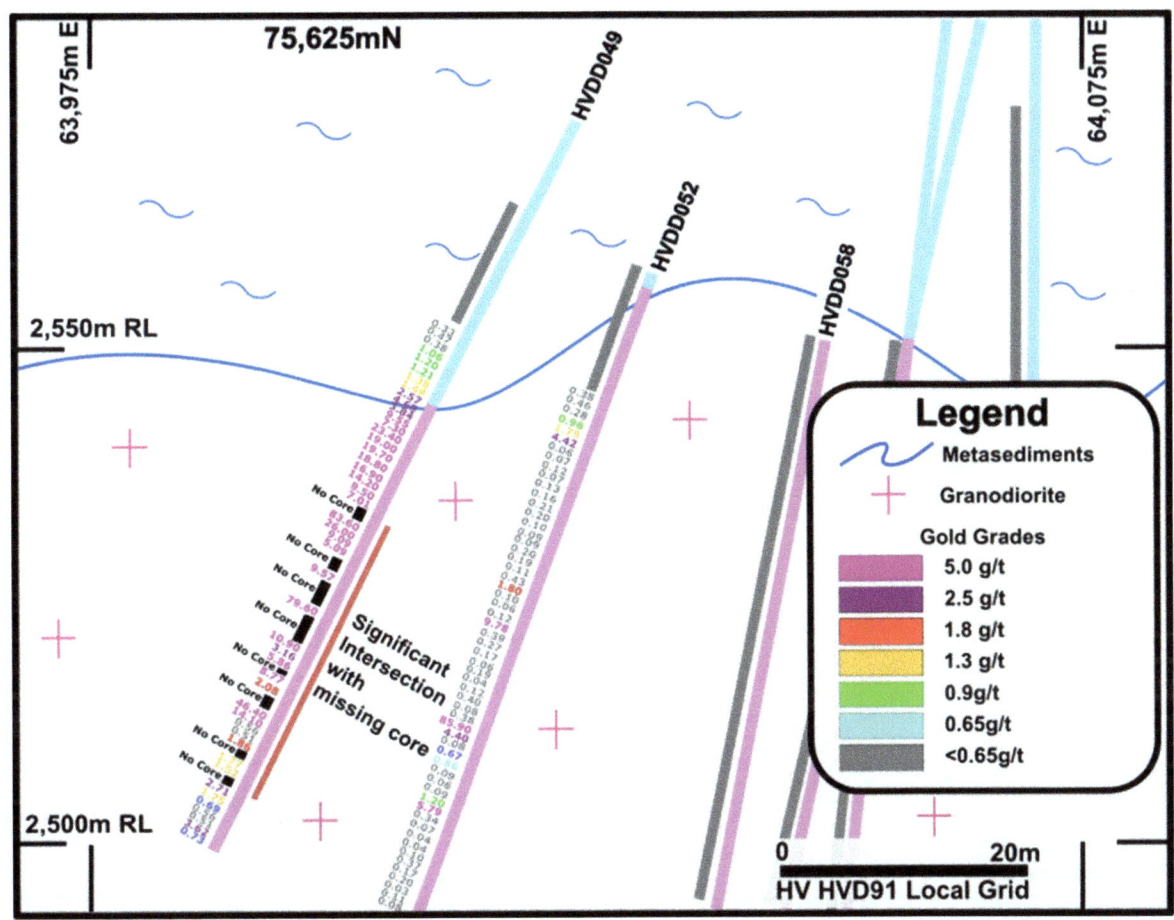

FIG 2 – Section through the Kaveroi deposit showing the Big Red structure intersected in HVDD049.

GEOLOGY

The Hidden Valley-Kaveroi deposit is a low-sulfidation carbonate-base metal epithermal gold-silver deposit contained within the Miocene-aged (12–14 Ma) Morobe Granodiorite (Bodorkos *et al*, 2013; Pascoe, 1991). The Morobe Granodiorite intrudes the upper-greenschist Kaindi Metamorphics, comprising psammites, pelites, metaconglomerates and marbles. Recent shrimp-based U-Pb dates indicate the Kaindi Metamorphics are Lower Permian in age (Bodorkos *et al*, 2013), a much older sub-unit of the regionally extensive Owen Stanley Metamorphics which are generally considered Late Cretaceous in age (Bodorkos *et al*, 2013; Corbett and Leach, 1998; Dow, 1977; Pascoe, 1991; Rinne *et al*, 2018). The deposit sits on the south-east edge of the Morobe Goldfield, which is contained within the Bulolo/Wau Graben (Figure 1), a regionally significant extensional graben (Corbett and Leach, 1998). The Morobe Goldfield deposits have been correlated to the Edie Porphyry suite of intrusives and are dated at 3 to 4 Ma using K-Ar and U-Pb methods (Bodorkos *et al*, 2013; Corbett, 1994; Dow, 1977; Dow, Smit and Page, 1974).

The Hidden Valley deposit is dominated by a series of faults controlling the gold mineralisation, including a set of early north-trending normal faults and significant NW striking thrust faults (Figure 3). The orebodies within the Hidden Valley pit have a general north-west trend and are north-east-dipping. The NW trending, 30 to 35° NE dipping Hidden Valley Fault, with its associated breccia, is a significant planar structural base underlying the main mineralisation. The main Hidden Valley orebody is broadly tabular shallow-dipping lens, up to 100 m thick and is localised immediately above and parallel to the Hidden Valley Fault. The Kaveroi orebody sits in the hanging wall of Hidden Valley and has a steeper (60–70°), more easterly dip (Figure 4).

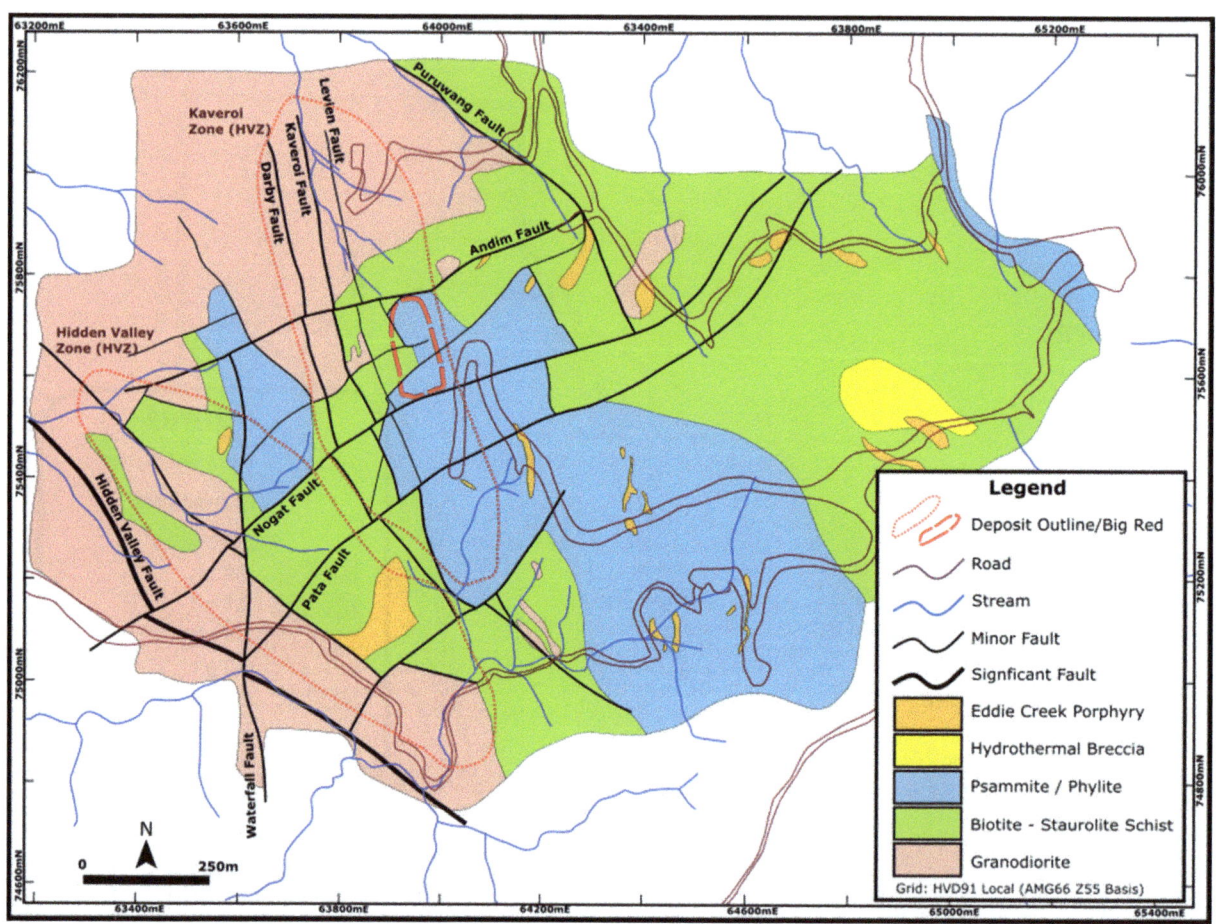

FIG 3 – Local pre-mining geology and structural setting showing the location of the Hidden Valley (HVZ) and Kaveroi (KVZ) orebodies and the approximate location of the Big Red Lens (heavy dashed red line).

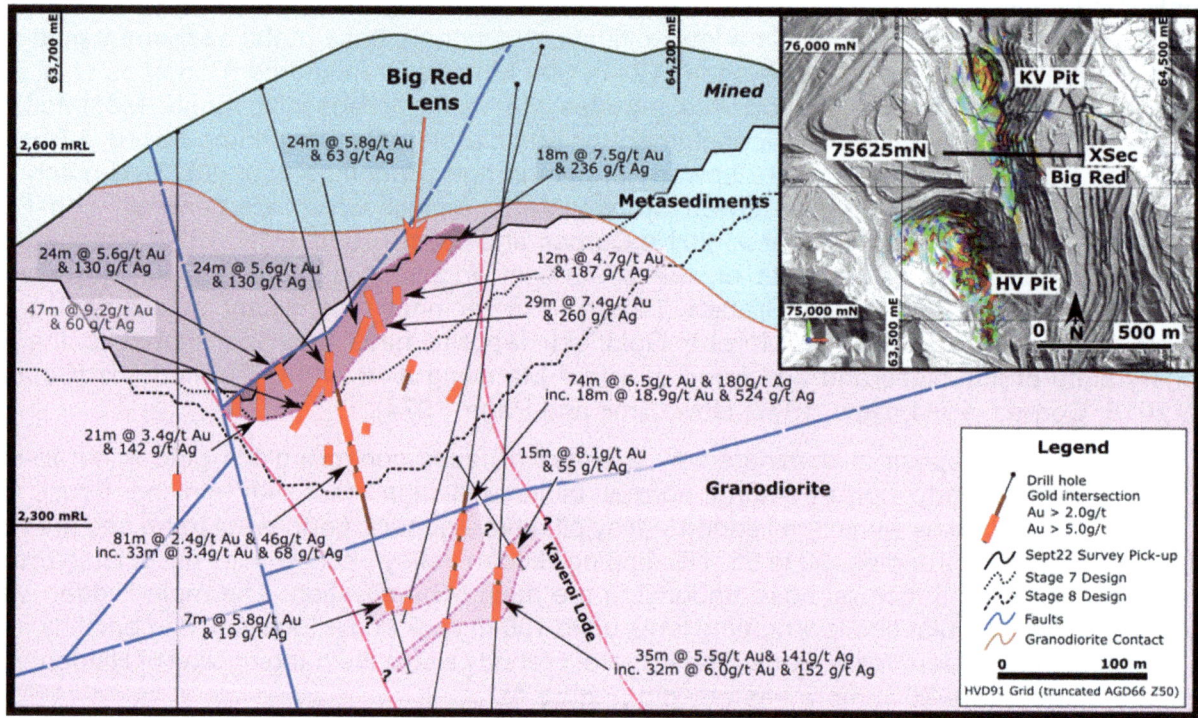

FIG 4 – Section of geology and structural Setting of the Big Red lens after initial drilling, also showing the location of some other Big Red-like intersections. The metasediment unit comprises the contact biotite schist and the overlying psammites and pelites. HV = Hidden Valley, KV = Kaveroi.

Most of the historical resource drilling was designed to target the east-dipping orebody with west/south-west-directed drilling. The resource drilling has an average spacing of 40–80 m, often more widely spaced at depth. Recent RC infill drilling attempted to close this spacing down to around 20–30 m ahead of the mining front. Grade control drilling is completed using RC rigs and is drilled as 30–60 m holes on an offset grid pattern of 8 × 6 m. This results in a significant data set of close-spaced, high-quality drilling that is used to estimate the grade control model, inform change of support for the Resource, assist in reconciliation and update the planning model.

BIG RED

Historically, drill targeting within the Kaveroi system has focused on known feeder structures such as the Levien and the Darby Faults (Figures 3 and 5). Subsequent follow-up drilling with a tighter grid pattern in the Levien structure's hanging wall revealed Big Red's up-dip mineralisation trend. One deep-probing grade control RC hole intersected 19.5 m at 8.9 g/t gold and 212 g/t silver in the same area. In response to this intersection, a concerted drill campaign commenced with the aim of understanding this area, this resulted in significant intersections, including:

- 21 m at 167 g/t gold and 638 g/t silver (including 1.5 m at 2187 g/t gold and 3476 g/t silver)
- 33 m at 106 g/t gold and 420 g/t silver
- 79.5 m at 11.7 g/t gold and 129 g/t silver
- 49.5 m at 17.9 g/t gold and 521 g/t silver.

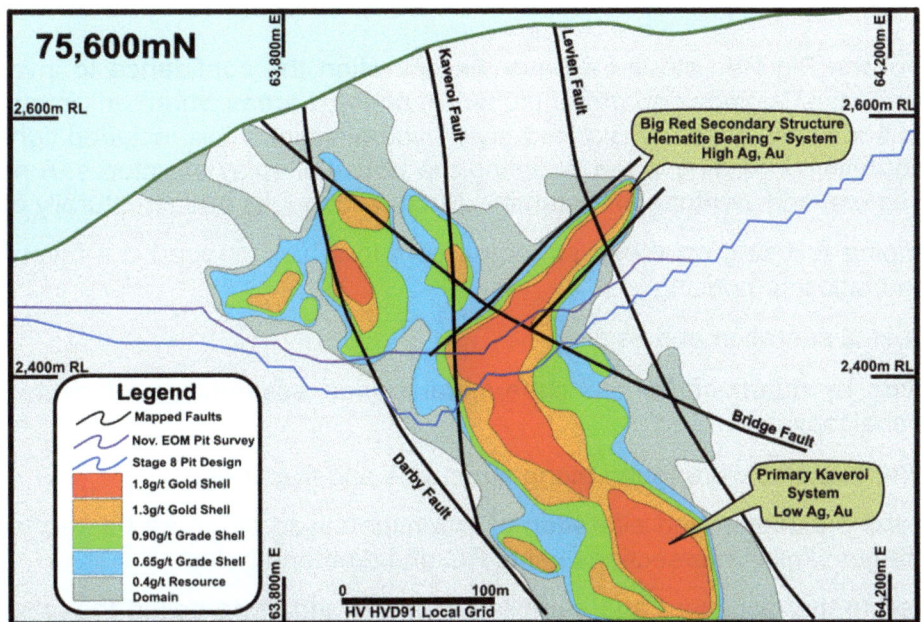

FIG 5 – Section through the Kaveroi deposit showing the Big Red structure. The bottom of the Big Red lens, below the Bridge Fault, is the high-grade Kaveroi load and is different in composition from BR.

Structurally, this lode dips 56° towards 238° and is recognised as a conjugate fault striking approximately 20° more westerly than the Kaveroi lode and dipping almost perpendicular to the 60° easterly dip of the main Kaveroi orebody (Figures 4 and 5).

The Big Red structure is believed to result from the reactivation of pre-existing Kaveroi structural conduits and the overprinting of NW normal faults (Andim and Nogat Faults – Figure 4). Andim Fault bounds the Big Red Lode to the north, Nogat Fault to the south, Levien to the east and Bridge Fault to the west (Figure 5).

Mineralisation in the Big Red lode comprises predominantly wide zones of silica-sericite ± Mn-Fe Oxide alteration with stockworks of quartz fragments ± pyrite ± base metals ± minor specular hematite. The high Au/Ag Big Red Lode results from reactivation and overprinting associated with multiple cycles of early banded quartz, pyrite + chlorite, hematite (minor Au/Ag) + sulfide epithermal vein system overprinted by hydrothermal maganocarbonate + base metal sulfide (pyrite, sphalerite, galena and chalcopyrite) veins.

CHANGING STRATEGIES

The discovery of such a high-grade lens in what was, until this point, a low-grade orebody presented significant challenges to the mining operations. Strict grade management was required to ensure that the high-grade was blended through the Mill at a rate that provided maximum recovery so as not to flood the circuit with high-grade gold and, especially, silver. The Mill was optimised for a 1.8 g/t gold and 32 g/t silver head grade; sending 10+ g/t gold and 200 g/t silver material to the Mill could adversely overwhelm the Mills capacity to handle the metal, resulting in a significant metal loss to tails.

The rugged nature of the topography means a relatively small ROM pad and limited stockpiling space, so stockpiling the material for any extended time adversely impacts the ROM's ability to operate efficiently by cluttering up the available space. Thus, ensuring the dig block design and scheduling were adequate for a timely and ongoing supply of high-grades to the ROM and subsequent blends without impacting the smooth operation of loaders and trucks on a tight ROM pad was important.

Further and importantly from a mine life perspective, were there more of these high-grade structures that had been missed and what could be done to find them?

Exploration

The discovery of the Big Red structure gave the operation the confidence to invest in additional drilling that allowed the Geology department to target and test areas where drill intersections implied the potential for additional lenses of Big Red-style mineralisation. This included tighter drill spacing for the resource definition drilling when targeting the potential splay structures. A new focus on pit mapping was commenced, centring on potential structural conduits and structurally disrupted zones.

Robust pit mapping focusing on these structural corridors has mapped out favourable host rock alteration assemblages comprising:

- Early hematite alteration and veining.
- Overprinting by quartz-carbonate base-metal sulfide assemblage of pyrite + sphalerite + galena ± chalcopyrite.
- Where Mn-Fe oxides were found along structures and fractures.

Understanding the orientation and alteration of this mineralised structure has allowed the geology department to target similar intersections using RC and diamond holes.

The mapping led to the discovery and sampling of a new structure called the Izzy Structure. The resulting map is shown in Figure 6, with results for the sampling shown in Table 1. The Izzy Structure sits outside the current Resource model but within the current pit design, making for an easy win for additional ounces.

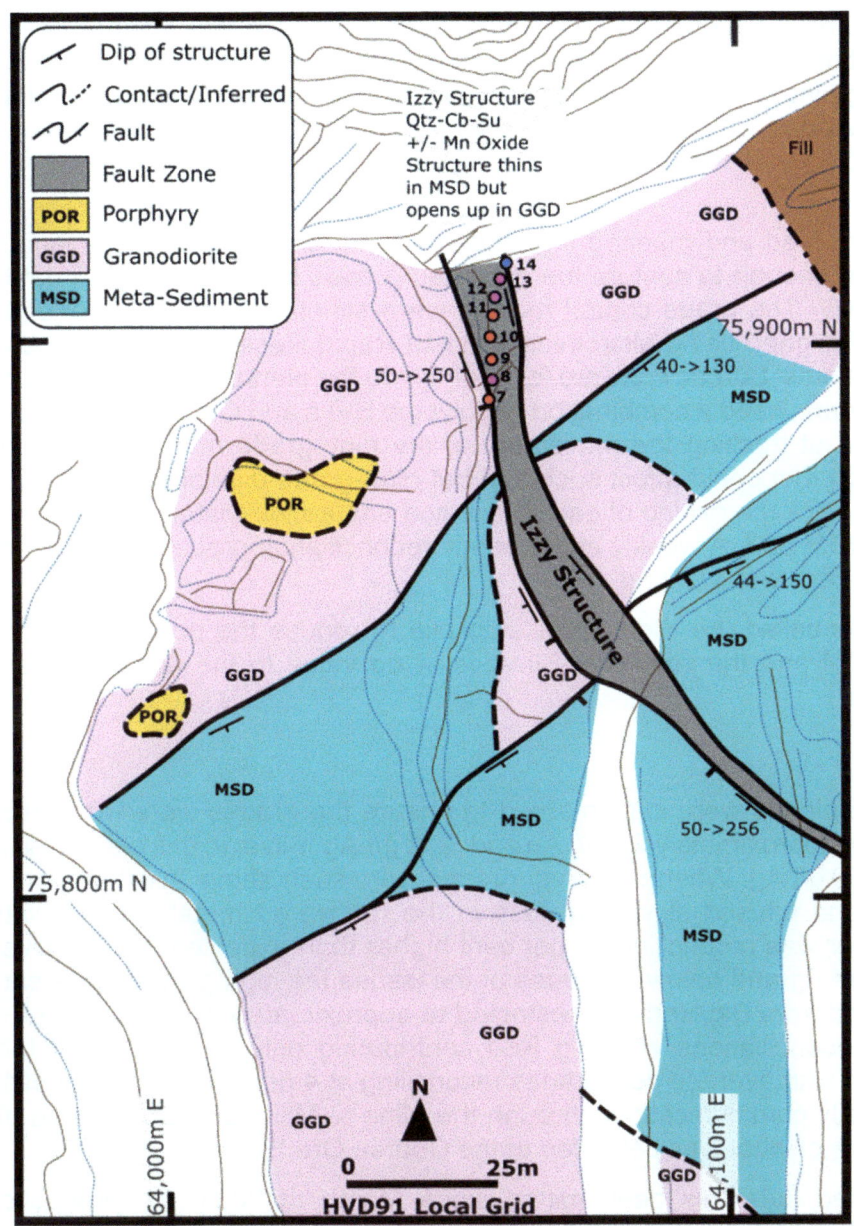

FIG 6 – Drafted Pit Map showing the newly identified Izzy Structure and location of channel samples.

TABLE 1

Channel sample results across the newly mapped Izzy Structure.

Sample #	SampleID	Au g/t	Ag g/t
7	0804135	0.40	28
8	0804136	0.25	29
9	0804137	47.49	487
10	0804138	0.61	42
11	0804139	0.42	45
12	0804140	14.37	428
13	0804141	6.28	278
14	0804142	0.25	18

The Izzy Structure is currently targeted with close-spaced grade control drilling and additional Resource Definition drilling and will be included in the mine plan over 2024 and 2025. Further, mapping and targeting are ongoing, with significant work on understanding the relationships between the Big Red lenses and the rest of the Hidden Valley-Kaveroi orebody.

Mining

To control the ore feed and manage the ROM pad, dig block sizes were managed to ensure the blocks were large enough to capture the expected grades but small enough not to dilute the block grades significantly. The grade control models were simulated using Golder's OBO software; the digblocks were designed using Micromine's Alastri Rapid Reserver software before being brought into Dassault Systems' Surpac software for coding into the planning and grade control models. Blast movement monitoring used a combination of polypipe BVI's and Orica's Orepro 3D; this combination proved very good at tracking the movement of the high-grade material and allowed the mine to reduce the dilution that could affect such thin but high-grade digblocks. Additionally, spotters were used in the pit to track the mining of each block and ensure the material made its way to the correct location on the ROM pad. A weekly and monthly reconciliation process on volumes was completed to ensure the process was working.

The stockpile was called the High Silver stockpile to reduce the potential for illegal miners to be drawn to the ROM and the stockpile. Silver is of no value to the artisan miners who attempt to frequent the area.

Blending

Strict controls on blends were implemented to ensure the grades were managed at a moderately consistent level of approximately 2.5 g/t Au and 52.2 g/t Ag. Silver in this blend was monitored closely to keep it below 60 g/t Ag. Whenever silver grades were seen above this threshold, it overloaded the leaching circuit and caused significant loss in the recovered metal. Reconciliation from the Mill indicated the silver was reconciling 13 per cent higher than expected, so the silver head grade fed was set at 52.2 g/t Ag and below. Because of the issues the high-grades were causing the Mill, the percentage of feed from Big Red was restricted to approximately 15 per cent per month of the total blend. Monthly reconciliations with Big Red contributing only 15 per cent of the blend feed still reconciled higher, with average gold grades reconciling at 4 per cent and silver at 12 per cent above expected. The daily communications through the Mine to Mill meetings became pivotal in ensuring the Mill was aware of what was being fed to the Coarse Ore Stockpile (COS).

While the Big Red lode has been mined since March of 2023, full production from the lens commenced in May 2023. The ore production from Big Red since May 2023, ex-pit was 1 157 617 t at 2.54 g/t Au and 76.0 g/t Ag for 94 691 gold oz and 2.83 Moz of silver. The additional ounces from the Big Red lens, over and above that expected from the Ore Reserve model, were 10 513 oz of gold and 519 480 oz of silver. The impact of Big Red on Hidden Valley production is shown in Table 2. A significant increase in ore tonnage and ounces produced was seen over the period the Big Red lens was mined and fed.

TABLE 2

Reconciliation of tonnes mined from Hidden Valley-Kaveroi between March 2023 and December 2023 when Big Red was fed through the Mill.

	Tonnes	Au g/t	Ag g/t	Au oz	Ag oz
Reserve Model (RSV)	2 010 223	1.99	37.0	128 686	2 391 892
Grade Control Model (GC)	2 292 652	2.00	46.3	147 260	3 415 899
Reconciled Mined (RM)	2 327 256	2.07	51.7	154 915	3 868 726
RM – RSV	317 033	0.08	14.7	26 230	1 476 834
	116%	104%	140%	120%	162%
RM – GC	34 604	0.07	5.4	7 656	452 827
	102%	104%	112%	105%	113%

CONCLUSIONS

The discovery of the Big Red lens from an isolated intersection 100 m above the orebody in the hanging wall generated significant excitement and problems for the Geology team in equal measure. The discovery of the Big Red lens stimulated additional mapping, leading to the identification of new lenses outside the current mine plan. The high-grades made the department rethink its dig block design and blending strategy. Overall, the high-grades and additional tonnes were positive for the operation in 2023. The grade control system continuously under-called the grades, particularly the silver, even with no top cuts applied to the composite files. Reconciliation of the milled material has shown a strong positive reconciliation against both the Ore Reserve and the grade control models; unfortunately, the high-grades proved difficult to manage despite the team's best efforts.

The unknown presence of Big Red in what was considered a well-drilled orebody highlights the risk of drilling a Resource based on a model with a strong bias in drill hole direction. Often, what is important in geology is not what is in the data, but rather, what is not. Trusting that the database is accurate and that the model is fully informed can mislead the geologist into mistaken assumptions, or even cause them to miss important clues that can lead to significant additions to the deposit, both of which will drive lost value. In this case, the geologists logging the original core noted the missing core, but never followed up on where it might have gone. When the intersection results were returned, the validation process did not identify that there was a high-grade intersection that was missing core, the final assays where not taken back to the core for review. These were the early steps that could have identified the structure when pit geometry would have allowed additional drilling to be conducted in a timely manner. Had these steps been done, would there have been changes to the mine plan, would subsequent periods of slowdown in mining and care and maintenance have been avoided? Unknowable, but perhaps the costs involved would have been mitigated had there been a better understanding of the location and tenor of the Big Red lenses. These are tasks the geologists on-site should routinely do, these relatively small tasks are the steps required to build a robust geological model. In an age where AI and Machine Learning is becoming common place, it is more important than ever to ensure that the underlying geological database is as robust and fully informed, as possible, else, how can you trust that the output is reliable?

ACKNOWLEDGEMENTS

The authors extend their thanks to the whole Geology Team for being available on-site to manage the long hours and good problems such a surprise brings to an operating mine. Particular mention goes to Michael Mondo for the long hours of mapping and detailed geological descriptions and Delilah Magautu for her outstanding management abilities in holding the fort together over the last several years. Thanks also to Harmony for allowing us to publish this good news story and supporting us in presenting at the 2024 International Mining Geology Conference.

REFERENCES

Bodorkos, S, Sheppard, S, Saroa, D, Tsiperau, C U and Sircombe, K N, 2013. New SHRIMP U-Pb zircon ages from the Wau-Bulolo region, Papua New Guinea, Geoscience Australia and Mineral Resources Authority, Papua New Guinea. doi:10.11636/Record.2013.025.

Corbett, G J, 1994. Regional structural control of selected Cu/Au occurrences in Papua New Guinea, in *Proceedings of the Geology, Exploration and Mining Conference* (ed: R Rogerson), pp 57–70 (The Australasian Institute of Mining and Metallurgy).

Corbett, G J and Leach, T M, 1998. Southwest Pacific Rim Gold-Copper Systems: Structure, Alteration and Mineralization, Society of Economic Geologists. doi:10.5382/SP.06.

Dow, D B, 1977. A Geological Synthesis of Papua New Guinea (No. 201). Bureau of Mineral Resources, Geology and Geophysics, Commonwealth of Australia.

Dow, D B, Smit, A J and Page, R W, 1974. 1:250,000 Geological Series – Explanatory notes, Wau, Papua New Guinea.

Harmony Gold, 2023. Mineral Resources and Mineral Reserves. Available from: <https://www.harmony.co.za/operations/mineral-resources-mineral-reserves/> [Accessed: 9 Jan 2024).

Nelson, R W, Bartram, J A and Christie, M H, 1990. Hidden Valley gold-silver deposit, in *Geology of the Mineral Deposits of Australia and Papua New Guinea* (ed: F E Hughes), pp 1773–1776 (The Australasian Institute of Mining and Metallurgy: Melbourne).

Pascoe, G J, 1991. Hidden Valley Gold project development summary, 1987–1991, in *Proceedings of the PNG Geology, Exploration and Mining Conference 1991* (ed: R Rogerson), pp 69–76 (The Australasian Institute of Mining and Metallurgy).

Porter GeoConsultancy Pty Ltd, 2003. Morobe Goldfield – Wau, Edie Creek, Hidden Valley, Kerimenge, Bulolo Papua New Guinea, Portergeo Database. Available from: <https://portergeo.com.au/database/mineinfo.asp?mineid=mn812> [Accessed: 5 January 2024].

Rinne, M L, Cooke, D R, Harris, A C, Finn, D J, Allen, C M, Heizler, M T and Creaser, R A, 2018. Geology and geochronology of the Golpu porphyry and Wafi epithermal deposit, Morobe Province, Papua New Guinea, *Economic Geology*, 113:271–294. doi:10.5382/econgeo.2018.4551.

Sapiie, B and Cloos, M, 2004. Strike-slip faulting in the core of the Central Range of west New Guinea: Ertsberg Mining District, Indonesia, *Geo, Society Am, Bull*, 116:277. doi:10.1130/B25319.1.

Opening the geologists tool box – building a grade control system

C Rowett[1], P Rolley[2] and D O'Rielly[3]

1. Geology Manager, Hillgrove Resources, Unley SA 5061.
   Email: caitlin.rowett@hillgroveresources.com.au
2. Chief Geologist and Exploration Manager, Hillgrove Resources, Unley SA 5061.
   Email: peter.rolley@hillgroveresources.com.au
3. Senior Geologist, Hillgrove Resources, Unley SA 5061.
   Email: daniel.orielly@hillgroveresources.com.au

ABSTRACT

The Kanmantoo Cu-Au-Bi-Ag deposit is hosted within Cambrian age metamorphosed sediments with copper mineralisation hosted by structurally controlled sulfide dominant veins and vein stockworks. This style of mineralisation provides underground (UG) grade control challenges like those experienced at many orogenic UG gold mines including variable geometry of the mineralised envelope at higher cut-off grades and the short scale variation in continuity of the mineralisation. The best mitigant to reduce the risk of incorrect stope outlines is a higher data density, which is a problem of both cost and time.

The grade control process implemented at Kanmantoo utilises an innovative toolbox of data science, geology and risk optimisation to optimise grade control costs and enable greater data density whilst quantifying risks.

- Use spectral scanning and Artificial Intelligence/Machine Learning (AI/ML) training of assayed drill core for direct Cu, Al, S and Fe grade estimation, and application of the scanning and ML model for Cu grades and geologic mapping of entire development faces at a pixel scale. Geological mapping is interpreted by geologists and interpreted vectors attributed for implicit modelling.

- UG grade control diamond drilling is logged and on-site analysed by Hand-held X-ray Fluorescence (HHXRF) of crushed material, calibrated through extensive trials to produce Cu, Fe, Al, S values. The rapid assay turn-around-time allows for areas of unexpected changes in Cu continuity to be highlighted for infill drilling.

- The integration of the spectral scanning, HHXRF Cu, Fe, Al, S values and interpolated geologic vectors, is used to build simulation models of the Cu grades. These simulation models are processed via a Maximum Profit and UG stope optimisation algorithm to identify the optimal stope volume given the grade uncertainty and the locations of specific stope uncertainty that may require additional data.

In conclusion, a novel approach to UG grade control has been implemented that is based on sound science, good geologic understanding and state-of-the-art technology.

INTRODUCTION

The Kanmantoo Copper Gold Deposit is situated 55 km south-east of Adelaide and the site has a long history of copper mining. Mining first commenced at the Kanmantoo Deposit in the 1800s as a series of small high-grade copper and copper-gold underground mines producing over 24 000 t of ore at around 8.5 per cent copper (Rolley and Wright, 2017). Between 1970–1976 (Verwoerd and Cleghorn, 1975) a single open pit was developed on the Kavanagh lode system and approximately 4.1 Mt at 0.87 per cent Cu, 0.06 g/t Au was extracted from the open pit before operations ceased in 1976 due to low copper prices. Between 2011 and 2019 Hillgrove Resources (HGO) mined three open pits at Kanmantoo for around 26.9 Mt at 0.58 per cent Cu, 0.1 g/t Au. In 2023 Hillgrove commenced an underground mine on the Kavanagh Cu-Au lode system and Cu production is expected in 2024.

The key geological objective during the development phase of the underground operation has been to develop a grade control system that provides geological information for stope definition whilst optimising costs and manages the risks of knowledge inadequacy and information turnaround time (TAT). To achieve this, a hybrid approach has been developed that utilises traditional methods such as underground (UG) diamond drilling, with newer technologies such as Hand-held X-ray Fluorescence (HHXRF) assaying of drill core and Near Infrared/Short wave Infrared (NIR/SWIR) spectral scanning of UG development faces.

The introduction of NIR/SWIR spectral scanning with Artificial Intelligence/Machine Learning (AI/ML) to convert the spectral responses to quantitative Cu, Fe and S values and for mapping of the alumino-silicate alteration of every UG heading provides a high density of images of the 3D spatial geometry of the structurally controlled mineralisation and can be used in the development of geological and mineralisation wireframes. The grade control drilling is optimised to provide information between the development levels for grade estimation.

DEPOSIT GEOLOGY

The Kanmantoo Trough of the Cambro-Ordovician Kanmantoo Province of south-east South Australia is host to numerous Cu, Pb, Zn, Ag, Au, As, Bi deposits over its 300 km length of which the Kanmantoo Cu-Au deposit is the largest known, albeit over 95 per cent of the Kanmantoo Province is under the cover of the Tertiary-Quaternary Murray Basin. The Kanmantoo Trough is a rift basin formed 521–511 Ma and the Cu-Au mineralisation at Kanmantoo is within the Tapanappa Formation of the Kanmantoo Group. The Tapanappa Formation is described as a sequence of turbidites, siltstones and pelites with minor Fe sulfide members (Rolley and Wright, 2017).

The Delamerian Orogen has affected the Kanmantoo Group with three episodes of deformation between 514 and 490 Ma (Foden *et al*, 2006). At the Kanmantoo deposit, D2 is the dominant deformation event with upright north–south folds with shallow southerly plunges and a dominant axial-plane parallel S2 fabric (Offler and Flemming, 1968).

Buchan-style regional metamorphism to upper amphibolite grade has affected the Kanmantoo Province during the Delamerian Orogen with peak metamorphism coinciding with peak deformation during D2. The Kanmantoo deposit is within the andalusite-staurolite metamorphic zone (Offler and Flemming, 1968).

Magmatism within the Kanmantoo Province can be correlated with each of the three Delamerian age deformation events, with S-type intrusives dominant with D1, S- and I-type intrusives with D2 and A-type intrusives post deformation (Foden *et al*, 2020). Recent work by The Geological Survey of South Australia (GSSA) indicates that post-deformation magmatism continues through to 465 Ma and possibly 390 Ma. There is minor I- and S-type felsic magmatism at the Kanmantoo deposit and local environs that is pre- and syn-peak deformation.

The main copper orebody at Kanmantoo (Kavanagh) is discordant both to bedding and to all mineral lineations and is partly sympathetic with the axial planar schistosity of the D2 Kanmantoo Syncline which averages ~65° -> 084°. Figure 1 shows the broad distribution of the copper zones at Kanmantoo and their general copper geometry relative to the S2 axial planar schistosity (Rolley and Wright, 2017).

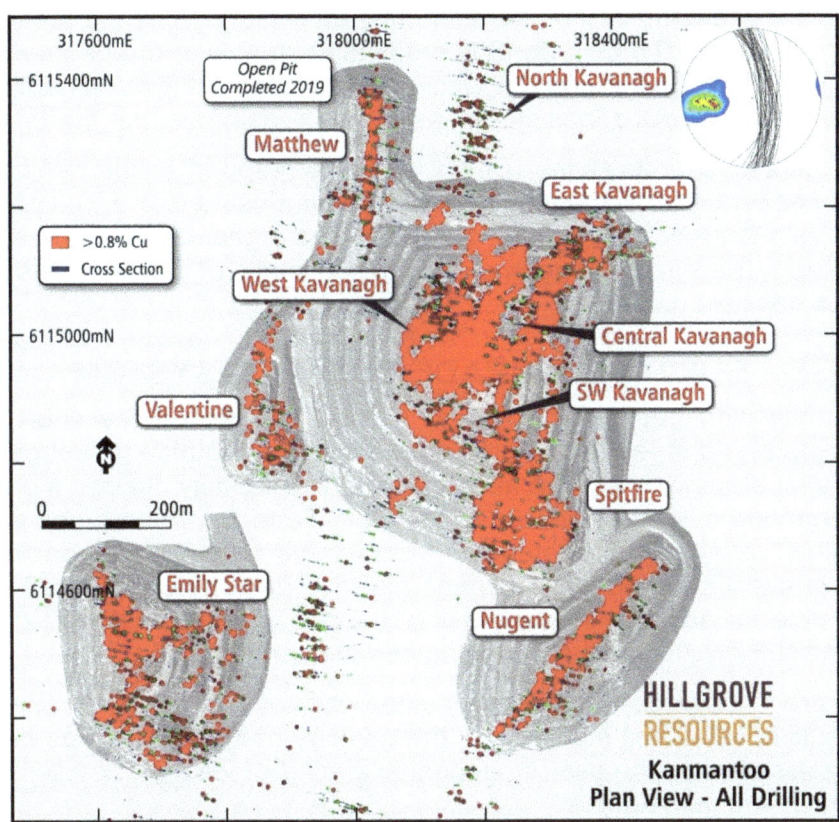

FIG 1 – Plan view of all drilling and open pit blasthole assays greater than 0.8 per cent Cu illustrating the various mineralised lodes and the stereonet of S2 cleavage.

Mineralisation consists of quartz poor sulfide veins and veinlets and is dominated by pyrrhotite and chalcopyrite with minor bismuthinite and pyrite. The sulfide mineralisation (seen in Figure 2) overprints all metamorphic assemblages and overprints or infills all peak deformation structures. Age dating suggests mineralisation occurs 490–485 Ma. Copper and silver are strongly correlated and gold and bismuth are strongly correlated and often overprinting chalcopyrite and pyrrhotite. Alteration mineralogy is dominantly garnet, biotite, chlorite with minor K-spar and magnetite. The mineralisation has been drilled to 800 metres below surface as a continuous structural system. Mineralisation at the Kanmantoo deposit is interpreted to be epigenetic (Oliver *et al*, 1968) and the result of basinal metamorphic fluids mixing with granite derived fluids channelled along pre-existing structures by magmatic driven geothermal fronts during syn- to post-peak deformation.

FIG 2 – Example of high-grade Cu-Au mineralisation from NQ core hole 23KVUG064 through the East Kavanagh Lode from 51.39 m to 55.40 m downhole.

On a mining scale, the geometry of the alteration halo as evidenced by the presence of garnet and chlorite and sulfur grades of ~0.6 per cent S is predictable, however the geometry and scale of the structurally controlled Cu-Au mineralisation within this corridor is highly variable. Variability of the geometry and scale of the Cu-Au zones increases with increasing copper grade.

There are natural copper cut-off grades (COG) which correspond to the density of S2 controlled copper (0.2 to 0.4 per cent Cu), epigenetic sulfide dominant veins (0.4 to 0.8 per cent Cu) and the brecciated sulfide vein stockworks (>0.8 per cent Cu). The highest grades are associated with the brecciated stockworks however the geometry and continuity of these zones is the hardest to predict from sparse drill hole data.

METHODOLOGY – BUILDING A GRADE CONTROL SYSTEM

The variable mineralisation geometry at higher cut-off grades and the short scale variation in continuity of the mineralisation requires a high data density to produce robust geological shapes at different Cu COG's for stope design. The need for high data density creates a problem of both cost and time for underground mining compared to open pit. To mitigate the problem of cost, always a high priority for Mine Management, strategies to reduce the cost per data point were investigated whilst maximising relevant knowledge of the mineralisation geometries. This led to the two pillars of the grade control data collection processes being the use of XRF of UG NQ drill holes for grade control analysis and NIR/SWIR spectral scanning of development faces.

Due to the simple ore mineralogy with coarse-grained chalcopyrite the only copper bearing mineral in the epigenetic sulfide veins and stockworks, the geology of the deposit has facilitated the use of HHXRF for drilling assays and the use of spectral scanning of washed development faces to map the 3D Cu distribution and vein density.

Use of ML/AI to train the spectral responses from assayed drill core to directly estimate Cu, Fe, S grades has been a break-through for the effective use of spectral scanning coupled with LiDAR positioning for direct 3D mapping of copper grades and silicate/sulfide veining.

The flow chart in Figure 3 illustrates the key steps and decision points with the grade control process. These steps are supported by robust quality assurance and quality control (QA/QC) processes specific to the data collection method.

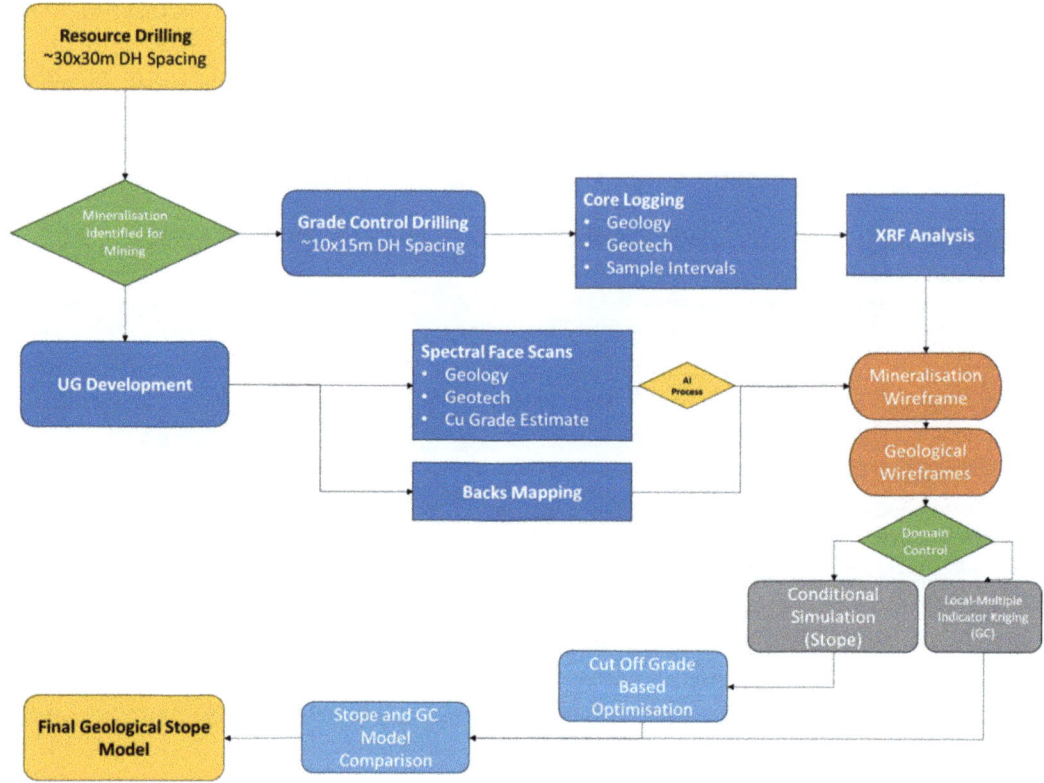

FIG 3 – Grade control system flow chart.

Grade control drilling

UG core drilling is conducted on approximately 10 m × 15 m spacing with the scheduling of each drill fan optimised for significant mine planning/design decision points. For example, the drill data required to facilitate the planning and design of development drives is completed as a first priority before the drilling for stope grade estimation.

Optimisation of the delivery of drill hole data results (facilitated by rapid HHXRF analysis of the drill holes) allows for dynamic changes in drill hole planning and drill hole sequencing. This provides opportunities to add additional drill holes as required or to remove redundant drill hole designs from the planned mine schedule prior to drilling. The TAT for drill hole data is optimised for on-site priorities and is not impacted by an external lab, with the opportunity for HHXRF results within 24 hours of hole completion for high priority drill holes.

Core logging

During core logging, basic geological information such as lithological boundaries, alteration type and intensity and location of mineralisation is collected. From this the sampling intervals are identified which are nominally 1 m and amended to reflect the geological boundaries. Prior to sampling, all core has high quality, lighting controlled, photographs taken for long-term records.

Alongside the logged geological information, geotechnical information is also collected that facilitates the creation of a Q value (rock mass quality index) for the entire drill hole. These Q values are used in the stope design to identify areas of poor ground conditions.

The logging procedures and core yard layout have been optimised to allow for safe and optimal material movement and for the maximum information to be collected in the minimum time possible.

XRF analysis

As a result of the successful application of HHXRF grade control processes for the open pit operations at Kanmantoo (Arbon and Rolley, 2019), HHXRF processes have been developed for UG grade control drilling.

During open pit operations, 2 kg samples were collected via a cyclone and riffle splitting from 3 m intervals from the percussion drilling of blastholes, the drilling of which produced drilling cuttings with a mixed particle size varying from 100 to 150 mm chips down to <1 mm fines. These 2 kg samples were then sieved and the fine fraction (<2 mm) retained for XRF analysis. Removing the coarse size fraction reduced the particle size heterogeneity and the sample error associated with coarse grained mono-mineralic chalcopyrite vein particles from the sample before presentation to the HHXRF for analysis (Arbon and Rolley, 2019).

UG grade control is based on NQ diamond drill core and consequently core samples for underground grade control must be wholly crushed prior to presentation to the HHXRF so the entire sample is homogenised and representative of the entire interval.

This required the development of a sample preparation laboratory on-site, where the whole core for the designated sample interval is prepared for the HHXRF analysis. At the on-site sample laboratory, the designated whole core interval is presented to the Orbis OM100 crusher and crushed to >70 per cent passing 2 mm. Once crushed the samples are rotary split into duplicates and can then continue to be prepared for HHXRF analysis using the same process as was developed for the open pit operations, including a high density of blanks and standards to monitor and capture crusher contamination.

A rigorous QA/QC process has been established for the operation of the on-site sample preparation and assay performance. A significant part of the QA/QC process is the comparison of 4-acid digest results (from an independent third party assay laboratory) to the HHXRF results for the same sample intervals. This work has allowed HGO to produce a robust regression factor for the HHXRF Cu, Fe, S results. This regression factor is continually reviewed as new QA/QC 4-acid digest results are returned for selected HHXRF samples. To facilitate the review of the regression used for the HHXRF samples, each month a selection of waste, low and high Cu grades from the various lodes are sent for 4-acid digest and assessment against the HHXRF results. The regression used is built to consider different grade bins and the variance seen at different copper grades. Based on this work the current Cu regression used has an R^2 of 0.96. Figure 4(top) shows the scatter plot of the regressed Cu grades against the 4-acid-ICP-MS (inductively coupled plasma mass spectrometry) Cu grades for the same intervals. Figure 4 (bottom) shows the QQ plot of the regressed Cu values and the 4-acid-ICP-MS Cu grades for the same sample intervals.

For grade estimation and geologic domaining, where 4-acid assays are available they are given priority over HHXRF assays for the same interval.

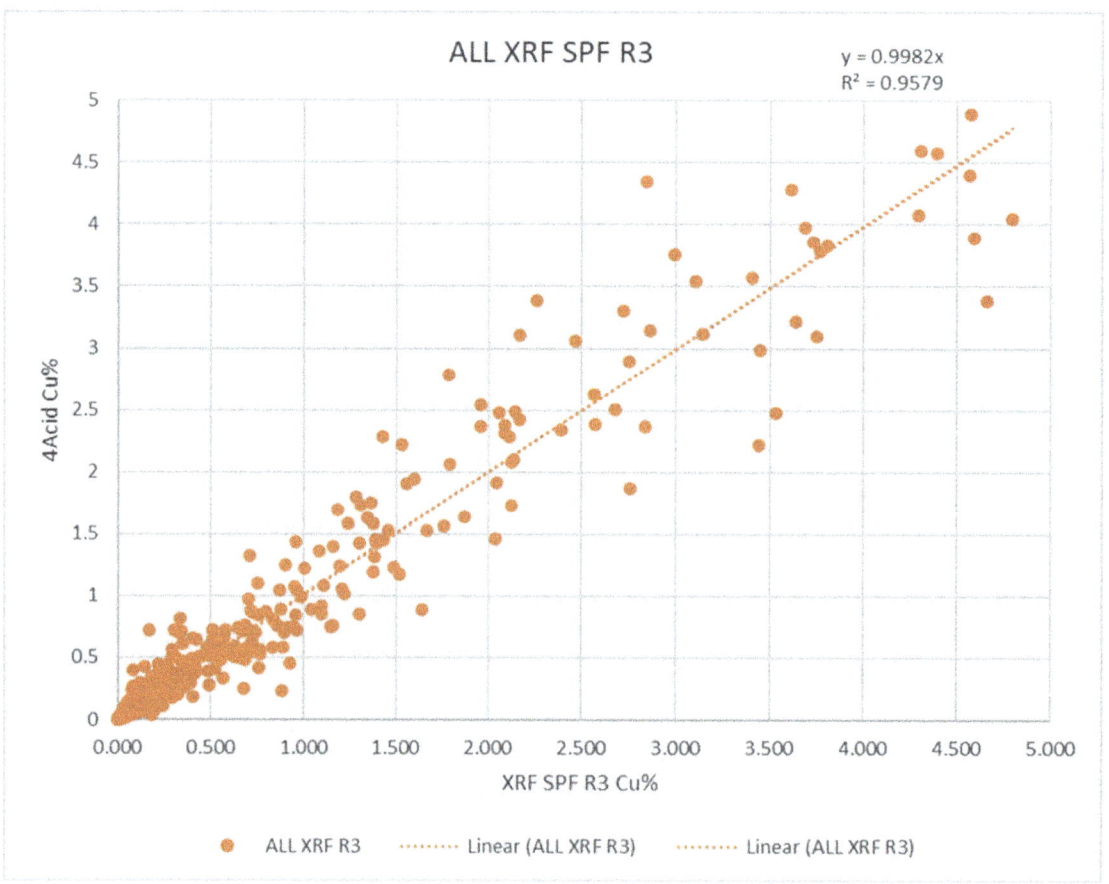

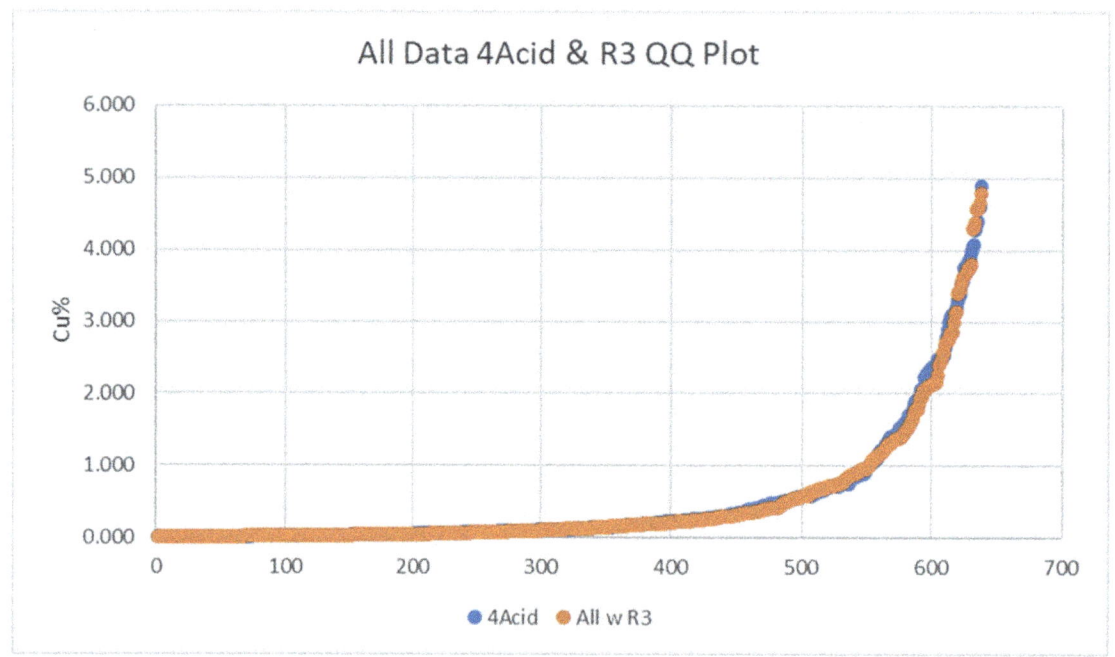

FIG 4 – (top) scatter plot of the regressed Cu grades against the 4-acid-ICP-MS Cu grade. (bottom) QQ plot of the regressed Cu values and the 4-acid-ICP-MS Cu grades.

The ability to continue using HHXRF for analysis of grade control core significantly reduces the cost of assaying, reduces the TAT for information and allows for optimal sample drilling/XRF prioritisation. The use of the HHXRF ensures that the benefits of the low cost and rapid TAT of data analysis can be captured to assist with local drill hole location optimisation and additional drill hole metres if justified.

Underground development

As development of ore drives commenced, the opportunity to collect a vast amount of geological information evolved using an innovative NIR/SWIR spectral scanning process.

Spectral face scans

Traditional face mapping and face sampling has several limitations ranging from safety concerns to sample quality and repeatability. Typically, when face mapping, a geology team member is usually given 30 mins in the mining cycle to map and sample a face. Within this time frame they are trying to safely create a detailed geological map including the mineralisation distribution on the face, any visible structures including joints, veins etc including their dip and dip direction measurements and take panel samples across the face that are representative of the geology. As a result each face is mapped with varied levels of detail based on time available, experience of the team member and ground support coverage over the face. The rock samples taken from the face also have a varied level of reliability based on what material can physically be chipped from face and often include an unconscious sampling bias of ore material versus waste.

Consequently, a process or system that will allow for the creation of unbiased and repeatable face maps and grade information within the scheduled face access window was required. Thus, increasing geological information from a structurally controlled mineral system that can be used in stope design.

Following previous work completed by HGO using the HyLogger™ (Mauger *et al*, 2018) the use of NIR/SWIR spectral data was identified as an opportunity to spatially map alumino-silicate assemblages and, using ML/AI of the NIR/SWIR spectra to map at a fine scale the Cu, Fe, S grades over entire consecutive development faces. Following research into new mining technologies, especially in the spectral space, led to partnering with Plotlogic and use of the OreSense® System UG. The OreSense® system (seen in Figure 5) combines hyperspectral data with LiDAR and red, blue green (RGB) imagery to deliver high resolution mineral maps and grade estimates using advanced ML/AI algorithms (Plotlogic, 2024).

FIG 5 – OreSense® scanner mounted on a 4WD vehicle safely deployed to an UG ore drive face.

To establish the calibrated algorithms to estimate Cu, Fe and S estimates from NIR/SWIR spectra, spectral scanning of ~6700 m of exploration drill core for which 4-acid digest ICP-MS assays were available was completed (reference training data). The drill hole data set for the ML/AI algorithm training was selected to ensure that all expected lithologies, alteration assemblages and copper abundances were included in the calibration data set.

The development testing of this ML/AI model was an iterative process with tests of the model regularly completed by scanning of drill holes not included in the reference training data set, until the spectral scan estimated copper grade correlated with the 4-acid digest assay results at numerous Cu COG's and as continuous intersection lengths. Calibration of the mineralogy was further assisted by a suite of X-ray diffraction (XRD) results also selected from representative intervals of common lithologies and mineral assemblages.

The OreSense® scanning unit is permanently mounted on the UG geology light vehicle. The unit is driven directly to the ore headings and offers a quick set-up for scanning. The vehicle is stationed under supported ground and no person is subject to any risk. The unit is operated by the UG Field Technicians and at each heading the location of the scanning unit is recorded against survey points. When at the face, the scan unit operators follow a standard operating procedure, ensuring repeatability and consistency of scans. Scans are processed and available for use within an hour of return to the surface.

The spectral scanner is calibrated prior to every scan against a known copper bearing standard and against a white background to ensure there is no instrument drift or bias.

Continual improvement of the field operating processes is ongoing with new information from the face scans to take into account the specific conditions of the UG environment. The development of these operating procedures and equipment optimisation will be ongoing as additional data is collected.

The scans produce a level of geological data that could not be captured by a geologist in the time frame allocated in the mining sequence. The mapping of the alumino-silicate alteration, identifies the micro-scale structures including bedding, folds and the overprinting of the copper rich mineralising fluids that is not visible to the naked eye. The copper images map at a pixel scale of the mineralised areas and this information is used to create a quantitative face grade estimate. Examples of the scan data are seen in Figure 6.

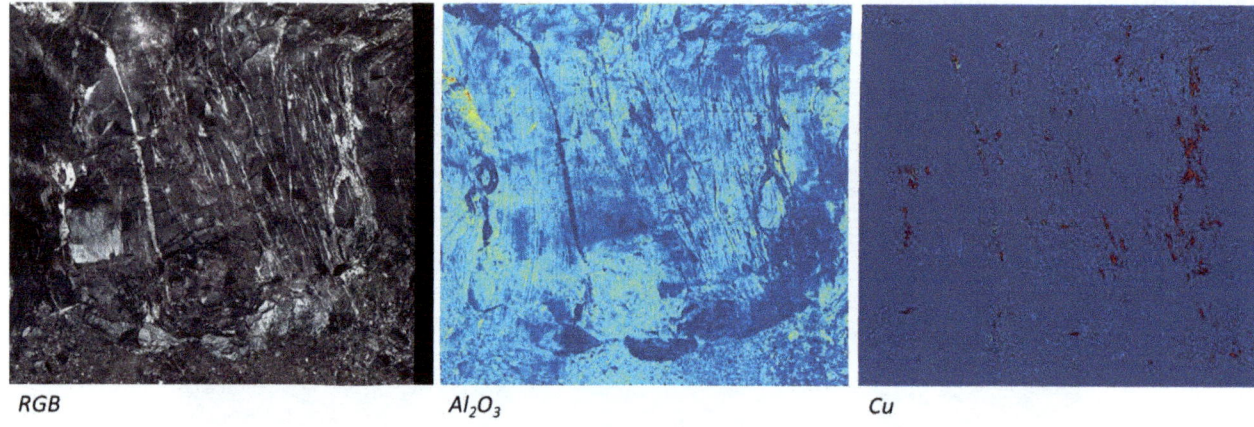

RGB Al_2O_3 Cu

FIG 6 – Face scan results.

The copper face estimates are used daily for designation of ore, waste and stockpile destination.

The compilation of the face scans for each ore drive (Figure 7) provides HGO with a detailed face map and copper distribution every 4 m horizontally along the drives and every 25 m vertically.

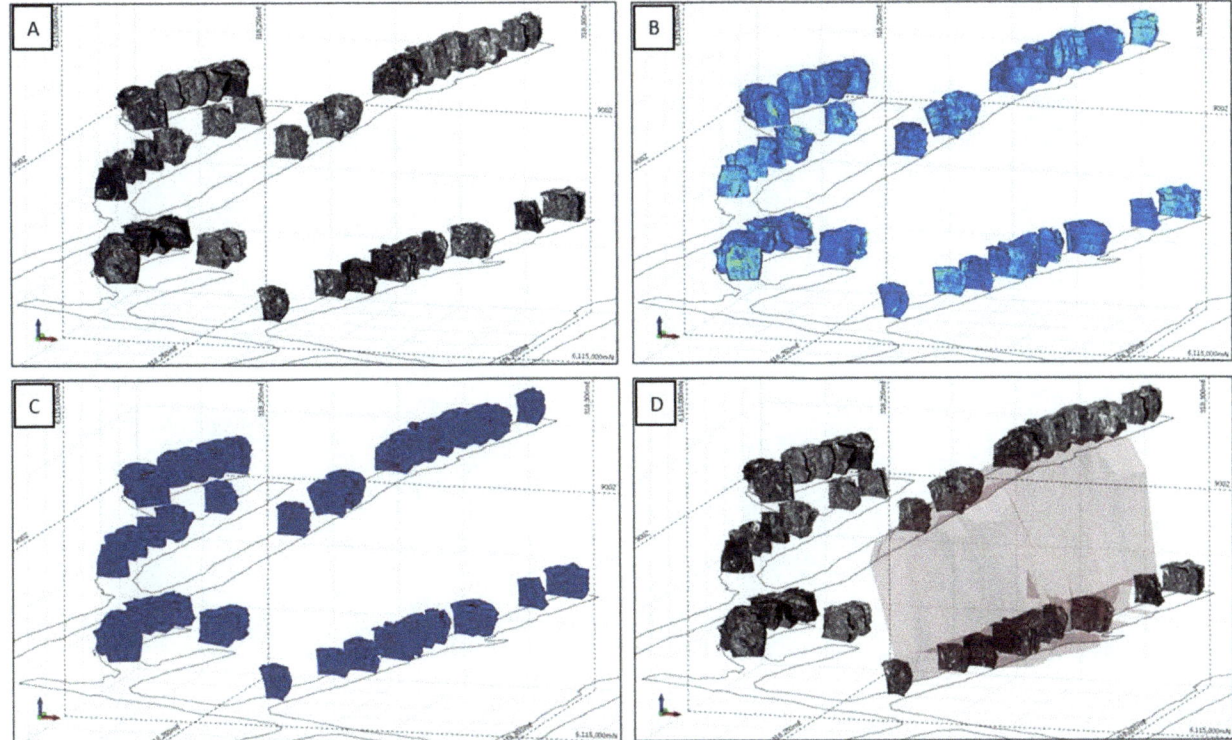

FIG 7 – Geolocated face scans as viewed within Micromine for two underground levels. (a) RGB images on geolocated LiDAR scans. (b) Al_2O_3 images on geolocated LiDAR scans (c) Cu grade distribution image on geolocated scans (d) RGB scans in relation to designed stope shape.

These detailed face images and copper distribution maps can then be utilised to assist with the creation of copper domain wire frames and mineral structure boundaries. These domains can be created both implicitly from coded strings or explicitly by manually modelling the copper or alteration boundaries as seen in Figure 8.

Further to the geological information collected from the scans, geotechnical information is also collected, including joints and from the LiDAR information their dip and dip direction. The mineralogical information from the scan also provides information on joint infill.

When synthesised, all the layers of information from the spectral, RGB and LiDAR face scans are processed to generate detailed maps of ground conditions, grade distribution for pillar placement and mineralisation geometry to assist the engineers with stope design. This mapping is on much higher data density than if relying solely on drill hole data. Due to the discontinuous nature of the structurally controlled mineralisation within the mineralised envelope, the detailed maps assist with the interpolation of localised mineralisation truncations and geometry changes. Face scanning provides frequent, repeatable and unbiased information in almost real time which allows for prompt decision-making.

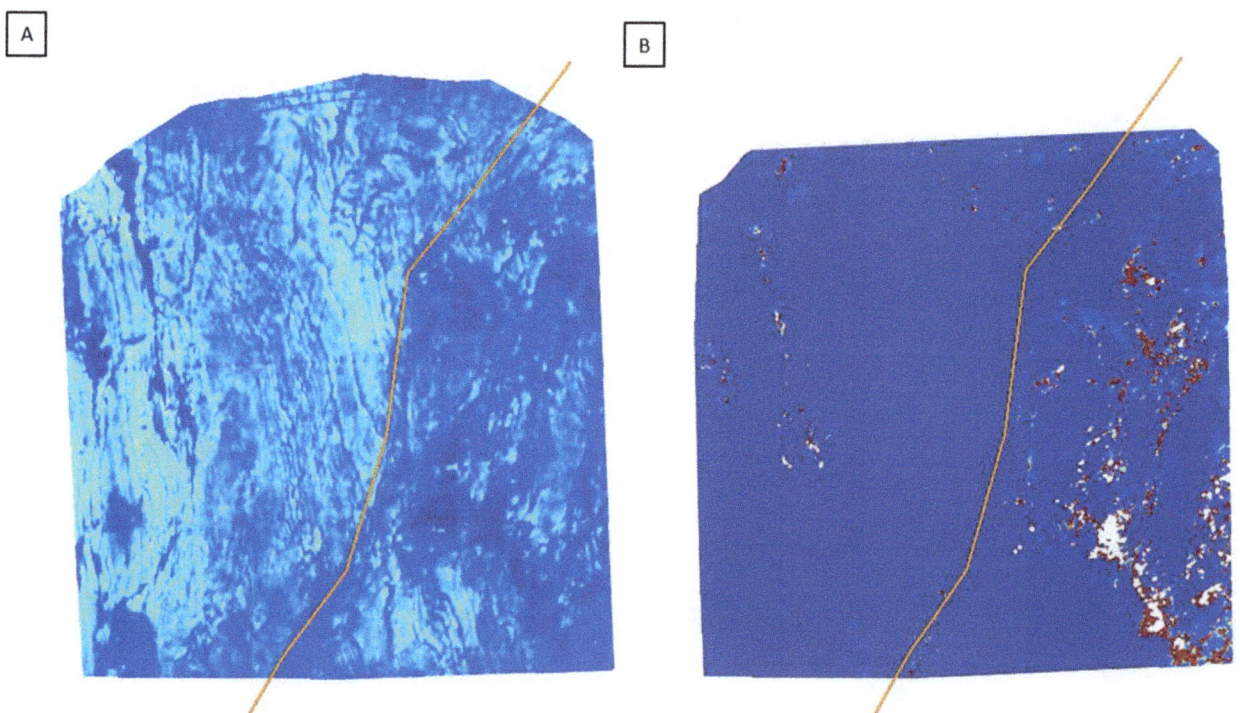

FIG 8 – (a) Al_2O_3 face image with digitised string of the alteration boundary. (b) Cu mineralisation image with the same digitised string identifying the high-grade mineralisation distribution boundary.

Mapping

To corroborate the spatial location of the observations made by the face scanning, qualitative visual backs mapping is also completed. This ensures a member of the geology team has inspected all headings, visually confirmed all the geological and geotechnical structures to build their own knowledge of the mineralisation system and ensure they do not only rely on processed information observed on the computer. This helps to build a healthy scepticism and curiosity which ensures that the underlying algorithms used in the face scanning will be routinely questioned and improved.

Estimation

Conditional simulation

From 2016 onward open pit operations utilised conditional simulation to build copper grade maps and a profit-loss optimisation function was implemented to select the optimal regions for mill processing. A detailed description of the process was described by (Reid and Arbon, 2017). This is also the preferred method for estimation of stope grades, following the excellent reconciliations seen during the open pit.

The open pit grade control data density (~3 m × 3 m) is not possible to replicate in the UG environment due to available drill locations, time and cost. To mitigate this lack of data in an UG environment, greater attention is given to the additional step of domain control within the conditional simulation process. These domains are created from the geological information collected by the face scanning and the diamond drill hole data. They allow for simulations to be created for each mineralised domain and within each alteration/waste halo.

The suite of simulation maps of copper grade is processed with an UG optimisation algorithm and a block model generated of expected copper grade and economic value given the copper grade uncertainty. The block model provides the planning engineer's guidance for optimal stope design and corporate objectives.

FUTURE WORK

Validation of the grade control process will be tested as mill production commences and reconciliation data is collected. With the reconciliation information further refinement will occur to the face scanning algorithms and the HHXRF regressions, along with review of the wireframe domain models.

Updates and refinements will be continuous to ensure that the grade control process continues to strive for the fastest possible TAT on information while maximising the information collected from every dollar spent.

CONCLUSIONS

The grade control processes developed have demonstrated that it is possible for practical integration of new and old technology to safely achieve high density of quantitative data in almost real time that is relevant to the demands of current mining processes. This paradigm shift in the operating efficiency for the geologic data acquisition has enabled data to be optimally collected for both the mine plan and for geological observations without being subject to qualitative observations and able to be dis-associated from the commercial objectives of third-party laboratories.

Data sets able to be provided by the integration of HHXRF and spectral scanning includes:

- Pixel scale copper grade maps of every development heading.
- LiDAR imagery of every heading.
- RGB photography of every heading.
- Geologic mapping of every heading at pixel scale.
- Drill core assays within 12 hours.

Information maps able to be generated include:

- Average copper grade of every heading.
- 3D alteration domains.
- 3D structural vectors and planes.
- Geotechnical structures, orientation and joint mineralogy.

The time/efficiency savings of the XRF/spectral scanning and optimisation are hard to quantify as the fixed cost of drilling contracts and labour has been re-prioritised to allow for additional targets or areas of geometric uncertainty to be analysed.

High resolution of quantitative data collection from underground development has been clearly demonstrated to be safely achievable and expedited through the use of the spectral scanning. This has also allowed for a greater pool of geological team members to collect unbiased, repeatable, high-quality data that is available for use within an hour of return to the surface unlike traditional face sampling techniques awaiting biased and limited sample information 12 to 24 hours after face collection. The scanning process facilitates optimal decision-making for the mining development cycle.

Other benefits observed by the geological team include:

- Low geological cost per sample providing the flexibility to take more samples as required.
- Efficient 3D classification and mapping of ore and waste material during development.
- High volume of data assisting with mineralisation geometry characterisation, geotechnical structures, lithological domains and mineralisation abundance.
- Additional time to analyse geological data as opposed to time spent collecting data.

The development of the grade control process is not static and will continue to develop and change. Review of the process will come with the addition of reconciliation information following production and will allow for determination of the robustness of estimations.

Once the robustness of the process is established a determination on any additional steps that may be required or steps that are redundant will be made. Illustrating that a geologist's toolbox is a dynamic place that with ongoing scientific curiosity, technology and a focus on optimisation (cost and time) new techniques and process will continue to develop.

ACKNOWLEDGEMENTS

The authors wish to acknowledge the hard work and continuing dedication of the Hillgrove Team from both the Open Pit and Underground Operations to engage and optimise new and innovative processes for grade control. Without their courage to engage in change processes and to 'think' about how these changes can be practically, safely and objectively applied to achieve these quality data sets, the new systems would not have been established. Thank you to the team at Plotlogic for their partnership in developing our underground spectral scanning systems.

We also thank Hillgrove Resources for permission to publish this paper.

REFERENCES

Arbon, H and Rolley, P, 2019. Continuous improvements in mine geology at Kanmantoo. in *Proceedings of 11th International Mining Geology Conference 2019*, pp 11–25 (The Australasian Institute of Mining and Metallurgy: Melbourne).

Foden, J D, Elberg, M A, Dougherty-Page, J and Burtt, A, 2006. The timing and duration of the Delamerian Orogeny: Correlation with the Ross Orogen and Implications for Gondwana assembly, *The Journal of Geology*, 156:10–33.

Foden, J D, Elburg, M, Turner, S, Clark, C, Blades, M L, Cox, G, Collins, A S, Wolff, K and George, C, 2020. Cambro-Ordovician magmatism in the Delamerian orogeny: implications for tectonic development of the southern Gondwanan margin, *Gondwana Research*, 81:490–521.

Mauger, A, Keeling, J, Rolley, P and Arbon, H, 2018. Spectral analysis of drill core from the Kanmantoo copper deposit, *MESA Journal*, 88(3):15–24.

Offler, R and Flemming, P D, 1968. A Synthesis of Folding and metamorphism in the Mount Lofty Ranges, South Australia, *Journal of the Geological Society of Australia*, 12(2):245–266.

Oliver, N H S, Dipple, G M, Cartwright, I and Schiller, J, 1968. Fluid flow and metasomatism in the genesis of the amphibolite-facies, pelite-hosted Kanmantoo copper deposit, South Australia, *Journal of the Geological Society of Australia*, pp 245–266.

Plotlogic, 2024. Plotlogic [online]. Available from: <https://www.plotlogic.com/> [Accessed: 10 January 2024].

Reid, M and Arbon, H, 2017. Setting new standards at Kanmantoo – effective grade control procedures, in *Proceedings of the Mining Geology Conference* (The Australasian Institute of Mining and Metallurgy: Melbourne).

Rolley, P and Wright, M, 2017. Kanmantoo Copper Deposits, *Australian Ore Deposits*, Monograph series 32 (ed: N Phillips), pp 667–670 (The Australasian Institute of Mining and Metallurgy: Melbourne).

Verwoerd, P and Cleghorn, J H, 1975. Kanmantoo copper orebody, *Economic Geology of Australia and Papua New Guinea* (ed: C L Knight), third edn, Monograph series 5, pp 560–565 (The Australasian Institute of Mining and Metallurgy: Melbourne).

Cosmic-ray muon tomography – new developments in mining applications at BHP's Leinster and Olympic Dam mines

D Schouten[1], J Townsend[2], E Western[3], M Owers[4], J Taylor[5] and A Tunnadine[6]

1. CTO and Co-Founder, Ideon Technologies, Richmond, BC, V6V 2K9, Canada.
   Email: doug@ideon.ai
2. Principal Innovation, Think & Act Differently, Powered by BHP, Perth WA 6000.
   Email: jared.townsend@bhp.com
3. Superintendent Geoscience Planning, BHP Nickel West, Perth WA 6000.
   Email: erin.western@bhp.com
4. Principal Geophysicist, BHP Nickel West, Perth WA 6000. Email: matthew.owers@bhp.com
5. Principal Geologist (Geological Modelling), BHP Olympic Dam, Olympic Dam SA 5725.
   Email: james.taylor3@bhp.com
6. Senior Consultant (Geology), SRK Consulting (Australasia), Sydney, NSW 2000.
   Email: atunnadine@srk.com.au

INTRODUCTION

Cosmic-ray muons are charged elementary particles that arise naturally from cosmic radiation interacting with the Earth's upper atmosphere. Particle showers of muons bombard Earth steadily and are attenuated by their interaction with matter along their trajectory. Muons can penetrate deep into the Earth's crust, up to thousands of metres and their attenuation in matter is proportional to the density of material the muon passes through.

By measuring muon flux through detectors positioned beneath the surface, the average density in the overburden within a wide field of view above the detector can be determined. Each detector produces a radiographic image of the rock mass above it and combining these images enables 3D tomographic reconstructions of subsurface density.

Rock density models derived via muon tomography can be used in a wide variety of settings from exploration and resource characterisation, to void mapping and monitoring. Where density is a proxy for mineralisation, muon tomography can complement existing exploration techniques in both near-mine and early-stage settings. Ideon Technologies, Inc has successfully deployed muon tomography to map MVT and VMS-type base metal deposits, as well as unconformity-hosted uranium deposits. Initial studies and forward modelling suggest both IOCG and magmatic nickel sulfide mineralisation types can present sufficient density contrast when compared to their host lithologies, making it amenable to using muon tomography.

BHP Nickel West has leveraged this capability to successfully image massive sulfide deposits over a 3 km strike extent at Cliffs and Leinster nickel mines in Western Australia, including in a multiphysics analysis that combined ground and airborne gravity data sets with muon tomography, in addition to other key insights ranging from regional geological mapping to valuable geotechnical information. The density models were extensively validated by drill information, where that was available.

BHP Innovation, in conjunction with Ideon Technologies and SRK Consulting, are also conducting a muon tomography survey at the world-class Olympic Dam deposit, utilising it to delineate near-mine exploration targets, constrain orE–Waste geometries between drill holes and characterise voids. Early results indicate that muon tomography can be a valuable source of 3D data early in the mining life cycle, facilitating better drill targeting, informing mine designs and enhancing orebody models.

MUON TOMOGRAPHY ANALYSIS METHODOLOGY

The muon tomography method is predicated around measuring the muon rate (number observed within a given exposure time) along each ray path between the surface of the earth and a detector. As described in detail in Schouten, Furseth and van Nieuwkoop (2022) and Schouten and Ledru (2018), the expected number of muons detected within a given pixel p corresponding to a 3D solid angle Ω is derived with the following equation:

$$N_{exp} = \Delta t \int_{\Omega_p} \left(\int_0^\infty D_\mu(E, \theta) \cdot p(E; \mathcal{O}) dE \right) \alpha(\hat{n}) \, d\Omega$$

Ideon has developed a robust, precise model that describes the expected intensity of muons at the surface $D_\mu(E, \theta)$ as a function of initial muon energy E and zenith angle θ, using data collected over the past six decades (Schouten, Furseth and van Nieuwkoop, 2022). The units of intensity are $cm^{-2}\ s^{-1}\ sr^{-1}$. As the muons propagate through rock, they lose energy at a rate that is heavily dependent on the density of the rock they traverse. Therefore, the probability $p(E; \mathcal{O})$ that a muon survives through a certain rock opacity ($\mathcal{O} = \int_{\text{path}} \rho(x, y, z) \, d\ell$) is calculated with Monte Carlo simulations and multiplied by the muon flux at sea level to determine the muon intensity at any given rock depth. The muon survival probability depends on the initial muon energy at the surface of the Earth; hence, the contribution is summed overall relevant energies. Finally, the detector active area and performance is accounted for in the $\alpha(\hat{n})$ term. The integration is performed over all angles corresponding to a cone Ω_p above the detector defined for each pixel p in the radiograph and multiplied by the exposure time Δt.

A geological model is developed based on *a priori* geological information about the rocks and other features in the study area and the statistical comparison of the expected number of muons to those observed in the field reveals high- and low-density with significance correlating with regions of more and less muon attenuation. The statistical significance in each pixel is defined as:

$$z = \frac{N - N_{exp}}{\sigma_N}$$

where N is the observed number of muons and σ_N is the statistical uncertainty of the expected number, which follows the Poisson distribution. In the following, the prior geological model consists only of an assumed uniform density ρ_0 in the subsurface, underneath a digital elevation model taken from airborne LIDAR data.

The expected and observed muon data sets are subsequently inverted into a 3D density model $\rho(x, y, z)$ by minimising a global function ϕ that incorporates a data misfit term for the muon tomography data and/or other data sets (eg ground and airborne gravity) with complementary sensitivities compared to the density model and a model objective function that ensures model smoothness and that can incorporate optional additional information, such as spatial weighting terms and a reference model:

$$\min_{\rho(x,y,z)} \phi = \min_{\rho(x,y,z)} (\phi_D^\mu + \lambda_g \phi_D^g + \beta \phi_M), \text{ where}$$

$$\phi_D \sim \sum_i (z_i - \sum_j G_{ij} \rho_j)^2, \text{ and}$$

$$\phi_M \sim \sum_{w \in \{x,y,z\}} \alpha_w \int_V \left|\frac{\partial \rho}{\partial w}\right|^{q_w} dV + \alpha_r \int_V \rho^p \, dV$$

where ϕ_D^μ, ϕ_D^g are the data misfits for the muon tomography and gravity data compared to an initial model, respectively, λ_g scales the relative contribution of muon and gravity data (if any) in the inversion, β is a regularisation trade-off parameter that controls the relative importance of model complexity and data misfit and ϕ_M is a model objective function that ensures smoothness (Davis *et al*, 2011; Tarantola, 2005). G_{ij} is a (sparse, for the case of muon tomography) sensitivity matrix that relates z_i of the ith pixel in the radiographic images or gravity measurement station to the anomalous density ρ_j of the jth voxel in the image volume, α_w is a constant that penalises roughness in each of the $w = x, y, z$ coordinates and α_r is another constant that penalises deviations from a reference model. In cases where only muon or gravity data is incorporated into the inversion, the other term is simply excluded from the objective function.

The 3D density model that is derived is relative to the assumed uniform density – ie the absolute density is simply $\rho_{\text{abs}}(x, y, z) = \rho_0 + \rho(x, y, z)$. This is only a calculational tool to ensure the initial model is 'close' to the true geology and has no other bearing on the final result. The measurement is properly an absolute density measure, without any prior constraints (except where otherwise noted).

LEINSTER MINE

Leinster Mine is situated in the Agnew-Wiluna greenstone belt in Western Australia and is wholly owned by BHP and operated by BHP's Nickel West group.

At Leinster, the Perseverance Ultramafic Complex hosts several sulfide orebodies including Perseverance, Venus and Rocky's Reward. The 2 km long drive between Perseverance and Venus was an ideal location to deploy in-mine muon detectors and effectively search the volume above the drive for undiscovered massive mineralisation from within readily accessible underground infrastructure. Further to the North, near Rocky's Reward, mineralisation was hypothesised to recur down-dip, which was the basis of the Balboa exploration target. Drill testing of the Balboa exploration target was undertaken in 2021 and these drill holes were identified as an opportunity to deploy borehole muon tomography.

As indicated, two different muon-tracking detectors were used: a gallery-style detector and a narrow cylindrical borehole detector (shown in Figure 1). The properties of the gallery detectors are discussed in greater detail in (Schouten and Ledru, 2018). The borehole detectors are 3.5 m long, 89 mm in diameter and suitable for deployment in uncased HQ holes. Multiple borehole detectors were positioned in each of three drill holes at the Balboa exploration target site, connected to one another and to the surface via communications and mechanical cables in a daisy-chain configuration.

FIG 1 – The surface box (left) and borehole muon detector (right) employed in this case study. The equipment is designed, manufactured and deployed by Ideon Technologies.

The Leinster muon tomography survey (depicted in Figure 2) proceeded in three phases:

- In Phase 1 and 2, gallery detectors collected data at nine locations along an underground drive that connected mining operations for the Perseverance and Venus deposits, at vertical depths below 800 m.

- In Phase 3, borehole detectors were installed to the north to explore the Balboa exploration target zone at downhole depths ranging from 200 to 400 m.

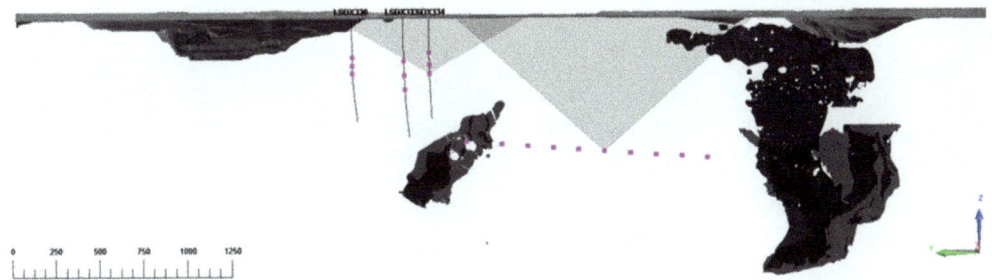

FIG 2 – Survey layout at Leinster Mine showing gallery and borehole detector locations, two example fields of view and existing mineralisation (Venus on the left, Perseverance on the right). The strike length explored in the survey was larger than 2.5 km. The total volume of rock imaged exceeded 1.5 billion m³.

The average rock density was directly measured in each direction within the fields of view of the detectors within approximately 110° opening angle and is shown in muon radiographic images in Figure 3. Each of the approximately 10 000 pixels in these radiographs is an independent, directional measurement of rock density within fields of view encompassing over 1.5 billion m³ of the subsurface.

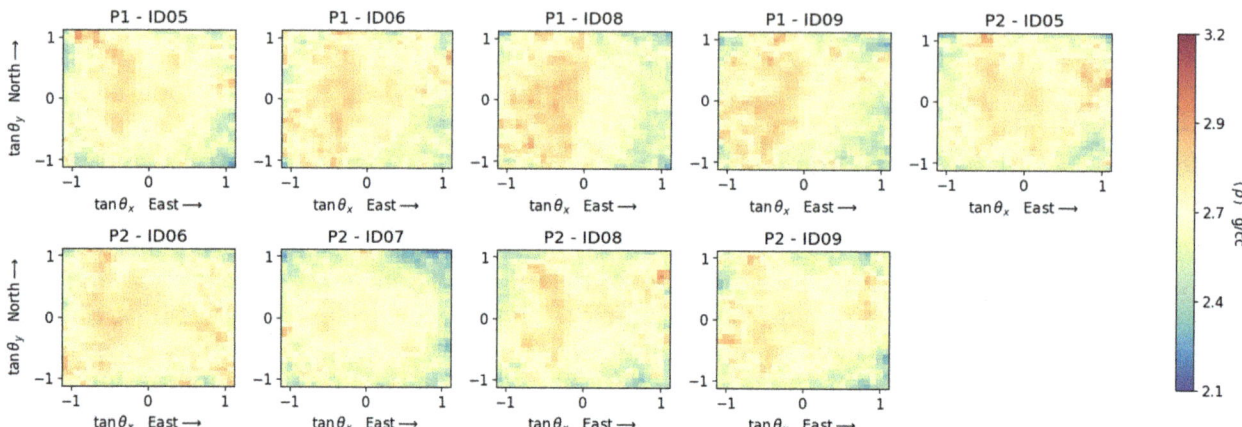

FIG 3 – Average density measurements for Phase 1 and Phase 2 detector data. High average density bands are observed in the West. In detector ID 07 data, there is a significant low-density anomaly observed to the North-East direction – which was identified after the survey was completed as due to voids from mining operations around the Venus deposit.

The data were integrated in an unconstrained (blind) 3D density inversion to yield a density model with voxels of 10 × 10 × 10 m³ over the 3 km strike extent and down to 830 m depth. As shown in Figure 4, a horizontal section of the density model near the surface yields interesting features, in particular a low-density contour that aligns very well with a large waste dump of unconsolidated rock, which unsurprisingly has significantly lower density than the country rock. There is also a very good correlation with a shear zone that is the conduit for sulfide mobilisation. This provided good confidence in the veracity of the model.

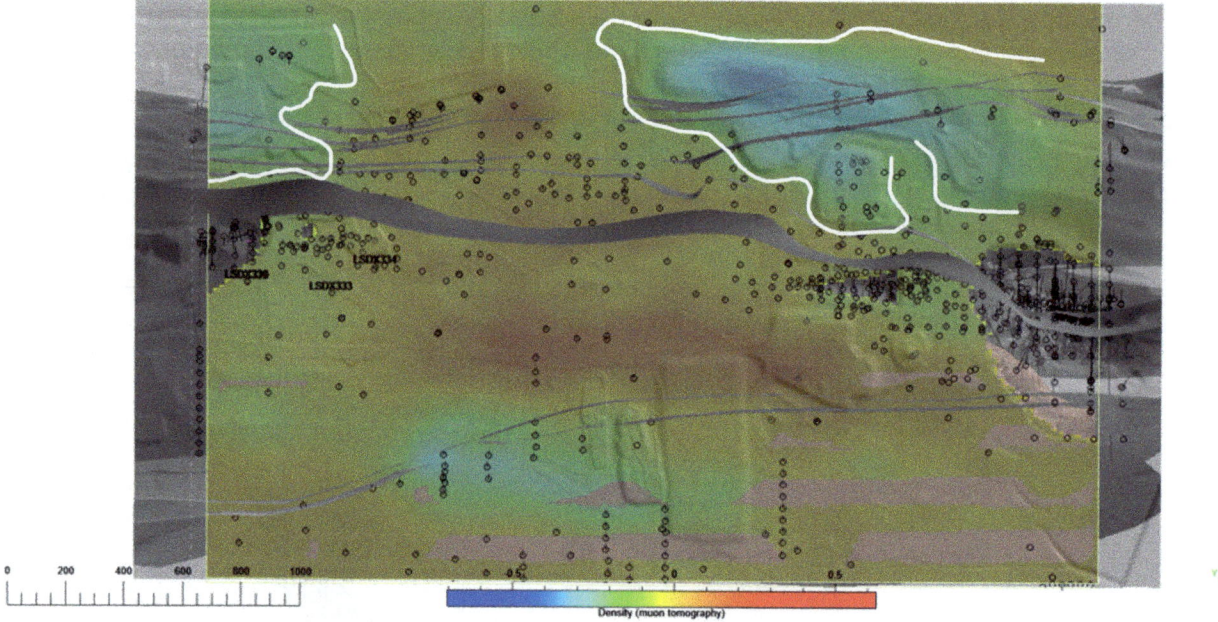

FIG 4 – Shallow horizontal section of inverse density model, derived from muon tomography data in Leinster Phase 1 and Phase 2. The surface topography and main geological contacts, as well as drill hole collars, are shown in grey scale. The high density of drilling in the North and South extents of the survey area are proximal to the Venus and Perseverance deposits.

In Figure 5, another horizontal section at 200 m depth indicates very interesting features. In particular, a North–south trending mafic intrusion is mapped very well by the muon tomography survey, especially in regions where there is significant drill data – and in regions where there is slightly less drilling density, indicates a possible adjustment to the geological model. The ultramafic UPC is seen as a clear density low, also aligning with the prior geological model very well. Finally, a discrete high-density anomaly to the East of the UPC was identified in drill data as a massive sulfide unit, shown in Figure 6. This structure was tightly resolved by the muon detectors from a distance of >750 m away.

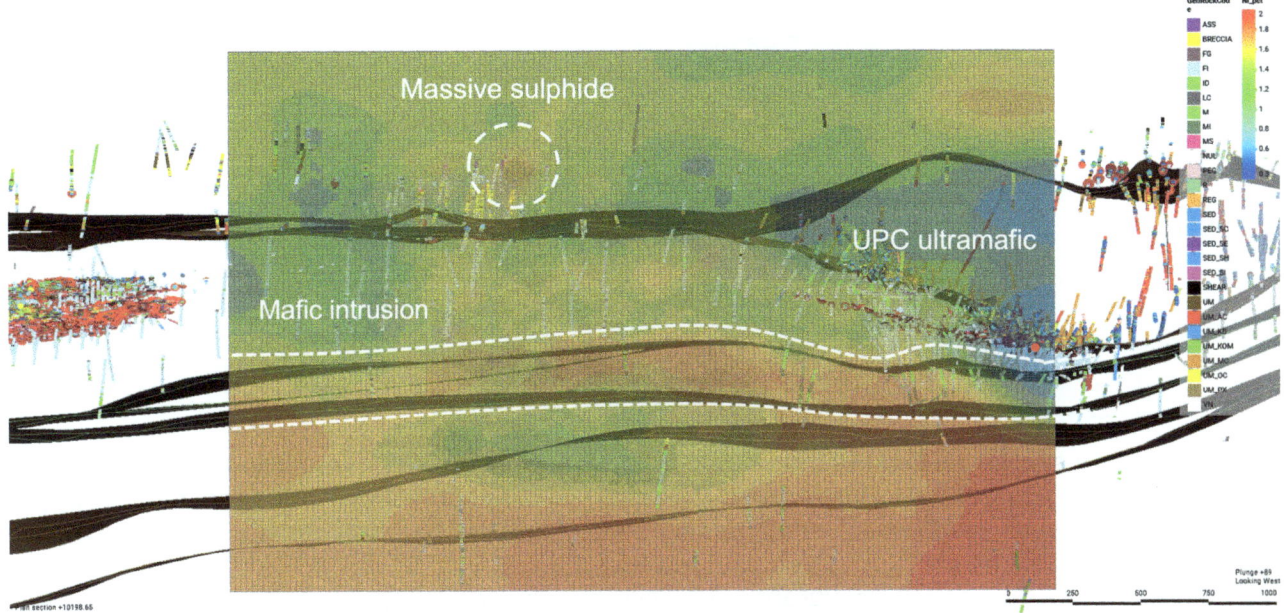

FIG 5 – Horizontal section of the muon tomography inverse model at 200 m depth. The sub-vertical mafic intrusion indicated by two geological contacts is confirmed by the muon tomography model with good resolution (the contour of the intrusion is indicated by the dashed white line), as is the high-density massive sulfide and the lower density ultramafic.

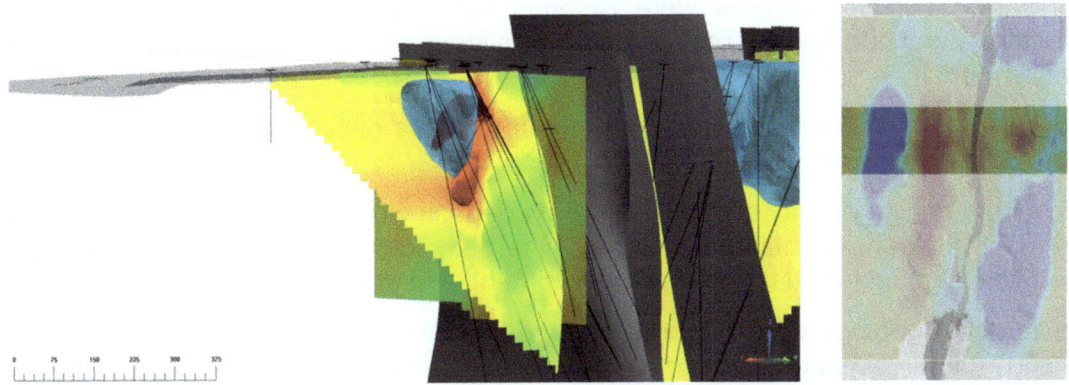

FIG 6 – (Left) massive sulfide structure, revealed in drill data and shown in the interpolated density model derived from drilling with East–West and North–South slices shown along with a red iso-surface corresponding to >3.5 g/cc. The density anomaly seen in the muon tomography data is shown by the semi-transparent blue iso-surface corresponding to a 0.3 g/cc anomaly. The massive sulfide was mapped from >750 m away by the muon detectors. (Right) – a plan view of the survey showing the density model at -50 m TVD, with the slice corresponding to the left view highlighted. The massive sulfide is clearly seen adjacent to the North–South trend proximal to the shear zone.

The gallery and borehole muon tomography data were integrated into a comprehensive density model encompassing both target areas. The inversion indicated continuity of the main geological features and alignment with an airborne gravity survey. Subsequently, a joint inversion of airborne, ground gravity and the full muon tomography data set was performed, yielding a high-resolution density model over the full volume, shown in Figure 7. The gravity data was found to be particularly helpful in constraining the model very close to surface, whereas the muon tomography data was far more powerful at depth.

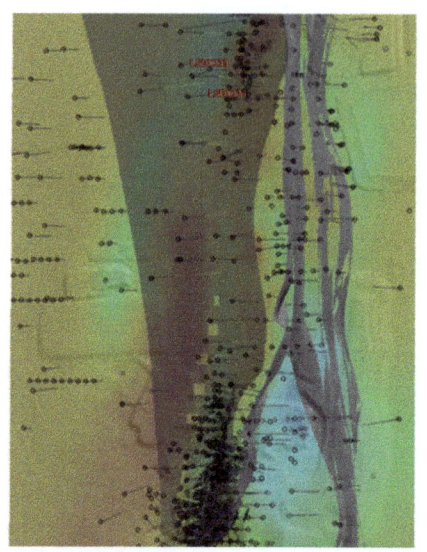

FIG 7 – Shallow section (100 m) of the unconstrained inverse density model from Phase 1, 2 and 3 muon tomography data at Leinster Mine. The main geological features are seen in as largely continuous between the Phase 1 and 2 gallery survey and the borehole survey in Phase 3. The borehole collar locations are highlighted with red and the many other drill holes near the Venus and Perseverance deposits are also indicated.

Key results from the Leinster muon tomography program were:

- Multiphysics integration of borehole and gallery muon tomography data, along with ground and airborne gravity, was demonstrated to yield a high-resolution model of subsurface density.
- Well-known surficial features were mapped with few-metre lateral resolution from muon detectors many hundreds of metres away, providing intrinsic confirmation of the validity of the density model.
- The ability to map relatively small regions of massive sulfide, from detectors installed more than 750 m away, was demonstrated.
- Voids from mining activity around the Venus deposit were clearly seen in the inverse models and in the 2D density images from one of the detectors that had this region within its field of view. This demonstrated the geotechnical potential for the technology.
- The regional geology was mapped with very good resolution from a very sparse detector density and with almost no ground disturbance and minimal operational overhead on the mine.
- The whole analysis was performed without imposition of geological constraints – indicating the capability of leveraging muon tomography for mine-scale geological mapping, ore delineation and brownfield exploration, even in contexts where the geological data is sparse.

OLYMPIC DAM MINE

Olympic Dam is an iron-oxide copper-uranium-gold deposit, located about 500 km north of Adelaide in central South Australia. It occurs within a large breccia complex within the Mesoproterozoic Roxby Downs Granite. The deposit is overlain by more than 250 m of overlying Neoproterozoic and Cambrian sedimentary rocks. It is worth noting here that geophysical measurements were responsible for its discovery.

The deposit has a large spatial extent and highly variable, complex geological structure. BHP Innovations worked with Ideon Technologies and the Olympic Dam geology team to investigate how muon tomography could be used to: (a) provide detailed mapping of the orebody and geotechnical structures; and (b) support in-mine searches for additional lobes of mineralisation with a significantly reduced level of required drilling.

In the first discovery phase of the program, gallery muon detectors were deployed at three locations in the so-called Greens area of the mine (see Figure 8), at depths down to 840 m, along with one detector being placed close to the unconformity, just below the overburden.

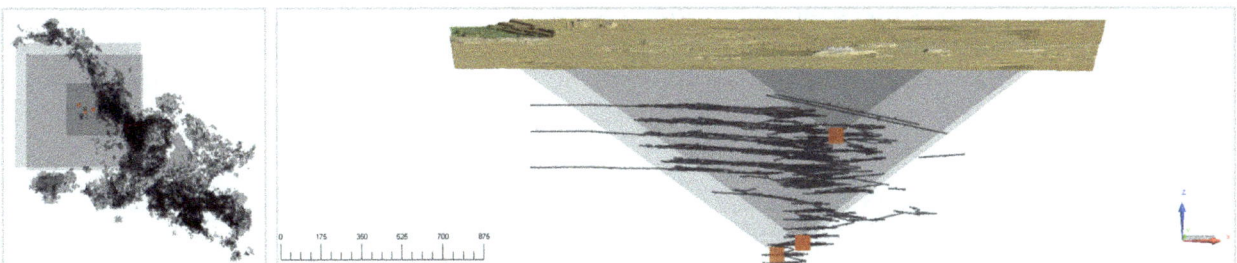

FIG 8 – Plan view (left) and zoomed in oblique view (right) of the muon survey layout at Olympic Dam. The detector locations are shown as orange squares and the fields of view are indicated by the semi-transparent pyramidal cones.

Deployment of the gallery muon detectors was delayed due to the logistical complexity of installing the sensors within a busy active mining area. The detectors are themselves robust against electromagnetic noise, vibrations, temperature variations and water flow. Nevertheless, blasting and equipment movement for mine operations understandably took precedence over the activities required to install and commission the muon detectors.

Since commissioning in March 2023, the detectors operated independently (ie without inbound or outbound internet connection) whilst power was supplied. Due to inaccessible network connectivity in the deployment locations, data were offloaded to a laptop by Olympic Dam personnel approximately every two weeks and relayed from the surface to Ideon Technologies' cloud platform for processing.

In all the following, the prior geological model consists only of an assumed uniform density ρ_0 in the subsurface below the overburden (with a more complex overburden model derived from prior drill data) and a digital elevation model taken from airborne LIDAR data. The overburden is known to be horizontally continuous in the survey region and was used to constrain the model in the top ~300 m of the survey region. Radiographic images of anomalous density significance z from one of the three locations are shown in Figure 9.

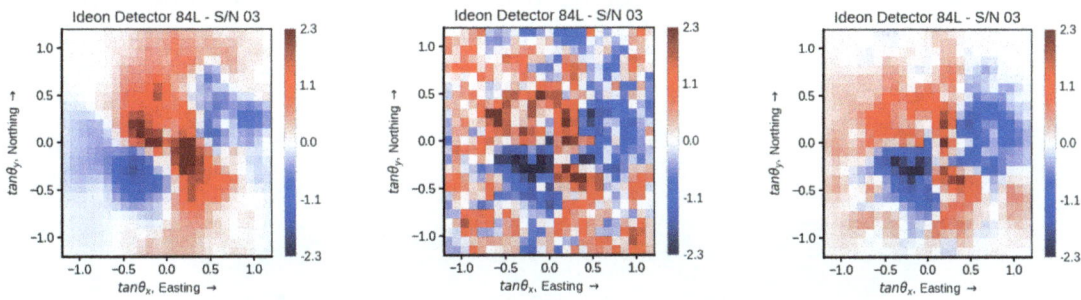

FIG 9 – Expected radiographic image of z (in units of standard deviations) from a forward simulation of the block model (left), the measured field data (middle) and the forward model of the recovered (unconstrained) density model (right). There is good alignment between the data and the forward models, both for the ground truth – indicating correct measurement of the geology, and for the inverse model – indicating good data fitness in the inversion.

As in the Leinster data analysis, the muon data were inverted to yield a model of subsurface density, with some notable differences:

- The spatial co-location of fewer detectors meant that the expected spatial resolution, especially along the depth dimension, would be significantly lower.

- Constraints were placed on the inverse model to align it with the known overburden within the survey region and the density below the overburden was constrained to be >2.6 g/cc.

In Figures 10 and 11, East–West sections of the inverse model from field data are compared to a course block model. Only the parts of the model that are within the fields of view of the detectors (ie with direct data support) are shown. A few noteworthy findings are apparent:

- The backfilled stopes are seen quite clearly as regions of lower density in the field data, albeit with coarser resolution; however, at the extremes of the fields of view, where the geometric triangulation that would normally be afforded by detectors dispersed more widely apart, the average density of mineralised rock and backfilled stopes is seen only as a region of overall higher-than-background density.

- The mineralised regions are mapped quite well, from only two detector locations, within a volume of more than 700×10^6 m^3 of the subsurface.

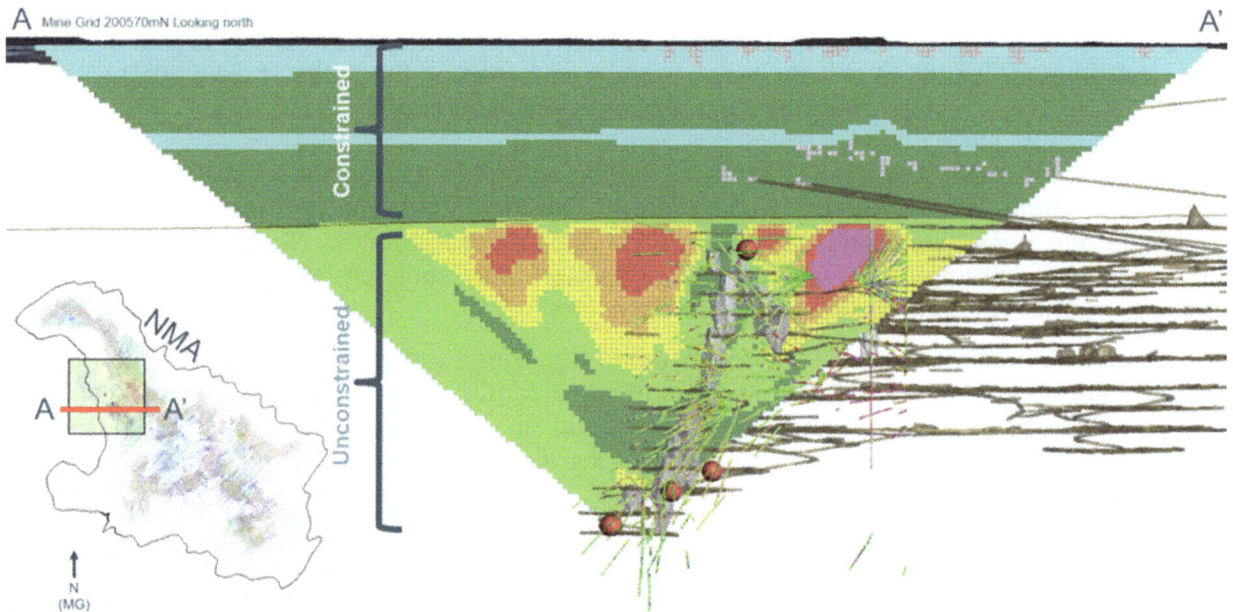

FIG 10 – E–W section through the density model in the survey area, showing the recovered density model from muon tomography, with overlain stopes (gray surfaces) and some of the drilling data. The high-density mineralisation is seen clearly in the muon tomography density reconstruction, as is very good definition of the lower density mine backfill and stopes to the West of the mineralisation.

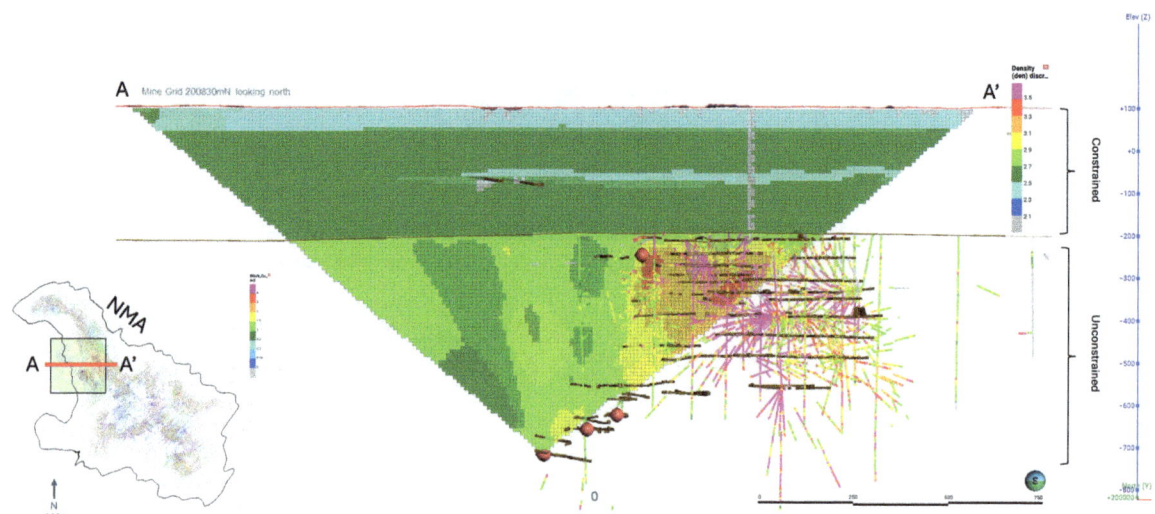

FIG 11 – E–W section through the unconstrained density model in the survey area. The high-density mineralisation is seen clearly in the muon tomography density reconstruction and confirmed in detail with extensive drilling information in the area.

Finally, in response to the deployment challenges posed by the gallery detector installation at Olympic Dam mine, Ideon Technologies has developed a thin panel detector, with equivalent spatial resolution and general robustness criteria, for easier widescale deployment on the backs of drifts or instrumentation cuddies. Parameters of this panel detector are indicated in Figure 12 and Table 1.

FIG 12 – Ideon's Mine Muon Hub installed with ceiling mount option inside a mine in North America (note: this is not BHP Olympic Dam).

TABLE 1
Ideon's Mine Muon Hub system parameters.

Panel detector parameters	
Power	120–240V A/C, 50/60 Hz, 25 W
Size	120 cm × 120 cm × 25 cm
Low maintenance	Extended deployment capabilities
Constant data flow	Fibre, ethernet, wi-fi connection (mine network), Bluetooth
Non-operating temp.	-30°C to +60°C
Operating temp.	-10°C to +55°C
Durability	Corrosion, dust, and water-resistant housing

CONCLUSIONS

BHP has led the way in leveraging a wholly new geophysical method for in-mine exploration, resource delineation and geotechnical applications. Muon tomography, given its inherent directional imaging capability and insensitivity to external influences such as electromagnetic noise, vibrations and other mine operations-induced noise sources, is uniquely positioned to provide valuable insights and high-resolution density mapping over large volumes of rock from within active mines. This has been demonstrated recently at two sites in Australia, namely Leinster and Olympic Dam and expanded use of this technology will bring more value mine operators around the world.

REFERENCES

Davis, K, Oldenburg, D, Kaminski, V, Bryman, D, Bueno, J and Liu, Z, 2011. Joint 3D inversion of muon tomography and gravity data, presented at the International Workshop on Gravity, Electrical and Magnetic Methods and Their Applications, Beijing, China, SEG.

Schouten, D and Ledru, P, 2018. Muon Tomography Applied to a Dense Uranium Deposit at the McArthur River Mine, *Journal of Geophysical Research: Solid Earth*, 123(10):8637–8652. doi:10.1029/2018JB015626.

Schouten, D, Furseth, D and van Nieuwkoop, J, 2022. Muon tomography for underground resources, in *Muography: Exploring Earth's Subsurface with Elementary Particles* (eds: L O Hiroyuki, K M Tanaka and D Varga) (American Geophysical Union).

Tarantola, A, 2005. *Inverse Problem Theory and Methods for Model Parameter Estimation* (SIAM).

Paving the way for future value – confidence in Olympic Dam's structural models utilising digital workflows

J Sidabutar[1], J Taylor[2] and M Goldman[3]

1. MAusIMM, Senior Modelling Geologist, BHP Olympic Dam, Adelaide SA 5000.
   Email: jelita.sidabutar1@bhp.com.
2. MAusIMM, Principal Modelling Geologist, BHP Olympic Dam, Adelaide SA 5000.
   Email: james.taylor3@bhp.com.
3. MAusIMM, Senior Mine Geologist, BHP Olympic Dam, Adelaide SA 5000.
   Email: matthew.goldman@bhp.com.

ABSTRACT

Structural geology is well understood to be a critical component of exploration and successful mine development. Within the mine environment, structural geology and subsequent 3D structural models play a vital role in supporting safe operations and mine production. Robust structural models improve rock mass characterisation, improve predictive capabilities, optimise stope shapes, result in strong mining compliance, reduce ongoing rehabilitation costs and reduce risk of fall of ground events.

To ensure adequate characterisation of structures within the mining footprint, it is crucial for mining operations to invest in effective data acquisition and robust, up-to-date structural models. Mine Geologists go to great effort collecting structural data, with mature operations having hundreds of thousands of discrete point observations collected down orientated drill holes or within the underground workings. It is common however, for the workflow to cease at data-capture, as it can be difficult to integrate new observations into a coherent 3D structural framework. This was largely true for legacy structural models at BHP's Olympic Dam Mine, where updates to fault wireframes were predominantly reactive and prompted by overbreak in stopes or other geotechnical events. Previously, the majority of historical structural models were 'static' and either generated by external consultants and/or previous generations of geologists, with little transparency on source points. This made it difficult to integrate new interpretations into existing structural models as the mine continued to expand.

To address this problem, the resource and geomodelling group was able to leverage the recent transition towards automated underground mapping workflows. Utilising Deswik MDM and Deswik CAD, mapping observations are entered in a standardised format and are rapidly available for integration into structural models within other software packages. This paper summarises Olympic Dam's new structural modelling workflow, which integrates data from Deswik MDM and AcQuire into one Leapfrog workspace. From here, faults can easily be wireframed, seeking to maximise the benefits of each respective software whilst reducing the challenges of working across multiple platforms. This has allowed geologists to produce robust structural models, integrating orientated underground mapping, drill section interpretations, digital mine-area interpretations, drill core logging observations and final LiDAR stope scans, in one location. This allows for timely, robust, and nimble structural models, has improved the ability to communicate risk and uncertainty, whilst also characterising the variability of faults, empowering decision-making supported by greater understanding of structural risk.

INTRODUCTION

Located approximately 580 km north-north-west of Adelaide, near the eastern margin of the Gawler Craton, Olympic Dam is one of the world's most significant deposits of copper, gold, silver and uranium. The deposit is located at the confluence of regional-scale NW- and ENE- faults and fault splays, which produced dilation zones supporting breccia formation and tapping into deep trans-lithospheric faults (Hayward and Skirrow, 2010; Clark *et al*, 2018). The Olympic Dam Breccia Complex (ODBC) has been overprinted by a complex system of cross-cutting and interconnecting brittle structures. During development of underground workings, these structures have the potential to form wedge failures in drive walls and backs and additionally can cause overbreak/underbreak in stopes. This poses a risk to production, causing dilution to the processing stream or affecting the stability of adjacent development (Widdup et al, 2004). These structures can be broadly categorised into three different scales based on their continuity (Figure 1):

- 'Primary' Deposit-scale faults have broad damage halos and often extend over 100–1000s of metres.
- Secondary' Mine-area-scale faults are continuous planes over 10s to 100s of metres and are observed in logging and typically correlate well across multiple development drives.
- 'Tertiary' Stope-scale faults are typically discontinuous planes and are projected tens of metres from their point observations.

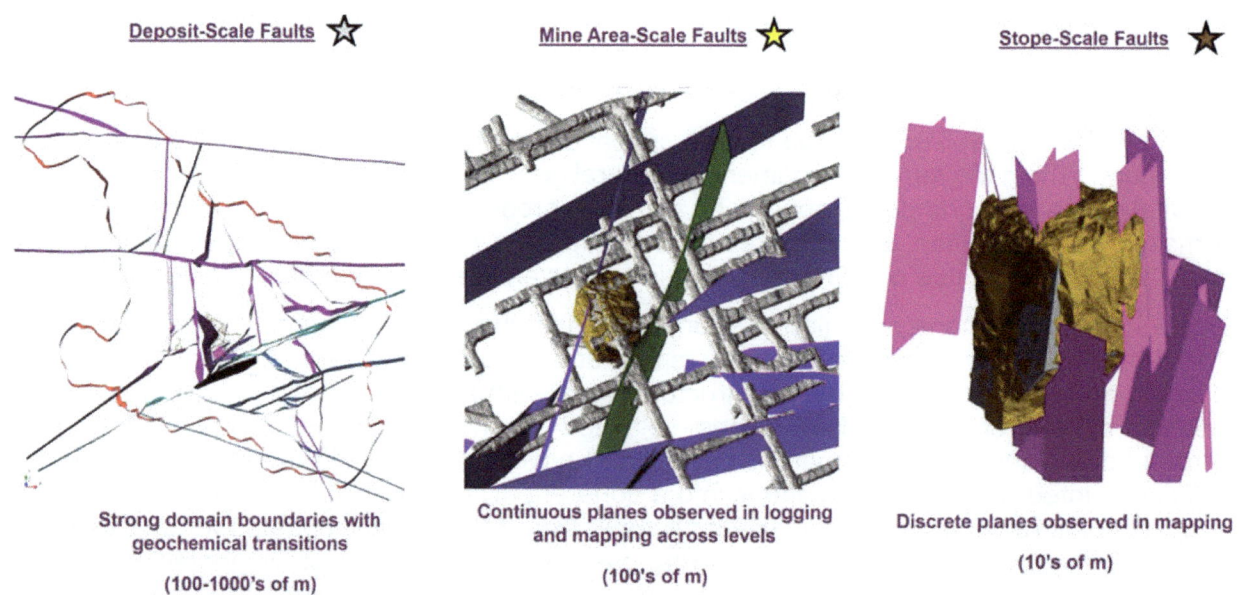

FIG 1 – Fractal relationships between structures at different scales.

Olympic Dam consists of two major mine areas: the historical Northern Mining Area (NMA) and the rapidly expanding Southern Mining Area (SMA). The SMA is far more structurally complex than the NMA and during initial expansion into this mine area, a number of rock mass related events occurred. To ensure the continued mining compliance, a dynamic structural model was deemed essential to support the execution of ~35 kms of underground development and extraction of approximately 40–50 stopes annually; Following this, the SMA was subject to an in-depth geological review to better understand the structural complexity. The ability to predict potential overbreak is important to enable better decision-making and improve mining performance.

Figure 2 shows the final stope shape, demonstrating how structures can induce significant overbreak in stope walls, leading to suboptimal performance relative to the designed shape. The middle image showcases the impact of modelled structures on the distribution of grade. Poorly understood structures can result in overbreak and dilution of grade within stopes but may also present opportunity if the structural story can be unravelled and produce exploration upside adjacent to current mining zone. In the image on the right, a numerical model of RQD data evaluated onto development solids effectively highlight geotechnical risk. Notably, the narrow band of low-RQD is governed by the modelled structures and the intersection of structures of that scale can cause poor rock mass conditions within this area.

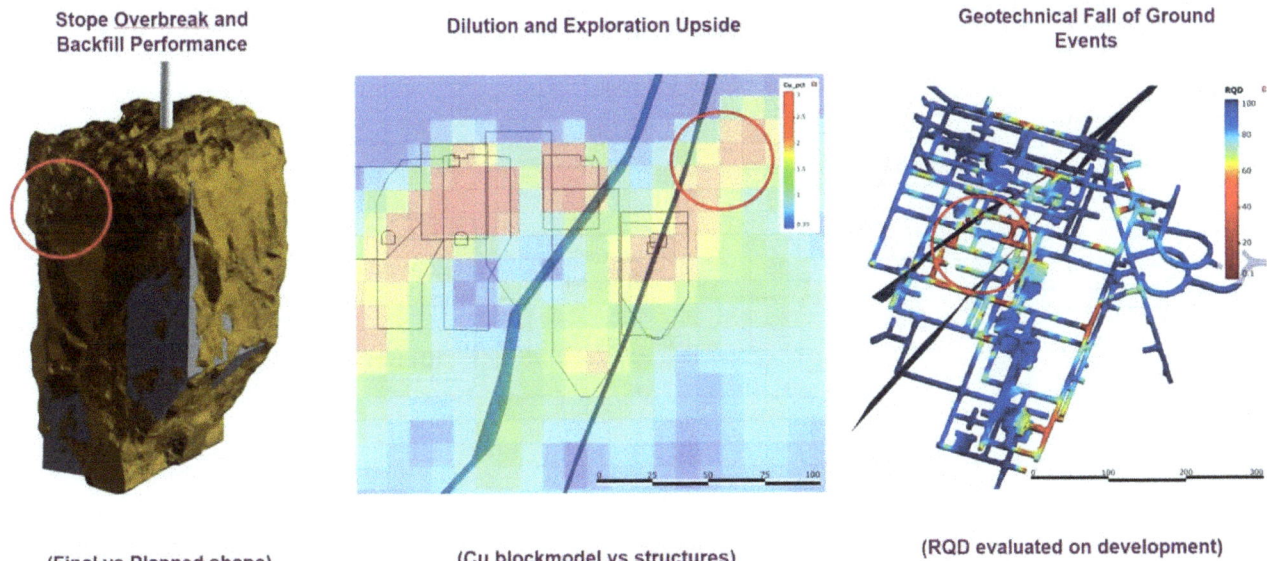

FIG 2 – Structural influence on stope performance, grade distribution and rock mass conditions.

Historically, the static nature and lack of source inputs of structural wireframes rendered subsequent updates to structural models impossible. Adjusting and refining existing structures during the acquisition of new data was difficult, rendering them obsolete as the mine progressed. As the mining footprint expanded, new structural models were largely reactive to rock mass related events and unavailable at the correct planning horizons, causing poor mining compliance. As a result of the reactive nature of modelling efforts, new areas of the mine progressed without detailed structural models, removing the ability to be proactive and 'predictive' in altering mine plans.

To address these challenges, data from MDM and AcQuire was combined into a single Leapfrog workspace from which faults could easily be wireframed. This key change allowed the structural model to be easily updated with new oriented mapping and drill hole observations. This paper provides insight into the problem-solving journey, including software integration to assist the optimised workflow. The result of the optimised structural workflow links fault wireframes to the raw observations recorded by geologists, allowing for the advancement of dynamic structural models within the mine over time.

DYNAMIC STRUCTURAL FRAMEWORK

Overview

The new dynamic structural framework has been built by linking the source inputs from Deswik MDM and AcQuire directly into a unified Leapfrog workspace. This amalgamation ensures ease of updates for future geologists. The new dynamic workflow is shown in Figure 3 and is summarised further.

Source inputs: Structural data is collected based on geological observations within underground mapping (Deswik MDM), drill core logging (AcQuire) and final LiDAR stope CMS. Modelling process: complete 2D-polyline interpretation in Deswik CAD from underground mapping, subsequently integrating all the source inputs (mapping, 2D interpretation, structural logging) into one leapfrog workspace, conducting stereonet analysis and 3D interpretation in Leapfrog. Product generation: generate three different scale structures with their accompanying source inputs and make both wireframes and input points (relative confidence) available in Deswik MDM. Product validation with LiDAR scans, new drilling and ground-truthing in underground to support continued refinement of structural models.

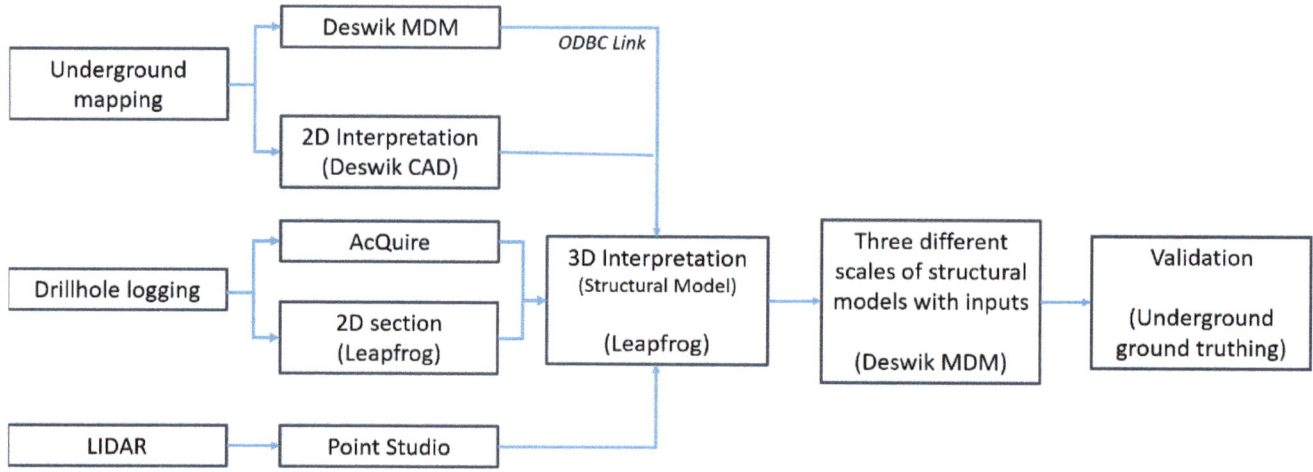

FIG 3 – The development of a dynamic consolidated structural framework (add existing models).

The dynamic structural workflow is designed to speed up the process of updating the structural model when the new data becomes available, providing transparency in model creation, showcasing source inputs, wireframe confidence and allowing for information to be accessed quickly and easily by all stakeholders.

Methodology

To define zones of structural modelling priority, a buffer of 100 m around 'Issued for Construction' development (IFC) and stopes within the five-year mine plan (5YP) was established (Figure 4). This ensured that the initial structural models generated has an immediate impact on the mine plan. The intent is to achieve a comprehensive dynamic structural model covering the entire future mining footprint, which is incredibly labour intensive and requires continuous refinement over an extended period. Given these challenges, prioritising initial modelling efforts in critical areas of the mine is paramount.

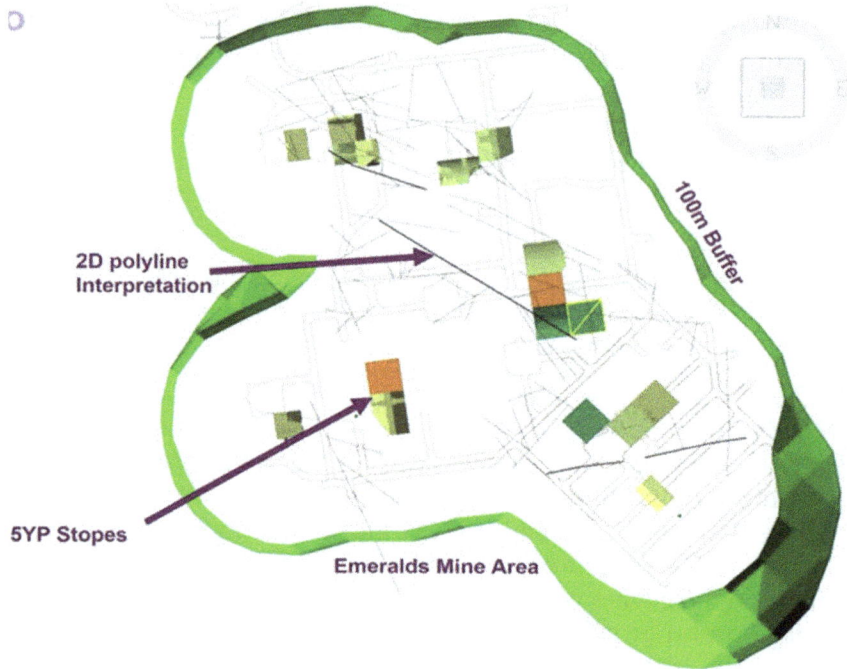

FIG 4 – The 2D polyline project has initially focused interpretation efforts inside 100 m buffers around the five-year plan stopes.

The generation of the 2D structural interpretation is the foundation of the dynamic structural framework, linking physical underground observations to coherent, structural interpretations (Figure 5). Faults are interpreted from underground mapping observations within a particular mine area, and geologists aim to correlate measurements and structures across adjacent development drives. Collaboration of the Geomodelling and Deswik MDM teams was required to conceptualise a new fit-for-purpose tool within Deswik to facilitate the linking of mapped structures. The physical geological properties from the mapped structures are attributed to the polylines and published to Deswik MDM for key stakeholders to utilise during stope, development and other infrastructure reviews.

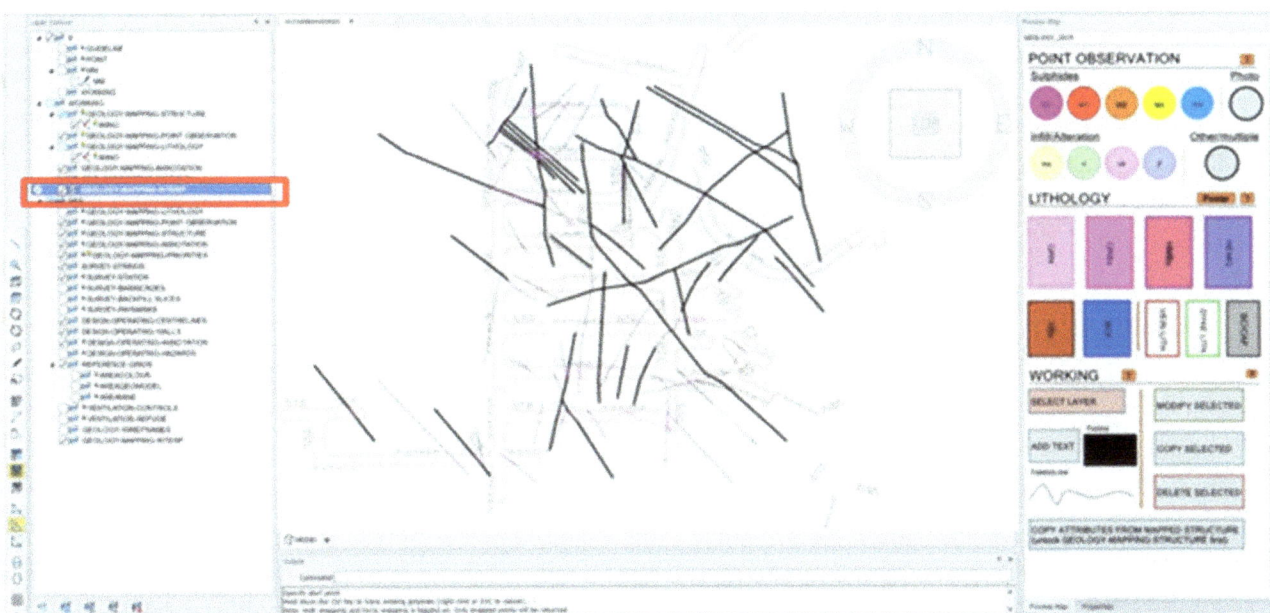

FIG 5 – The generation of the two-dimensional structural interpretations (2D polyline) in Deswik.

The easily updatable 2D interpretation layer enhances transparency in the Mine-Area and Deposit-Scale structural models through clearly defining chronologies, linking interpretations to physical observations. Subsequent 3D wireframes are anchored to these 2D interpretations, enabling the mine geology team to project structures within active headings and prompting ground-truthing by the geotechnical department.

Mine geologists at Olympic Dam routinely provide monthly interpretation cross-sections on completed underground drill fans, summarising geological observations. This includes lithology, mineralisation and structures which are then communicated to downstream customers. Inferring fault relationships across individual drill holes and fans can be difficult, however once several sections from the same area are georeferenced, the geological story begins to unfold. Sections enable subsequent interpretation to be anchored to recent drill core observations. Achieving this involves overlaying georeferenced cross-sections with surrounding drilling to pinpoint correlations in fault continuity through the area (Figure 6).

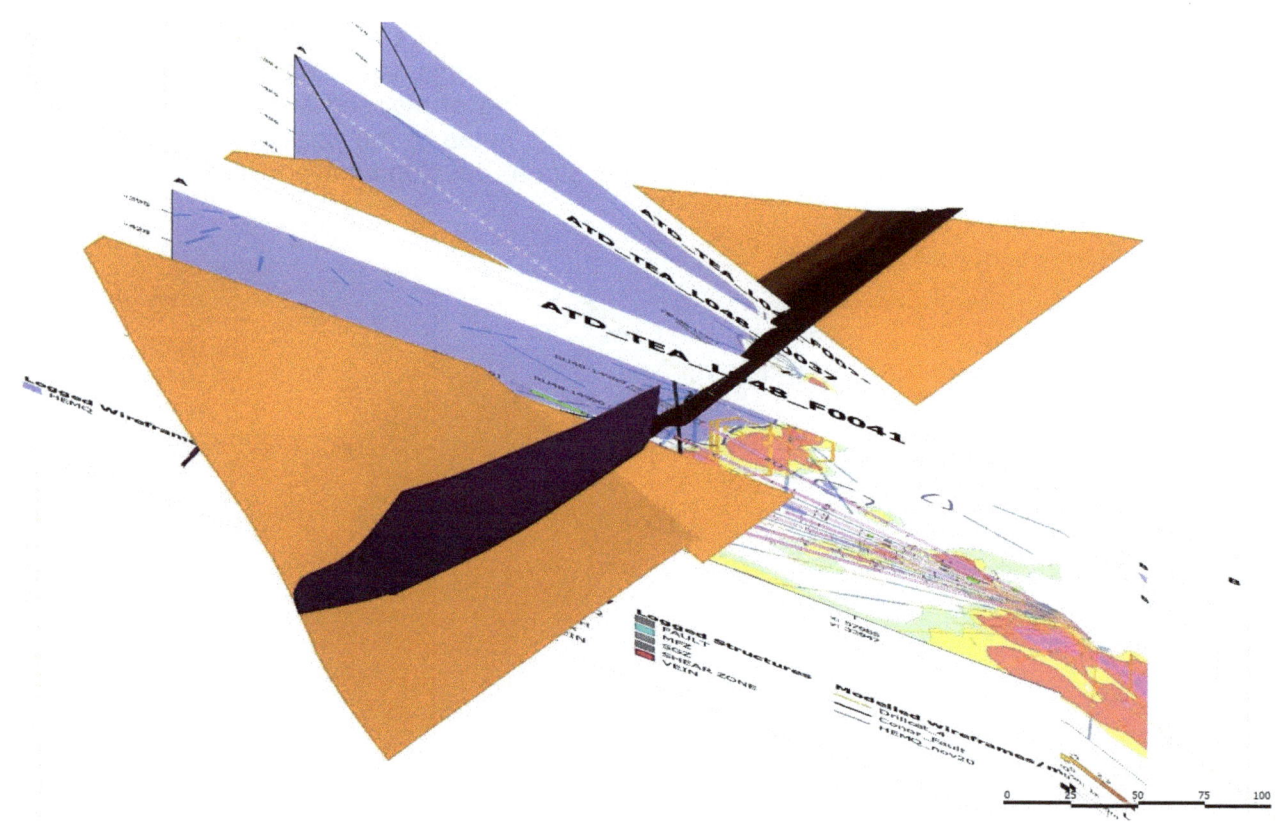

FIG 6 – A structure overlaid against georeferenced sections add confidence.

To generate 3D structural wireframes, all available data including underground mapping, drill hole logging, polyline interpretations and drill sections is used in an integrated workflow. Software integration has linked underground mapping in Deswik MDM into Leapfrog, enabling a seamless, automated process for reloading mapped structures into the existing dynamic structural models. The Deswik MDM – Leapfrog link ensures up-to-date mapped structures are readily available for the modelling team. A basic stereonet analysis is completed to validate the orientations identified during the preliminary 2D interpretation, allowing for the identification of 3D trends (Figure 7). The resulting 3D interpretations are locally constrained structural models within areas of relatively high confidence. These live (dynamic) models can be easily updated by future generations of geologist and can be refined as new data becomes available. The overarching goal is to create a streamlined 'one-stop-shop' for structural modelling across multiple scales.

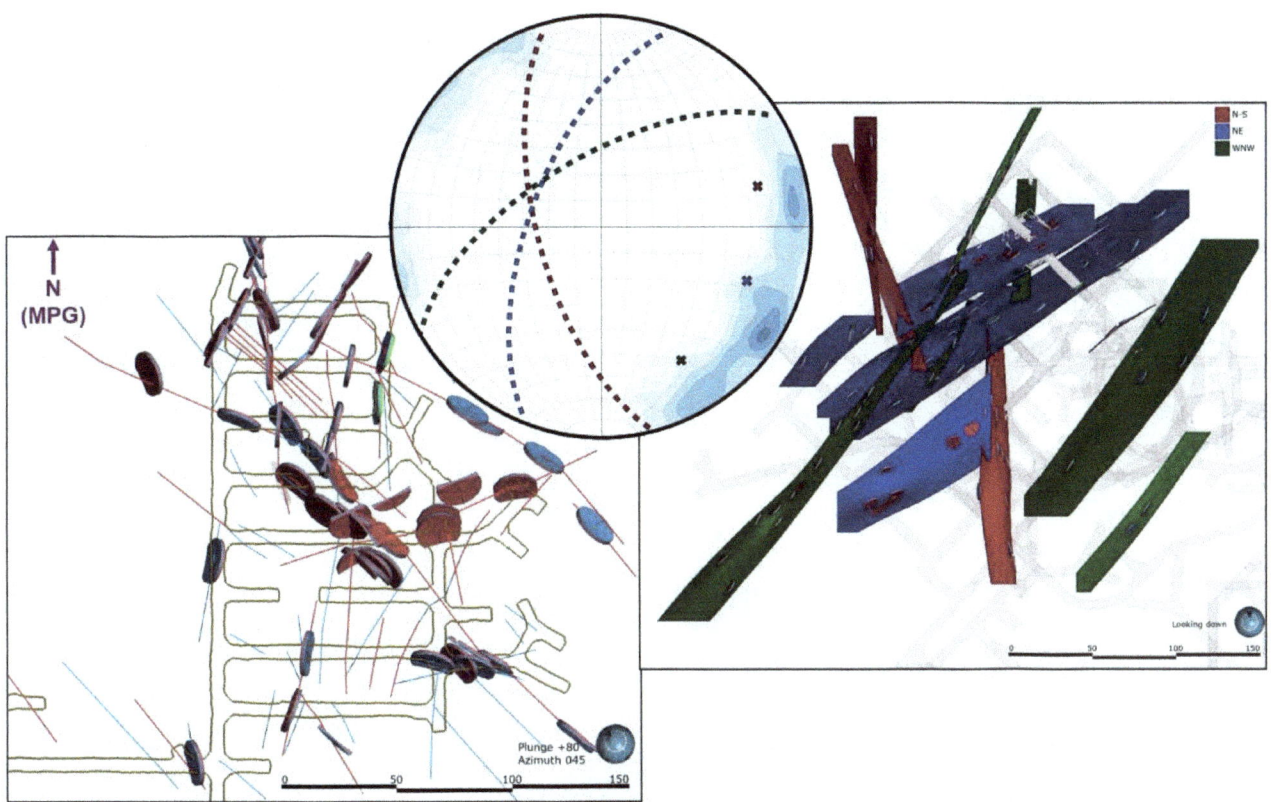

FIG 7 – Fault selections driven by 2D interpretations and an example of structural interpretation in three-dimensional software.

The source points within the new dynamic structural framework are now linked to each individual fault wireframe. It is the integration of physical observations during logging and mapping, complemented by stope CMS data, which builds multi-layered geo-confidence (Figure 8). Each of these individual observations are edited within different input layers within Leapfrog Geo, ensuring transparency through the entire process. These individual structural interpretation points are subsequently published to Deswik MDM, attributed to their respective fault, to demonstrate the exact location of each observed point (Figure 9).

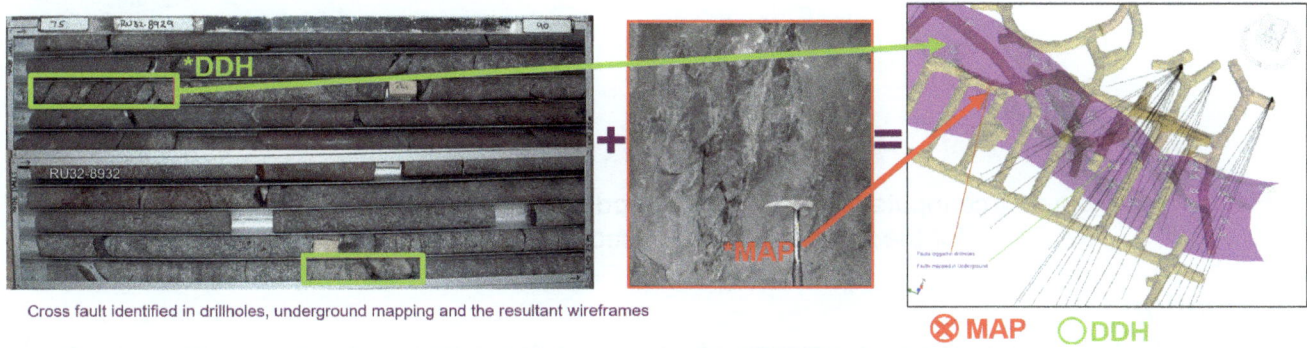

FIG 8 – The example of transparent source inputs fed into the structural wireframe.

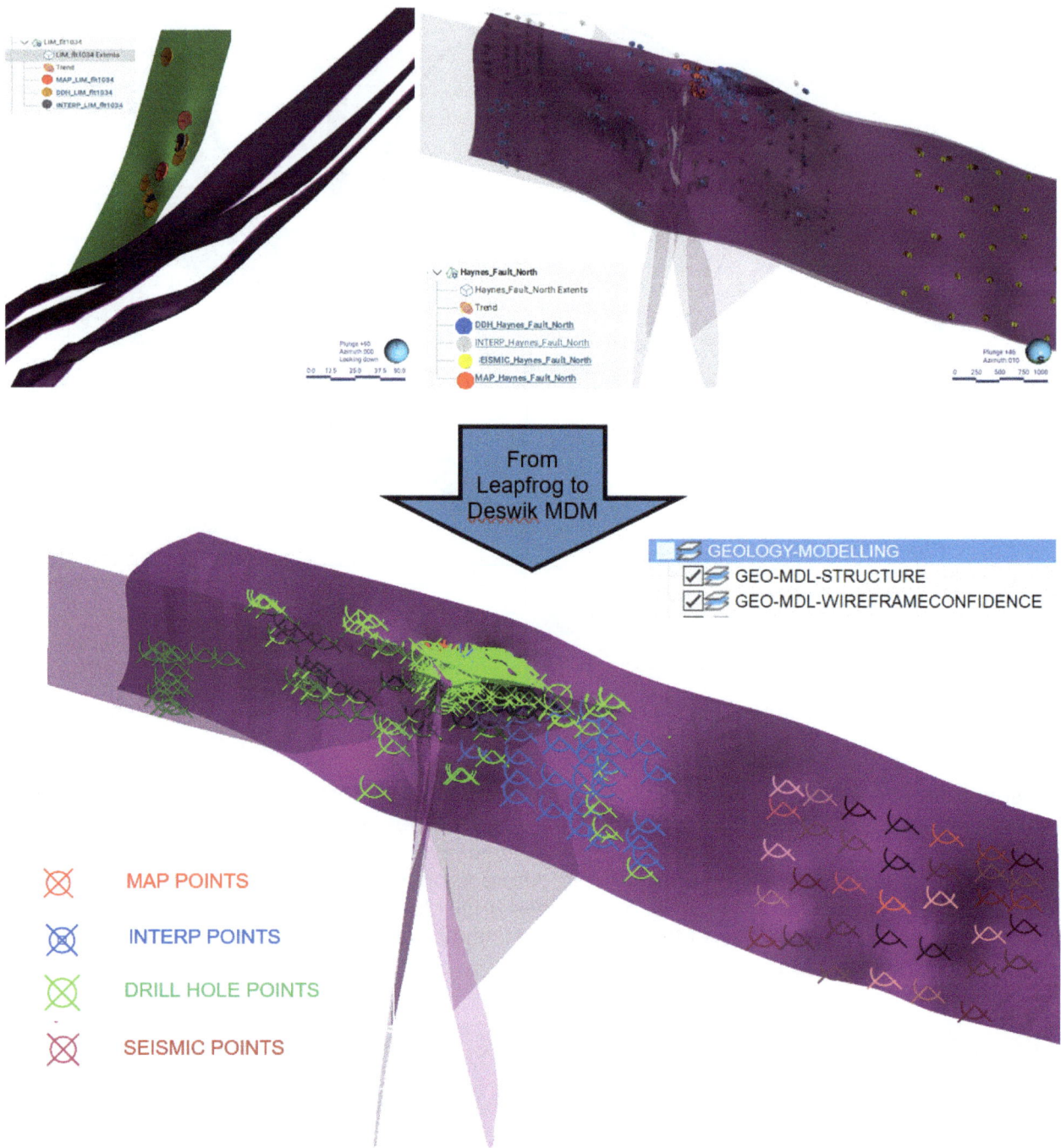

FIG 9 – Example source inputs of the structural models flow to Deswik MDM (the 'source of truth' for the wireframes) with accompanying attributes.

Validation

To validate the fault models, a process of ground truthing was necessitated to enhance the confidence level. In instances of stope overbreak, LiDAR scans can be utilised for the validation of modelled faults, aligning them with the observed overbreak within a stope. This validation technique ensures a more robust and reliable assessment of the model's fidelity (Figure 10).

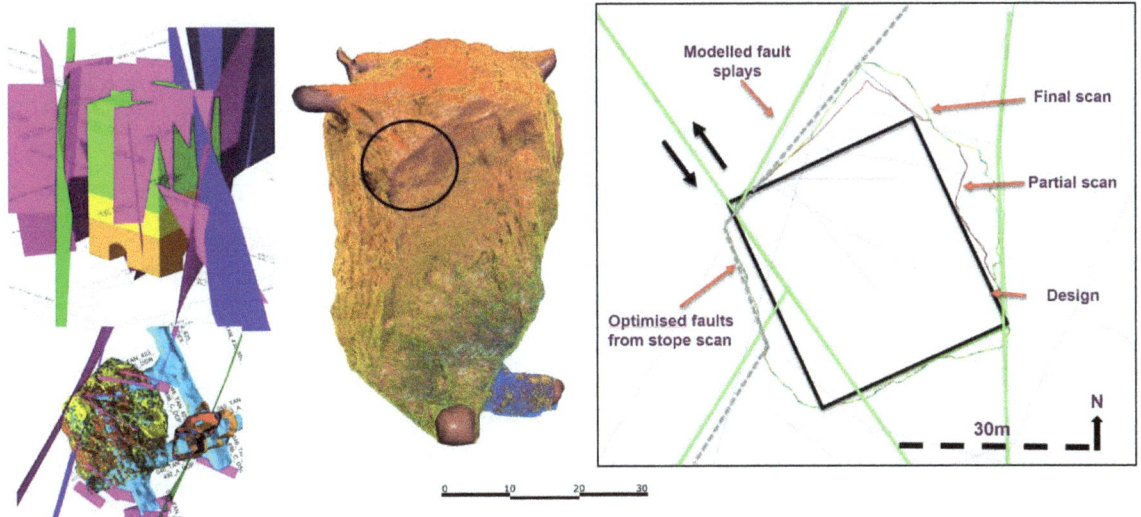

FIG 10 – Comparing current stope scale model with LiDAR scans.

The historical static models were used as reference points in the new structural modelling process. Subsequently, when the new model (dynamic model) with accompanying source inputs is established, the old model will become obsolete. This practice aims to reduce confusion and improve confidence. Ideally, all historical models should be either upgraded to a dynamic model with accompanying source inputs or rendered obsolete.

These structural models will continue to evolve with the increasing amount of input data as new development and drilling progress within the mine. In cases where a structural model is flagged as a potential risk, geologist typically requests ground-truthing to enhance the confidence level of the modelled structures. Not all models undergo ground-truthing due to excessive number of models and priority is based on potential impacts to mining operations.

DISCUSSION

Olympic Dam continues to refine and improve its confidence in structural models, paralleled by an expanding knowledge of fault characteristics and their implications. By distilling the impact of specific faults into concise summaries, we encapsulate key details on a one-page attachment for customer use. These summaries, seamlessly integrated into the wireframes in Deswik MDM, encompassing fault location, relevant lithologies, mine areas and wireframes confidence levels. These summaries also include a comprehensive overview of faults and their associated characteristics and are supported with example pictures of the fault from drill core and underground exposures. This systematic approach not only refines our models but also enriches our knowledge base and provide operational insights.

There are several challenges arise during the transformation of the historical structural models into a dynamic structural model. This challenge includes the large volume of data and the fact that underground mapping is the major source of oriented structural data. A larger presence of oriented drill hole data would facilitate its incorporation and enhance the confidence of the models. Furthermore, there are other difficulties related to software limitations that involve manual selection and assigning processes.

To address these challenges, we could consider the following actions:

- When possible, obtaining kinematic data set to supplement the existing data set.
- Collaborate with software developers to address limitations and enhance the functionality of modelling tools, reducing the need for manual intervention.

Looking ahead, the future improvements may involve:

- Investing in advance software solutions capable of handling large volumes of data and automating complex tasks.

- Conducting research and development efforts to enhance the capabilities of underground mapping technologies for capturing structure data.
- Implementing training programs to upskill team members in utilising advanced modelling techniques and software functionalities.
- Establishing robust data management protocols to ensure the integrity and reliability of input data for modelling purposes.

By implementing these strategies and pursuing continuous improvement initiatives, we aim to overcome existing challenges and elevate the effectiveness and efficiency of our modelling process.

CONCLUSIONS

In conclusion, this paper has summarised the process of developing a consolidated and comprehensive structural framework for Olympic Dam. Over the past two years, the integration of new data has enabled rapid progress in generating robust models throughout the mine and enhances our understanding and confidence in the structural framework of Olympic Dam across multiple mine areas.

The approach involves modelling structures at multiple scales to influence targeting, drilling, planning and production strategies. Striving for transparent source inputs, with the addition of confidence points added to fault wireframes to convey the local accuracy of data, ensuring the framework can be easily updated. These enhancements align with our commitment to foster stakeholder partnerships, deliver high-quality geoscience products, explore value and characterise risks. This comprehensive review has yielded a structural model crucial for informing the mine plan, marks a significant step toward achieving geoscience excellence in the Olympic Dam mine operation.

ACKNOWLEDGEMENTS

The authors extend their gratitude to the Olympic Dam Mine Geology team and underground mapping geologists for their commitment to safely gathering robust structural measurements. Special appreciation is also extended to the modelling and resource teams for their valuable contributions, which have influenced this paper. It is hoped that this case study will serve as a catalyst for positive change within the broader mining industry.

REFERENCES

Clark, J M, Ehrig, K, Poznik, N, Cherry, A R, McPhie, J and Kamenetsky, V, 2018. Syn- to post-mineralisation structural dismemberment of the Olympic Dam Fe-oxide Cu-U-Au-Ag deposit, in *Society of Economic Geologists Annual Conference: 2018*, extended abstract and poster (Metals, Mining and Society: Denver).

Hayward, N and Skirrow, R, 2010. Geodynamic setting and controls on iron oxide Cu-Au (±U) ore in the Gawler Craton, South Australia, in *Hydrothermal Iron Oxide Copper-Gold Deposits: A Global Perspective*, ver 3, Advances in the Understanding of IOCG Deposits, pp 119–146 (PGC Publishing: Adelaide).

Widdup, H, Fouet, T, Hodgkison, J, McCuaig, T C and Miller, J, 2004. A three-dimensional structural interpretation of the Olympic Dam deposit – Implications for mine planning and exploration, in *Proceedings of PACRIM 2004*, pp 417–426 (The Australasian Institute of Mining and Metallurgy: Melbourne).

Strategies for enhancing ore control and production at Martabe Gold Mine Indonesia

A Triyunita[1], N Saala[2], N Khariyah[2], M I Hidayat[2] and J Hertrijana[2]

1. Senior Mine Geologist, PT Agincourt Resources, Martabe Gold Mine, South Tapanuli, North Sumatera, Indonesia. Email: ade.triyunita@agincourtresources.com
2. PT Agincourt Resources, Martabe Gold Mine, South Tapanuli, North Sumatera, Indonesia.

ABSTRACT

Martabe Gold Mine is an open pit mine located in the North Sumatra province, Indonesia (Figure 1). Martabe project managed by PT Agincourt Resources (PTAR), a member of ASTRA, a well-established Indonesian mining company. It is known for its substantial gold and silver deposits with 6.2 million ounces gold and 59 million ounces silver in resource model, and 3.7 million ounces gold and 33 million ounces silver in reserve model. Martabe Gold Mine is in the western Sunda-Banda magmatic arc within the West Barisan physiographic fore-arc region. This region is dominated by high sulfidation epithermal type hosted in tertiary volcanic dome complex and sedimentary rocks (Figure 2). There are three active pits in Martabe Gold Mine namely Purnama, Barani and Ramba Joring with a distinctive ore characteristic.

FIG 1 – Martabe Gold Mine location.

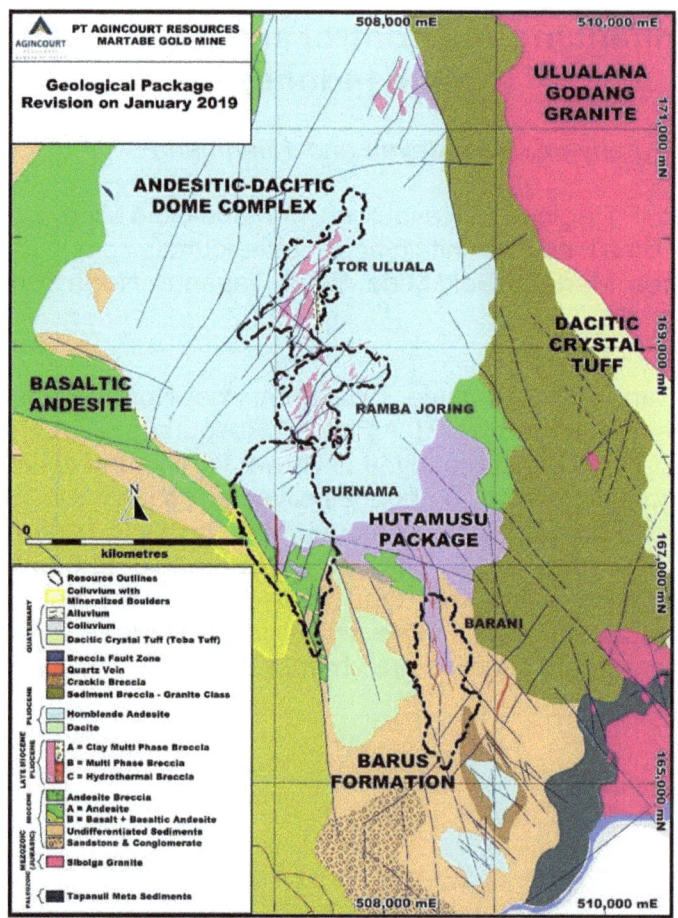

FIG 2 – Martabe Gold Mine regional geology overview.

The mine geology team has a significant role to provide data and maintain excellent quality control for every step of the mining process. The first step of the ore control process is collecting detailed samples from drilling to provide additional data to build a grade control model that has already been updated by some of fundamental attributes of the reserve model. This model is the initial guideline to derive a comprehensive classification and delineation for ore and waste to accommodate plant and waste dump placing requirements. Considering the challenging geological conditions where various classifications of ore and waste are blasted together, it is essential to mitigate unplanned ore loss and dilution by modelling the after-blast profile before digging activity. To optimise process plant feed and throughput, supplied ore must be in accordance with planned geological data and plant ability to received ore material. Because of the diverse types of physical and chemical condition of ore material, it is best to manage and blend the ore to achieve the most optimal value.

INTRODUCTION

Martabe Gold Mine consists of three active mine pits characterised by high-sulfide epithermal gold-silver deposits in Purnama and Ramba Joring pit and one low-sulfide epithermal gold-silver deposit in Barani (Crispin, Hertrijana and Albert, 2015). There are variations of characteristics in those three pits, wherein Purnama is a stratiform epithermal deposit that shows examples of both high sulfide vuggy disseminated gold and low sulfide banded quartz veins, dominantly controlled by shallow dipping lithological features and cut by feeder systems with dominant mineral assemblages of silica + dickite ± alunite and abundant pyrite, enargite and luzonite. Ramba Joring was dominated by volcanic lithofacies with coherent hornblende phyric, feldspathic porphyry of dacitic to andesitic composition, emplaced as a high level sub-volcanic intrusion. Meanwhile Barani is dominated by silicified sandstones and conglomeratic beds that host low sulfidation quartz veins and pyrite cemented hydrothermal breccia (Harlan *et al*, 2005). The various material types and complexity of the mineralisation style and oxidation profile requires an estimation not only of the contained metal but also the recovery of contained gold and silver metal through the carbon-in-leach (CIL) process. The complexity of the mineral deposit is depicted in numerical block model which will be classified into several material types. It can significantly impact processing recovery achievement if the mining process is not professionally managed.

Mine geology has the main responsibility to carry out grade control processes to ensure the quality of mined material is the most appropriate for mill requirements. Grade control is the process used to identify and delineate the tonnage and grade of ore for one or more cut-off grades. It is required to identify and describe the tonnage and grade of various ore types. The main purpose of grade control is to maximise the mine's profitability by minimising the amount of misclassification during mining (Deka Dynamics, 2019). Misclassification is the incorrect assignment of ore and waste resulting in dilution and ore loss. Mine Geology team in Martabe Gold Mine has collective target that refers to Indonesian Government Regulation No. 26/2018 (Minister of Energy and Mineral Resources, 2018) regarding good mining practices in mineral and coal mining.

To achieve good mining practice, grade control team ensures the quality of delivered materials by increasing data quality. A systematic workflow was designed to achieve collective target of better reconciliation and value of mining recovery >90 per cent. All work components must be synergised and all supporting data must be valid.

Mine reconciliation at the Martabe Gold Mine is a comparison of the tons, grade and metal as predicted from the Ore Reserve against the tons, grade and metal predicted by subsequent grade control activities, against the tons, grade and actual metal produced by processing plant. Mine reconciliation encompasses a multi-disciplinary approach across geology, mining engineering, operations and metallurgy to deliver benefits throughout the mining value chain.

GRADE CONTROL WORKFLOW

From the earlier stage of production, Martabe Gold Mine constantly faced many challenges. Throughout the years, evaluation and improvement are actively carried out throughout all the work steps to shift the reconciliation trend closer towards the threshold value. Some of the significant issues are lack of infill drilling data and poor QA/QC in sample caused by an overestimate reserve model and poor control on mining practice (see Figure 3).

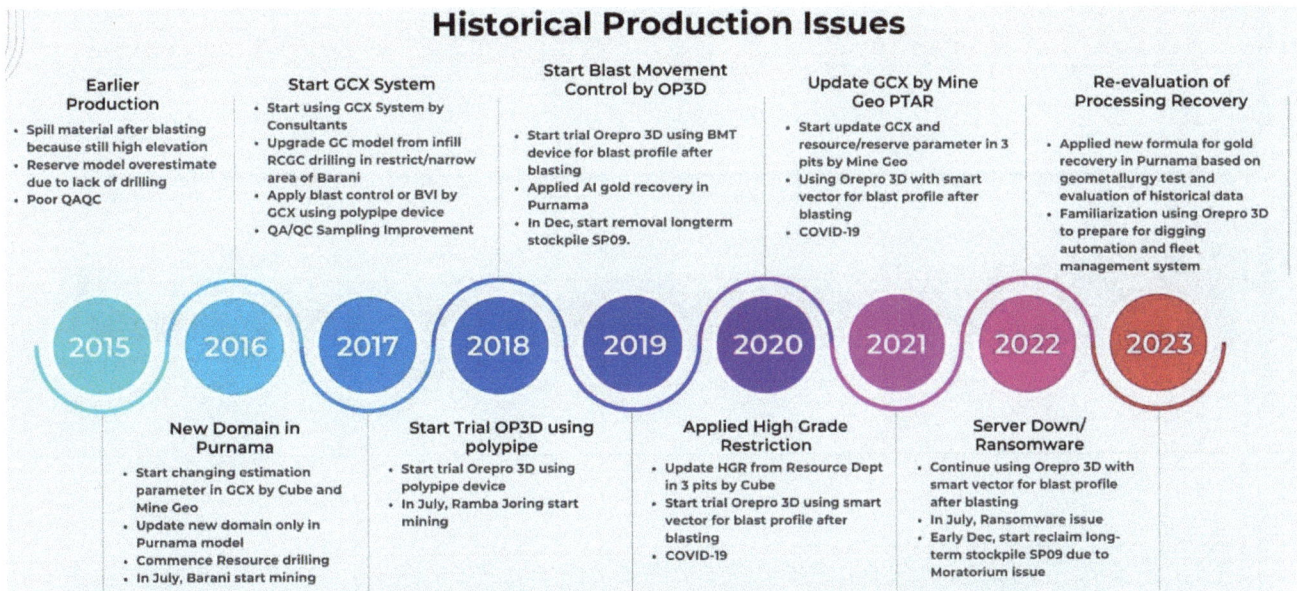

FIG 3 – Historical production issues and solutions from year to year in Martabe Gold Mine.

Corrective steps are taken by collecting more data, improving QA/QC and sampling methods, updating estimation parameters and carrying out good control over mining activities until a systematic work process is developed. Workflow of Mine Geology team in Martabe Gold Mine can be explained by the following steps.

Data collection and QA/QC data

Initial data collection is the fundamental process that must be carried out with good supervision and control to ensure the data quality. Martabe is using a reverse circulation rig to collect grade control samples (RCGC). RCGC drilling is a closed off space drilling, carried out to append data between sparse resource drilling data. The selection of RC method considers the geometry of orebody as the rig can be angled for optimal intersection widths of the target.

This step ensures the correct assessment of grade across the orebody to identify ore and waste zones. Each pit has distinct characteristics, therefore, drilling azimuth for each pit is different. Throughout the drilling process, quality control and quality assurance are continuously applied. RCGC drilling samples will be sent to on-site laboratory to test the constituent elements for every sample. Element suites for every pit are different as the mineralogy of each pit is also distinct. The crucial part of the sampling process lies in quality control and quality assurance.

As part of QA/QC process, standard sample and duplicate sample are included in every batch of sample before submitted to on-site laboratory. Certified reference material (CRM) with a known grade and variability is included in every 50-sample number of grade control sample. Field blank samples and samples are included. There were several types of CRM that were bought from Ore Research and Exploration (OREAS) Pty Ltd. In 2011 to July 2014, significant outliers of gold analysis happened, caused by systematic error occur in laboratory. The previous on-site laboratory used an average result to calculate the standard deviation instead of using the standard deviation of CRM from certificate. It was making the tolerance of outlier higher than it should. Over time, accuracy results improved since the on-site management laboratory changed (Figure 4). The laboratory improved quality control and mine geology improved in recording of selected standard. Mine geology recorded the selected standard by pulling off the label and stuck the label on the standard book (Table 1).

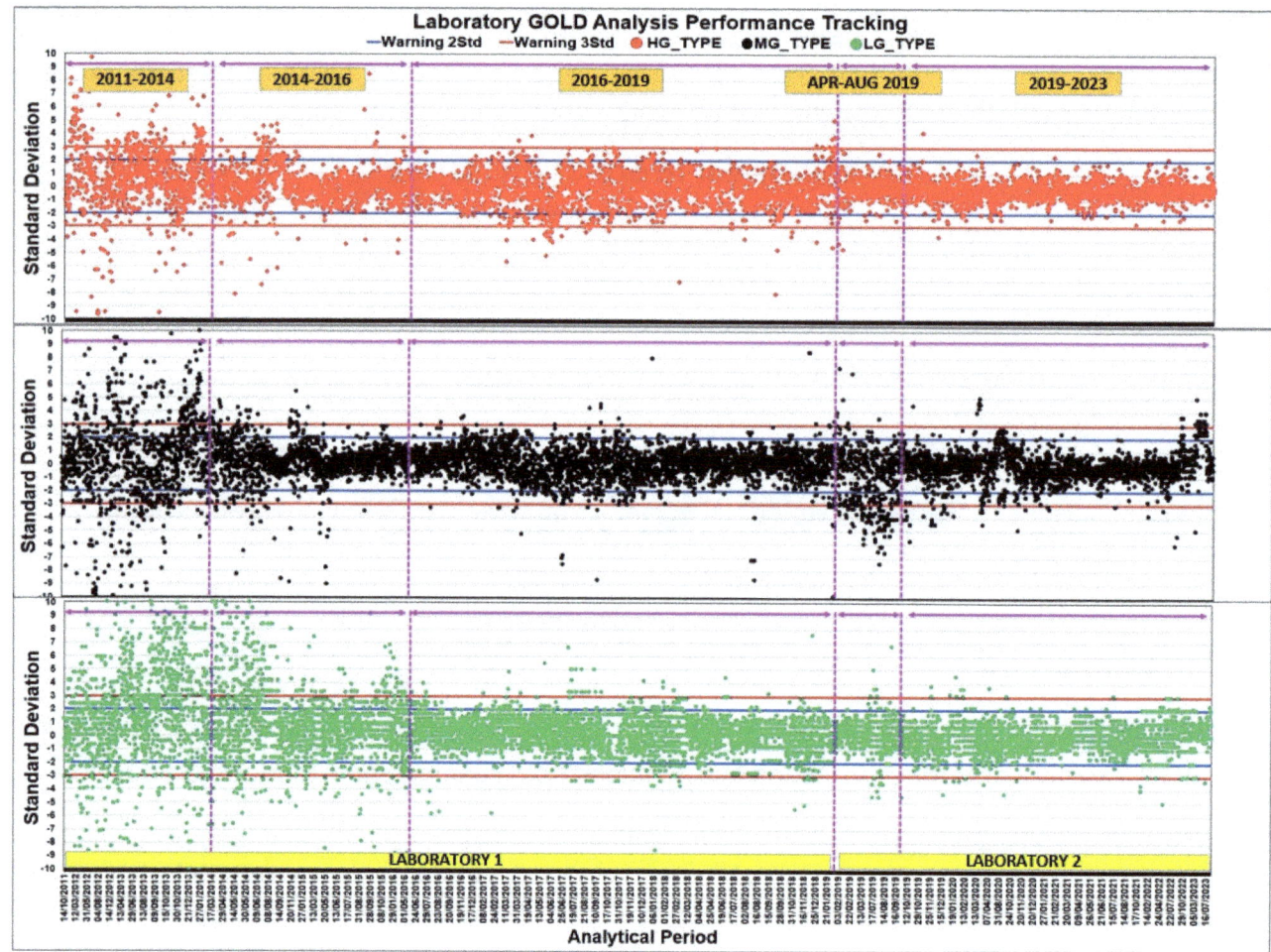

FIG 4 – QA/QC graphic of gold analysis for each type of CRM grade shows a decreasing number of outliers from 2011–2023.

TABLE 1

QA/QC performance and issues in on-site laboratory throughout the years.

On-site lab	Periods	Notes
Laboratory 1	2011 to July 2014	33.38 per cent CRM failed. There were systematic errors occurring and lab was in huge problem.
	August 2014 to July 2016	9.05 per cent CRM failed. Lab improves the system of QC. Mine geology recorded the selected CRM, then removed the label and stuck the label on to standard record book.
	August 2016 to March 2019	3.41 per cent CRM failed. Lab performance improved since management lab changed.
Transition from Lab 1 to Lab 2	April to August 2019	13.53 per cent CRM failed. Lab transition from Lab 1 to Lab 2
Laboratory 2	August 2019 to December 2023	2.22 per cent laboratory performance very well.

Blank samples are considered as single value with value known as zero or less than the detection limit. This sample requires to indicate contamination during preparation sample also reflect thoroughly cleaned equipment, furthermore, to avoid overestimation of grade control model. These samples were inserted into sample batch every 20 and 80 sample number. Outlier in highlighted area happened as blank material from vendor has not standardised yet (Figure 5). From 2018 onwards, Martabe already uses standardised sandy quartz material as blank sample (Table 2).

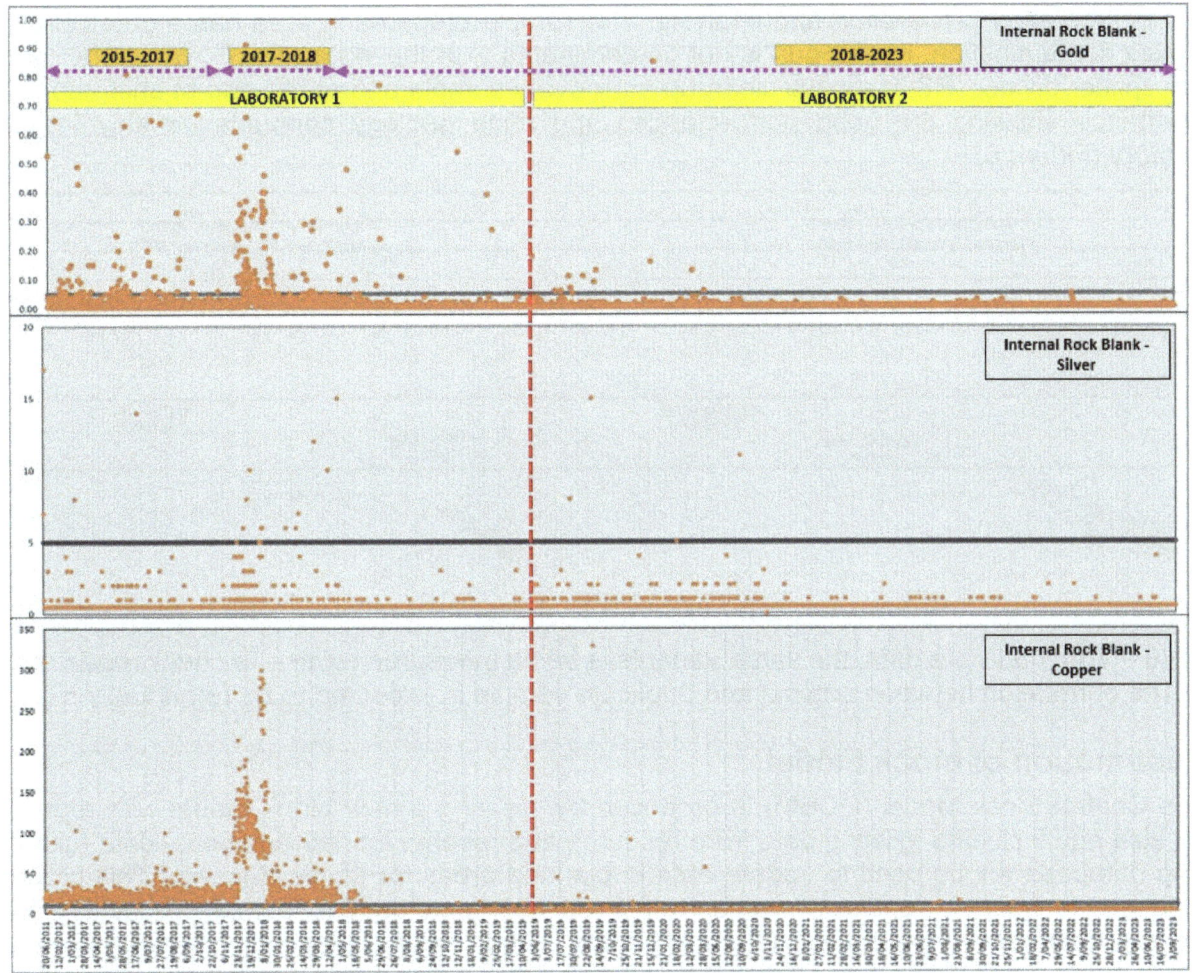

FIG 5 – Internal rock blank plotted for gold, silver and copper.

TABLE 2
Historical list of blank sample sources used in Martabe Gold Mine.

Periods	Blank material
January 2015 to October 2017	Basalt (from Batunadua, Padangsidimpuan)
November 2017 to April 2018	Basalt (from basalt quarry, Martabe)
May 2018 to October 2023	Sandy Quartz

Field duplicates is the second sample taken from the original material from the drill hole using a half or rifle splitter. These samples are used to check the contamination during sample preparation. In addition, it can be used to analyse sample homogeneity, sampling technique and analytical accuracy. Field duplicates are mostly informative for estimation of the overall precision of samples (Abzalov, 2011). The field duplicate will be submitted as a unique sample number. In mine geology the field duplicate were taken from every 22 samples.

Statistical analysis of duplicate data uses various tools to compare the duplicate sample to the original sample. The difference between the duplicate pairs can be calculated directly or indirectly. Initial analysis of duplicate data should be completed using X-Y scatterplots to gain a general view of the repeatability of results and to identify samples with obvious errors (Greig and Cook, 1998). There are some improvements implementations in the sampling practices that can minimise error from field such as replacement of the drill cuttings collection bag in every hole, flushing or clean up rod done in every rod to avoid the clogged up, the drilling contractor should be minimising the dust losses and the riffle splitter should be on level ground.

The management changes of the on-site laboratory also make many improvements in minimising errors in laboratory preparation and analysis. The sample preparation area has a dust collector to minimise contamination, the air gun is not contaminated with the water, the Boyd crusher and the rotary splitter divider (RSD) always clean up after every sample and at the end of shift, the QA/QC performance showing the good performances and mine geology conducts monthly inspection laboratory (Figure 6).

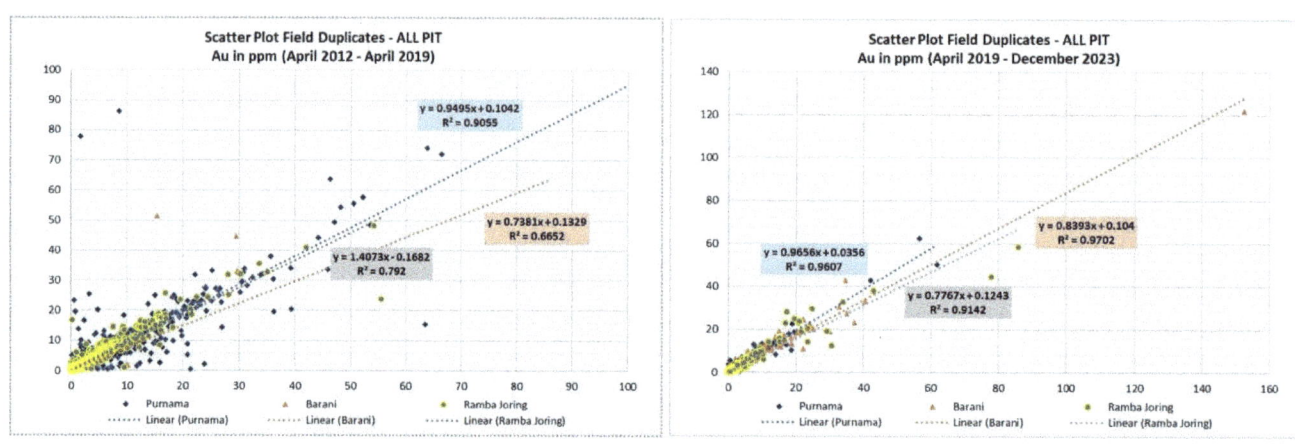

FIG 6 – With duplicate data, the same variable is being measured twice – to check repeatability. The correlation between original and duplicate sample is expected to be remarkably high.

Re-estimation of block model

Grade Control Block Model (GCBM) is produced by creating a new blank model with regularised block size and including existing data from resource and reserve model. Additional data taken from drilling database will be used to update data in planned areas for digging activity. This process is carried out to keep a consistent value from initial resource data to actual production. Before resource parameters were applied to grade control model, several statistical studies were carried out to evaluate and fit the grade control estimation result.

Each pit has a certain set of parameters and a different method of estimation. GC parameters will be updated when there is a change in resource or reserve model. Data from pit mapping, resource development and grade control drilling data will be incorporated into the re-domaining step when resource model update is performed. Then, newest resource model will be used for multi-domain updating and interpretation in grade control system (see Figure 7).

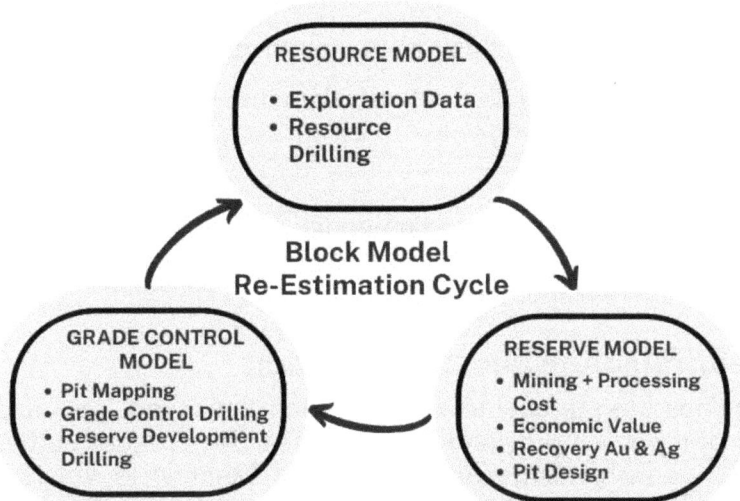

FIG 7 – Block model re-estimation process in Martabe.

Martabe is using grade control system module, an integrated grade estimation and reporting system, designed to facilitate routine grade estimation and reporting requirements for most open cut and underground mining operations. This is one system of process to create localised grade control block model with a set of steps and restrictions to ensure consistent results for every user.

The entire process starts begins with drilling database extraction which then will become guidance to create domain interpretation. Currently, there are three types of geological domain and five types of mineralisation domain (see Table 3). Re-interpretation of resource domain has significant impact on the quality of the model. A denser data set contributes to creating a more refined domain boundary.

TABLE 3

Several types of domains that should be re-interpreted in Grade Control Estimation process.

Pit	Geology domain	Mineralisation domain					
		Au	Ag	SxS	Cu	Ca	ANC
Purnama	Lithology + Alteration + Oxidation	v	v	v	-	-	-
Barani		v	-	-	-	-	v
Ramba Joring		v	-	v	v	v	-

ANC: Acid Naturalisation Capacity.

Drilling data inside the domain boundary will be composited to ensure the samples have comparable influence on the statistics. If the samples are collected over variable lengths, then there is a risk of introducing a bias into the data analysis. The effect of compositing is to weight sample grades according to the corresponding interval length. A top cut value will be applied to any composite grades higher than the value.

The estimation process in grade control is using a system as the estimation engine, allowing multi-pass estimation across several techniques including ordinary kriging and dynamic kriging (Table 4). This process estimates all composite assay data inside a block model campaign.

TABLE 4

Estimation process in Martabe estimation engine system.

No	Estimation process
1	Estimation of top cut value (both adjusted and unadjusted) using ordinary kriging/dynamic kriging
2	Estimation of HG restriction code using Inverse Distance Weighting (IDW)
3	Estimation of uncut value (both adjusted and unadjusted) using ordinary/dynamic kriging

The challenge is to combine qualitative understanding of geological processes with the patterns exhibited in the quantitative data to create accurate spatial predictions that help mining engineers plan, design and extract mineralisation economically.

Ore and waste determination and classification

Mine geologist determines ore material using a calculated recovery gold and silver value, also an economic value (EV) of the profit provided after reduced by revenue, processing cost and selling cost according to current gold prices assumption in $/Oz. Meanwhile, waste material determination is based on the total value of Net Acid Producing Potential (NAPP). Three running pits in Martabe have such distinctive ore characteristics, therefore, each of those has different formula for calculating EV, recovery and NAPP (see Table 5).

TABLE 5

Calculating parameters for ore and waste definitions.

Pit	Recovery Au	Recovery Ag	Economic value	NAPP
Purnama	Calculation using Al + OSR adjustment Metallurgical test work (Lewis's formula) + OSR adjustment	Lewis's formula + metallurgical test work (Darryn M)	EV 1_A and EV 2_A	MPA – ANC calculation
Barani	Metallurgical test work (Peter Colbert Formula) + OSR Adjustment			MPA – ANC titration
Ramba Joring	Metallurgical test work (Michael Liu and Darryn M Formula)			NAGpH (not using NAPP)
Notes	OSR	Oxygen Shear Reactor		
	NAPP	Net Acid Producing Potential		
	MPA	Maximum Potential Acid (sxs × 30.63)		
	ANC	Acid Naturalisation Capacity (ca × ((100.09/40.08) × 10 × 0.98))		
	NAG	Net Acid Generating		

Further classification of ore and waste is based on other associated elements beside gold (Au). There are 21 types of ore and six types of waste material in Martabe (Table 6). Detailed classification on ore and waste materials is done to consider plant processing specification. Ore classification is based on value of gold (Au), copper (Cu), silver (Ag) and sulfide sulfur (SxS). The processing plant was designed to accommodate maximum value of 3 g/t Au, 36 g/t Ag, 200 g/t Cu and <3 per cent SxS. If the average value of certain material is above threshold tolerance, the material will be classified into 'problematic ore' (PO). Besides geochemical composition of ore material, physical properties also become important aspects to consider in ore delivery process. These physical properties are affected by alteration and mineralisation. Clay dominant and hard ore fall in the same category as problematic ore. Therefore, those materials underwent a specific management before delivered to milled.

TABLE 6

Ore classification code, values and flagging tape colour for each type.

Values	Ore Type	Code	Flagging Tape
EV1_a > 0 ; Au > 10	Super High Grade	HS	
EV1_a > 0 ; Au > 2.5 - 10	High Grade	HG	
EV1_a > 0 ; Au > 1.5 - 2.5	Medium Grade	MG	
EV1_a > 0 ; Au > 0 - 1.5	Low Grade	LG	
EV1_a > 0 ; Au > 2.5 - 10 ; Ag > 30 or Cu > 300	High Grade Problematic Ore (High Ag, Cu)	HG-PO	
EV1_a > 0 ; Au > 1.5 - 2.5 ; Ag > 30 or Cu > 300	Medium Grade Problematic Ore (High Ag, Cu)	MG-PO	
EV1_a > 0 ; Au > 0 - 1.5 ; Ag > 30 or Cu > 300	Low Grade Problematic Ore (High Ag, Cu)	LG-PO	
EV1_a > 0 ; Au > 2.5 - 10 ; CuCN > 500	High Grade High CuCN	HG-CUCN/ HG-POS	
EV1_a > 0 ; Au > 1.5 - 2.5 ; CuCN > 500	Medium Grade High CuCN	MG-CUCN/ MG-POS	
EV1_a > 0 ; Au > 0 - 1.5 ; CuCN > 500	Low Grade High CuCN	LG-CUCN/ LG-POS	
EV1_a > 0 ; Au > 2.5 - 10 ; SxS > 3	High Grade High SxS	HG-SXS	
EV1_a > 0 ; Au > 1.5 - 2.5 ; SxS > 3	Medium Grade High SxS	MG-SXS	
EV1_a > 0 ; Au > 0 - 1.5 ; SxS > 3	Low Grade High SxS	LG-SXS	
EV2_a > 0 ; EV1_a <= 0 ; Au < 1.5	Mineralized Waste	MW	

* EV1_a : Economic Value base on the current gold price and current silver price
* EV2_a : Economic Value base on the upgrade 20% of the current gold price and upgrade

Ore mine control to reduce misclassification

Mining activity in Martabe Gold Mine consists of drilling, blasting, digging, hauling and dumping activity. Ore mine activity will go in accordance with daily production plan and blending ratio will be the guidance to determine ore proportion that need to be sent to crusher or stockpiles. Digging activity is conducted in every flitch with 2.5 m height. Mine geology personnel in the field will record and report the total truckload for every material and destination. Throughout the ore mine step, there are several tools and procedures to control the quality such as ore and waste boundary set-up, temporary benchmark (TBM) and laser installation to control the floor digging achievement. Mine geology personnel in the field also control ore and waste spotting specifically in contact area with expertise that has been trained by geologists.

Throughout the mining process, there are many possibilities of bias occurrence and material misclassification. Ore loss is economic mineralisation that has been lost during the mining activity and dilutions is defined as the incorporation of waste material to ore due to the operational incapacity to efficiently separate the materials during the mining process, considering the physical processes and the operating and geometric configurations of the mining with the equipment available.

Martabe distinguishes planned and unplanned misclassifications based on the cause. Planned misclassifications occur in the ore/waste delineation process when a little amount of waste below selective mining unit (SMU) size included in ore boundary or vice versa. Unplanned misclassifications occur in active mining activity, notably after blasting activity conducted. There are several approaches to control blasting activity, such as fitting a suitable tie up pattern with ore geometry, using poly pipe to monitor blast movement and creating a post blast model (Figure 8). By managing the cause, the amount of ore loss and dilution can be reduced and lead to increased revenue. Currently Martabe is using OREPro™ 3D. Since 2019, study about blast movement monitoring is continuously done with over 70 projects in all three active pits. Over the past five years, the use of OREPro™ 3D has led with a total reduction of 53 per cent throughout this period, subsequently increase substantial revenue up to 174 per cent (see Figure 9).

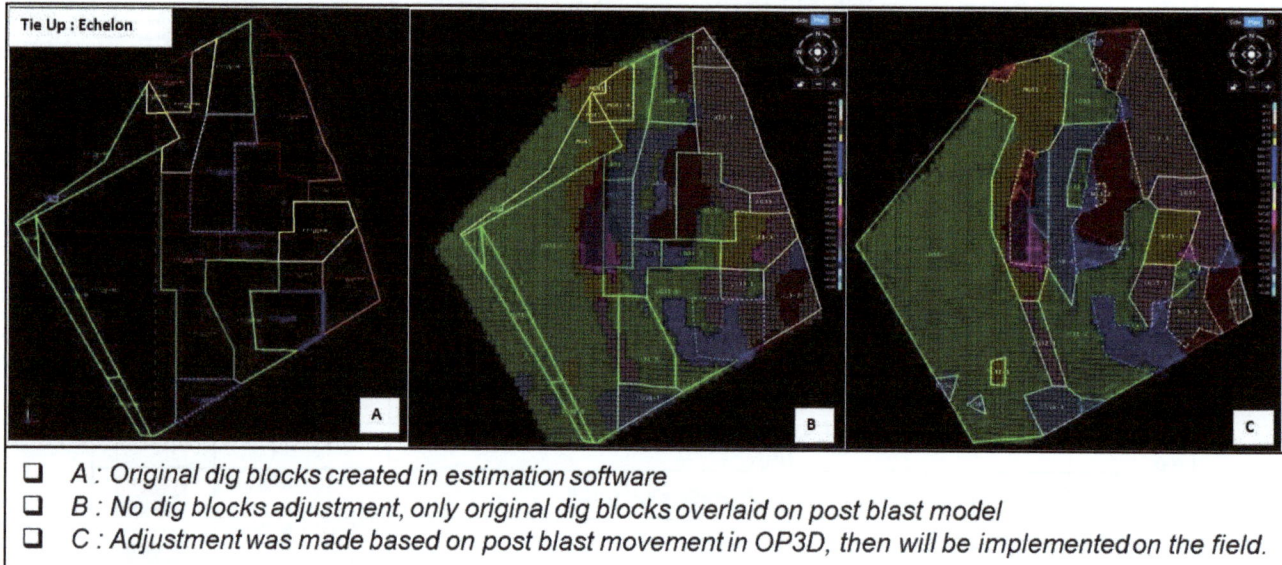

- A : Original dig blocks created in estimation software
- B : No dig blocks adjustment, only original dig blocks overlaid on post blast model
- C : Adjustment was made based on post blast movement in OP3D, then will be implemented on the field.

FIG 8 – Comparison of digging block before and after post-blast model adjustment to minimise misclassification.

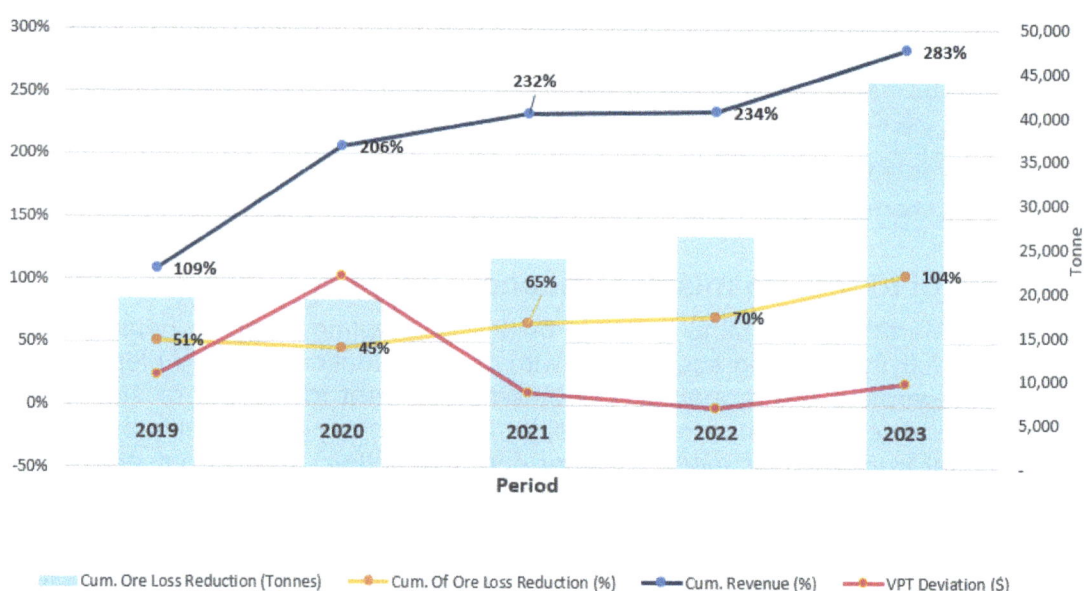

FIG 9 – Comparison graph between revenue value and total ore loss reduction.

Results from blast movement monitoring software play a significant role in mining practice improvement, but it has not directly used yet in the reconciliation process.

Ore blending and stockpile management

Blending and stockpiling is an effective way to treat the problematic ore to ensure streamlined ore delivery to process plant. Short-term plan mining will provide proposed blending ratio every day in accordance with monthly target (see Figure 10). This ratio will help to maintain the amount of material that will be directly sent to crusher, dump to and feed from ROM Stockpile. Mine geology team will control and ensure daily target achieved by direct supervision in the pit and ROM area.

Proposed Blend Ratio PIT - ROM Finger - December 29th, 2023			Estimated Average Direct Tip:		89%							
Source	Material Type	Ratio	Primary / backup Source	Est. Direct Tipping	PLI	Est. Au (g/t)	Est. Ag (g/t)	Est. Cu (g/t)	Est. CuCN (g/t)	Est. sxs (%)	Rec_Au (%)	Rec_Ag (%)
PUR - RL 327.5 (P3); BAR - RL 370 (B1)	HG	0.04	Primary	100%	5.2	3.31	4.29	84.64	14.61	1.04	0.82	0.61
PUR - RL 327.5 (P3); BAR - RL 370 (B1)	MG	0.05	Primary	100%	4.1	2.08	1.67	50.88	12.28	0.39	0.91	0.41
PUR - RL 327.5 (P3); BAR - RL 370 (B1); PUR - RL 362.5 (P3); PUR - RL 320 (P6)	LG	0.80	Primary	100%	3.9	0.81	4.09	90.82	13.78	1.40	0.74	0.68
RF04_LG			BackUp		4.7	0.90	7.48	77.90	25.91	1.16	0.75	0.72
RF02_PO	PO	0.04	BackUp		4.5	1.22	19.28	322.50	140.65	2.47	0.65	0.55
PUR - RL 305 (P4)	PO-S	0.07	Primary	50%	5.4	3.04	23.85	1161.07	676.71	2.08	0.63	0.58
Estimate ore grade feed to crusher					4.17	1.15	6.22	173.87	67.39	1.41	0.74	0.65
Crusher Throughput		ton/hour			1,100							
		Daily Tonnage			23,000							

Note: 1#EX1250 at Purnama RL 327.5 (P3), 1#EX1250 at Barani RL 370 (B1 & B2), #1x40T at Purnama RL 305 (P4), RL 362.5 (P3), RL 320 (P6)

Sticky and viscous material : 2,531 tonnes #Temporary Finger 2 ADT/hrs
RF 04_LG rehandled from ROM 2 ADT/hrs
RF 02_PO rehandled from ROM 1 ADT/hrs

*) Based on historical and clay domain by Minegeo

FIG 10 – Daily proposed blend.

Martabe has three types of stockpiles consist of active, long-term and crushed ore stockpile. Active stockpile or Run-of-mine (ROM) Stockpile has four ore categories and will be used for temporal dumping of ore to maintain grade target of delivered ore (see Figure 11). Problematic ore supply can also be maintained by split dumping to crusher and to ROM Stockpile. A two-finger stockpile system for each category is utilised whereby there is an active stockpile for crusher feed (closed for tipping) and an active stockpile for tipping (closed for crusher feed). A stockpile finger must be fully depleted before new material is added again on to it.

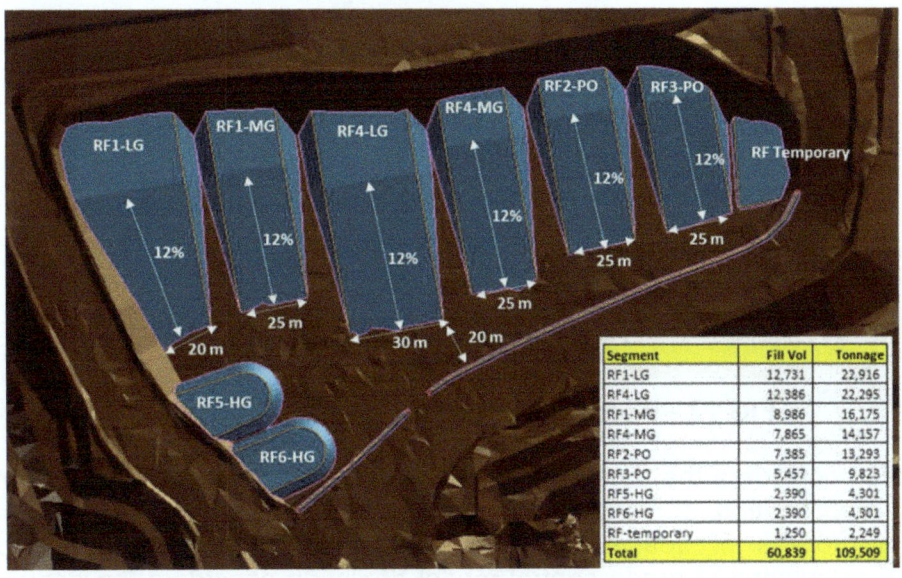

FIG 11 – Run-of-mine (ROM) Stockpile as active stockpile.

There are several long-term stockpiles in Martabe and it is specifically designed to accommodate very low-grade or marginal ore. From 2020 until now, one of the biggest long-term stockpiles in Martabe is actively supplied to crusher as gold price increase, this material then categorised as economic.

Crushed ore stockpile (COS) is the stockpile cone of ore produced after primary crushing prior to grinding through the mill. This is a continuous stockpile as material is constantly fed onto it and taken from it.

Ore mine calculation by as-mined method and reconciliation flow

The purpose of ore mine calculation by using the as-mined method is to record the progress of ore being mined daily before being input to a production database. Either ex-pit to crusher or to stockpile and stockpile to crusher. Daily ore calculation is the preliminary calculation of volume and tonnage based on total number of tally truck records from personnel in field. The trucking database is also used to estimate the crusher feed grade by proportioning the number of direct tipping and stockpile rehandle that sent into the crusher. The result of these records is the fundamental data source for creating reports and mining reconciliation.

Production database was set-up by using SQL database. The mining recovery is a number that shows the comparison between mining ore production from the pit with total of mineral reserves estimated for a certain period and location, expressed in percent. This process is a part of a bigger value chain that is called mining reconciliation that usually evaluates the whole mining flow process; from Resource Block Model to Declared Ore Mined (see Figure 12). Mining reconciliation is essentially the process of identifying, analysing and managing variations between planned and actual results in such a way that it highlights opportunities to improve a mine's performance (Hargreaves and Morley, 2010).

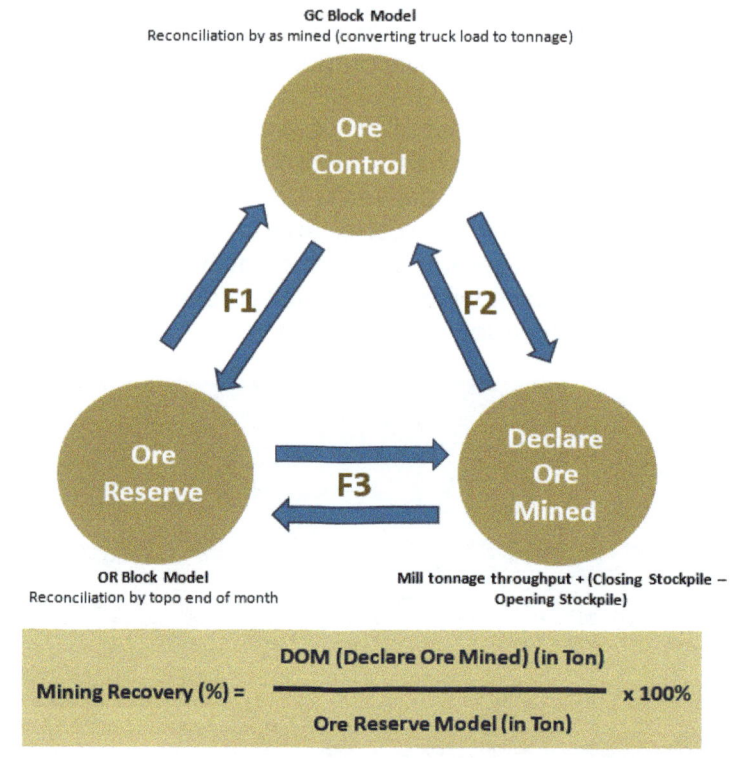

FIG 12 – Reconciliation flow in Martabe Gold Mine.

Mine reconciliation considers the data flow for the entire production process (mining value chain) and allows for identification of variation between planned versus actual measurements. Ore calculation is a process of spatial reconciliation between grade control model and actual topography. The calculation of grade control model depletion considers the actual field condition. Poor mining practice can directly be depicted by comparing the total truckload with actual condition captured in topography. Planned data refers to total depletion of reserve model based on actual topography but not considering mining practice factor. Actual measurement data refers to total milled ore tonnage and total balance of all the stockpiles. These three data will be reconciled to assess the mining performance monthly and annual. An industry guideline for reconciliation, 5 per cent variance is considered good while 10 per cent is reasonable with a requirement to investigate and proper justification. Reconciliation outside of 10 per cent variance requires further detailed investigation.

Mine reconciliation in Martabe Gold Mine is showing a positive trend compared to earlier production stage where technical and operational issues have not been resolved yet. But throughout the years, each of the issues is solved by conducting extensive study, analysis and evaluation. This is the result of continuous improvement in every key process (see Figure 13).

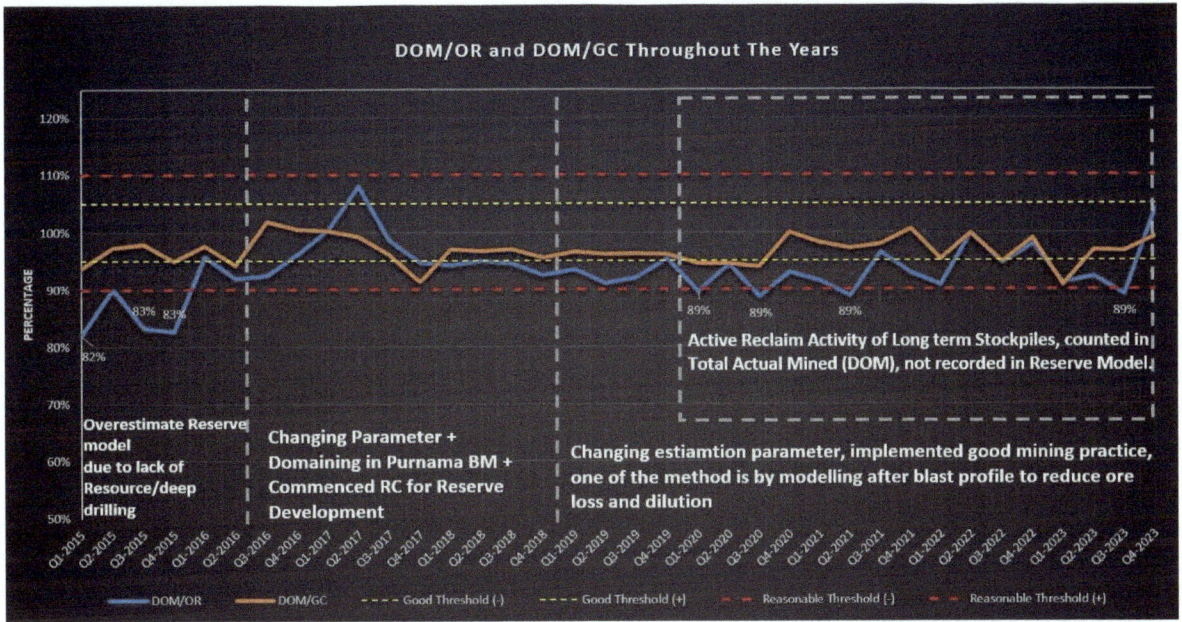

FIG 13 – Reconciliation trend from earlier production in 2015.

Contributing factor of mine reconciliation result will be calculated and become a guidance to determine which aspect needs to be improved. Ore tonnage in grade control block model is 426 Kt (5.9 per cent) lower and gold metal content is 25.4 Koz (6.3 per cent) lower (see Figure 14). Overall variance caused by mining practice is still below the threshold tolerance at 1–2 per cent bias. Based on 2021 data, variance topography including spilled material, boulder, dilution and ore loss issue are becoming the highlight that must be improve hereafter.

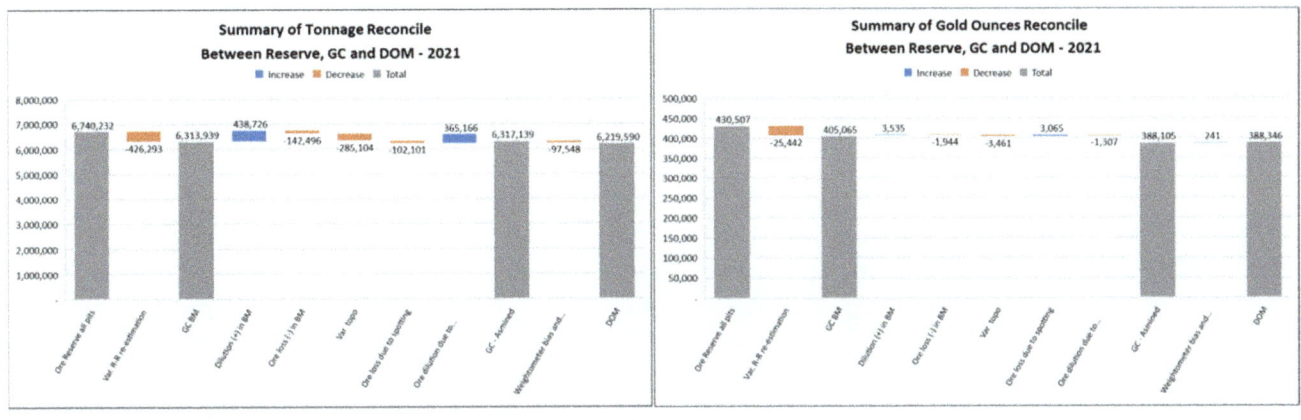

FIG 14 – Water fall graph that are showing the reconciliation result between Reserve, Grade Control Model and Declared Ore Mined (DOM) in 2021.

Based on data in Table 7, comparison between Grade Control and Reserve Block Model shows 6 per cent bias in tonnage, 4 per cent bias in gold grade and 10 per cent variance in total gold metal content. Actual mined (Declared Ore Mined: DOM) tonnage and Grade Control Model is showing a small number of variances of 2 per cent.

TABLE 7

Data reconciliation for all pit in 2021.

Reconciliation All Pit - YTD : December 2021					
	Tonnes	Grade Au (g/t)	Grade Ag (g/t)	Au (Oz)	Ag (oz)
Declared Ore Mined (DOM)	6,219,590	1.94	11.50	388,346	2,299,498
Grade Control (GC) Asmined	6,317,139	1.91	10.89	388,105	2,210,793
Ore Reserve (OR)	6,740,232	1.99	10.89	430,507	2,360,105
DOM/GC %	-2%	2%	6%	0%	4%
DOM/OR %	-8%	-2%	6%	-10%	-3%
GC /OR %	-6%	-4%	0%	-10%	-6%

Throughout the years, Grade Control (GC) model data shows a satisfactory performance when compared to the actual material received in the processing plant (Figure 15). Beside the model performance, mining activity and stockpile reclaim activity also impact the total GC model data. Reserve model data shows an 8 per cent variance when compared to DOM. One of the causes is the active reclaim activity in long-term stockpiles throughout 2021. Data from long-term reclaim is captured in total amount of DOM and was not accounted in Reserve Model.

Reconciliation trend in 2021 shows an exceptionally good trend in comparison between GC model and DOM. It means, the mining practice factor is constantly controlled. Meanwhile, Reserve Model Data shows a higher value compared to GC and DOM. It denotes the re estimation process need more deep drilling data from resource development. It means the domain dimension and the variogram model need to be evaluated.

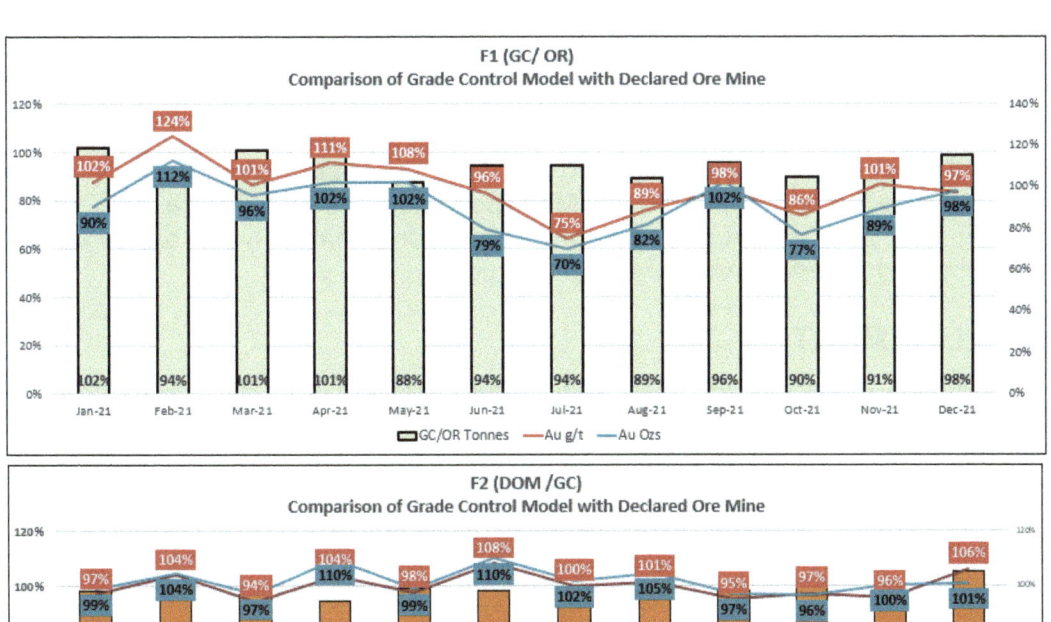

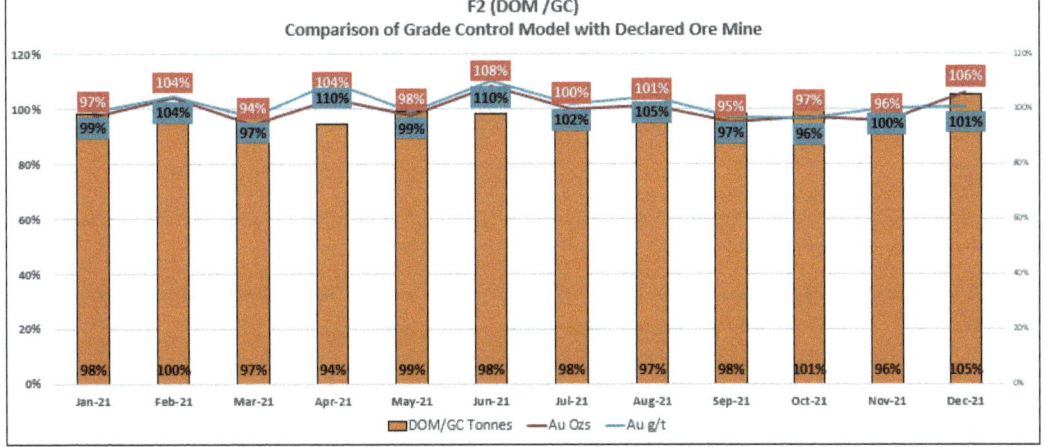

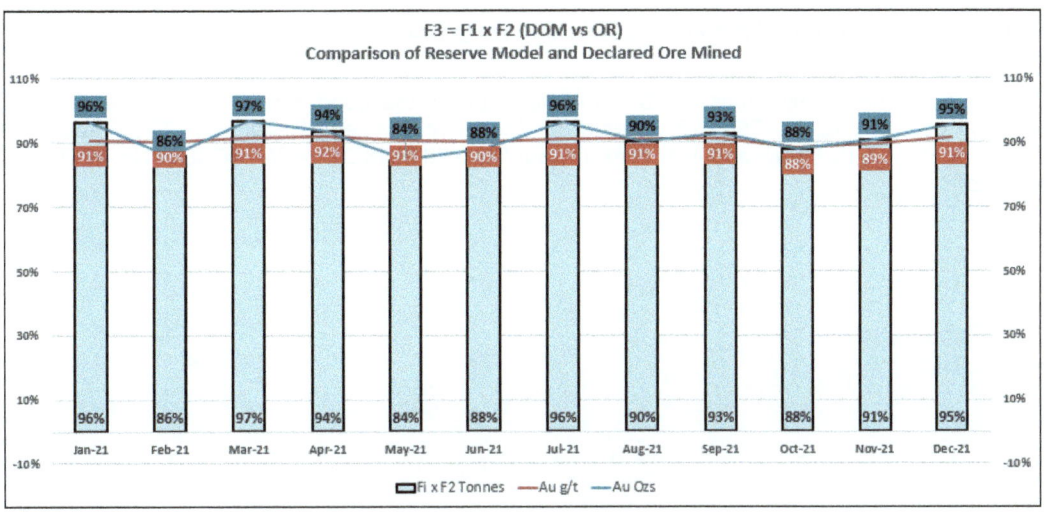

FIG 15 – Comparison graph between the grade control, reserve model and actual mined data during 2023.

CONCLUSIONS

Since production first commenced in 2012, Martabe Gold Mine has faced many issues regarding operational and technical aspects. These issues prompt ideas that lead to proper and effective solutions to emerge. The Mine Geology team in Martabe Gold Mine continue to preserve a systematic work and constantly seek room for improvements and innovation. The results are proven to have a positive impact, seen by reconciliation trend that move towards threshold target.

The reconciliation process carried out at the Martabe Gold Mine shows good improvement from year to year because improvements continue to be made. By ensuring that data quality is maintained at every step in the mining cycle, it has a very influential impact on the results being better and closer to the target. QA/QC carried out when taking and testing samples in the laboratory is the initial preparation for building a credible data set. Regular studies, test work and evaluations are also particularly important aspects for maintaining data quality.

Ore control and management in the field is a crucial aspect in the mining cycle to ensure delivery of well extractable ore. Accurate determination of the quality of supplied ore will be depicted on the reconciliation result at the end of the month. Martabe Gold Mine aims to maintain a system that is already running well and continue to explore innovations that can be applied relevantly and effectively to support the performance of the production.

ACKNOWLEDGEMENTS

The author would like to convey gratitude to Nurafrianti Saala, Nur Khariyah, M Imam Hidayat and Janjan Hertrijana as co-authors for all discussion session and contributions throughout the preparation of this paper. The author would like to thank all personnel of the Mine Geology team in Martabe site for their encouragement and support. The author would also like to thank all administrative officers at Martabe Site and Jakarta Head Office for preparing all necessary documents. Lastly, the author would like to thank the management of PT Agincourt Resources (Member of ASTRA) for their support and permission to publish this paper.

REFERENCES

Abzalov, M, 2011. Sampling Errors and Control of Assay Data Quality in Exploration and Mining Geology, in *Applications and Experiences of Quality Control* (ed: O Ivanov). doi:10.5772/14965.

Crispin, S, Hertrijana, J and Albert, P, 2015. Exploration Success at the Martabe Gold Mine, Indonesia (PACRIM 2015 Conference).

Greig, D and Cook, W, 1998. Methods for Assay Database Validation, in *More Meaningful Sampling in the Mining Industry*, Australian Institute of Geoscientists, Bulletin No. 22, pp 31–37.

Hargreaves, R and Morley, C, 2010. Mining Reconciliation – An Overview of data Collection Points and Data Analysis, in *Mineral Resource and Ore Reserve Estimation – The Aus IMM Guide to Good Practice*.

Harlan, J, Jones, M, Sutopo, B and Hoschke, T, 2005. Discovery and Characterization of the Martabe Epithermal Gold Deposits, North Sumatera, Indonesia.

Minister of Energy and Mineral Resources, 2018. Implementation of the good mining practices and supervision of the mineral and coal mining, No 26/2018, Minister of Energy and Mineral Resources, The Republic of Indonesia. Available from: <https://www.apbi-icma.org/uploads/files/old/2018/06/PerMen-ESDM-No.-26-Thn-2018-English-Version-.pdf>

Black rock cave flow model reconciliation and optimised draw strategy

M C Vermaak[1] and C Curtis-Morar[2]

1. Project Geologist, Glencore, McArthur River Mine, Borroloola NT 0852.
   Email: mathew.vermaak@glencore.com.au
2. Senior Geometallurgist, Glencore, Brisbane Qld 4000. Email: catherine.curtis@glencore.com.au

ABSTRACT

The Black Rock orebody, which forms part of the Mount Isa Copper deposit, is situated in north-west Queensland, Australia. It is a key component of Glencore's Mount Isa Mines operation, which will celebrate a century of Cu-Zn-Pb-Ag extraction in 2024. This extensive mining history precipitates unique technical challenges that influence the current extraction of copper from Black Rock (Shiels, 2022). Ore was originally mined from the Black Rock Open Cut (BROC) between 1957 and 1965 before wall instability led to early pit closure. The remaining Indicated Resource is currently extracted by sub-level caving, at the Black Rock Cave (BRC), located beneath BROC.

The challenges to reconciliation at BRC include a difficult-to-simulate mining method, heterogeneous supergene mineralisation (Smith, 1967), a demanding geotechnical environment (Shiels, 2022) and the presence of historic backfill within BROC. Backfill is a common diluent at BRC drawpoints and poses a grade control challenge – over and above waste rock ingress. Reconciliation work conducted at BRC showed variances between modelled forecasts and actual outcomes from the cave. Analysis of the variances indicates that a faster than predicted flow mobility of backfill contributes significantly to the observed discrepancies, necessitating a refinement of this parameter.

Utilising a staged approach, the BRC team determined the optimal backfill flow mobility parameter. This methodology involved testing different backfill flow mobility parameters against visual backfill dilution estimates and drawpoint assay data. More specifically, multiple backfill flow mobilities were tested and regression analyses were carried out to determine the optimal dilution flow parameter whose outputs aligned with observations. The results were then used to revise the BRC flow model.

This updated flow model, currently implemented on-site, demonstrates improved accuracy in production forecasts and increased predictability of material flow in BRC. This has assisted the BRC operational teams to optimise ore extraction. This paper aims to describe the process used to reconcile and refine the BRC flow model and the impact this had on draw strategy and production forecasting.

INTRODUCTION

Black Rock Mine is situated in the north-west Queensland Minerals Province and is part of the Mount Isa Mines Copper operation (Figures 1 and 2). The copper mineralisation consists of supergene enrichment above leached, primary chalcopyrite ore (Figures 2, 3 and 4) and is interpreted to be outside of the primary Silica-Dolomite alteration (Glencore, 2023). The orebody had a remaining Indicated Resource of 2.5 Mt at 5.49 per cent Cu, pre-cave mining (Glencore, 2019) and a mine life of four years.

Chalcocite ore is currently extracted from Black Rock by sub-level caving (SLC). Block caving and an open pit extension were investigated as potential mining methods for BRC. Shiels (2022) states that the orebody geometry, variable cave flow and propagation and stability risks meant that the orebody is not amenable to block caving. An open pit extension was also not suitable due to the cost associated with relocating critical surface infrastructure. The SLC method was thus chosen, also offering a potentially higher chance of total resource extraction versus the other methods. The cave sequence targeted the highest-grade domain first so that more metal was recovered and processed early (Shiels, 2022).

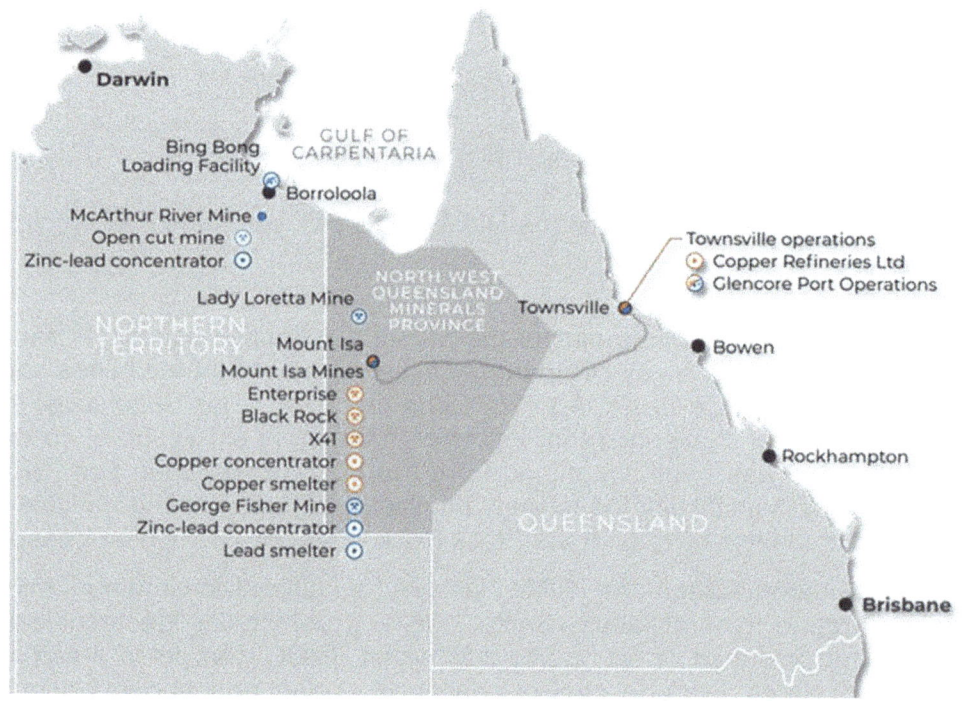

FIG 1 – Location of Mount Isa Mines in north-west Queensland, Australia. Black Rock Cave is an integral part of Glencore's Queensland copper business.

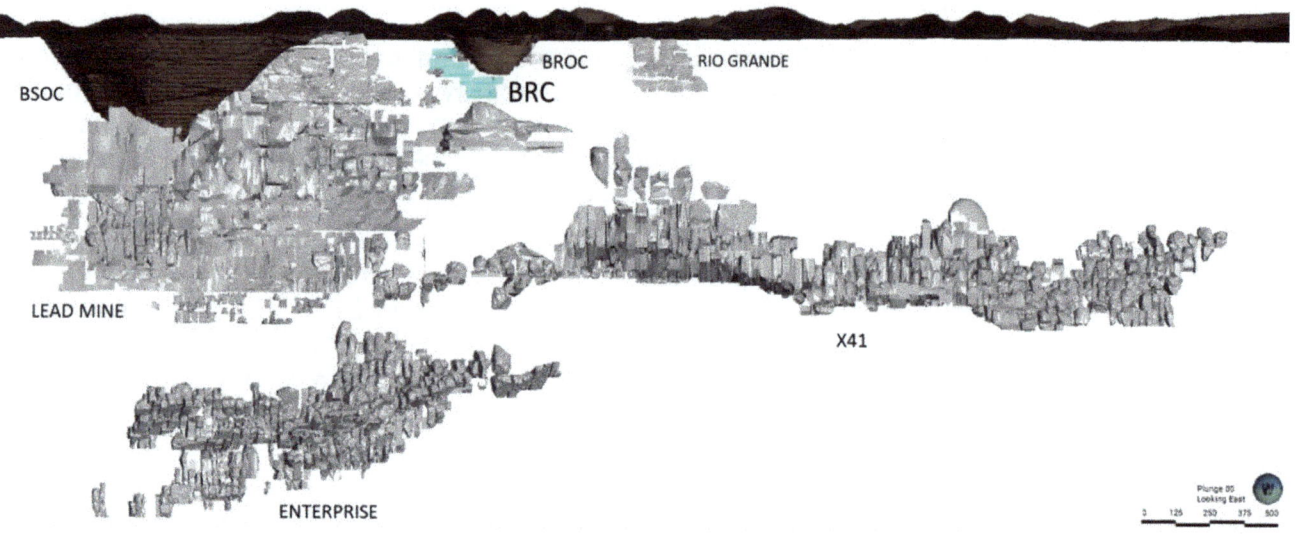

FIG 2 – Long section looking east, showing Black Rock Cave (BRC) and the Mount Isa Mines complex.

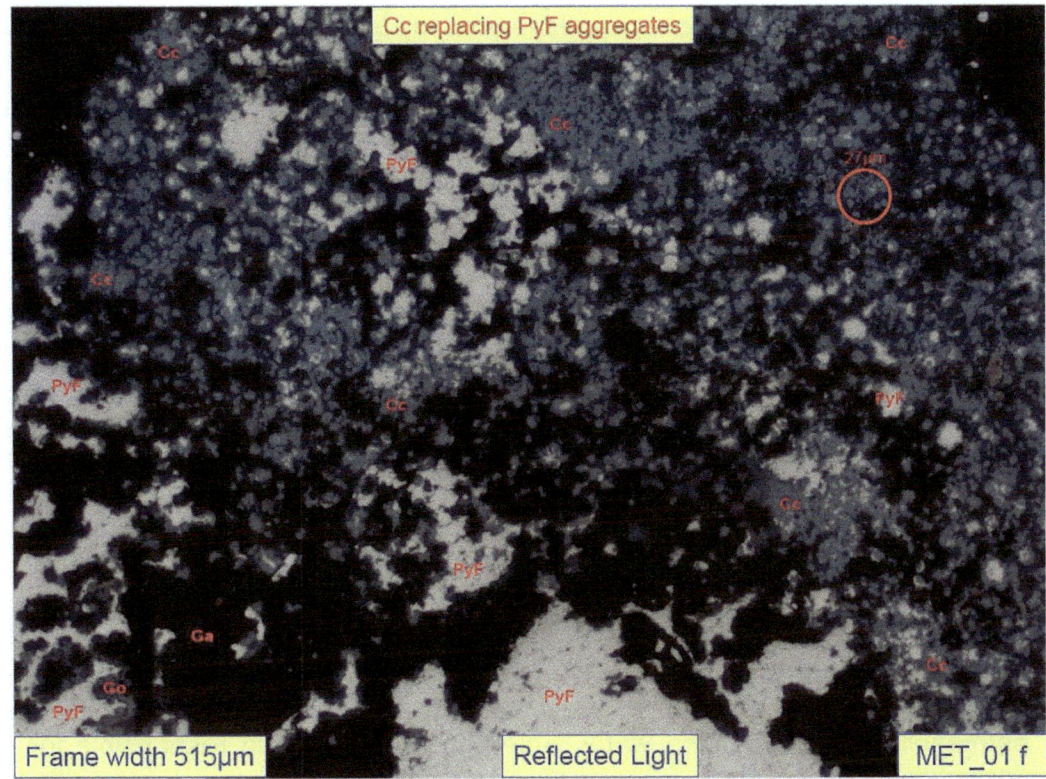

FIG 3 – Photomicrograph showing chalcocite replacement of fine-grained pyrite. Cc – Chalcocite, PyF – Remnant Framboidal Pyrite, Ga – Gangue.

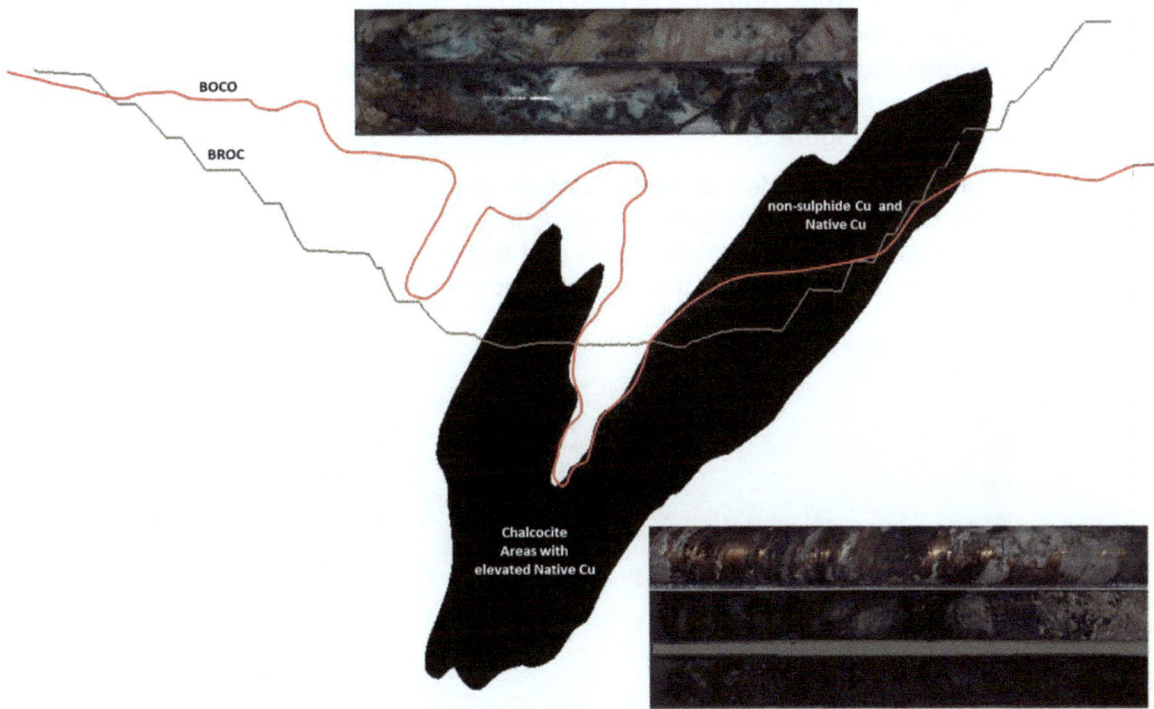

FIG 4 – A schematic showing the modelled supergene mineralisation with the generalised distribution of sulfide and non-sulfide copper species in relation to the base of complete oxidation (BOCO) and the historic Black Rock Open Cut. The core photo above the BOCO shows the variety of carbonate, silicate and oxide copper species present in the oxide domain. The core photo below the BOCO shows sooty chalcocite and Native copper.

The following is a brief description of the SLC mining method, adapted from Cokayne (1982): it is a top-down mining method where ore is drilled and then blasted from regular sub-levels developed into the orebody. The orebody and rock around the orebody are allowed to collapse naturally, where caved material is allowed to flow down as ore is drawn out through the sub-levels.

Material movement within a SLC can be simulated by flow modelling which uses several adjustable parameters to model cave flow. In the case of BRC, we utilise Power Geotechnical Cellular Automata (PGCA). The results derived from the flow model then inform the production schedule, targets and forecasts. Therefore, optimisation of the flow model parameters is crucial as cave production progresses from start-up to a mature operational phase. The mechanisms used to refine and optimise the flow model constitute various forms of reconciliation (Campbell, 2020).

GEOLOGY OF THE MOUNT ISA COPPER DEPOSIT

The Mount Isa Copper Deposit was estimated to be 248 Mt at 3.3 per cent Cu before mining (Conaghan, Hannan and Tolman, 2003). Mineralisation is hosted exclusively with the Urquhart Shale Formation of the Proterozoic Mount Isa Group. Copper occurs as chalcopyrite within Silica-Dolomite alteration, a complex sequence of microfacies resulting from progressive metasomatic alteration (Perkins, 1984). This is divided into two broad zones – an inner silicious core and an outer dolomitic halo.

Extensive late-stage cross faulting enabled the circulation of acidic groundwater through part of the deposit. This resulted in a localised area with a depressed base of complete oxidation (BOCO) and extensive, deep (>800 m), anoxic leaching and removed an estimated 20 per cent to 50 per cent of the dolomitic groundmass (Hewett, 1968). Smith (1967) showed that the resultant porosity, increased permeability and fluctuating water table controlled the fluid responsible for supergene alteration processes at Black Rock.

THE GEOLOGY OF THE BLACK ROCK CHALCOCITE OREBODY

The formation of the Black Rock orebody is described in detail by Smith (1967) and summarised here. Black Rock supergene mineralisation is hosted by the Urquhart shale, in the hanging wall of the Silica-Dolomite alteration. The orebody was formed by shallow secondary copper enrichment where cupriferous fluids associated with deep leaching, reacted with fine-grained pyrite to precipitate secondary copper species (Figure 3).

Supergene mineralisation is complex consisting of an assemblage of non-sulfide copper species occurring above the BOCO and high-grade chalcocite below the BOCO (Figure 4). Chalcocite is predominantly a replacement of framboidal pyrite and largely occurs as sooty chalcocite (Figures 3 to 5). The chalcocite is the target ore for the Black Rock sub-level cave.

FIG 5 – 'Protolith' (ie unaltered host rock) versus Black Rock Cave (BRC) Ore – showing chalcocite replacement of framboidal pyrite layers to form 'sooty chalcocite.'

CHALLENGES ASSOCIATED WITH MINING THE BLACK ROCK OREBODY

The supergene mineralisation is heterogeneous and complex but can be subdivided into two main domains – a chalcocite domain and an oxide domain. These were modelled using diagnostic leaching, geochemistry and mineralogy (Hollitt, 2017; Barnes, 2018). Nevertheless, due to the heterogeneity of the deposit, non-sulfide copper can report to the drawpoints. These species are unrecoverable in the sulfide-optimised flotation circuit (O'Donnell and Begelhole, 2023), however; BRC has come full circle and the high-silica tails are used as a flux for the copper smelter. The repurposing of the BRC tails as a flux has introduced an innovative pathway to recover previously unrecoverable Cu species (O'Donnell and Begelhole, 2023).

Black Rock sulfide ore was originally mined underground by room and pillar and cut-and-fill stoping in 1931 (Berkman, 1996). Silica flux was mined from the Black Rock Open Cut (BROC) for the smelter, from 1957 to 1965, with chalcocite ore being extracted between 1963 and 1965 before instability in the western wall closed the pit (Edwards, 1968; Hewett, 1968). Near-pit fixed infrastructure was at risk due to ongoing pit-wall instability and an estimated 5 million cubic metres of backfill was placed into the pit to mitigate the risk (Shiels, 2022).

This complex history has created several challenges for BRC. The first is historic voids which affected the upper levels of the cave. These required extensive prop drilling to ensure the upper levels could be produced safely. The historic drives, stope fill and old material introduced oxidised sulfides into the ore stream, which had a negative impact on metallurgical performance when the ore was processed through the concentrator. The variable cave propagation, chimneying and caving into BROC also allowed the ingress of historic backfill to reach the drawpoints. It is this backfill material that is the focus of this paper.

BLACK ROCK RECONCILIATION

The flow model reconciliation process constituted three parts:

1. Dilution quantification.
2. Assay reconciliation.
3. Sequential copper analysis.

Reconciliation at BRC showed variances between modelled forecasts and actual outcomes from the cave.

DILUTION QUANTIFICATION

Dilution was quantified through geological inspections and visual estimates. The backfill is oxidised and bright orange whereas the BRC ore is grey (Figure 6). The colour contrast combined with experienced observation, allowed for an estimate of dilution. The data was collated and compared with dilution estimates derived from the flow model.

FIG 6 – Example of backfill ingress in a BRC drawpoint.

The process to optimise the backfill flow mobility parameter was iterative. Multiple flow model simulations were run testing different backfill mobilities. The simulation results were compared to the actual estimated dilution. The relationship between the simulation results and actuals was analysed using regression modelling which determined the optimal backfill flow mobility parameter (Figure 7). The updated flow model shows good alignment between the predicted and observed dilution.

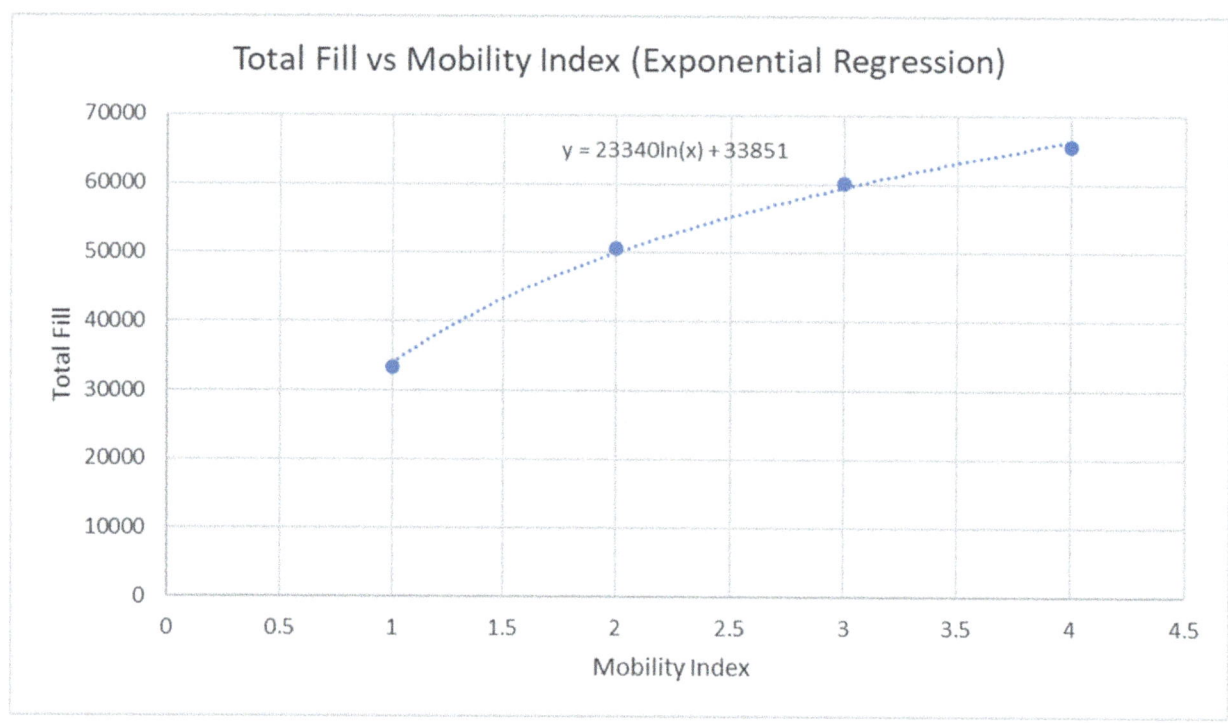

FIG 7 – Regression modelling showing predicted total backfill tonnes as a function of the flow mobility index. Note that all other domain mobilities were held constant for this exercise.

It was thus found that the backfill flowed ~3.4× faster than previously thought, in-line with observations around the backfill rilling over the top of freshly blasted drawpoints. Thus, analysis of the variances indicates that a faster than predicted flow mobility of backfill contributed significantly to the observed discrepancies, necessitating a refinement of this parameter.

TESTING MODEL-ASSAY ADHERENCE

Assay grade adherence was tested between the original and updated flow models. Reconciliation work was carried out to test the representativity of BRC drawpoint (grab sample) assays versus the mill. This work is outside the scope of this paper but found very good adherence between drawpoint XRF assay data and mill XRF assay data.

It was found that the updated flow model aligned better with the production assays than the original flow model. Figure 8 shows that the grade variance between forecast and assay was variable across the sub-levels. Understanding the spatial distribution of the variance allowed for deeper analysis of the discrepancies. Areas with variance unaccountable by flow modelling are assessed using other geological and mining factors outside the scope of this paper.

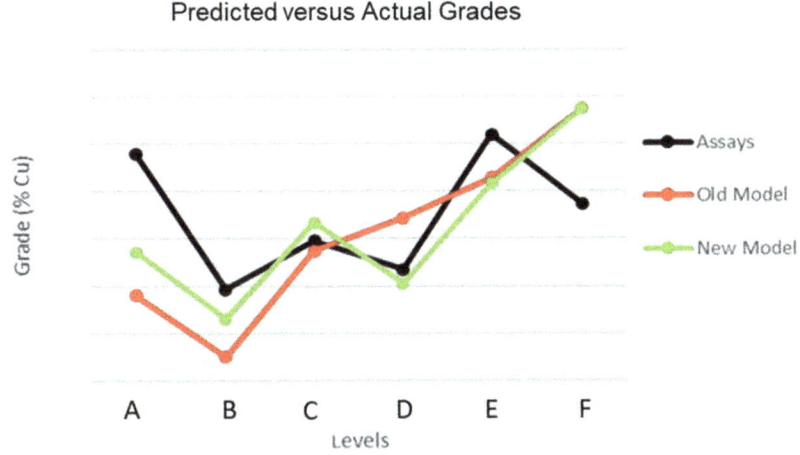

FIG 8 – Assay grade adherence between the original and new flow models. The 'Old Model' is the Original Model, whereas the 'New Model' is the Updated Model.

As can be seen in Figure 8, the levels with the greatest discrepancies are 'A' and 'F.' The remainder of the levels are in relative agreeance. What this work allowed the BRC team to do, was to dig further into the sources of the discrepancies, aiding in mine scheduling and optimisation.

GEOMETALLURGICAL DOMAIN FLOW MOBILITY

The geometallurgical domains derived by Hollitt (2017) and Barnes (2018) are flagged in the BRC Block Model. The domains assist with forecasting the proportion of chalcocite and oxide copper in the Mill feed. Considering the BRC orebody constitutes a variety of unrecoverable copper species, this last step helped us refine geometallurgical domain flow parameters, overall refining the flow model to better predict material from, as well as set up a mechanism to better understand and report unrecoverable copper species proportions to the metallurgy teams. This is especially pertinent considering the overdraw strategy combined with the outcomes achieved by O'Donnell and Begelhole (2023).

OUTCOMES

The results of this reconciliation effort were multifaceted:

- Data-backed refinement of flow model parameters has led to improved grade and dilution predictability and forecasting, at both a whole mine and a sub-level-by-sub-level basis (Figure 9). This has assisted production planning across the whole value chain.

- The improved flow model confidence and material flow understanding has enabled the implementation of an optimised operational draw strategy. This has assisted the operational teams to optimise ore extraction, which allows us to better recover planned metal tonnes and benefit from greater overall mine efficiencies. This optimised draw strategy has enabled the BRC team to implement a flexible production schedule which has made the mine more resilient to recent extreme weather events.

- The process has highlighted additional challenges for cave flow simulation in heterogenous and complex orebodies such as BRC. Assay-assisted reporting strategies and reconciliation are crucial for accurate production forecasting. This has produced a relatively robust end of month (EOM) reconciliation and has helped identify sources of error through the value chain.

- Given the more accurate forecasting and reporting strategies, this work also enabled the implementation of a suitable (and dynamic) cut-off grade and allowed sensitivity and scenario analysis for mine optimisation, especially with regards to throughput grade targeting. A hypothetical illustration of this analysis is provided in Figure 10.

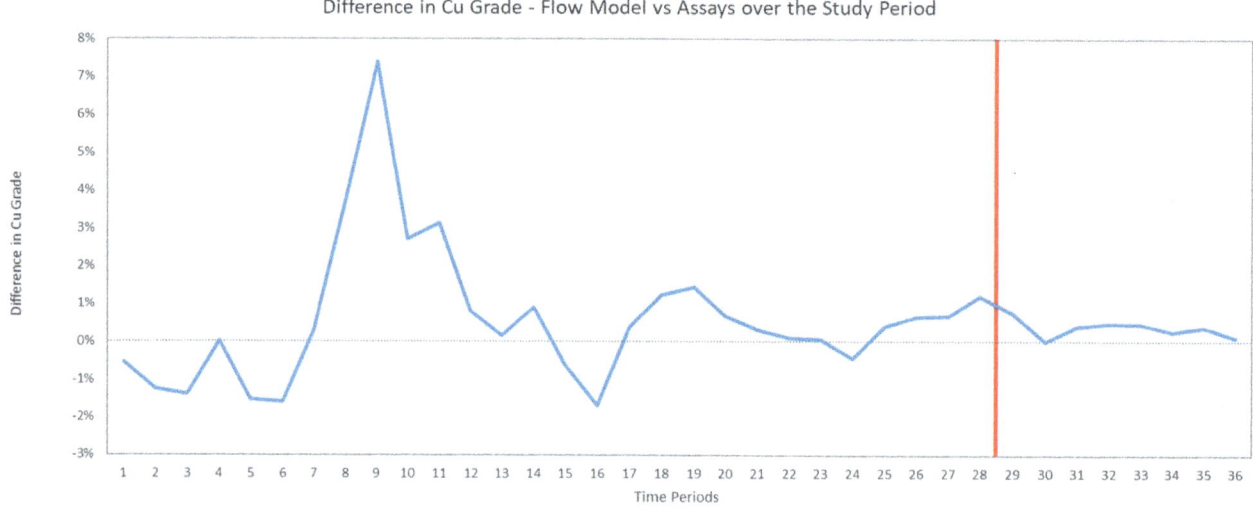

FIG 9 – Difference between modelled/forecasted and actual grades. Highlighted periods: (i) Period 28: implementation of the new flow model into the mine schedule (note the sustained drop in variability between the forecast and actuals). (ii) Period 34: implementation of sub-level specific factoring into the schedule, leading to further improved alignment between the plan/forecast and actuals.

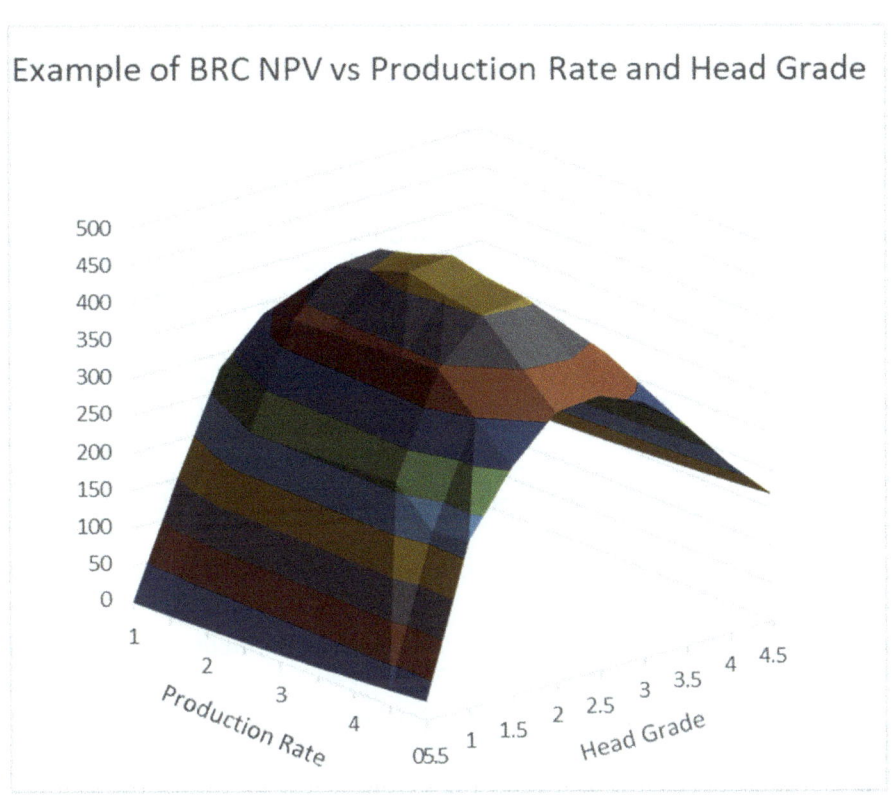

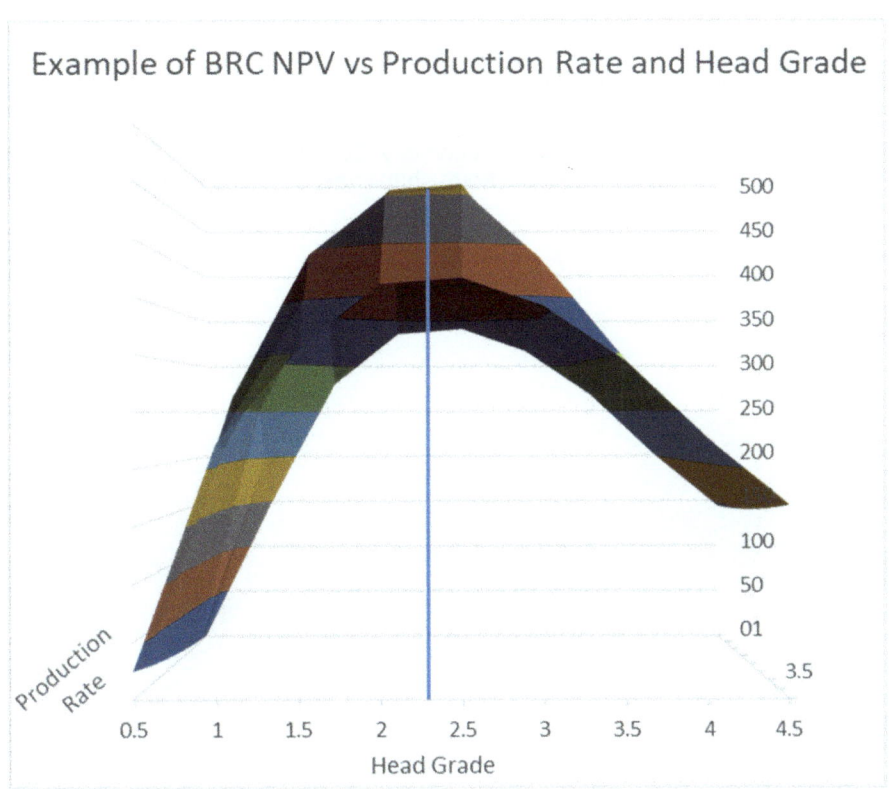

FIG 10 – Hypothetical example of BRC Three-Parameter Value Optimisation showing targeted head grade to be the major driver in terms of Net Present Value (NPV).

REFERENCES

Barnes, H, 2018. Copper Sequential Data, BROC ore domains. Unpublished Report.

Berkman, D, 1996. *Making the Mount Isa Mine 1923–1933*, Spectrum series, p 9; p 90 (The Australasian Institute of Mining and Metallurgy: Melbourne).

Campbell, A D, 2020. Recovery and flow in cave mining: current knowledge gaps and the role of technology in the future, in *UMT 2020: Proceedings of the Second International Conference on Underground Mining Technology* (ed: J Wesseloo), pp 77–104 (Australian Centre for Geomechanics: Perth) doi:10.36487/ACG_repo/2035_0.04.

Cokayne, E W, 1982. Sublevel caving Chapter 1: Introduction, in *Underground Mining Methods Handbook* (ed: W A Hustrulid), pp 872–879 (Society of Mining Engineers of The American Institute of Mining, Metallurgical and Petroleum Engineers, Inc.: New York).

Conaghan, E L, Hannan, K W and Tolman, J, 2003. Mount Isa Cu and Pb-Ag-Zn deposits of NW Queensland, Australia, *Regolith Expression of Australian Ore Systems, CRC LEME Geochemistry Special Monograph Series*, p 3.

Edwards, D B, 1968. Ground Stability Problems Associated with the Black Rock Open Cut at Mount Isa, in *AusIMM Proceedings*, no 226, pt 2, pp 61–72.

Glencore, 2019. Resources and Reserves as at 31 December 2019, report. Available from: <https://www.glencore.com/.rest/api/v1/documents/9d3cb62e5188ec6d35cb69b57679eb9e/GLEN_2019_Resources_Reserves_Report--+%281%29.pdf>

Glencore, 2023. Resources and Reserves as at 31 December 2023, report. Available from: <https://www.glencore.com/.rest/api/v1/documents/static/a53e27b1-6025-4ef2-9be8-f3be543dfb26/GLENCORE-Resources-and-Reserves-report-2023.pdf>

Hewett, R L, 1968. Deep Leaching and Accelerated Post-mine Oxidation in the 500 Ore Body at Mount Isa, in *AusIMM Proceedings*, no 226, pt 2, pp 73–88.

Hollitt, M, 2017. BROC Cutback project: Assistance with the Development Program – Black Rock Open Cut Geomet sampling and analysis from 2015 drilling (II). Unpublished Report.

O'Donnell, R and Begelhole, J, 2023. One Plant's Trash is Another Plant's Treasure; a Synergistic Approach to Novel Uses for Tailings Streams, in *Proceedings of the 11th MetPlant Conference*, pp 364–381.

Perkins, W, 1984. Mount Isa silica dolomite and copper orebodies; the result of a syntectonic hydrothermal alteration system, *Econ Geol*, 79:601–637.

Shiels, A 2022. Unearthing the Black Rock orebody with sublevel caving, in *Caving 2022: Proceedings of the Fifth International Conference on Block and Sublevel Caving* (ed: Y Potvin), pp 1443–1458 (Australian Centre for Geomechanics: Perth).

Smith, W D, 1967. The Geology of the Black Rock Secondary Copper Orebody, The Department of Geology and Mineralogy, University of Queensland. Available from: <https://espace.library.uq.edu.au/view/UQ:00a0962>

Technological solutions to remnant mining ore control at the Martha Underground Mine

A Whaanga[1], W Vigor-Brown[2], A Harvison[3] and J Maxted[4]

1. Superintendent Mine Geology, OceanaGold, Waihi 3610, New Zealand.
   Email: abe.whaanga@oceanagold.com
2. Senior Geologist Mining, OceanaGold, Waihi 3610, New Zealand.
   Email: william.vigor-brown@oceanagold.com
3. Project Geologist, OceanaGold, Waihi 3610, New Zealand.
   Email: adrian.harvison@oceanagold.com
4. Project Geologist, OceanaGold, Waihi 3610, New Zealand.
   Email: jared.maxted@oceanagold.com

ABSTRACT

Several new and emerging technological advances were trialled and implemented at the beginning of the Martha Underground remnant (MUG) mine project, which has been in production for three years. This paper presents a review of the effectiveness of these technologies in a remnant mine and an update on future work.

Three-dimensional (3D) scanning utilising photogrammetry has enabled more accurate and detailed geological mapping. This has led to more optimally positioned ore drives, with reduction of operational re-work, increased efficiency and improved stope reconciliations. In addition, every drive that intersects the vein skin or historical void is 3D scanned, with the data points used to update the vein and historical void models to assist with stoping and defining historical depletion.

Digital mapping has given the geologist a full suite of digital data to base decisions on. The ability to accurately geo-reference and ensure the current and future face positions will be within caution buffers has increased the speed of the drive direction process whilst ensuring that safety protocols in and around historical workings are adhered to.

Continuous improvement of grade control geological modelling and estimation techniques have reduced turnaround times of wireframe and intermediate model updates, including information on historical workings such as size and spatial location. This has enabled the geotechnical and engineering departments to make financial decisions with the latest information, reducing re-work and decreasing the time from drive completion to stope design and ultimately production.

Ore definition of remnant skins around voids is problematic, with the inability to strike drive along the vein. Footwall drives with transverse fill and extraction drives provide a drilling platform to grade control drill short holes from a bobcat mounted LM30 mobile drill rig. Cavity Auto-Scanning Laser (C-ALS) surveys down these holes into the historical stope voids assist in determining depletion for grade control models. 3D scanning of historical voids, once broken through, allow remnant stope skin definition.

The introduction of a new generation of Light Detection and Ranging (LiDAR) scanning technologies now provide the ability to create 3D geo-referenced scans at the active heading, thus removing a further step of registration and post-capture processing on the surface. Detail captured at the drive or face allows geologists to identify and map historical workings openings and interpret stope fill from collapse and *in situ* rock and veining. These interpretations are used to refine the historical void model, allowing accurate depletion of the historically mined vein.

However, no technological revolution is perfect. Time pressures at faces have resulted in incomplete or poor quality data at times. The learning curve with new and bespoke software and the requirement for new geologists to master multiple pieces of technology to complete a task requires specific and targeted training. An over reliance on technology means that core geological skills of mapping, spatial orientation and structural measurements require constant attention to ensure they do not become diminished. However, the benefits of these new technologies outweigh the challenges.

INTRODUCTION

Mining has played an important role in the history of Waihi since Au was first discovered in 1879. Since then, the Martha deposit in the heart of the Waihi town has produced close to 6.9 Moz of Au.

Historical underground mining of the Martha veins occurred from 1882 to 1952 and an open pit mine extracted ore from the upper portions of the vein system from 1988 to April 2015. When mineralisation was discovered approximately 2 km east of the Martha deposit, the Favona underground mine was developed and extraction of ore commenced in 2004 (McAra, 1988).

This underground mine expanded and is still in operation, having extracted ore from numerous vein systems in between the Favona and Martha orebodies, including the Trio and Correnso orebodies (producing approximately 1.1 Moz Au).

Mining beneath the existing Martha open pit, known as the modern Martha Underground (MUG), commenced in 2021. The MUG project consists of the remnant material left behind from historical mining and differs from the early orebodies, which were largely virgin veins and not historically mined; stope fill, stope vein skins and intact crowns and pillars make up the bulk of the current resource.

This paper reviews several new and emerging technologies implemented at the Waihi mine, with a discussion on how effective these technologies have been for remnant mining at the MUG project.

3D SCANNING

Currently at the MUG mine, every cut in an ore drive is 3D scanned using photogrammetry, with a geo referenced.obj file produced for importing into mining software (Whaanga, Vigor-Brown and Nowland, 2019).

With remnant mining, variability of the vein skin width can occur. Small inaccuracies in the vein position or void location has previously led to skins being created in the model where they do not exist or the absence of a skin where they should exist. To counter this, every fill/extraction drive that intersects the vein skin or historical void is 3D scanned. This gives many data points along the vein and historical void so that both models can be updated. Grade control drilling is typically done on a 12–20 m spacing depending on the orebody and location within. Adding scans of fill drives gives a 5 m × 5 m polyline of where the vein and void are spatially located. Combined with the drill hole grades for estimation, this greatly increases the accuracy of the model before stoping by helping to define the historical depletion (Figure 1).

FIG 1 – Plan view 3D scan with actual historical drive position captured against void model (light blue) with modelled vein position (dark blue).

Challenges exist with accurate 3D scans of unsupported historical voids; geologists must take enough overlapping photos with sufficient light to capture the void openings and minimise shadowing.

As the process is completely digital, instances have arisen where technology has failed. To mitigate this, a backup camera is kept in the geology underground vehicle and a spare tablet on the surface. As the process requires all data to be processed underground, situations of completely failed face data are rare. There is no backup sketch taken so in these cases face channel data is not able to be spatially located or used. However, with the ability to scan the backs between these areas of data loss, the net effect on the interpretation for the area is minimal.

Stope reconciliation

Optimal drive positioning is vital to improving stoping performance alongside minimising dilution through overbreak and ore loss through underbreak. The presence of historical drives passes and stopes in current development has presented a challenge to position ore drives that create a stable economic stope shape with sufficient geotechnical pillars. 3D scanning of each ore drive has allowed the geology and geotechnical teams to review the drive position cut by cut, comparing the modelled position of both vein and historical mining to the measured positions. This has enabled drive positions to be constantly updated and modified in real time. With MUG being high-graded through historical mining, modern stope grades are lower and closer to economic cut-off. Optimal positioning has resulted in significantly less re-work getting drives back online, translating to lower costs and higher recoveries. Prior to the introduction of 3D scanning, drives went offline due to directional issues up to several times per month, or 10–20 per cent of the ore drive metres mined, requiring some form of stripping or re-work to re-align. After 3D scanning technology was implemented, this was reduced to less than 5 per cent. This is especially significant when considering the bulk of the MUG deposit consists of thinner, lower grade, more structurally complex veins that were not historically mined due to being uneconomic at the time. Compared with the earlier, previously unmined, higher-grade orebodies, Favona, Trio and Correnso, which were less structurally complex and easier to follow, mining of the MUG deposit is challenging.

DIGITAL MAPPING

Improvements to digital mapping processes using Deswik.Mapping™ have increased the level of information available to track and review the presence of historical voids.

Hardware upgrades

The introduction of upgraded Panasonic Toughpad FZ-G2 notebooks from the FZ-G1 has increased the processing power, allowing for faster and more efficient use of the Deswik Geology Mapping software whilst underground. This has reduced the need for re-interpretation and 'tidying up' of underground mapping data. The geologist can now use the mapping software to select a pre-determined working plane (Whaanga, 2022), interpret the lithology they see at the face and effectively map lithologies underground, without the use of the textures function which previously acted as a stopgap due to hardware limitations. This has reduced the workload by removing the cleaning up stage of importing geological mapping, as the sections are already set to plan view and mapped to the appropriate lithologies, requiring only a simple import through Deswik process maps (Whaanga, 2022). The historical void model is large and manipulation is data intensive. The upgraded hardware allows more information to be displayed, facilitating better decision-making at the face rather than waiting until the geologist is back on the surface.

Structural measurements

Structural measurements are now being recorded electronically into the geologist's workflow at the active work face, using Deswik Geology Mapping software and a compass. This occurs after a face section has been created, which in turn geo-references a location based on the distance from a laser station, offsets and azimuth of the face (Whaanga, 2022). This process has allowed more accurate measurement of structures in the complex underground environment, as well as allowing the validation of the wireframe and vein orientation between drill holes while at the face. Reconciling the historical working positions with new drilling and vein interpretations is often difficult. Structural measurements at the face combined with the void model assist in determining which vein was historically mined and how to integrate new data to place drives in optimal positions.

Template updates

Remnant mining has increased the complexity and frequency of development and production interactions with historical voids and material types. Jumbo probe data, design stops, caution buffers, vein and void model wireframe updates, C-ALS surveys and grade control drill information are all considered at the design signoff stage. Once approved, attributed information is incorporated into master design projects with critical safety information available on digital survey memoranda for operations personnel and digital templates for technical services personnel. The geology team has incorporated this information into the latest Deswik templates, which are automatically refreshed overnight with completed and approved data from the previous day (Whaanga, 2022). Once spatially located, geologists can interpret the historical void/material type at the active face within the context of the design template notes. If a drive has broken through into an interpreted filled stope that contains a void, geologists contact the geotechnical and survey teams requesting respective surveys/inspections. A review of logged data from on level grade control drilling can offer insight into the accuracy of modelled void positions. Due to the unpredictable nature of remnant mining it would not be possible to print out a series maps covering every area of the mine and every eventuality that may be encountered at active headings. Updated digital data in a 3D mapping suite is an efficient and powerful way of organising a wealth of information, enabling fast, safe and confident decisions at the face (Doyle and Whaanga, 2017).

Mapping training

Since the introduction of a complete digital mapping workflow in 2019 at Waihi, several new geologists have been trained and moved through the department. Younger geologists are receptive to digital technology and the time taken to become competent with the Waihi workflows is notably fast. Mapping in 2D, both plan and section, are not required to the same extent and with implicit modelling is no longer required at the wireframe construction phase. The quality of these skills has noticeably and predictably declined. However, this does not appear to be a problem with the increase in the ability to work, think and communicate in 3D. Good quality mapping training must continue, including taking structural measurements, identifying kinematic indicators, mineralogical features and timing relationships.

HISTORICAL VOIDS

A historical void model for the MUG mine was created from historical map digitisation and reserve drilling of the orebodies. These wireframes are then updated and adjusted as new information becomes available (Muir, 2023).

Cavity auto scanning laser (C-ALS)

With a diameter of 50 mm, the C-ALS cavity monitoring system is easily deployed downhole or uphole through boreholes to survey inaccessible spaces. A system of hinged, lightweight 1 m rods provides a fixed azimuth capability and enables C-ALS deployment down boreholes typically up to 50 m in length. A nosecone camera gives full visibility of the borehole during deployment, so operators can see any obstructions and judge when C-ALS has entered the void. The C-ALS probe has sensors to ensure it can be tracked down the borehole and that the scan is automatically georeferenced to fit into existing 3D mine data. Once in the void, the laser-scanning head rotates on two axes, measuring the 3D shape of the void with full 360° coverage and with a range up to 150 m. A load-bearing cable attached to the probe transmits all the measured data back to the surface unit.

Grade control drilling

A bobcat mounted LM30 and more recently a mobile drill rig (MDR) have been utilised to drill short holes from footwall drives, ideally less than 30 m in length. Holes are designed to bridge open stopes if possible and continue on the opposite side of void, with the void and depth noted for poly insertion at completion of the drill hole. Designing holes in this way gives the option of obtaining a C-ALS of an open void if intersected, while still collecting core and samples of remnant vein skin around the void.

If an open void is intersected, the drill hole is lined with poly when the hole is completed and a C-ALS scan taken. Using the C-ALS scan and lithological logging of vein skin gives a much more accurate model for remnant vein material along with size and location of the void (Figure 2). A common issue with the scanning of voids is the collapse of drill holes. Due to the fragmented broken nature of the ground around these historical voids, the drill holes often collapse before poly can be inserted to the void depth. An improvement to this process would be to scan through the drill string itself before pulling the rod string out.

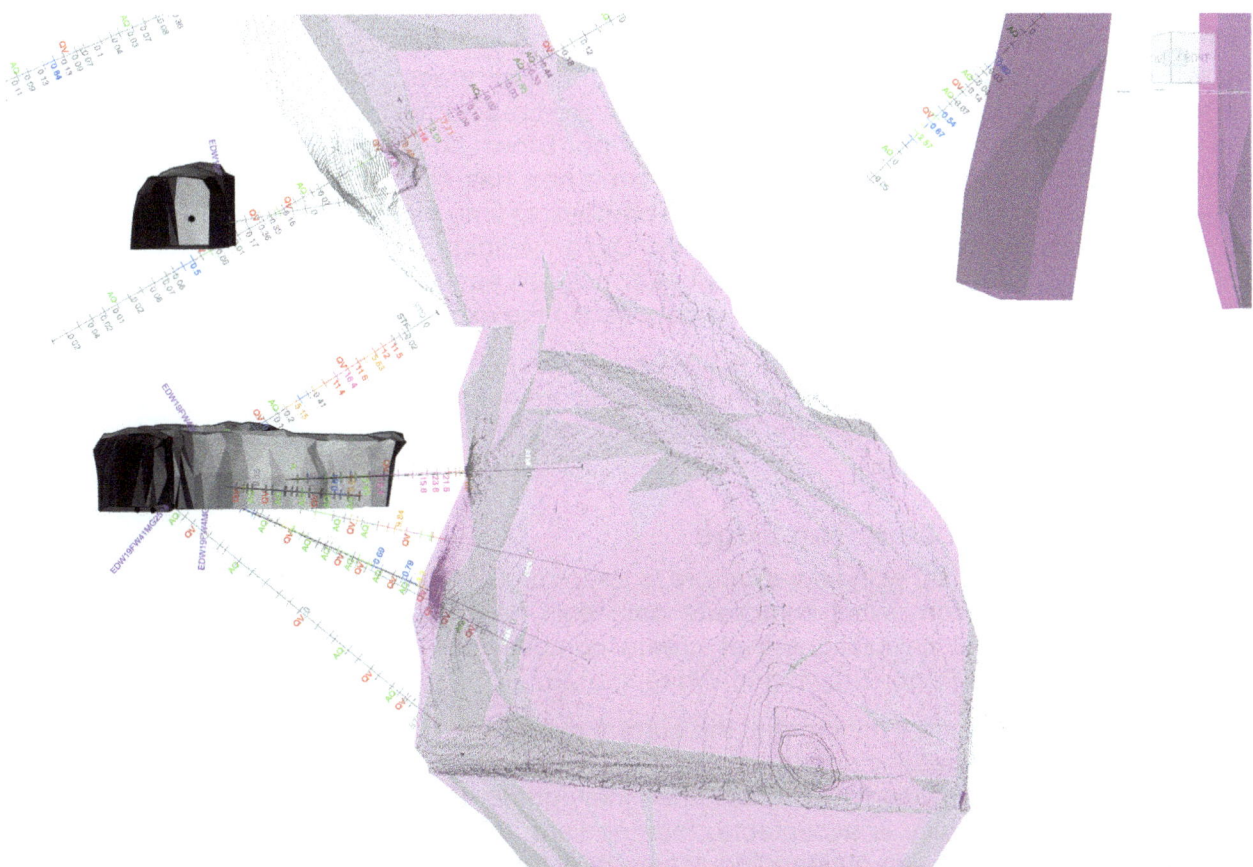

FIG 2 – Short grade control drill holes from footwall development targeting skins and pillars, with point cloud C-ALS scan (grey) and adjusted historical void shape (pink).

Due to the variable vein skin width, a denser spacing of drill holes is required around historical stopes. Without high precision of both the hanging wall/footwall and the historical void position, economic material may be overestimated or depleted.

If the drill hole intersects pillar, crown, collapse or fill material, digitised shapes can be manually adjusted to match the drilling. In areas with filled stopes, drill holes are designed to ensure at least one hole drills through a pillar or crown and near the edge of the historical stope, to best confirm its location.

GRADE CONTROL MODELLING AND ESTIMATION

Importing data

The inputs required for geological model wireframes and grade control model estimations are diamond drill hole data and underground ore control channel data. Previously, data was exported from the two acQuire databases and combined in an Access database, which is then connected to Leapfrog Geo via an ODBC link (Vigor-Brown and Whaanga, 2022). Software updates to Leapfrog

Geo have made this work around redundant. Multiple databases can now be imported, with data compiled into a combined table and combined merge table to be used for modelling and estimation.

Interval selection

Grade control modelling is done solely in Leapfrog Geo using face channels, drill hole data and polylines, generated from registered 3D scans (Whaanga, Vigor-Brown and Nowland, 2019). Interval selection is more complicated with the presence of open stopes and historical voids. Void information must be carefully selected as the original vein no longer exists in many areas. The relationship of the vein position as it exits the open stope has a large influence on vein skins.

Model relationships

The MUG mine can be grouped into several principal areas, each with distinct geological characteristics. Previously, grade control models were created for these mine areas and dynamic links established to the master geological model (Vigor-Brown and Whaanga, 2022). A single master geological model is no longer used. Each mine area now has its own geological and grade control model. This approach has significantly reduced processing time for geological models, as geologists can work on different mine areas concurrently. Dynamic links between the geological and grade control models within single projects make better use of Leapfrog's functionality; for example, the use of wireframes for variable orientation as opposed to the export and import of reference surfaces.

Seequent model branching functionality is still used to branch off alternative interpretations. Kriging Neighbourhood Analysis (KNA) models and estimates are manipulated for purposes outside drill and blast design; for example, life-of-mine and reserve models with modified variables for mineable shape optimisation processing.

Grade estimation in Leapfrog Edge

Leapfrog Edge has a similar workflow structure to Leapfrog Geo. Each step in the estimation process is dynamically linked, from initial composite data built from assays within a wireframe domain, through to block model calculations that create combinations of first and second passes and resource classifications (Vigor-Brown and Whaanga, 2022).

Three years of MUG remnant mining with the current dynamic grade control systems has embedded a workflow with the engineering department that allows the use of best available data to make timely decisions, whilst also allowing for accurate and timely grade estimation validation and final release to ensure the model quality is acceptable. Intermediate working grade models are released with correct geological models and outstanding assay data, for use by drill and blast engineers to begin stope designs. Final validated models are released and stope authority documentation updated to reflect any changes (Vigor-Brown and Whaanga, 2022). This has proven to be a fast and efficient method to reduce the delay from ore drive or drilling completion to stope design and mining. Occasionally, more major changes are made to the geological model and estimation with the addition of grade information, which creates re-work for the drill and blast engineers; however, production delays are still minimised.

Intermediate models with incomplete drilling information are also released to the long-term planning engineers. Mine Stope Optimiser (MSO) is used to generate stope shapes and determine the initial economics of an area and allow a head start to long-term designs. As discussed, final validated and released models are evaluated against the designs completed and minor changes made where required. In situations where stopes are generated in unexpected areas or there is an absence of shapes in expected grading material, there is a focus on review and validation of geology and estimation prior to the final model release.

This iterative approach to model updating and engineering design has greatly assisted with the dynamic nature of mining remnant material, where every historical void interaction can change the short-term mining plan, requiring flexibility from all departments.

Modelling and estimation of historical voids and remnant ore

Extensive historical underground workings in the MUG mine area pose a unique challenge with geological modelling and estimation. Historical drives, passes, open, filled and collapsed stopes need to be modelled to a high spatial accuracy, added to the grade control estimate and each dealt with uniquely. The historical void model is used to deplete and code material in grade control estimates.

In remnant areas, stope skins, crowns and pillars are targeted for extraction and often contain economic Au grades. Accurately representing these in a block model estimate is challenging (Figure 3). Areas where the wireframes and void model overlap indicate the mined ore to be depleted and what remains is modelled and estimated. Where the relationship between the two is inaccurate there is a high risk of modelling and estimating skins, crowns and pillars that do not exist or depleting ones that do (Figure 4).

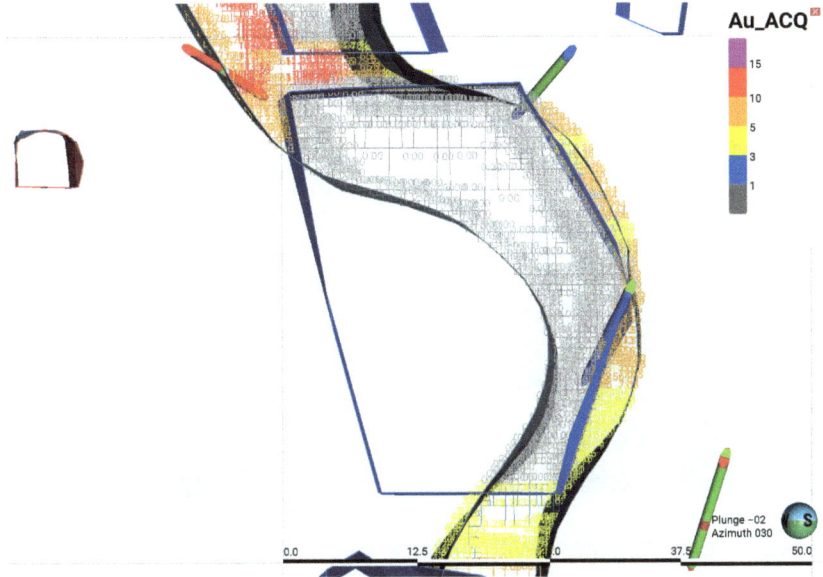

FIG 3 – A cross-section of a modelled historical open stope with no data to inform skins on the left had side of the void and incorrect depletion and skins forming on the right-hand side. The historical void is not informed by C-ALS.

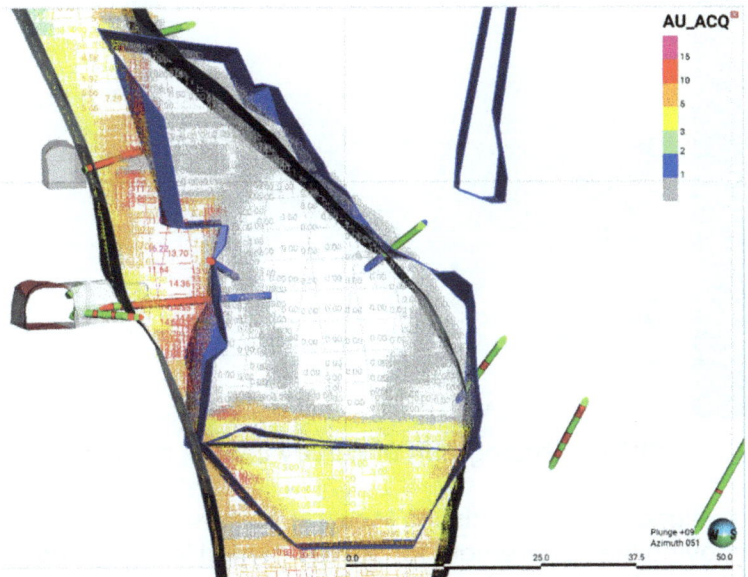

FIG 4 – Section view with footwall drive drilling to target skins and historical void. The historical void model is adjusted based on drilling and C-ALS. Updated wireframe interpretation and excluding portions of selected segments and midpoints then re-estimated giving better representation of reality.

The ability to update grade estimates quickly based on new data, including information on historical workings such as size and spatial location, supports the need to model remnant material as accurately and timely as possible. Development breaking into historical workings for filling and extraction provides additional data and resolution of the location of skins and voids. These breakthrough cuts can be scanned, the wireframe and voids updated, re-estimated and provided to engineers for finalising drill and blast design on remnant ore.

LIDAR

OceanaGold is currently trialling Mine Vision Systems (MVS) FaceCapture™. This features a 90° × 360° field of view LiDAR with a 20 000 lumen light, combined onto a sensor head with integrated vest and containing fast GPU processing, standard batteries and a 4 TB hard drive. It is connected to a wireless tough tablet (Figure 5).

FIG 5 – FaceCapture™ system with sensor head, integrated vest and tough tablet.

Georeferencing is achieved by the operator using imported survey control. A photo is taken of the survey markers and correlated with the existing survey database (Figure 6).

FIG 6 – Photograph of survey control marker.

Subsequent scans are aligned with the existing location base map and georeferencing is passed to the localised scan, removing the requirement for survey control in every scan. Painting the LiDAR at the face/exposure of interest achieves higher resolution through over-sampling and allows a 3D registered mesh to be created live (Figure 7). The LiDAR has a working range of 50 m, allowing point clouds of historical drives and voids to be generated. 3D meshing has a range of 1–20 m, exceeding the current working range of photogrammetry of ~10 m. Drift accuracy of the system is 1 per cent, giving centimetre rather than millimetre accuracy. From a mining perspective the accuracy is on par or greater than Waihi's existing system and prioritises speed of use and processing, with file sizes in the megabytes rather than gigabytes. Other LiDAR systems require post processing and down scaling of the point cloud data sets generated to be immediately useable for a higher level of accuracy and detail not required for mining applications.

FIG 7 – Drive point cloud (blue) with 3D geo-referenced face meshes and survey control (yellow).

FaceCapture™ and Deswik Mapping are operated on the same tablet. The 3D georeferenced meshes are imported directly into Deswik Mapping, via the import.obj feature; point clouds can also be imported. Geologists map directly onto the meshes as per the existing photogrammetry process; however, geo-referencing or conversion of mapped features from Metashape to Deswik Mapping and post-processing is no longer required. Data captured underground only needs to be synchronised with the master mapping layers on surface. Detail captured at the drive or face allows geologists to identify and map historical workings openings and interpret stope fill from collapse and *in situ* rock and veining (Figure 8).

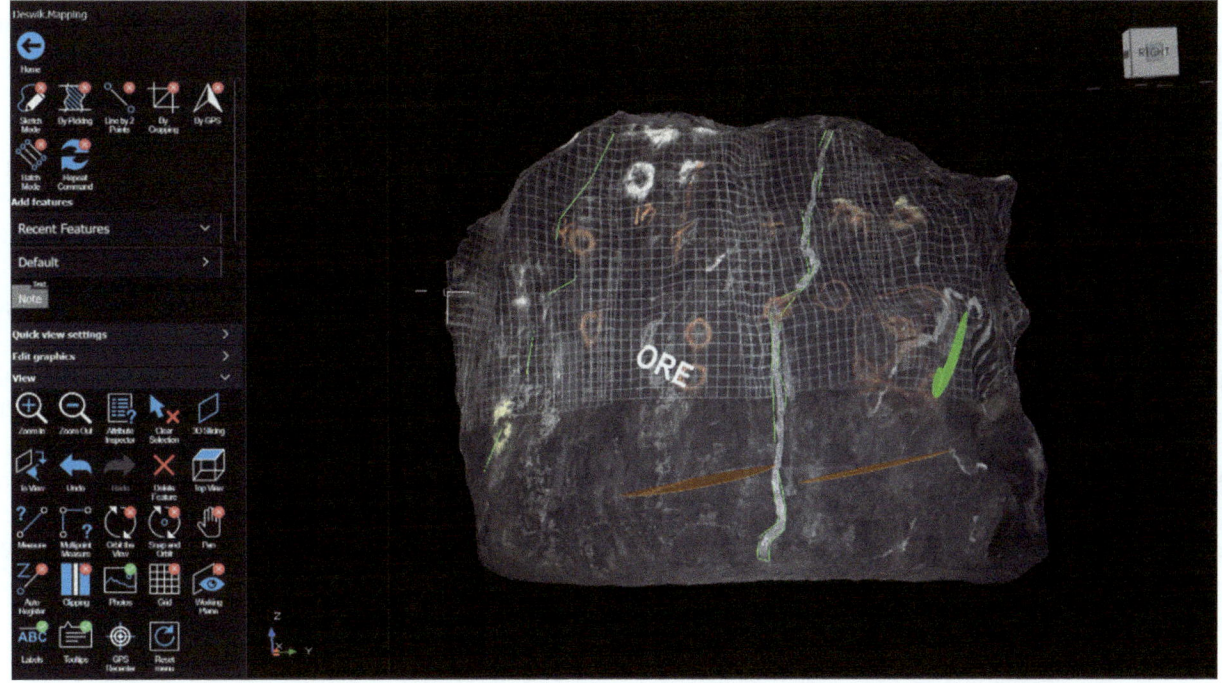

FIG 8 – Face sketching directly onto geo-referenced meshes in Deswik Mapping.

These interpretations are used to refine the working void model, allowing accurate depletion of the historically mined vein.

Integration with Deswik.OPS™ allows calculation of advance, grade line, as built, volumetrics for overbreak/underbreak and volume mined from the generated point clouds. The end of month pickup process can be streamlined with actual face positions and scans, assisting survey and increasing the accuracy of the mine reconciliation.

CONCLUSION

The objectives of this review were to look at the real-world application of several technological innovations and grade control workflow improvements developed over the last seven years at the Waihi underground mine, share learning learnings on the successes and highlight areas with room for improvement.

The challenge presented by the MUG remnant mining project has meant that the authors have had three years of implementing and rigorous testing of the workflows and technologies developed. Every aspect of ore control in a remnant mine is more difficult and challenging: 3D scanning of unsupported historical voids; managing diamond drilling programs through existing voids and categorising the different aspects of historical fill, collapse and vein skin; ore drive direction with the complexity of reconciling historical void positions with vein interpretations and optimal placement to achieve both economic and stable stope shapes; accurately representing depleted historical void shapes to generate grade control models; and managing large data sets on portable hardware devices to ensure timely decisions can be made underground rather than on the surface.

The technology advancements described have been utilised to overcome the challenges presented by the MUG remnant project and have directly contributed to improvements in ore drive development and stope placement, in turn lowering costs and improving profit margins.

Recent improvements to current scanning techniques with the introduction of LiDAR and real time geo-referencing will allow more work to be completed at the underground face. This will further improve information turnaround time, allowing the underground geologist to make correct, timely decisions with accurate data.

ACKNOWLEDGEMENTS

The authors would like to acknowledge the following people: Samantha Muir and the Resource Development team at Waihi for their work on drilling around historical voids; Waihi ore control geologists that have assisted with improving processes; OceanaGold for granting their approval to publish this work; Dr Angela Storkey for proof reading.

REFERENCES

Doyle, L R and Whaanga, A, 2017. Maximising profits from a challenging ore body: the optimisation of ore control at the Waihi underground gold mine, in *Proceedings AusIMM New Zealand Branch Conference 2017* (The Australasian Institute of Mining and Metallurgy: Christchurch).

McAra, J B, 1988. *Gold Mining at Waihi*, 352 p (Martha Press: Waihi).

Muir, S L, 2023. Managing and validating diamond drilling around complex historic workings, in *Proceedings AusIMM New Zealand Branch Conference 2023* (The Australasian Institute of Mining and Metallurgy: Christchurch).

Vigor-Brown, W and Whaanga, A, 2022. Dynamic grade control modelling processes at the Waihi Underground Gold Mine, in in *Proceedings 12th International Mining Geology Conference 2022*, pp 179–186 (The Australasian Institute of Mining and Metallurgy: Melbourne).

Whaanga, A, 2022. From paper to bytes – digital mapping implementation, in *Proceedings 12th International Mining Geology Conference 2022*, pp 506–513 (The Australasian Institute of Mining and Metallurgy: Melbourne).

Whaanga, A, Vigor-Brown, W and Nowland, S, 2019. Dynamic grade control modelling processes at the Waihi Underground Gold Mine, in *Proceedings 11th International Mining Geology Conference 2019*, pp 355–365 (The Australasian Institute of Mining and Metallurgy: Melbourne).

Futureproofing mine geology – planning ahead

Copper mineralisation at Mt Isa – unseen insights from seismic surveys after a century of geoscientific study

S Bright[1], G Turner[2], A Brown[3] and M Megebry[4]

1. Senior Structural Geologist, HiSeis Pty Ltd, Perth WA 6008. Email: s.bright@hiseis.com
2. General Manager Technical Solutions, HiSeis Pty Ltd, Perth WA 6008. Email: g.turner@hiseis.com
3. Consultant (previously Principal Geologist Mt Isa Mines, Glencore), Mt Isa Qld 4825. Email: alex.brown@lithinsight.com.au
4. Senior Geophysicist, HiSeis Pty Ltd, Perth WA 6008. Email: m.megebry@hiseis.com

ABSTRACT

The Mt Isa base metal mineral system is a world-class ore deposit. Lead, zinc and silver were initially discovered in 1923, with copper discovered shortly after. The mineral system has been the subject of intense study by company geologists, consultants and researchers over a long period of time. Despite this, recent 2D and 3D seismic surveys over the deposit and surrounding areas have brought new insights into the Mt Isa minerals system, challenged some of the previous understanding and validated some recent ideas about the deposit.

The surveys have provided evidence of strong controls on the mineralisation by a west dipping thrust system. The copper and silica-dolomite alteration show a good correlation with an antiformal zone of strong reflectivity seen on a north–south 2D seismic section. Correlation with east–west 2D lines suggest that this antiform is the longitudinal representation of a west dipping imbricate thrust stack. The Buck Quartz Fault forms the base to this stack and the Bernborough, J46 and other faults appear as break back thrusts with mineralisation proximal to the intersections. The Mt Isa Fault appears to be part of the west dipping thrust system rather than the previously interpreted more vertical feature. It is apparent from cross-cutting relationships that there has been later movement along this fault. The Leichardt Fault is also a west dipping fault that appears to have been more recently reactivated.

Two other fault sets have been mapped in 3D by the reflection seismic and the refraction seismic tomography and the relative timings of these structures can be deduced from cross-cutting relationships. This is highly beneficial for mine planning because it assists with understanding potential offsets in the mineralisation in addition to the potential geotechnical implications of these faults. The refraction seismic tomography is also very effective at providing a very detailed map of weathering depth variations and this can be directly used to optimise pit designs as well as mapping the sub-crop geology.

The 3D seismic reflection survey mapped the basement of the Mt Isa group sediments and offsets in its position due to both the west dipping and other faults. This information is guiding new exploration targets.

INTRODUCTION

The Mt Isa Copper orebody is a world-class deposit. Lead, zinc and silver were first discovered at Mt Isa by prospector John Campbell Miles in 1923 and copper was discovered in the late 1920s. Since that time over 8 Mt of Cu, 15 Mt of Zn and 11 Mt of Pb have been mined from these orebodies and current resources of approximately 2.5 Mt of Cu have been delineated (Glencore, 2022).

As a result of its high value mineral endowment, Mt Isa is one of the more highly studied geological volumes on the planet. Over 20 000 drill holes have been drilled within a 20 × 5 km area around the Mt Isa orebodies.

In recent years Glencore engaged HiSeis to complete seismic surveys and interpretation. Whilst drilling provides detailed geological data along each drill trace and conventional geophysical techniques such as Induced Polarisation (IP) and Electromagnetics (EM) used in mining can image the upper few hundred metres, active seismic techniques provide the only way to obtain detailed continuous 3D images of the subsurface overall economically mineable depths.

The seismic investigations included:

- core and wireline rock property measurements
- eight × 2D seismic surveys
- two × vertical seismic profile surveys
- a 12 km² 3D seismic survey.

Despite the high level of investigative work over the last 100 years, the seismic has brought some new insights into the Mt Isa mineral system. This paper highlights some of the key results that can guide mine geologists to best assist future efforts both to exploit the remaining mineralisation at Mt Isa, and use seismic as a tool to assist other mines efficiently extract their ore.

FAULT ARCHITECTURE

The north–south 2D seismic line acquired over the Mt Isa mineralisation showed a surprising correlation between a prominent antiformal zone of strong seismic reflectivity and the zone of silica-dolomite alteration as determined from drilling (see Figure 1a). The antiform is interpreted to be the longitudinal representation of a west over east imbricate thrust stack (Figure 1b). The thicker zone of the thrust stack between where it tips out to the north and south could have provided a region of increased dilation and permeability that may have been important in localising the Mt Isa Copper orebody.

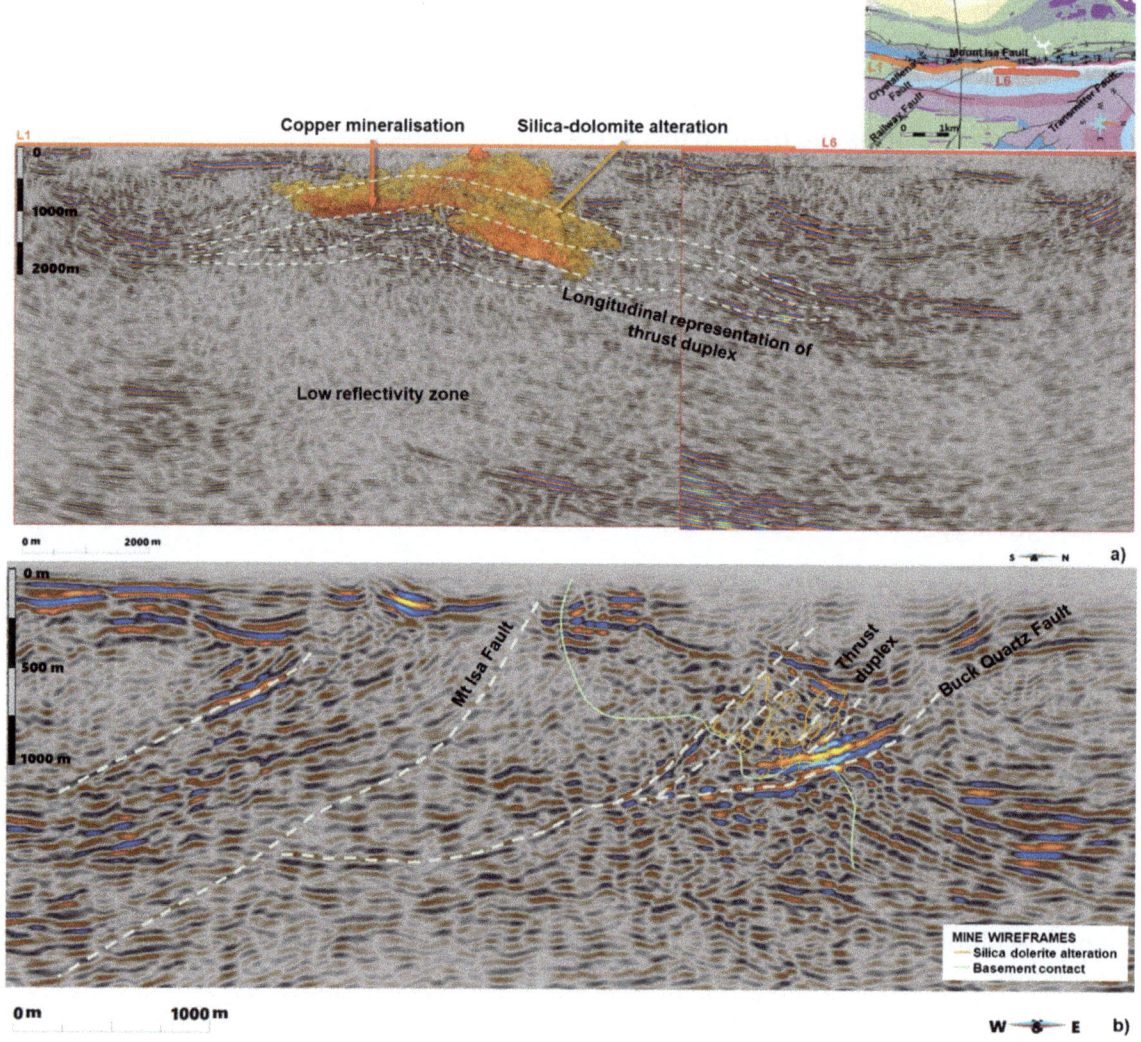

FIG 1 – (a) East–west 2D seismic section over the south end of the Mt Isa Copper mineralisation illustrating key features of the interpreted west–east imbricate thrust duplex. (b) North–south 2D seismic sections illustrating the longitudinal section through the interpreted E–W thrust stack and its correlation with the copper mineralisation and alteration at Mt Isa together with an underlying large low seismic reflectivity zone.

The E–W 2D seismic lines (eg Figure 1b) suggest the large scale N–S Mt Isa Fault which is mapped for many kilometres to the immediate west of the Mt Isa mineralisation, is part of this west dipping thrust system rather than the previously interpreted more vertical feature.

The Bernborough, J46 and Buck Quartz Fault which have all been mapped within the mine are representative surfaces of thrusts/break back thrusts and appear to be an important control on mineralisation with copper present at or proximal to the intersection of these thrust ramps (Figure 1b). The Paroo Fault, which is the name ascribed to the unconformable contact between the Mt Isa Group sediments and the Eastern Creek Volcanics, is likely to be an early fault (prior to the above thrusting) that underwent complex deformation subsequent to its formation including during the thrusting.

An extensive low reflectivity zone (Figure 1a) occurs at depth in all seismic sections and may represent an extensive alteration zone (with a volume of approximately 60 km^3). The presence of this zone suggests that the Mt Isa mineralisation may have been emplaced or enriched as a result of hydrothermal alteration homogenising the seismic properties.

We think that the longitudinal representation of the localised thrust stack associated with this large copper orebody and the zone of low reflectivity below represent important signatures to search for as the industry looks for new ways to focus deep exploration.

The 3D seismic survey completed to the immediate south of the 1100 Cu orebody in 2022 provided an additional level of detail on the 3D position, orientation and timing on the multiple generations of faults present at the mine. The different generations of faulting are visible in the detailed map of the depth to solid/fresh bedrock obtained from the seismic refraction tomography data (Figure 2). The authors find that because different geological units can be more or less resistant to weathering, the top of fresh rock image can be a very good mapping tool in covered or deeply weathered terrains such as is found at Mt Isa. In addition, deeper zones of weathering often follow faults due to the associated damage zones.

Figure 2 shows the sub-crop position of a series of faults interpreted from the full 3D seismic reflection cube overlain on the depth to bedrock map. Linear bedrock lows outline the Mt Isa and Leichardt faults together with a series of NW–SE faults. Other faults are delineated by distinct boundaries to bedrock highs which demarcate the extent of different geologic units with different levels of resistance to weathering.

By observing the cross-cutting relationships we can infer the timing of the faults. The Mt Isa fault and Leichardt faults terminate the NW–SE faults indicating that they have experienced later movement. The NW–SE Faults themselves terminate the NNE–SSW Faults and these in turn terminate the NNW–SSE Faults. In some cases the seismic also provides an indication of the direction of movement and offsets across the faults.

The ability to map out the positions of these faults in 3D, understand their timing and understand the sense of movement has important implications for mine planning since some fault sets localise mineralisation and others offset it. It can be difficult or impossible to confidently join drill hole intercepts into 3D surfaces and the drill data cannot usually provide information on fault throws.

An example of faults mapped by the seismic offsetting mineralisation at Mt Isa is shown in Figure 2. The extensions of the two NNE–SSW faults mapped at the north of the seismic data are coincident with significant offsets in the mineralisation as mapped by drilling.

A further benefit of refraction tomography is that it maps out detailed near surface variations in seismic velocity. Seismic velocity is related to rock competency – which is why refraction tomography is so good at mapping depth to bedrock. The seismic tomography data can thus be used to delineate areas of free-dig for open pit planning and can help with pit-slope design (by delineating changes in rock competency and at the very least the depth of weathered material). Accurate mapping of the depth of weathering assists with better predicting the extent of oxidised ore for optimising processing. Glencore also used this data to search for additional areas of shallow clays needed for tailings storage and rehabilitation activities. The tomography was able to identify several zones with similar near surface velocity profiles to known clay deposits indicating that they may host similar material.

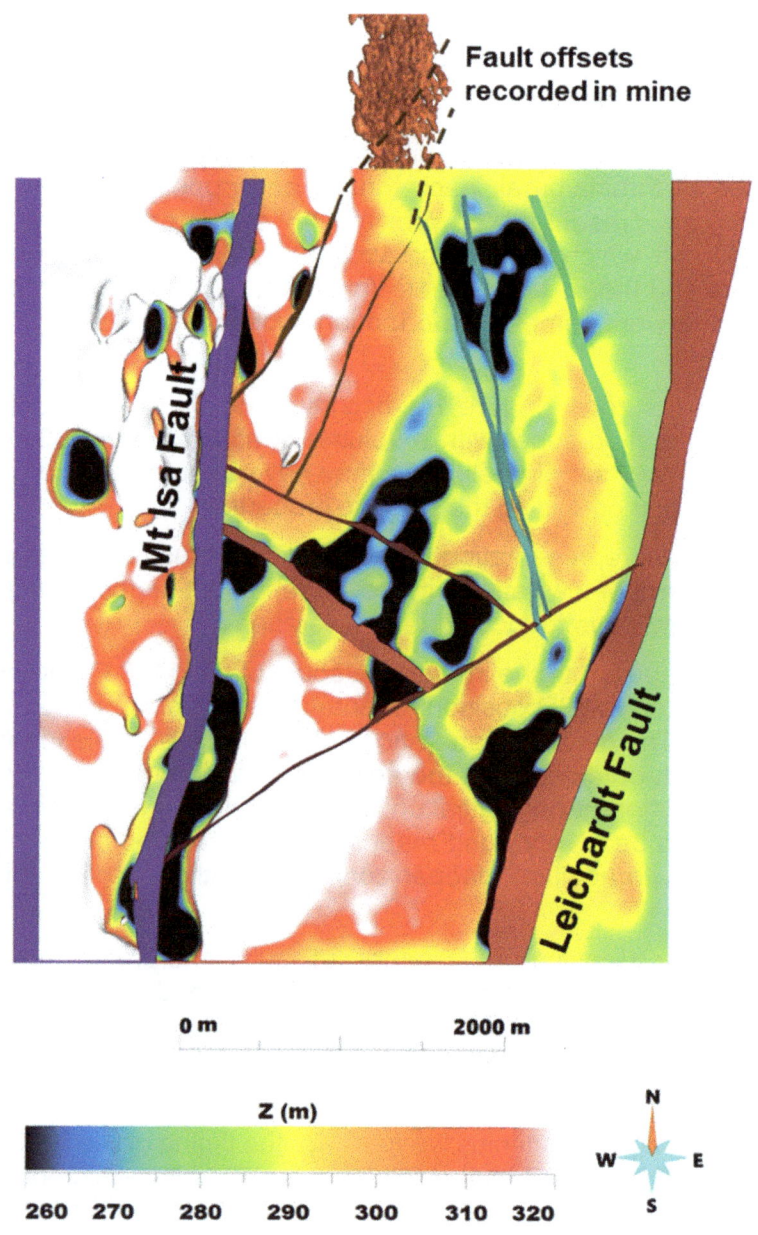

FIG 2 – Sub-crop position of faults observed in the 3D seismic reflection data set overlain on the depth to bedrock map extracted from the seismic tomography. The image also shows how two faults rapidly mapped within the 3D survey extent (where there is limited deep drilling) offset mineralisation as shown by the mine mineralisation wireframe.

SMALLER SCALE FAULTING

To enhance the ability to map the smaller scale faulting within the 3D seismic cube a machine learning approach, dubbed Fault-Mod, was used to map out small scale breaks/offsets in the data. The power of getting the computer to do this is that it can be a lot more effective and objective in assessing the continuity of structures in 3D. In contrast, a human interpreter will typically need to flick between multiple vertical (eg E–W and N–S) and plan sections to do this and this can be a very time consuming and somewhat subjective process.

Figure 3 shows a vertical section though the Fault-Mod cube overlain on the seismic data and compared to one of the relatively few deep drill holes through the seismic cube. The location of the interpreted faults correlate very well with the location of structures recorded in the drill hole and shown by the markers. The core photos shown to the side illustrate the broken core logged as structures. Importantly where no structure has been indicated by Fault-Mod there are long sections of unbroken core.

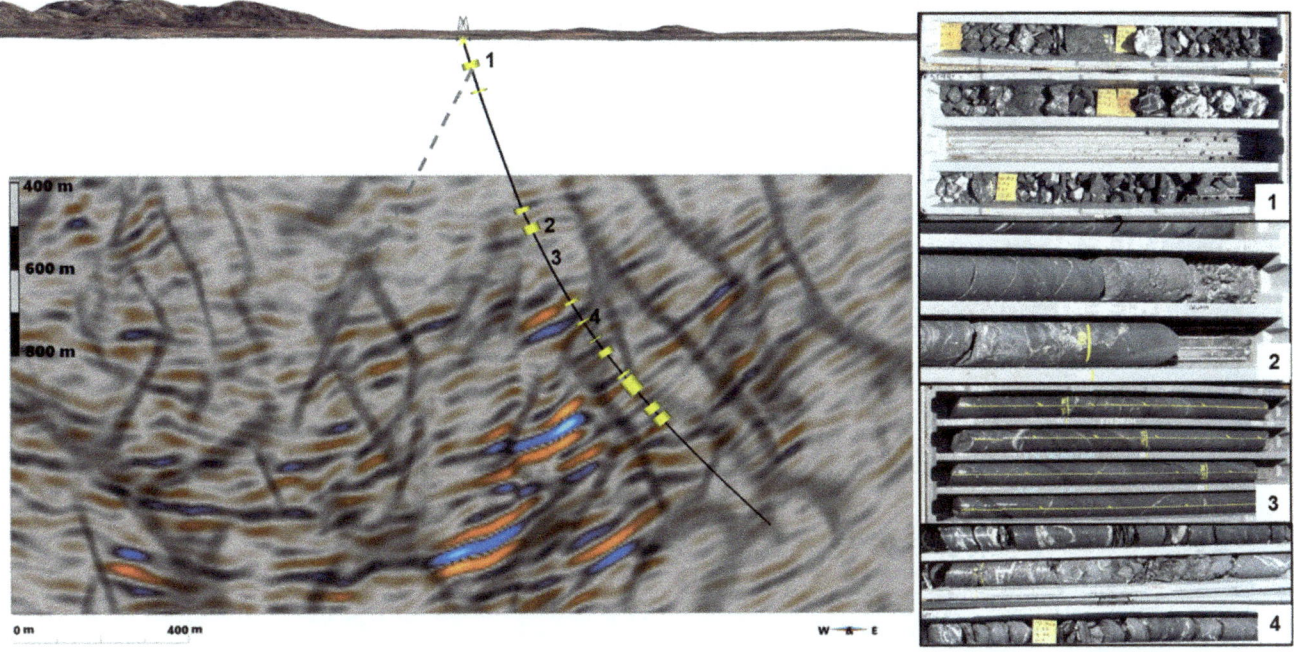

FIG 3 – Vertical section through the seismic cube showing faults (grey) picked using a machine learning process superimposed on the seismic data. The core photos demonstrate the correlation of broken and altered core (core photos 1, 2 and 4) and picked faults and large section of competent core where no faults were picked (core photo 3).

CONTACT BETWEEN THE MT ISA GROUP (MIG) SEDIMENTS AND THE EASTERN CREEK VOLCANICS (ECV)

The Mt Isa Group (MIG) - Eastern Creek Volcanics (ECV) (basement) contact to the western side of the Bernborough Fault appears as a high-amplitude sub-horizontal reflector in the north central part of the 3D seismic cube (Figure 4). It was possible to use automated horizon tracking tools available within Paradigm's GoCAD software package to track this contact. The process uses seed points selected along the reflector and generates a surface by following the seismic signal outwards from the seed points. The resulting surface shows excellent correlation with the intercepts of this interface logged in drill core.

The detailed nature of this surface made it possible to observe a number of small offsets/faults in the surface (Figure 4). The more extensive of these were interpreted into 3D wireframes using a combination of the steep dip attribute across the MIG-ECV contact and the automated Fault-Mod attribute discussed previously.

Based on the observations already made about the importance of the north–south striking imbricate thrust system; together with the independent interpretation that structures crossing through the MIG-ECV potentially provide a conduit for copper rich fluids leached from the volcanics to pass into the more reducing MIG (particularly the Urquhart shale) where they are precipitated (eg Andrew, 2020; Gregory, 2006); these offsets/faults provide new mine scale targets for zones of higher grade mineralisation. Four of the fault surfaces interpreted in this way are shown in yellow and blue in Figure 4 overlain on the seismic data together with other larger scale faults.

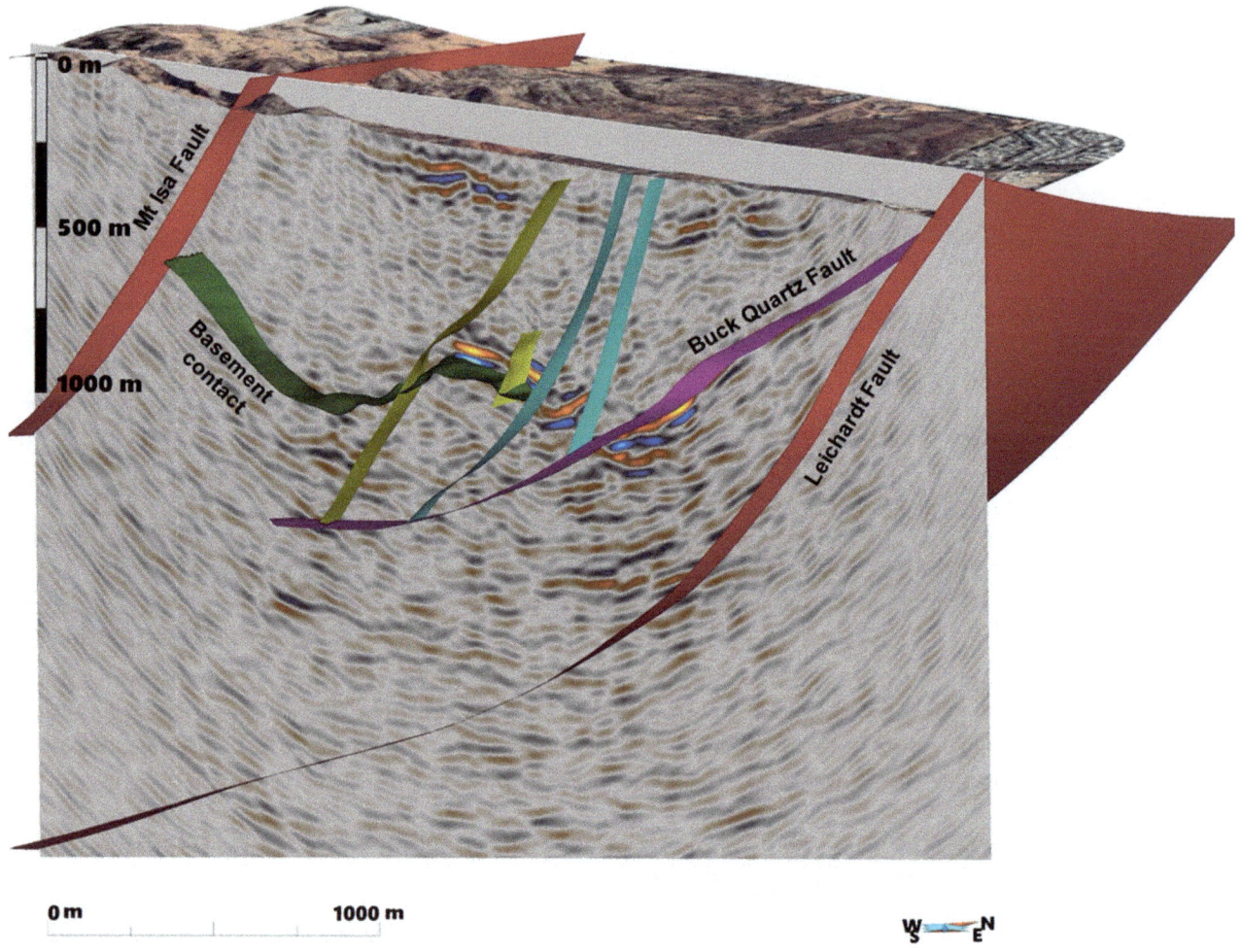

FIG 4 – Location of the basement contact together with large scale and small scale structures interpreted from the 3D seismic overlain on a vertical section through the 3D seismic cube.

SUMMARY OF IMPLICATIONS FOR MINING AT MT ISA AND ELSEWHERE

The 3D active seismic data is the only data set that can provide measured data with a resolution of tens of metres over cubic kilometres covering a mineral deposit and its immediate surrounds. As this paper shows, this can unlock new insights that cannot be obtained by surface and underground mapping, drilling, associated geochemical analyses or other geophysical techniques.

For the mine geologist and mine planning team this includes the ability to:

- Understand the significance of different faults and fault systems.
- Understand which fault systems (if any) control the deposition sites of mineralisation and therefore more effectively detect and delineate additional mineralisation.
- Understand which fault systems offset mineralisation and the extent and direction the mineralisation is offset.
- Continuously map the 3D location of faults to optimise the development and stopes from both an orebody extraction and a geotechnical perspective.
- Better estimate the amount of free-dig material for open pit mining.
- Better estimate the distribution of oxidised ore to assist processing optimisation.
- Locate new sources of construction materials for operations.

We see a growing role for this type of data as a key foundation data set which can frame mining decisions through the life of a mine and multiply the benefits of each new observation by providing a detailed framework of measured data which can be used to extrapolate in 3D away from the point of observation.

ACKNOWLEDGEMENTS

The authors would like to acknowledge Glencore for permission to publish these results.

We would also like to acknowledge the HiSeis team who acquired and processed the data and Dr Roric Smith whose insights guided the early 2D interpretation and set a foundation for some of the results presented here.

REFERENCES

Andrew, B S, 2020. Recognising cryptic alteration surrounding the Mount Isa Copper Deposits: Implications for controls on fluid flow, and mineral exploration, PhD thesis, The University of Waikato, New Zealand.

Glencore, 2022. Annual Report, Glencore: Switzerland.

Gregory, M J, 2006. Copper mobility in the Eastern Creek Volcanics, Mount Isa, Australia: evidence from laser ablation ICP-MS of iron-titanium oxides, *Mineralium Deposita*, 41:691–711.

Optimising workflows in post-pandemic operations – impacts and implications for mine geologists

S Edmond[1], S O'Brien[2], A Tod[3], D Yeates[4] and M Ravella[5]

1. Lead Geologist, Veracio, Salt Lake City UT, USA. Email: shauna.edmond@veracio.com
2. Geoscientist Manager, Veracio, Adelaide SA 5950. Email: shaun.obrien@veracio.com
3. Manager – Minalyzer, Veracio, Perth WA 6000. Email: angus.tod@veracio.com
4. Head of Global Marketing, Veracio, Adelaide SA 5950. Email: dave.yeates@veracio.com
5. Chief Innovation Officer, Salt Lake City UT, USA. Email: dave.yeates@veracio.com

ABSTRACT

In an era marked by the COVID-19 pandemic, the mining sector has confronted unprecedented challenges, prompting a critical reassessment of geologists' roles and workflows. The traditional, conservative approach in mine geology, symbolised by the adage 'if it ain't broke, don't fix it,' has been challenged by a new paradigm necessitated by global labour issues and the urgent need for progress in mining processes. Our research, underpinned by time-in-motion and work measurement studies across global mining sites, explored the transformative potential of integrating core scanning technologies, such as TruScan and MinalyzerCS, into traditional workflows.

A time-in-motion study across more than half a dozen mining operations at various stages in maturity noted that the application of core scanning technologies led to a dramatic reduction in manual handling and enhanced process efficiency. This technological intervention proved crucial in identifying key geological features, significantly reducing sample volumes sent to laboratories. This adoption of technology resulted in streamlined lab assays and improved processes.

To codify this improved process, our study further explored the implications of these technologies at additional sites, each presenting unique challenges and opportunities. Through collaborative analysis with clients, we have distilled the collective practices in core logging and contrast these against a potential future state of core logging and analysis processes. This evolution from traditional to optimised workflows was accelerated through the COVID-19 pandemic and simultaneously underscores the critical role of leadership and change management in successful operational transformation.

In summary, our research highlights the pivotal role of core scanning technologies in advancing and optimising post-pandemic mining operations. These technologies enable a strategic shift in geologists' roles, from traditional to more analytical and decision-influencing functions. The study serves as a blueprint for adapting to evolving operational needs in the post-pandemic era, emphasising the coexistence of specialist expertise and efficiency in the evolving mining sector.

The 'don't fix it' challenge

The advent of the COVID-19 pandemic aggressively challenged mining operations all over the world to adopt one of two operating paradigms. Complete self-sufficiency, or digitalisation; ushering in an era of unprecedented challenges and transformative changes, particularly for those in the field.

A series of projects that involved core logging workflow optimisation during this period of disruption found renewed and compelling reasons to re-evaluate of entrenched workflows, especially in core logging geology and core shed operations. The longstanding maxim of 'if it ain't broke, don't fix it' became a moot point in the face of evolving global labour challenges, and a heightened need for safety, connectivity, efficiency and progress in mine geology processes.

This operational study probes into how the crisis has offered an opportunity to rethink and revamp traditional methodologies. By integrating innovative core scanning technologies into the geologists' workflow, our research highlights the potential for significant advancements in safety, efficiency, data governance and integration.

The pandemic was responsible for limiting physical access to sites and demanding operational agility. Consequently, this accelerated the adoption of technology in mining operations. This has highlighted the necessity for a paradigm shift in how geological data is collected, analysed and utilised in decision-making processes.

Our investigation explores how these technological interventions can transform the geologist's role from a purely data-gathering function to a more strategic, analytical and decision-influencing capacity. In essence, exploring the intersection between the crisis-induced challenges we faced during the pandemic and the opportunities they present for technology adoption and workflow improvement in core logging and mine geology operations.

Methodology

Central to our approach was a combination of workflow observations, time-in-motion studies and work measurement analyses; tools traditionally used in industrial engineering to assess and optimise operational processes. This blend of methodologies from 2020 through to end of 2023 allowed us to capture both quantitative and qualitative aspects of workflow changes resulting from technological integration.

Conducting comprehensive site assessments, we documented existing workflows, and identified key areas for potential improvements to decision-making, time delays and manual handling, as displayed in Figure 1. This initial phase involved close collaboration with on-site geologists and technicians, ensuring an in-depth understanding of current practices and challenges.

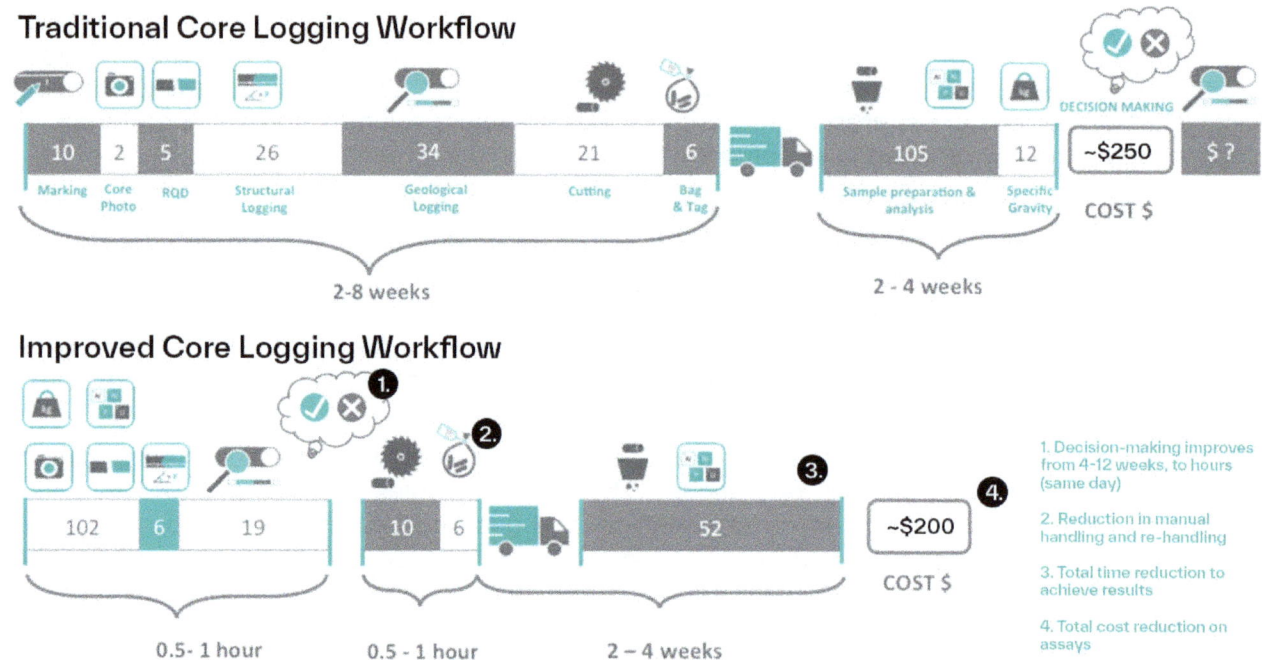

FIG 1 – Summary of improved workflow identified through an intricate study.

Subsequently, we deployed core scanning technologies, namely TruScan and MinalyzerCS, within these pre-existing workflows. Our focus was on understanding how these technologies could seamlessly integrate into and enhance the geological data collection and analysis process. This involved not just the installation of equipment but also training staff, adjusting processes and monitoring the adaptation period closely.

In analysing the data, we employed a comparative approach, juxtaposing the pre- and post-technology integration scenarios. This comparison allowed us to draw concrete conclusions about the efficacy of core scanning technologies in enhancing workflow efficiency, data precision and overall operational effectiveness in mine geology.

The culmination of our methodology was not only in quantifying the improvements brought about by these technologies but also in capturing the broader implications of such integrations in terms of strategic capacity building for geologists and the potential for long-term value generation in mining operations.

Site-specific implementations and results

Case studies in geology workflow optimisation

Our exploration into optimising workflows in post-pandemic operations spans across diverse geological formations and global locations. Each site presents a unique set of characteristics and challenges (Table 1).

TABLE 1

Sites, ore types and locations.

Site	Ore Type	Location
Site A	Orogenic Gold Deposit	Australia
Site B	Porphyry Copper Deposit	North America
Site C	Orogenic Copper-Gold Deposit	Australia
Site D	Porphyry Copper Deposit	South America
Site E	VHMS Deposit	North America
Site F	Sedex Deposit	Australia
Site G	Iron Oxide-Apatite Deposit	Europe
Site H	Complex Exploration Project	Australia
Site I	Orthomagmatic Nickel-Sulfide	Australia

Site A – Workflow optimisation in orogenic gold deposit

At Site A, an orogenic gold deposit in Australia, the integration of TruScan technology marked a significant departure from traditional core logging methods. The pre-existing process, labour-intensive and time-consuming, was streamlined dramatically by reducing manual movements from a staggering 17 down to a mere five. This optimisation not only resulted in enhanced efficiency but also reduced the physical strain on geologists.

TruScan's deployment brought forth an innovative era where geochemical data became a beacon for process change. Its utilisation in early 2019 at the mine's exploration core logging facility heralded a shift towards economic and targeted core analysis. By identifying arsenic-rich zones and distinguishing between pyrite and arsenopyrite, the technology enabled a focused approach to sampling. This precise differentiation was pivotal in reducing the volume of samples sent to laboratories by an impressive 80 per cent, underscoring the technology's profound impact on cost savings and operational efficiency.

The pandemic's onset further amplified TruScan's value, as its data streaming capability became instrumental in sustaining project momentum despite severe restrictions on site access. This adaptability prevented redundancies within the geology team and maintained the continuity of drilling projects, showcasing the technology's role in not only optimising processes but also in ensuring resilience in crisis situations.

Site B – Advancing porphyry copper deposit operations

Site B, hosting a porphyry copper deposit in North America, experienced a paradigm shift with the optimisation of drill core processing times. The project took a decisive lead in prioritising core samples for laboratory assays based on advanced geochemical information provided by TruScan, pre-empting traditional assay methods.

This strategic prioritisation was made possible through the establishment of a robust database infrastructure that seamlessly communicated with central systems, allowing for pre-emptive and informed decisions on mine design. The integration of live geochemistry reads during core logging transformed the workflow, significantly reducing reliance on laboratory assays for quality control and final results.

As a result, laboratory assays became a secondary step, primarily for validation rather than exploration, enabling a faster and more efficient design process. This shift not only occurred within the geology team, but also the operational workflow of the core shed, with an 'optimised workflow' (Figure 2) proposed and implanted, an increase in operational pace, as well as an enhancement in the accuracy and quality of geological data occurred, facilitating a more agile and informed exploration strategy.

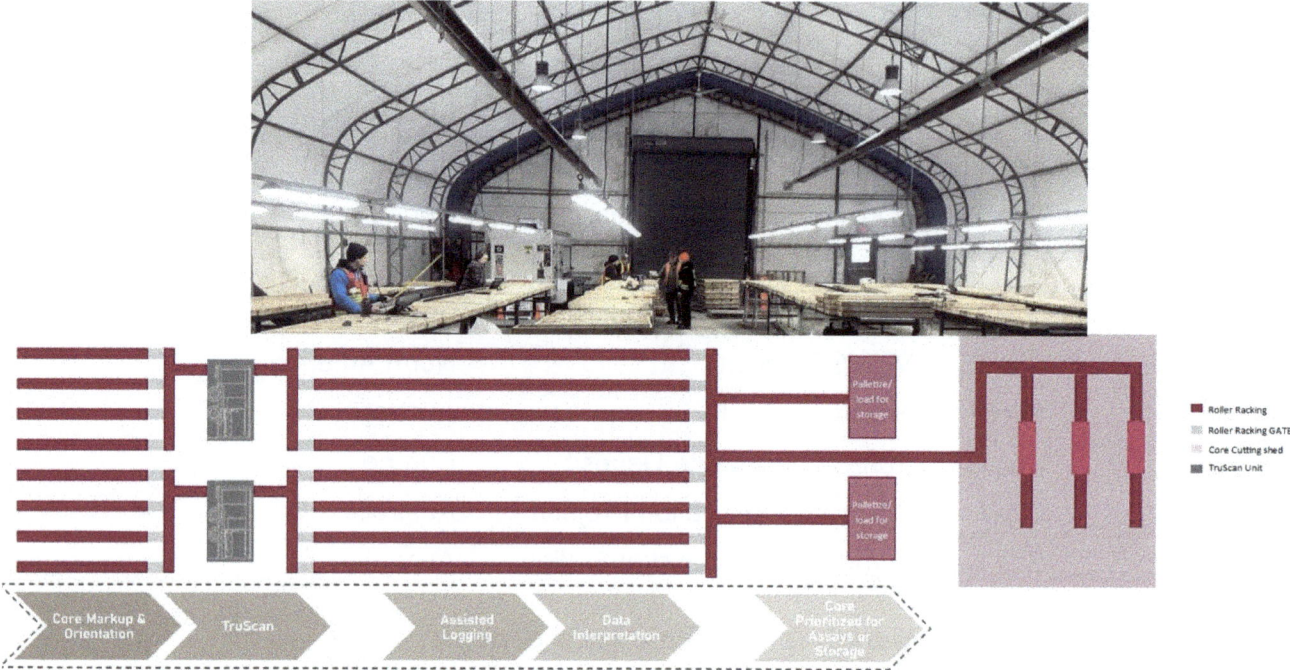

FIG 2 – Proposed 'optimised workflow' for Site B.

Site C – Enhanced geologist empowerment through core scanning integration

Site C's transformation of the core shed workflow epitomises innovation through the integration of TruScan technology. The shift from conventional 'factory logging' to a data-rich, selective approach marked a significant evolution in operational methodology. The transition was characterised by the automation of mundane logging tasks and the enrichment of geologist engagement with sections of the core of genuine interest or anomaly. This pivotal shift allowed geologists at Site C to pivot from time-intensive data collection to more nuanced and value-adding data interpretation and analysis.

The reimagined workflow, as depicted in Figure 3, reflects a meticulously crafted sequence of process improvements designed to maximise the benefits of TruScan technology. Core reception and initial preparation set the stage for the innovative scanning process, which includes in-depth QA/QC, data uploads to the cloud, and instant alerts for significant geochemical events. The geologists are then equipped with comprehensive data and assays, facilitating expedited and more accurate logging processes.

FIG 3 – Geologists empowered to auto-log core utilising Veracio's AutoLogger Platform on-site.

Site D – Digitisation and agile decision-making in porphyry copper exploration

At Site D, a porphyry copper deposit currently under study, the focus has been on establishing an optimised workflow that reflects the unique challenges and opportunities of the site. Our engagement with Site D is centred around developing a nuanced process that acknowledges the specific talent available, the geographical constraints, the operation's requirements and the crucial role of technology adoption.

Our initial assessments have highlighted the need for a digitised process that accelerates decision-making and integrates technology into the workflow in a manner that respects the site's current practices. This approach is rooted in observing incumbent practices, understanding the core challenges, and then carefully introducing technological solutions that can be adopted effectively. The goal is to create a workflow that not only enhances productivity but also ensures the accuracy and governance of geological data, aligning with the broader strategic objectives of the mining operation.

Site E – Enhanced exploration strategy at VHMS deposit

The implementation of the TruScan system at Site E has spearheaded a technological revolution in the core logging activities of the VHMS deposit in North America. This ED-XRF system brought forth a high-resolution elemental analysis that significantly advanced the project's chemostratigraphy understanding. The data gleaned from over 13 000 m of historical core, coupled with the 2500 m of freshly explored core, was integral in forging a new path in exploration strategy. The ongoing pilot study aims to assess the feasibility of weaving this data into a semi-automated logging process. The anticipated outcome is a considerable reduction in laboratory dependence, enabling both time and cost efficiencies. The implications of this integration extend beyond immediate operational enhancements; they promise a transformation in the exploration and mining cycle, offering a richer, more accessible data set that could revolutionise the way the deposit is understood and managed.

The benefits of TruScan at Site E are multifaceted. Firstly, it ensures the continuity of core logging activities during extreme weather conditions by decoupling from drill core assays. This strategic relocation of core logging activities to a regional centre approximately 700 km south ensures the safety and comfort of the geology team while maintaining uninterrupted throughput. Secondly, this relocation strategy has mitigated the limitations imposed by a consistently full camp and the inability to accommodate additional personnel on-site. The digital logging techniques and long-term storage of historical core have enhanced the geology team's productivity, allowing them to maintain a consistent output remotely.

Site F – Operational efficiency through core scanning technology

At Site F, a Sedex deposit in Australia, the introduction of core scanning technology directly after drilling represented a significant leap forward in operational efficiency. Working in close partnership with the client, a bespoke decision-making algorithm was developed to utilise lead-zinc values identified during the scanning process. This algorithm underpinned the secondary use of the technology, accelerating the identification and validation of geological domains, which in turn expedited the logging process on-site. The emphasis on consistent logging practices has been paramount in enhancing the geological modelling process, ensuring that models are not only accurate but also reflect a standardised understanding across different geologists.

The MinalyzerCS technology at Site F has been transformative in several ways. By reducing sample volume and enabling rapid domain validation, the technology has notably decreased the turnaround time for assay data, thus facilitating quicker updates to the geological models. The capacity to deliver more precise modelling of ore thickness and grade has translated into more efficient and profitable mining operations. Moreover, the data controls methodology, data management, reporting and quality control activities have all been elevated through the MinalyzerCS set-up. This system's robust QA/QC procedures, aligned with international standards, ensure the reliability and governance of the data produced. The streamlined workflow has not only reduced sampling costs but also enabled a more rapid response to geological findings, instilling a proactive approach to mine planning and exploration.

Site G – Iron oxide-apatite deposit, Europe – expanding throughput with core scanning

Site G's integration of core scanning technology was a strategic initiative to handle the burgeoning demands of an extensive drill program. In a region where labour availability and expertise were limited, the adoption of this technology allowed for a significant increase in geological throughput without the need for proportional increases in headcount, which wasn't feasible due to the remote location and specialised skills required. The core scanning technology, as seen in Figure 4, not only ensured the consistent and accurate logging of the increased core volume but also facilitated the rapid training and upskilling of new geologists. This training was critical, as the technology provided uninterrupted, high-quality data that maintained the integrity and governance of the geological information system.

FIG 4 – Digital scanning technologies at work on-site in Europe.

Furthermore, the real-time data acquisition from core scanning allowed for more dynamic and responsive drill program management. The technology's capability to process high volumes of core efficiently meant that the operation could keep pace with the ambitious expansion plans. This efficiency was not just a matter of cost but was pivotal in ensuring that the mining operation could exploit the window of opportunity presented by favourable market conditions and strategic operational goals.

Site H – Complex exploration project, Australia – streamlining processes with MinalyzerCS

The MinalyzerCS at Site H represented a significant advancement in the exploration project's capability to process geological data. The system's daily geochemical output dramatically accelerated the selection and prioritisation of samples for lab assays. This speed was not merely an operational benefit but a strategic advantage, enabling the geology team to make quicker, more informed decisions about drill progression and resource allocation. The rapid turnaround of geochemical data meant that drilling could be adapted on the fly, a crucial factor in a complex exploration environment where every day and every drill metre could lead to a different strategic direction.

The technology also had a substantial impact on the efficiency of core logging, reducing the time taken to log, assay and interpret geological data. The MinalyzerCS allowed geologists to validate domains in real-time, which is critical in the early stages of exploration when understanding the geology is rapidly evolving. This capability ensured that the project could maintain momentum, adapt to new findings quickly and refine its exploration model with a level of precision and speed previously unattainable.

Site I – Orthomagmatic nickel-sulfide deposit, Australia – advancing metallurgical domain validation

At Site I, the focus was on enhancing the efficiency of core logging processes and speeding up metallurgical test work for an orthomagmatic nickel-sulfide deposit. The introduction of the MinalyzerCS technology (Figure 5) marked a significant shift in workflow. Traditionally, the site faced delays of six to eight weeks for quarter-core assays to return from laboratories, hindering the pace of prioritising samples for metallurgical test work. The MinalyzerCS unit transformed this process, allowing for immediate scanning and decision-making post-drilling, thereby accelerating the validation of assumed metallurgical domains.

FIG 5 – Minalyzer technology being utilised on-site to increase efficiency of core logging.

The technology's impact extended beyond just improving core logging efficiency. It played a crucial role in confirming metallurgical domains quickly, enabling the project to proceed at a much faster rate. This acceleration was crucial in a context where metallurgical understanding directly influenced the exploration strategy and mine planning. The ability to rapidly confirm geological assumptions through immediate geochemical data fundamentally changed the project's trajectory, allowing for more agile and informed decision-making in a field where time is often a critical factor.

Technology integration and process optimisation

The journey of integrating core scanning technologies into mining operations begins with an understanding of existing practices, as depicted in Figure 6's workflow diagram.

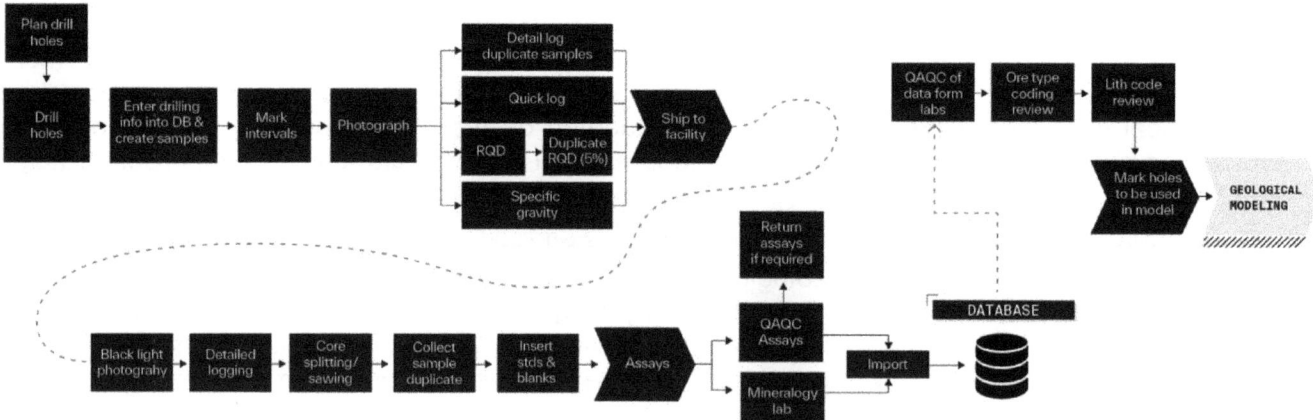

FIG 6 – A common overview of incumbent core logging processes, typically associated with site works globally.

Figure 6 represents the traditional methodologies that have been the bedrock of geological workflows, characterised by manual, time-intensive tasks.

- Data-driven decision-making – Core scanning technologies facilitate a significant shift towards data-driven decision-making, enabling geologists to focus on strategic analysis rather than manual data collection.

- Streamlined workflows – The integration of technologies like TruScan and MinalyzerCS streamlines workflows, enhancing efficiency and reducing the scope for human error.

- Collaborative process improvement – Customisation of technological solutions through collaborative analysis with site personnel ensures seamless integration and maximises operational benefits.

- Change management and leadership – The transition to technology-driven processes necessitates strong leadership and effective change management to ensure successful adoption and adaptation by the workforce.

The culmination of these benefits and the transformative impact of technology integration are encapsulated in the follow-up workflow that distils the advanced workflows, highlighting how the integration of core scanning technologies has reshaped core logging processes.

Figure 7 synthesises the key improvements observed across various sites, such as enhanced data precision, increased productivity, reduced manual handling and a strategic shift in geologists' roles.

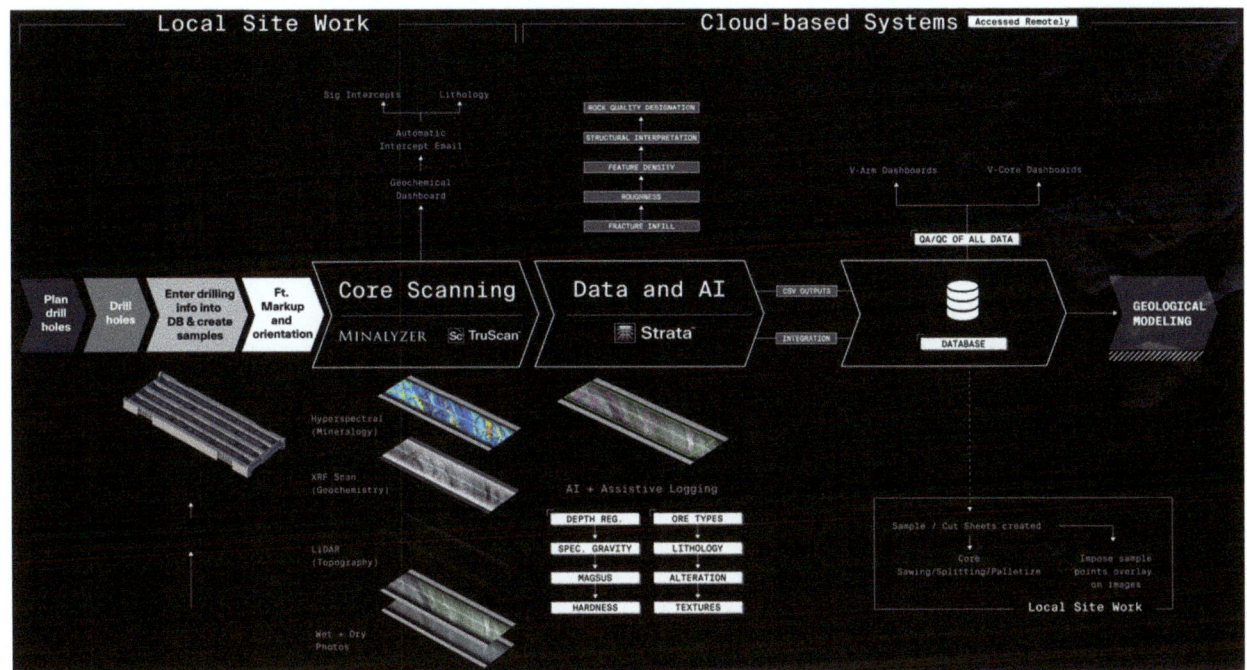

FIG 7 – A distillation of advanced core logging process proposed by Veracio.

This visualisation serves as a blueprint for the future of geological workflows, demonstrating a more automated, rapid, and well-governed approach to core logging and analysis, pivotal in the post-pandemic era of mining operations.

CONCLUSION

Our study across various global mining sites demonstrates that the integration of core scanning technologies like TruScan and MinalyzerCS fundamentally enhances geological workflows, particularly in the context of post-pandemic operations. This technological integration has proven to be a key driver in transitioning geologists' roles from traditional, manual-intensive tasks to more strategic, analytical functions. The shift is not just in process efficiency and data precision but also in the overall operational philosophy, emphasising the importance of data-driven decision-making and streamlined workflows.

The successful implementation of these technologies, as illustrated through our diverse site experiences, underscores the necessity of a collaborative approach, change management and strong leadership. The integration transcends mere technological adoption; it involves rethinking and redefining traditional practices to embrace a more efficient, safe and strategic operational model. As the mining sector continues to evolve in the post-pandemic era, the insights from this study offer a blueprint for leveraging technology to achieve operational resilience and efficiency, underscoring the vital coexistence of specialised expertise and innovation in mine geology.

Machine learning-based silicate alteration – optimising spectrographic core imaging, multi-element geochemical, and dynamic rebound hardness data set to improve the objectivity of alteration domaining

M H Rahadi[1], A Wiguna[2], M Heriyanto[3] and R Taube[4]

1. Geology Lead, Merdeka Copper Gold, South Jakarta 12910.
   Email: mochammad.rahadi@merdekacoppergold.com
2. Data Lead, Merdeka Copper Gold, South Jakarta 12910.
   Email: aldaka.wiguna@merdekacoppergold.com
3. Data Scientist, Merdeka Copper Gold, South Jakarta 12910.
   Email: mohammad.heriyanto@merdekacoppergold.com
4. Geoscience Manager, Merdeka Copper Gold, South Jakarta 12910.
   Email: rob.taube@merdekacoppergold.com

ABSTRACT

In today's era, the integration of technology-based geological data collection and machine learning analysis plays a pivotal role in revolutionising the field of geoscience. This paper explores the application of unsupervised machine learning for alteration classification and supervised machine learning for alteration prediction. A critical objective of this study is to significantly reduce the uncertainty of visual alteration logging by maximising the role of spectrographic core imaging (SCI), multi-element geochemical (MEG), and dynamic rebound hardness (DRH) data in classifying and predicting the alteration facies.

The unsupervised machine learning technique has initiated reliable alteration classification. The K-means clustering algorithm was employed to categorise the alteration classification harnessing SCI and DRH data. Nine alteration major mineral compositions combined with hardness from the latest drilling activity were used in this analysis which unveiled patterns that are undetectable by human vision. These clustering results provide valuable clarity into the diversity of alteration facies within the project, serving as key drivers for resource estimation, geotechnical and metallurgical purposes.

Thereafter, a supervised machine learning technique was performed to predict the alteration facies from historical drilling activity, which is devoid of SCI and DRH data, based on MEG data. Deploying Neural Network (NN) and AdaBoost (ADA) models to absorb knowledge from the pattern between alteration facies and MEG, enabling them to predict the alteration facies where SCI and DRH data were absent. Through this approach, it was possible to augment the spatial coverage of the deposit and provide sufficient data density to support the development of a robust alteration model in three-dimensional space which is one of the foundations in supporting the development of reliable mine/processing design and operation.

INTRODUCTION

Understanding the alteration processes of a porphyry deposit has a critical role to support the development of mining and processing design. Reducing uncertainty through technology application in the data collection and analysis allows us to increase the objectivity alteration classification.

The Tujuh Bukit deposit is a complex porphyry Cu-Au deposit, where faulting and overprinting processes have destroyed the initial texture of the rocks. Conventional logging methods are often unable to precisely separate the alteration facies due to intense overprinting. Understanding this complex alteration requires a quantitative approach in terms of separating the initial porphyry alteration and the overprinting alteration. Since human observation is limited to a qualitative approach, technological assistance is being used to overcome these limitations, including spectrographic core imaging (SCI), dynamic rebound hardness (DRH), multi-element geochemical (MEG) and machine learning algorithms.

This paper will outline the development of an unsupervised machine learning (UML) algorithm to improve the alteration classification based on a data set comprising spectrographic core imaging (SCI) and dynamic rebound hardness (DRH) data taken after 2019. On historical drill holes which lack spectrographic core imaging (SCI) and dynamic rebound hardness (DRH), a supervised machine learning (SML) algorithm, specifically Neural Network (NN) and AdaBoost (ADA) were trained using multi-element geochemistry data to predict the unsupervised machine learning (UML) alteration classification.

DATA COLLECTION

A quantitative approach has been used to try to improve the separation between the initial porphyry alteration and the overprinting alteration. Conventional alteration logging methods require geologists to make a qualitative observation of the drill core which has the potential to introduce bias and variability to data sets. Since human observation is limited in the qualitative approach, technology assistance is being used to overcome these limitations, including spectrographic core imaging (SCI), dynamic rebound hardness (DRH) and multi-element geochemistry (MEG). A brief introduction to these tools is provided as follows:

- Spectrographic core imaging (SCI) technologies can quantify the mineral composition of the core surface based on the electromagnetic signature of the minerals under specialised sensors. The data generated by spectrographic core imaging allows for the mineral composition of the core to be analysed quantitatively.

- Dynamic rebound hardness (DRH) testing tools drop an engineered tip under spring force against the test surface while measuring the impact and rebound velocities. The hardness of the core exterior can be measured with these devices. This approach correlates well with conventional field hardness estimates using a geological hammer and provides a more objective measure of rock hardness.

- Multi-element geochemistry (MEG) analysis refers to the simultaneous analysis of multiple chemical elements in geological samples. MEG involves determining the concentrations of various elements via Inductively Coupled Plasma Mass Spectrometry (ICP-MS) and Inductively Coupled Plasma Optical Emission Spectrometry (ICP-OES).

During unsupervised machine learning (UML) data set construction, all the data available of SCI and DRH within the Tujuh Bukit project has been collected. 287 597 rows of data with 1 m composite interval have been gathered from ten variables (aspectral, alunite, pyrophyllite, dickite, muscovite, kaolinite, illite, montmorillonite, chlorite, and DRH mean) which still contain 51.3 per cent missing values, mainly from old drill holes without SCI data. By removing all the missing values, 134 068 records were ready to be used as inputs for the new alteration classification using K-means clustering algorithms.

Combined with the UML result, MEG data was also extracted during supervised machine learning (SML) data set construction. By removing all the missing values and replacing the negative value with half detection limit, 131 952 rows of data can be used as a training data set for the SML algorithm to predict the UML alteration result.

The following subsections will provide descriptions of the spectrographic core imaging and dynamic hardness testing data used as inputs for the new alteration classification from K-means clustering machine learning algorithms.

Spectrographic Core Imaging (SCI)

Historically, alteration logging and subsequent model development relied on visual interpretations and geologists' limited field testing of mineral assemblages. Modern core imaging technologies provide an opportunity for greater objectivity in the determination of alteration mineral assemblages along the drill core. The data set generated by spectrographic core imaging (Corescan Technology) at the project site supplied the data for this paper, providing estimates of the modal percentages of key minerals along the core surface on 1 m interval lengths.

For unsupervised machine learning (UML), nine major minerals from spectrographic core imaging data are being used as the key features of K-means clustering: alunite, pyrophyllite, dickite, muscovite, kaolinite, illite, montmorillonite, chlorite and aspectral. Aspectral response is also used in this research because aspectral represents minerals that could not be detected by SWIR (short wave infrared) and may include quartz, opaline silica, sulfides or partially altered feldspar minerals. Aspectral responses occur in a wide range of lithology and alteration types at Tujuh Bukit.

Histograms of the nine minerals (Figure 1) display specific correlation to alteration logging, except aspectral mineral which presents almost in every alteration facies.

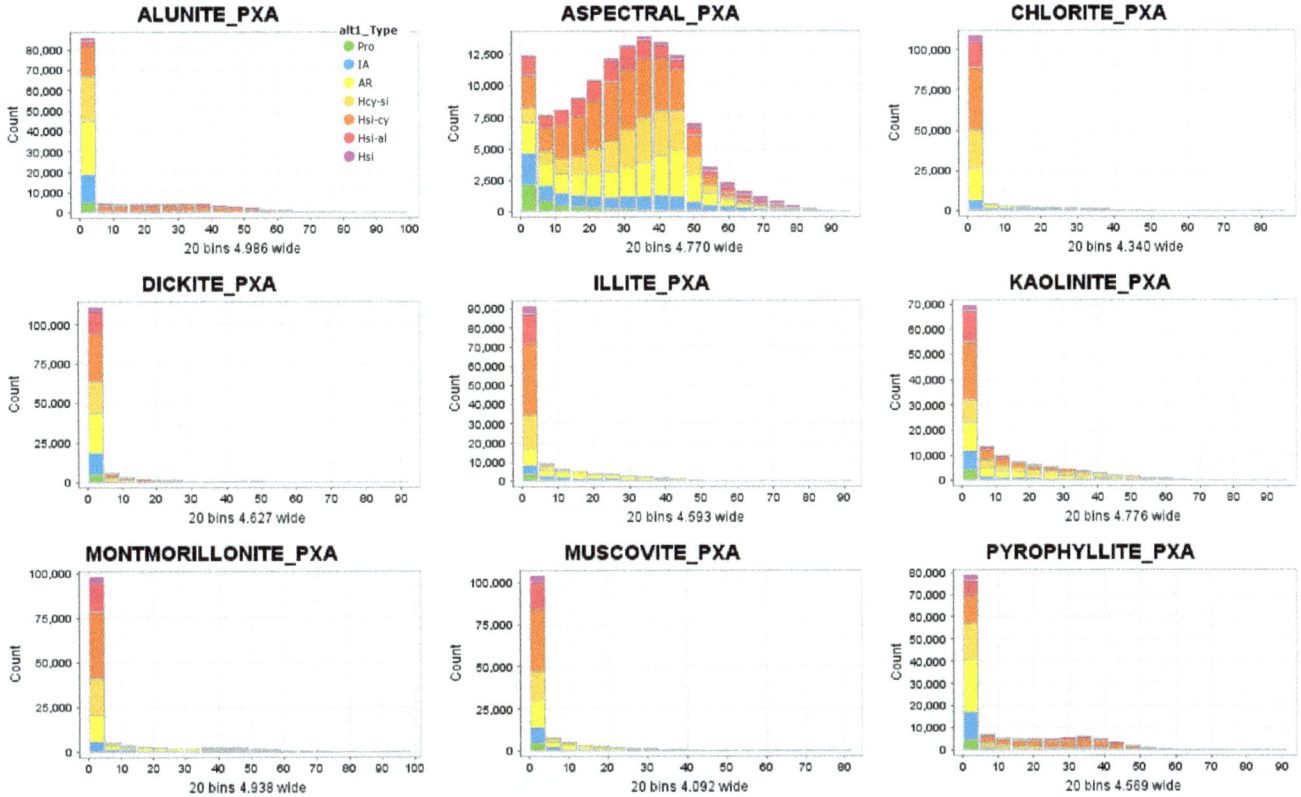

FIG 1 – Histogram of nine spectral minerals.

Figure 2 exhibits cumulative probability plots of the data after normalisation. A normal score of 0 is the mean of the data distribution. The plots show a strong correlation between mineral composition from spectrographic core imaging and alteration logging based on geologists' observation. Along with Figure 3 which also displays a correlative relationship of each mineral to a specific alteration facies.

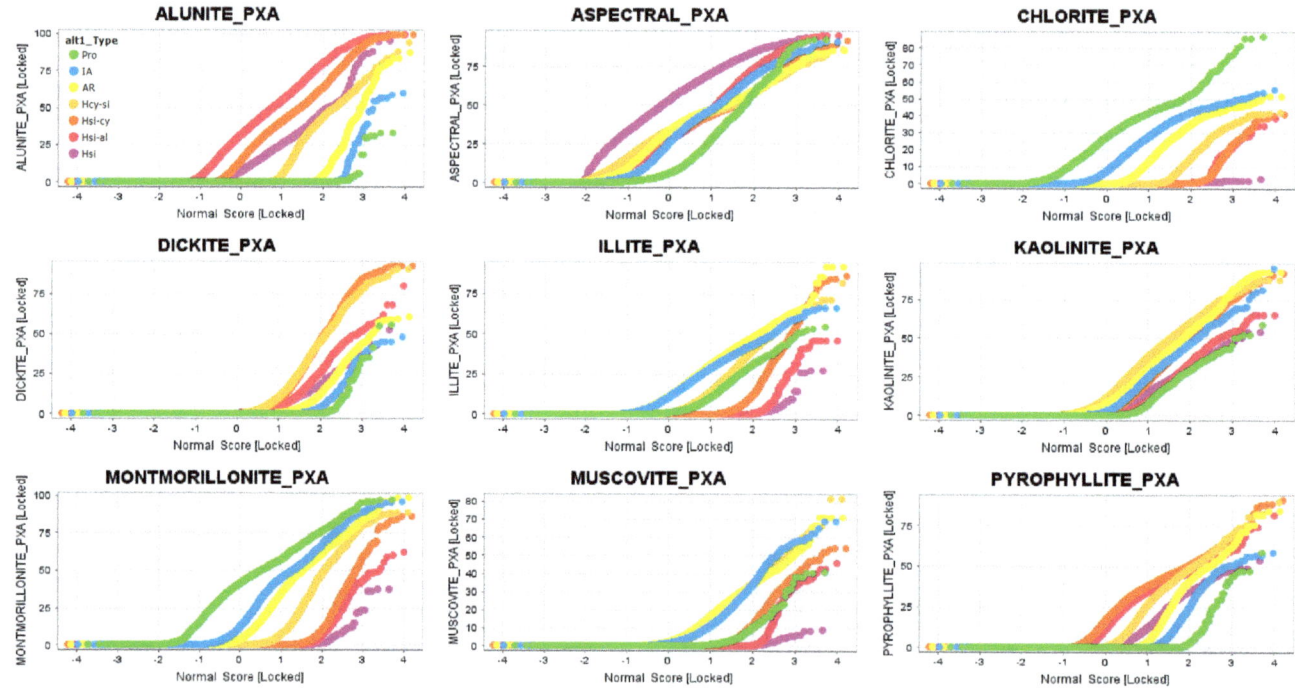

FIG 2 – Split probability plot of spectral minerals in all alteration facies.

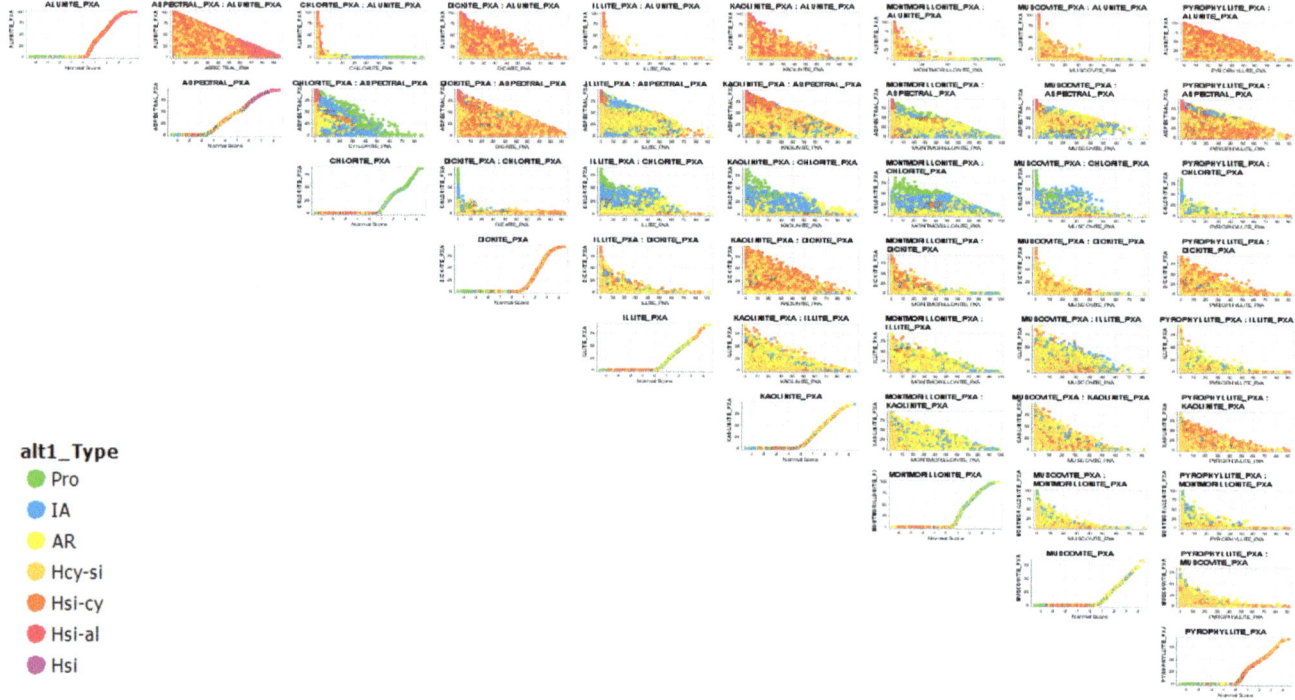

FIG 3 – Scatter plot matrix of each pair of minerals.

The key findings from the review of the spectrographic core imaging data are as follows:

- Alunite, pyrophyllite, and dickite show a strong relationship to the advanced argillic alteration likes Hcy-si (clay-silica), Hsi-cy (silica-clay), Hsi-al (silica-alunite), and Hsi (silica). Alunites are mostly localised in Hsi-al and tends to gradually decrease from Hsi-cy to Hsi and Hcy-si. Pyrophyllites are predominantly localised in Hsi-cy and Hsi-al while some are also detected on Hcy-si and rare on Hsi. Dickites are associated with Hsi-cy and Hcy-si while some are still observed on Hsi-al and even weaker on Hsi.

- High aspectral contents are specifically associated with the Hsi alteration. However, aspectral response can be found moderately in almost every alteration except propylitic which is normally located in the very distal area from the source hydrothermal activity.

- Kaolinites are normally associated with the zone of transition between the argillic and advanced argillic. It can be situated in Hcy-si and AR (argillic) while some still exist in Hsi-cy and IA (intermediate argillic) alteration.

- Illites and muscovites are the key minerals of AR (argillic) and IA (intermediate argillic) alteration. However, muscovite can also be associated with phyllic alteration. The possibility of separation between AR alteration and phyllic alteration using machine learning algorithm will be demonstrated in this documentation.

- Chlorites and montmorillonites are specifically associated with IA and Pro (propyllitic) alteration which is less affected by hydrothermal activity.

Each of these nine minerals was considered as the clustering signal which demonstrates a specific correlation to the alteration classification. Since alteration classification is based on these minerals, it has a correlative relationship to the alteration classification. Minerals that don't have a strong correlation with alteration are considered as the clustering noise and have been eliminated from the data set.

Dynamic Rebound Hardness (DRH)

Dynamic rebound hardness testing provides a non-destructive estimate of the rock hardness by measuring the impact and rebound velocity of the DRH tool's impact tip against the core surface (ie Equotip rebound hardness test). At Tujuh Bukit site, hardness values (Leeb) are recorded on approximately 10 cm intervals and then averaged over fixed 1 m interval lengths to provide a mean hardness value.

The histogram and split probability (Figure 4) plot show unreliable data which has below 200 Leeb. This condition commonly occurred during measurement in very soft material. Since the principle of this testing is measuring the rebound velocity of the tools against the surface, in the very soft material the tool's impact tip wouldn't produce any rebound impact. Therefore, Leeb values below 200 are considered the detection limit of Equotip testing.

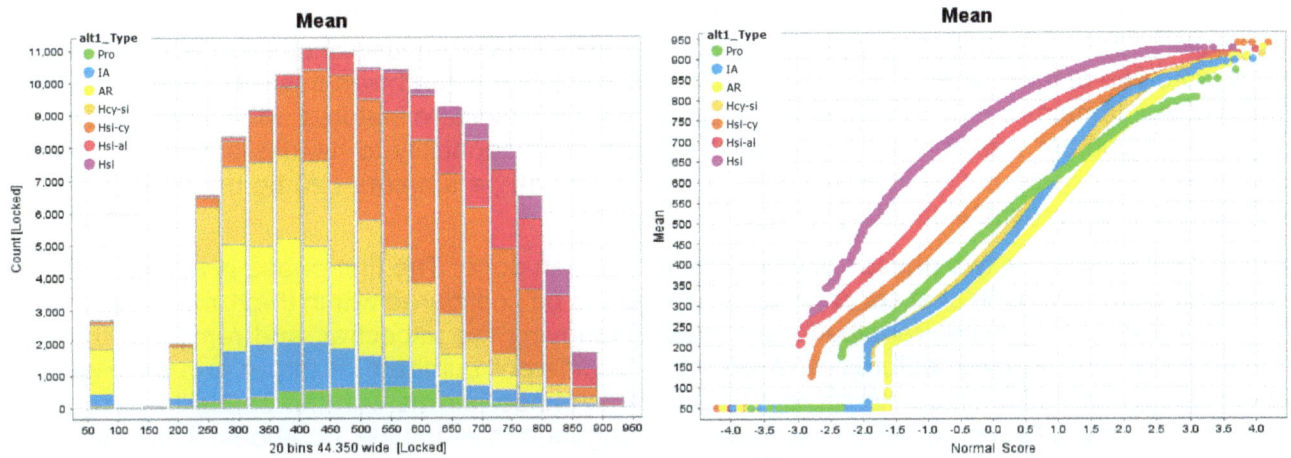

FIG 4 – Histogram and split probability plot of Equotip Leeb values for all alteration facies.

Both histogram and split probability plot show that the distributions of mean dynamic rebound hardness value vary between alteration classifications and reflect the expected hardness of the alteration minerals. The key findings from the review of the hardness testing data are as follows:

- The hardest materials are commonly associated with Hsi (silica) followed by Hsi-al (silica-alunite) and Hsi-cy (silica-clay). This condition aligns with the development of the advanced argillic alteration which tends to increase the silica replacement from the distal to proximal area.

- Pro (Propylitic) alteration is characterised by moderate hardness due to the weak impact of hydrothermal activity. It is commonly localised in the most distal part of hydrothermal activity hence it is still preserving the initial hardness of the fresh lithology.

- The softest materials are commonly associated with clay-dominant alterations such as Hcy-si (clay-silica), IA (intermediate argillic) and AR (argillic) alteration.

Although dynamic rebound hardness testing is not the basis of the alteration logging at Tujuh Bukit, the Leeb values demonstrate a strong correlation to the alteration classification. Therefore, in this study, dynamic rebound hardness data is still useful as an input to machine learning.

APPLICATION OF MACHINE LEARNING

Machine learning (ML) is a subset of artificial intelligence (AI) focused on the system that learns from data/environment, from the concept of AI initially introduced by Turing (1950). It is a method of data analysis that automates analytical model building, either for classification, regression, or patterns extraction, through algorithms that iteratively learn without being explicitly programmed. The key objective in machine learning is to build a model that in general will behave intelligently in totally new data/environment.

The geoscience field has witnessed rapid advancements in the progress and adoption of ML. Within geoscience, ML serves as a valuable tool for data preprocessing, enabling the application of previous knowledge beyond theoretical and synthetic scenarios and facilitating unprecedented applications. Applied ML has evolved into an established tool in computational geoscience, offering the potential to provide further insights (Dramsch, 2020).

In this study, an unsupervised machine learning (UML) algorithm, K-means, was used to improve the alteration classification based on a data set comprising spectrographic core imaging (SCI) and dynamic rebound hardness (DRH) data. For historical drill holes which lack spectrographic core imaging (SCI) and dynamic rebound hardness (DRH), a supervised machine learning (SML) algorithm, specifically Neural Network (NN) and AdaBoost (ADA) were trained using multi-element geochemistry data to predict the unsupervised machine learning (UML) alteration classification.

Unsupervised Machine Learning (UML)

K-means clustering, an unsupervised machine learning algorithm, is a method of vector quantisation, originally from signal processing, that aims to partition n observations into k clusters in which each observation belongs to the cluster with the nearest mean (cluster centres or cluster centroid), serving as a prototype of the cluster. The target of using the K-means algorithm in this research is to explore unrevealed domains based on the relationship between multivariate features within the data set. The K-means algorithm was chosen for the analysis due to its capability to handle large data sets and run an efficient computation. It is versatile, as the algorithm parameters can be modified to suit the needs of the analysis (Celebi, Kingravi and Vela, 2012).

In this UML, there are two phases of work that have been done. The first phase is algorithmic trials which include finding the best set of hyperparameters to be configured within the K-means algorithm, this included adding aspectral albedo and slope, removing the aspectral, and removing the DRH means from the selected features, and testing with different numbers of target groups (Figure 5). Each test generated a silhouette score. Once the first phase returns with the best silhouette score, the same processes will be rebuilt using python language in a Jupyter notebook. This stage is to overcome the limited processed data of Orange Data Mining software (Demsar *et al*, 2013), an open-source data visualisation, data mining, and machine learning tool, which is only able to process a maximum of 5000 rows of data when conducting K-means clustering. For other than clustering processing, Orange has no limitation on the amount of data processed.

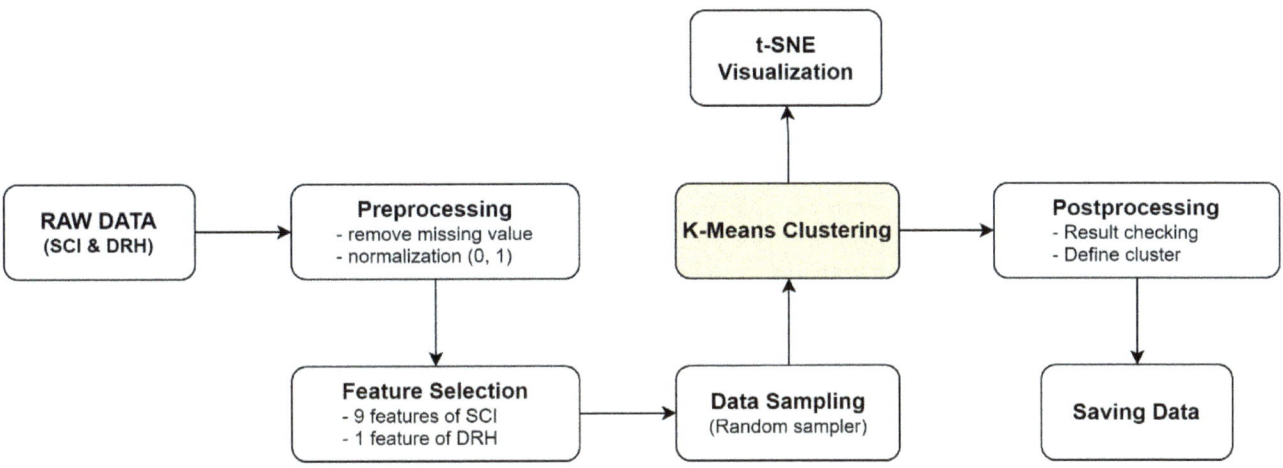

FIG 5 – Unsupervised Machine Learning (UML) workflow.

Data and preprocessing

K-means clustering process starts with raw data input, which is the merged data comprised of SCI, DRH and visual alteration logging data. To achieve more robust and reliable clustering results, the data set was prepared by removing rows with missing values, including failed DRH readings and drill holes without SCI data (mostly historical holes drilled from surface). This step removed 51.3 per cent of the records. Based on the needs of the analysis and the data characteristics, prior to further data processing, all the features were standardised to a mean of zero and a variance of 1 by using the StandardScaler function. Utilising preprocessed data, the ten features that are recognised by geological logging as the dominant signals for the alteration classification were selected as input. These are aspectral, alunite, pyrophyllite, dickite, muscovite, kaolinite, illite, montmorillonite, chlorite and DRH mean. To hasten the processing random data sampling was conducted using 5000 data to fit within the limit of Orange Data Mining software.

Parameters and metrics

The Optuna hyperparameter optimisation framework (Akiba *et al*, 2019) was adopted to help find the best set of hyperparameters to be configured within the K-means algorithm ranging from 2 to 10 clusters. The hyperparameters include the number of initialisations, number of iterations and random number for centroid initialisation. The best hyperparameters are evaluated through the accuracy value and the best clusters are statistically evaluated through the silhouette score.

The silhouette score ranged between -1 and 1 which provides a measure of how close the points in one cluster are relative to the points in neighbouring clusters in multivariate space. A value close to 1 indicates that points are far away from neighbouring clusters. A value close to 0 indicates points are close to the decision boundary between two neighbouring clusters. Negative values may indicate points have been assigned to the 'wrong' clusters.

The application of K-means in this research used Manhattan distance as a distance metric. Manhattan distance is arguably a distance metric that is robust to outliers and a better choice compared to Euclidean distance when the data set consists of high-dimensional data (Verdhan, 2021). We set K-means++ as the method for initialisation of the K-means algorithm. K-means++ is an improved K-means method that picks the initial centroid based on the distance-squared metric and has at least 10 per cent accuracy improvement over K-means and faster running time (Arthur and Vassilvitskii, 2006).

Supervised Machine Learning (SML)

Neural Network (NN) and AdaBoost (ADA), supervised machine learning (SML) algorithms, are algorithms which learn from labelled training data to build models from which predictions or decisions can be sourced. The training data consists of input (features) along with corresponding output labels (targets or classes) (Kotsiantis, 2007). The objective is to learn a mapping or function that can generalise from the training data and accurately predict targets from unseen input data. NN and ADA were chosen after exploring various algorithms such as Gradient Boosting, Random Forest, Tree, Logistic Regression, Naïve Bayes, k-Nearest Neighbours (kNN), Support Vector Machine (SVM) and NN and ADA itself.

Neural networks (NN) are computational techniques inspired by the structure and function of the human brain. These networks have gained immense popularity and have become a foundational concept in modern machine learning. Neural networks lead at learning from data and making predictions or decisions based on complex patterns and relationships within that data (Krose, 1996).

AdaBoost, short for Adaptive Boosting, is a ML algorithm that is used for binary classification tasks. It is a popular ensemble learning method that combines multiple 'weak' classifiers to create a strong classifier. The key idea behind AdaBoost is to iteratively train a series of weak classifiers on different subsets of the training data, where each subsequent weak classifier focuses more on the examples with which the previous classifiers struggled (Schapire, 2013).

In this SML encompasses two distinct working phases, the first phase involves SML algorithm selection using the same data set. Experiments were conducted to select the best SML algorithm using three variations of data sets and performance metrics were employed to evaluate the SMLs. The metrics used to measure SML performance, commonly used in machine learning, include accuracy, precision, recall and F1-score. The second phase involves training and executing NN and ADA algorithm to predict K-mean clustering output using the entire data set without data separation during the training (Figure 6).

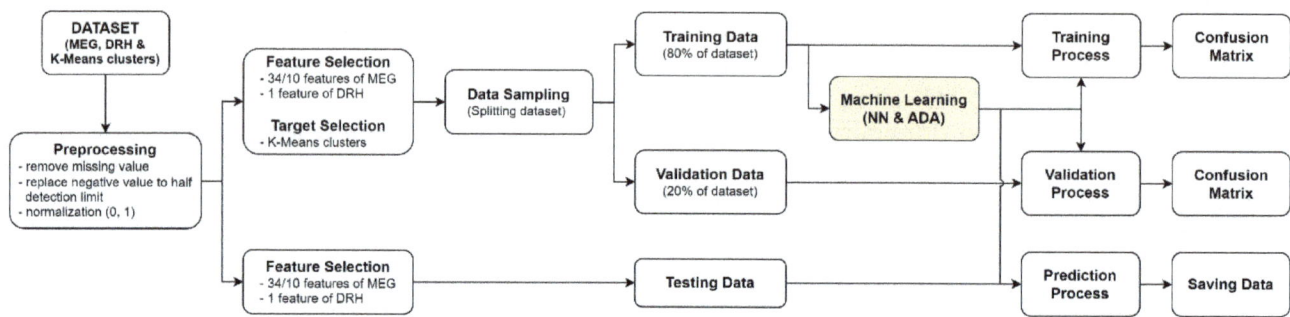

FIG 6 – Supervised Machine Learning (SML) workflow.

Data and preprocessing

The SML process starts with raw data input, which is the merged data comprised of K-means clustering output, DRH and MEG data. The merged data or data set was prepared by removing rows with missing values, replacing negative value to half detection limit and normalising feature input data to interval 0 to 1. This data set was divided to two types, one for the selection SML algorithm that is partitioned based on the features used and the other for the final prediction. The utilised features are varied to assess the impact of features number on the SML performance that are denoted as MA+E (many assays + Equotip) for the data with 35 features from MEG and DRH measurements, MA (many assays) for the data with 34 features from MEG and LA (limited assays) for the data with ten features from MEG.

Replacing negative value of MEG data using half below detection limit as follows:

- Elements in ppm including Ag (0.05 ppm), As (0.5 ppm), Au (0.0025 ppm), Ba (0.5 ppm), Bi (0.025 ppm), Cd (0.025 ppm), Co (0.5 ppm), Cr (1 ppm), Li (0.05 ppm), Mn (0.5 ppm), Mo (0.05 ppm), Nb (0.05 ppm), Ni (0.5 ppm), Pb (0.5 ppm), Sb (0.05 ppm), Sc (0. 5 ppm), Sn (0.05 ppm), Sr (0.25 ppm), Ta (0.025 ppm), Te (0.05 ppm), V (0.5 ppm), W (0.05 ppm), Y (0.05 ppm), Zn (0.5 ppm), and Zr (0.25 ppm).

- Elements in percent including Al (0.005 per cent), Ca (0.0025 per cent), Cu (no negative or 0 value), Fe (0.005 per cent), K (0.001 per cent), Mg (0.001 per cent), Na (0.001 per cent), Ti (0.0005 per cent), and S (0.005 per cent).

MEG data used in this research was divided into two types, 34 features or elements and ten elements. The 34 elements data consist of two components:

- Concentration in parts per million (ppm) of elements including Ag, As, Au, Ba, Bi, Cd, Co, Cr, Li, Mn, Mo, Nb, Ni, Pb, Sb, Sc, Sn, Sr, Ta, Te, V, W, Y, Zn and Zr.
- Percentage composition of elements including Al, Ca, Cu, Fe, K, Mg, Na and Ti.

On the other hand, the ten elements data consist of two components:

- Concentration in ppm of elements including Ag, As, Au, Ba, Mo, Pb, Sb and Zn.
- Percentage composition of elements including Cu and S.

The K-means clustering outputs were utilised as targets or classes, where the category labelled Pyrophyllite-Advanced Argillic (C0) exhibited the highest number of data samples, comprising approximately 23 783 samples, constituting approximately 18 per cent of the entire data set. In contrast, Dickite-Advanced Argillic (C7), demonstrated the lowest number of data samples, totalling approximately 6859 samples (~5 per cent). The distribution of the remaining classes is as follows:

- Silica Flooding (C2), encompasses 21 764 samples (~17 per cent).
- Alunite-Advanced Argillic (C4), comprises 21 187 samples (~16 per cent).
- Kaolinite-Argillic (C1), includes 18 813 samples (~14 per cent).
- Illite-Intermediate Argillic (C6), contains 15 373 samples (~12 per cent).
- Montmorillonite/Chlorite-Propyllitic (C3), holds 14 669 samples (~11 per cent).
- Muscovite-Phyllic (C5), consists of 9504 samples (~7 per cent).

The data set was divided into training, validation and test sets (Table 1) to facilitate the training and evaluation of SML models. The training set is utilised for training SML, allowing it to capture complex patterns effectively. During training, the SML adjusts its parameters to minimise discrepancies between predictions and true labels in the training data. The validation set is employed to fine-tune hyperparameters and prevent overfitting. Hyperparameters, such as learning rate and network architecture, are optimised during this phase. Finally, the test set assesses the SML's overall performance and generalisation ability on unlabelled or unseen data, ensuring its effectiveness in real-world scenarios.

TABLE 1

Summary of Supervised Machine Learning (SML) algorithm usage.

No.	Features		Data set		
	Input	Target	Train	Validation	Test
1	MA+E	K-means cluster	105 562	26 390	141 603
2	MA	K-means cluster	105 562	26 390	231 392
3	LA	K-means cluster	105 562	26 390	272 396

Parameters and metrics

Each SML algorithm utilised default parameters and hyperparameters from the Orange Data Mining software. In Orange, normalisation is applied within the NN/ADA widget. In order to compare the results of normalised and non-normalised data entered into the SML widget, we carried out experiments that yielded equivalent results. Thus, before or after normalising data will result in same performance. NN consists of one hidden layer with 100 neurons. The rectified linear unit (ReLU) activation function is utilised in this layer to introduce non-linearity and enhance the NN ability to capture complex patterns. The optimiser used is Adam, which is a popular optimisation algorithm known for its efficiency and effectiveness in updating the NN parameters based on the gradients of loss function. ADA was utilised with base estimators in the form of decision trees, specifically with a total of 50 estimators.

Comparison was conducted between Gradient Boosting, Random Forest, Tree, Logistic Regression, Naïve Bayes, kNN, SVM, NN and ADA. Accuracy, precision, recall and F1-score were used as metrics to measure SML performance. Precision measures the accuracy of positive predictions, recall gauges the SML model's ability to identify all positive instances and the F1-score combines both metrics for a balanced evaluation of overall performance. SML models are systematically evaluated and compared through quantitative metrics, facilitating the identification of the optimal model for a specific task. In this assessment, NN and ADA emerged as top performers, with NN demonstrating superior metrics in precision, recall, F1-score, and accuracy for MA+E and MA data sets (Tables 2 and 3), while ADA excelled in handling smaller LA data set (Table 4).

TABLE 2
Training result using MA+E (35 features) data set.

SML algorithm	Accuracy	F1-Score	Precision	Recall
Neural Network	0.736	0.735	0.735	0.736
AdaBoost	0.724	0.724	0.724	0.724
Gradient Boosting	0.719	0.718	0.718	0.719
Random Forest	0.718	0.718	0.718	0.718
Tree	0.657	0.657	0.658	0.657
Logistic Regression	0.650	0.647	0.646	0.650
Naïve Bayes	0.581	0.571	0.578	0.581
kNN	0.573	0.569	0.573	0.573
SVM	0.370	0.393	0.454	0.370

TABLE 3
Training result using MA (34 features) data set.

SML Algorithm	Accuracy	F1-Score	Precision	Recall
Neural Network	0.715	0.715	0.715	0.715
Gradient Boosting	0.7	0.699	0.699	0.7
AdaBoost	0.699	0.698	0.698	0.699
Random Forest	0.698	0.697	0.697	0.698
Tree	0.656	0.656	0.658	0.656
Logistic Regression	0.633	0.631	0.630	0.633
Naïve Bayes	0.568	0.557	0.565	0.568
kNN	0.517	0.514	0.519	0.517
SVM	0.33	0.353	0.446	0.33

TABLE 4
Training result using LA (ten features) data set.

Supervised ML	Accuracy	F1-Score	Precision	Recall
AdaBoost	0.633	0.632	0.631	0.633
Random Forest	0.622	0.620	0.619	0.622
Tree	0.56	0.56	0.563	0.56
Gradient Boosting	0.533	0.527	0.529	0.533
Neural Network	0.53	0.524	0.525	0.53
Naïve Bayes	0.433	0.418	0.427	0.433
kNN	0.421	0.417	0.421	0.421
Logistic Regression	0.419	0.393	0.413	0.419
SVM	0.105	0.099	0.245	0.105

Each SML model was subsequently validated using the data set from the validation column in Table 1 to ensure the absence of overfitting. The results confirm that the optimal SML model for each feature demonstrates non-overfitting characteristics. This is supported by the performance metrics of the SML models where slight differences between the training and validation phases are observed. The training stage uses the training data set while the validation stage uses the validation data set, results are presented in Tables 5 and 6.

TABLE 5
The best SML algorithm after training in ML selection phase.

Feature	Supervised ML	Accuracy	F1-Score	Precision	Recall
MA+E	Neural Network	0.736	0.735	0.735	0.736
MA	Neural Network	0.715	0.715	0.715	0.715
LA	AdaBoost	0.633	0.632	0.631	0.633

TABLE 6
Validation results of SML algorithm in ML selection phase.

Feature	Supervised ML	Accuracy	F1-Score	Precision	Recall
MA+E	Neural Network	0.739	0.737	0.739	0.739
MA	Neural Network	0.72	0.72	0.722	0.72
LA	AdaBoost	0.654	0.58	0.606	0.556

An experiment where all MEG values were modified to parts per million (ppm) were also compared. The findings indicate that the accuracy of MEG converted to ppm is lower than the best SML model (see Table 7). ML training also was conducted to all data set includes 61 413 rows. If compared to the best SML, the accuracy is lower. MEG configuration related to silicate minerals, including Al, Ca, Cu, Fe, K, Mg, Na, and Ti, performed less well with NN compared to the accuracy achieved by 34 other MEG.

TABLE 7

Performance of All MEG in ppm.

Feature	Supervised ML	Accuracy	F1-Score	Precision	Recall
MA+E	Neural Network	0.738	0.738	0.738	0.738
MA	Neural Network	0.719	0.718	0.718	0.719
LA	AdaBoost	0.658	0.657	0.657	0.658

Accuracies achieved for best SML algorithm in Table 5 are satisfactory for modelling objectives with restricted SCI data. This can be observed in the spatial validation, where the initial simplified modelling becomes more detailed. Finally, the best SML algorithm selected for the 34 MEG+DRH (MA+E) data and the 34 MEG (MA) were Neural Network and for ten MEG (LA) data, AdaBoost was chosen.

RESULT AND SPATIAL VALIDATION

Unsupervised Machine Learning (UML)

The purpose of clustering is to build an alteration model that is quantitative and 100 per cent data-driven. Utilising spectrographic core imaging mineral percentages and dynamic rebound hardness mean data, the K-means clustering algorithm was tested with different numbers of target groups. Each test generated a silhouette score. The best silhouette score in this processing is 0.280 with eight clusters output which indicates the clearest multivariate separation of the clusters.

To visualise high-dimensional data and validate the K-means clustering performance, dimensionality reduction is required, as the data sets consist of numerous features. The curse of dimensionality is a problem that occurs when there are too many input features within the data set and will reduce the model's performance (Verdhan, 2021).

In this research, PCA (Principal Component Analysis) and t-SNE (t-distributed stochastic neighbourhood embedding) were performed to visualise the data and to reduce the data dimensionality. PCA is an unsupervised linear dimensionality reduction and data visualisation technique for very high dimensional data, while t-SNE is also an unsupervised non-linear dimensionality reduction and data visualisation technique. Unlike PCA which tries to preserve the global structure of the data, t-SNE is one of the best dimensionality reduction techniques which tries to preserve the local structure of data. Dealing with outliers, PCA is highly affected by outliers while t-SNE can handle outliers effectively. For these reasons, t-SNE was selected for dimension reduction (Van der Maaten and Hinton, 2008).

A t-SNE transformation was performed and the data points were plotted in the space defined by the first two t-SNE variables (Figure 7). Dots that plotted close to each other demonstrate feature similarity. Under the t-SNE plot, the K-means alteration clusters show solid classification from each cluster. In the same plot, the conventional visual alteration logging displays a high degree of overlap between domains, indicating that the visual logs are unable to precisely separate the actual alteration based on the mineralogy composition.

The automated alteration class from SCI mineralogy produces a higher structured classification under the t-SNE plot, however, the K-means alteration provides several enhancements especially in the sub-domaining of the advanced argillic alteration into three different clusters (alunite-advanced argillic, pyrophyllite-advanced argillic, and dickite-advanced argillic) and phyllic alteration into two different clusters (muscovite-phyllic and illite-intermediate argillic).

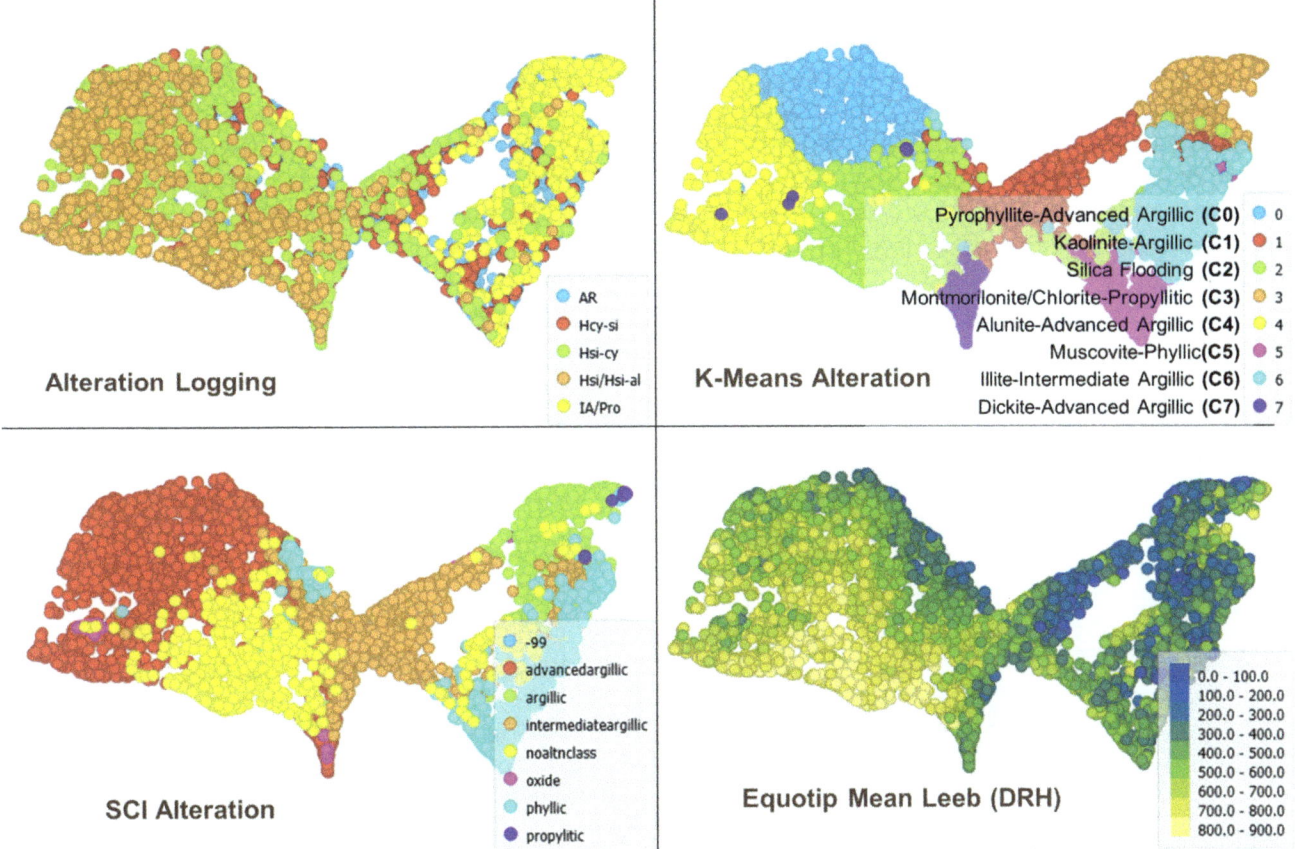

FIG 7 – t-SNE plot of Alteration Logging, K-means Alteration, SCI Alteration and Equotip Mean Leeb (DRH).

Figure 8 displays the average mineral compositions of samples in each cluster and the summary 'Mineral Assemblage' interpreted for each cluster (the silicate alteration facies). Each cluster has a characteristic clay mineral. These clay minerals seem to be proven as the main signals for the K-means clustering. On the other hand, the aspectral component is present to some level in almost all clusters. This condition was anticipated as aspectral can be present as several minerals including silica and feldspars.

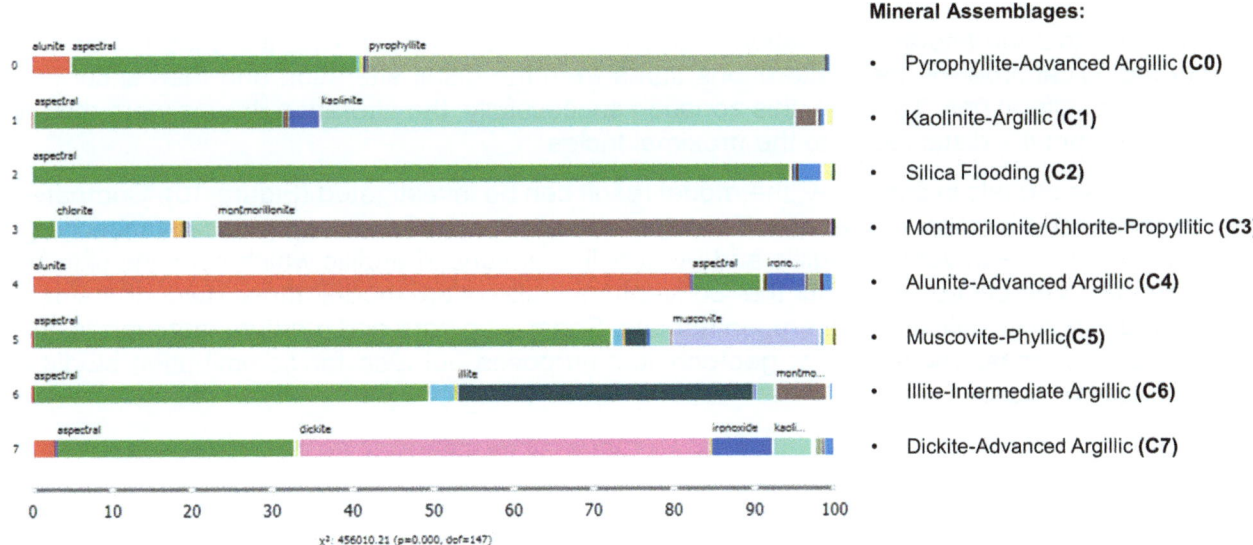

FIG 8 – Box plot showing mineral compositions within each cluster.

Figure 9 shows box plots of mean Leeb value for each K-means alteration domain. The K-means alteration domains clearly display hardness variation. Commonly, silica flooding and alunite-advanced argillic are the hardest materials in this deposit, followed by pyrophyllite-advanced argillic and dickite-advanced argillic. The soft material is commonly associated with non-advanced argillic alteration such as kaolinite-argillic, illite-intermediate argillic, muscovite-phyllic and montmorillonite/chlorite-propyllitic. Compared to the conventional visual logging data, the K-means alteration is more effective at separating hard and medium materials while conventional alteration logging shows higher overlaps between hard and medium materials. Although there are still some overlaps between medium and soft materials, the K-means alteration classification demonstrates a significant improvement from the visual logging data.

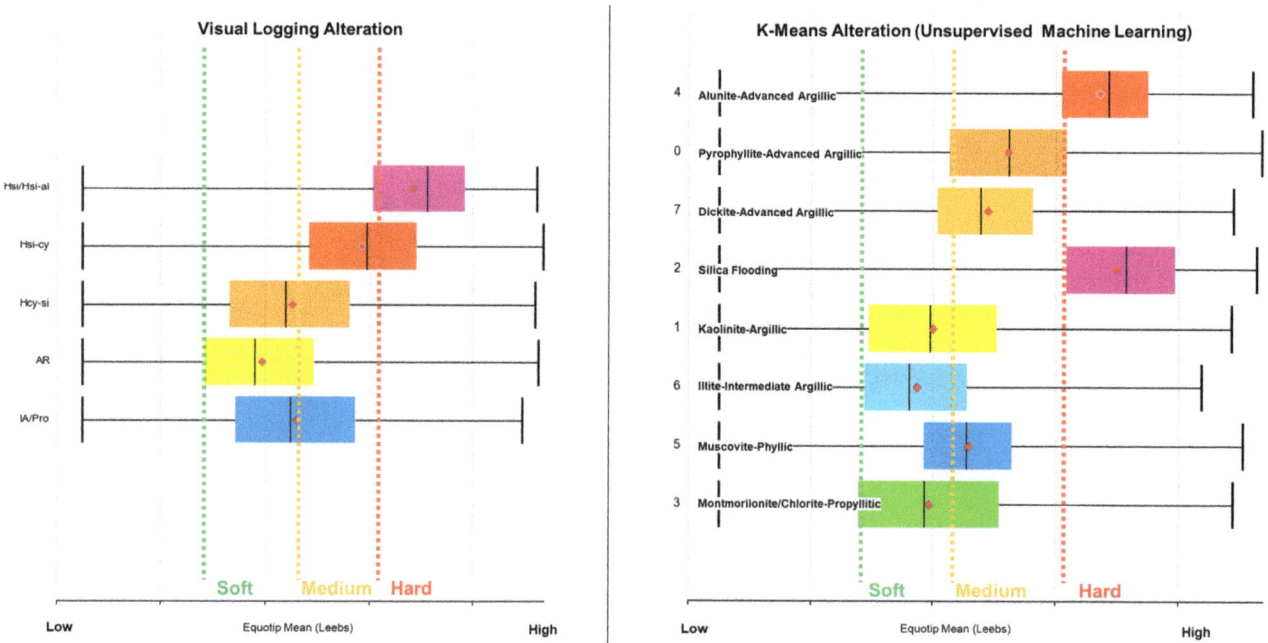

FIG 9 – Boxplot of Equotip mean (Leeb) versus Visual Alteration and K-means Alteration classification.

To examine the classifications in three dimensions (Cartesian coordinates), implicit modelling using Leapfrog software was performed using visual alteration logging, SCI alteration class and K-means silicate alteration data. The data being used in this modelling covers almost the whole mineralisation area at Tujuh Bukit. The same Leapfrog parameters were applied to each of these variables during implicit modelling.

It is observed that the alteration zonation is concentric, naturally mimicking the shape of an onion. The chronological order of these alterations starts from the distal alteration and then is cut by the more proximal alterations. To model this zonation successfully, the alteration groups were modelled in sequence from the distal facies to the proximal facies.

By slicing these models in plan-view, the model result can be investigated (Figure 10). Compared to the conventional alteration model, the K-means alteration model can separate advanced argillic alteration into alunite-advanced argillic and pyrophyllite-advanced argillic which contains significant rock strength differences, while in the conventional alteration model those two domains are generalised in a single silica clay alteration domain. Separating hard and medium materials is critical not only for rock mass modelling for geotechnical purposes but also for comminution studies for metallurgical purposes.

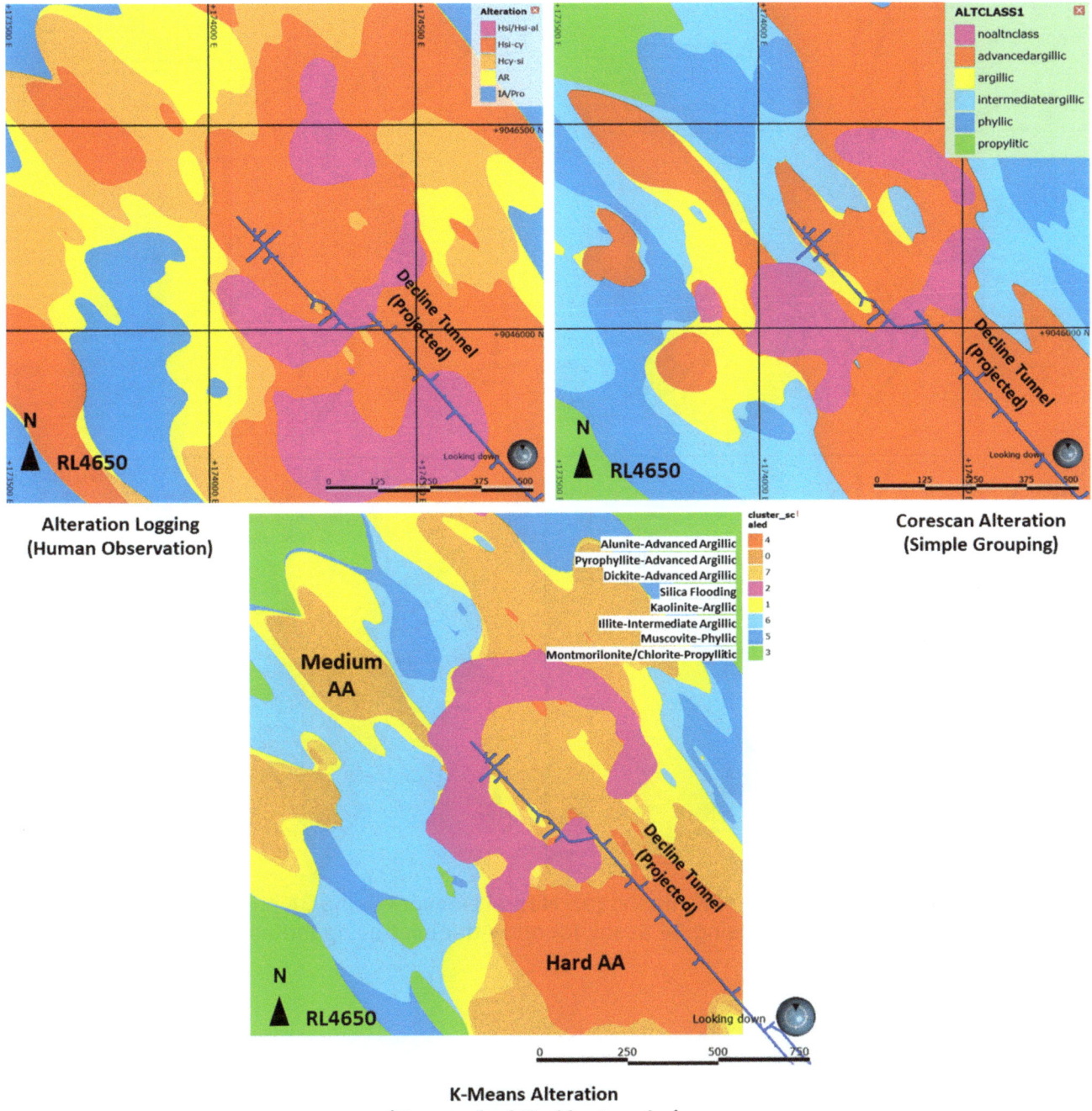

FIG 10 – Model comparison between the alteration logging, SCI alteration and K-means alteration.

Supervised Machine Learning (SML)

In this study, Leapfrog Geo software was employed to visualise the spatial distribution of machine learning results. Various machine learning techniques, namely K-means, Neural Network using 34 assay and Equotip (MA+E), Neural Network using 34 assay (MA), and AdaBoost using ten assay (LA) data, were plotted and visualised in both plan view (Figure 11) and elevation view (Figure 12). Through a comparative analysis with equal data support, it was determined that the results were highly satisfactory. Overall, no notable discrepancies were observed among the machine learning outcomes, particularly when modelling a wide range that can be effectively represented using an implicit model.

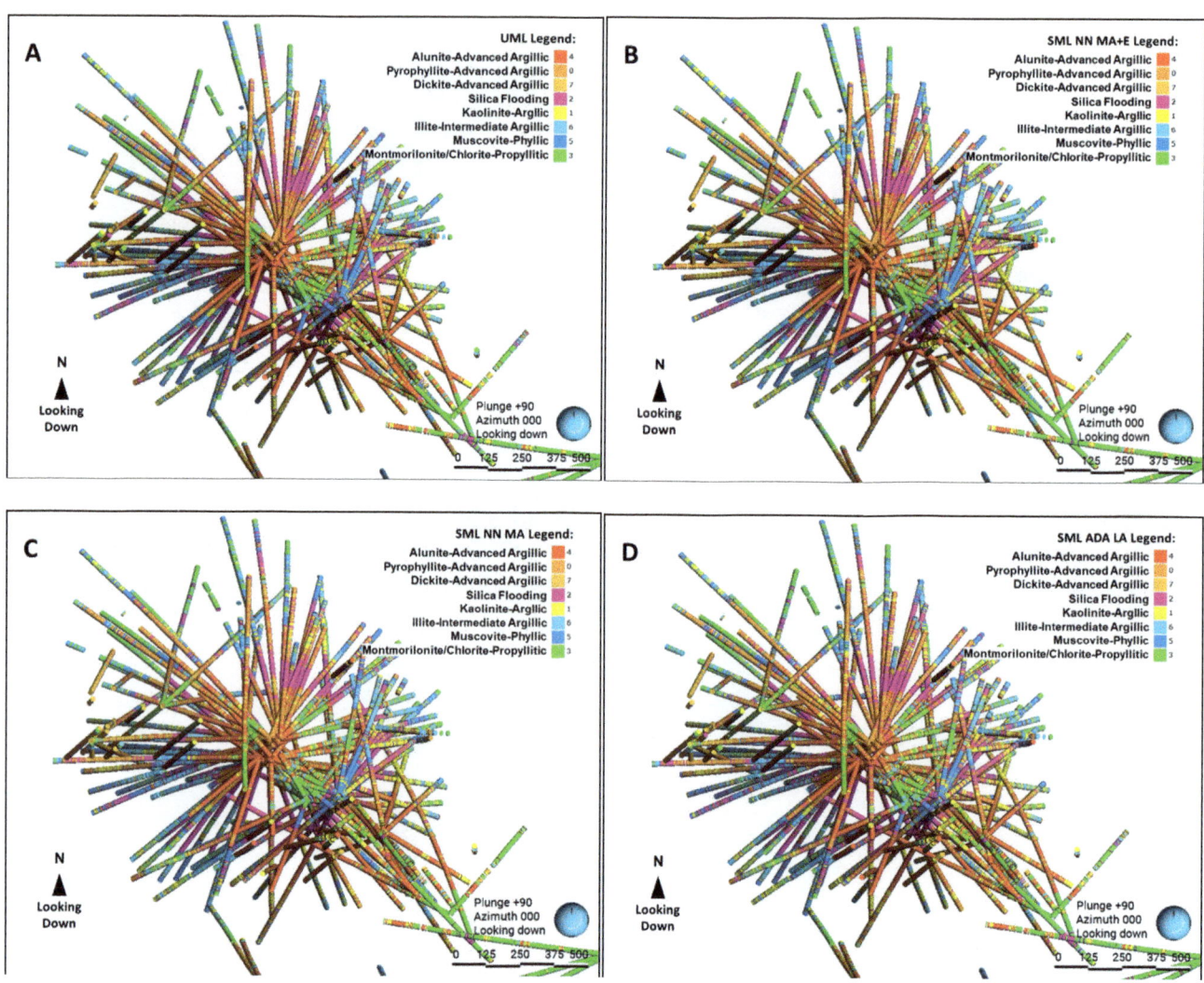

FIG 11 – Spatial visualisation of K-means (a), Neural Network using 34 assay and Equotip (b), Neural Network using 34 assay (c) and Adaboost using ten assay (d).

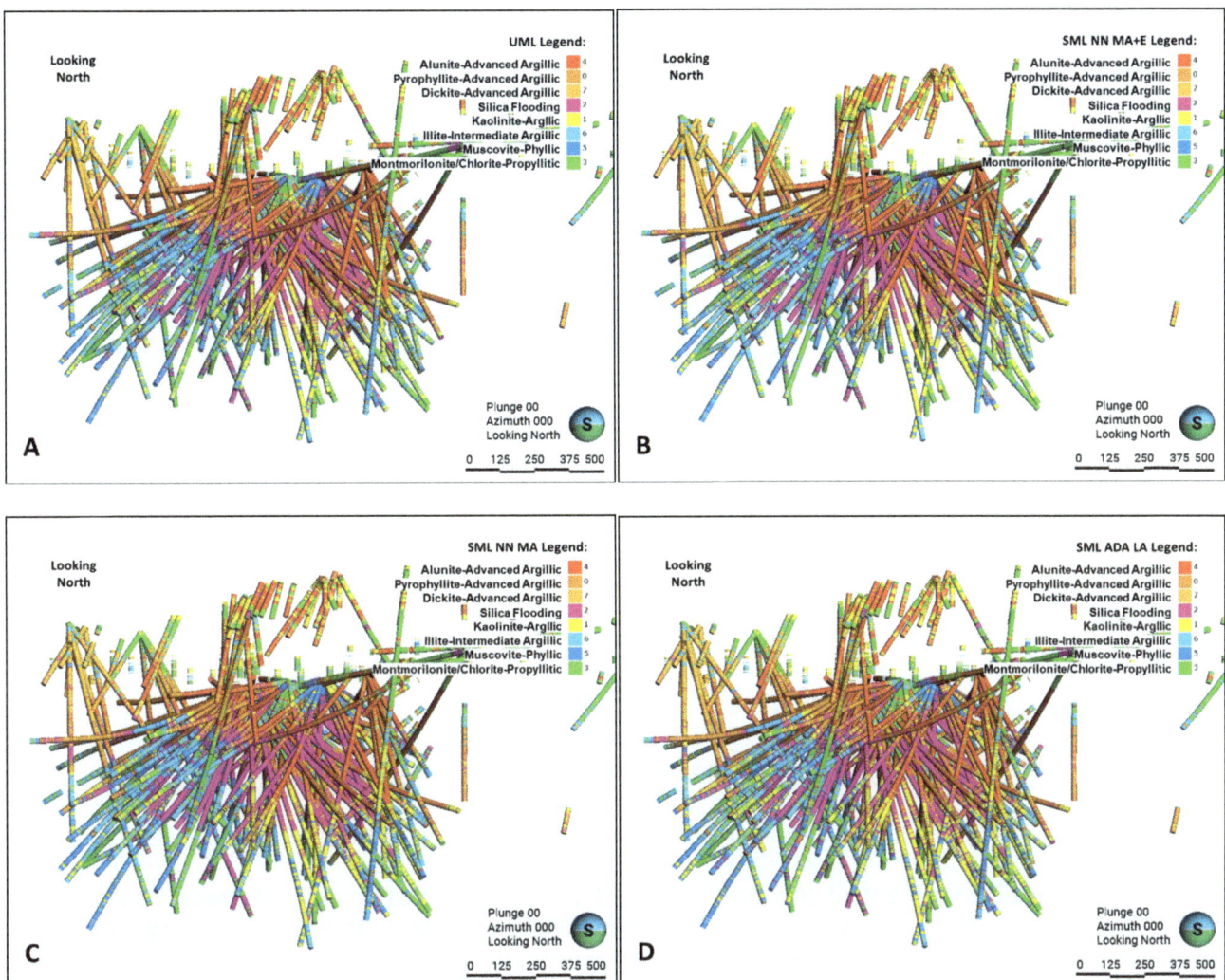

FIG 12 – Spatial visualisation of K-means (a), Neural Network using 34 assay and Equotip (b), Neural Network using 34 assay (c) and AdaBoost using ten assay (d).

During the modelling phase, all of the machine learning results were utilised to enhance the coverage in areas where there was a lack of SCI data. Although the results showed similarities, the machine learning validation revealed varying performance rates among them. Consequently, in intervals where multiple results overlapped, the need arose to identify the best-performing outcome. To address this, a merging strategy was employed to enhance the overall quality of the model (Figure 13).

The following data order was employed during the modelling phase:

1. The K-means result utilising SCI and Equotip data, which served as the target for the predictive model.

2. The Neural Network result using 34 assay and Equotip data, exhibiting the highest prediction performance.

3. The Neural Network result using 34 assay data, demonstrating the second highest prediction performance.

4. The AdaBoost result using ten assay data, presenting the third highest prediction performance.

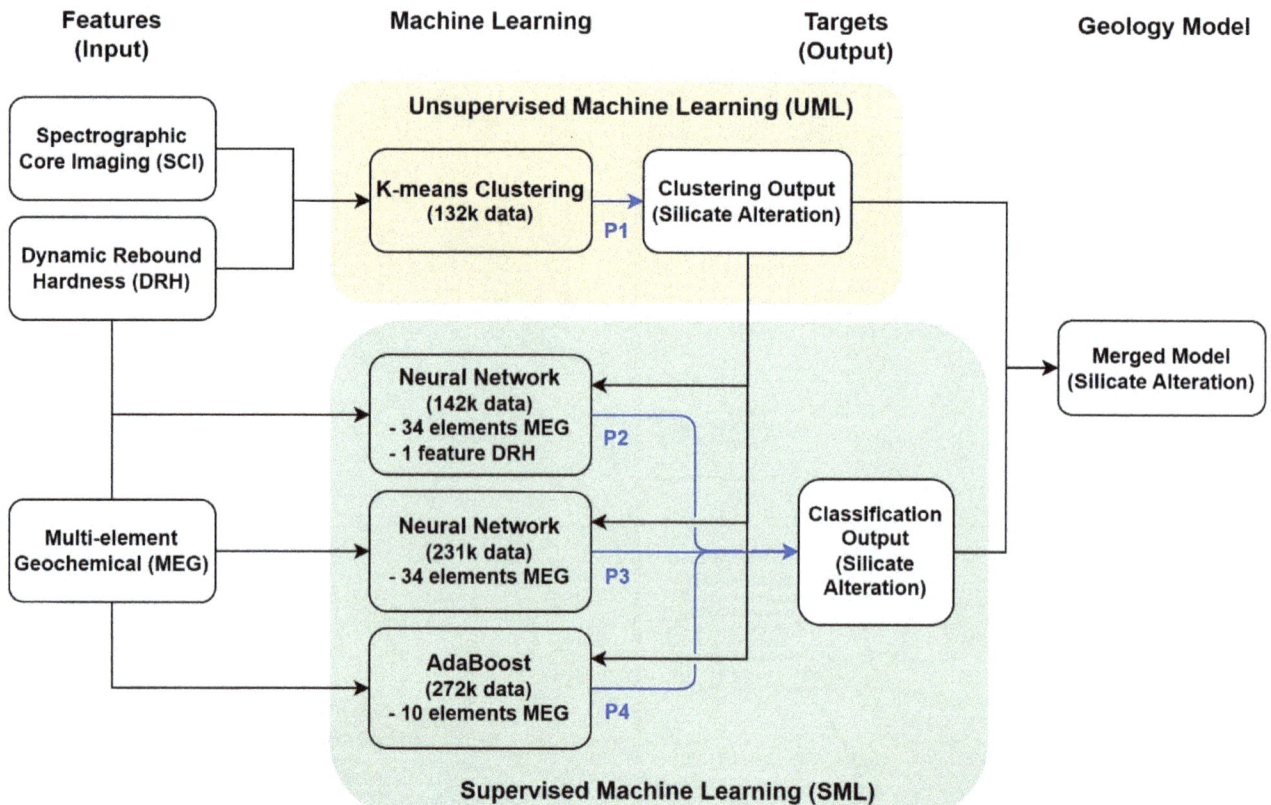

FIG 13 – Machine Learning data merging strategy.

This sequential arrangement of the data ensured a systematic approach to model construction, taking into account the varying strengths and capabilities of each machine learning technique. By prioritising the K-means result as the target and incorporating the superior prediction performance of the Neural Network results using 34 assay and Equotip data, the model benefited from the most accurate and reliable information available.

The inclusion of the Neural Network result using 34 assay data and the AdaBoost result using ten assay data further contributed to the comprehensive understanding of the domain. Despite their slightly lower prediction performances, these results added valuable insights and increased the overall robustness of the model.

By adhering to this data order, the modelling process followed a logical progression, maximising the effectiveness of each technique and enhancing the final model's predictive capabilities. This approach ensured that the model captured the most significant patterns and trends present in the data set, enabling informed decision-making and reliable analysis.

The consolidated data from all those results were utilised to construct the 3D model. Employing similar modelling parameters as the previous UML only data. The model exhibited noteworthy enhancements, particularly providing artificial data in regions lacking SCI data (Figure 14).

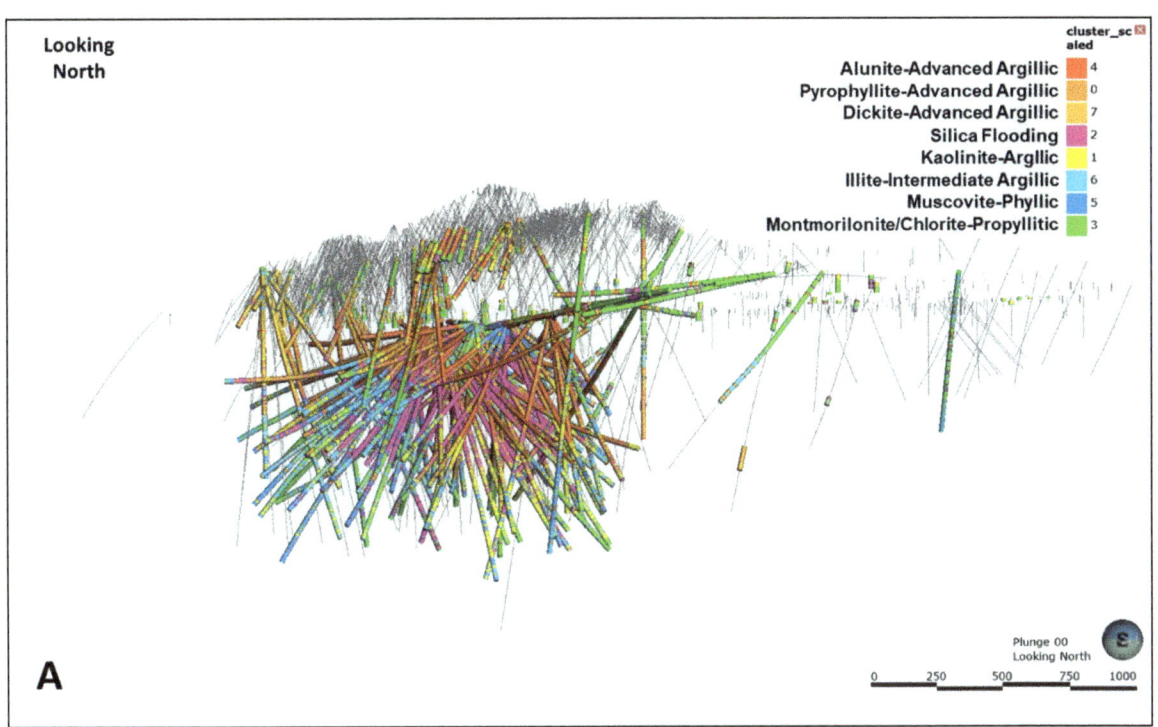

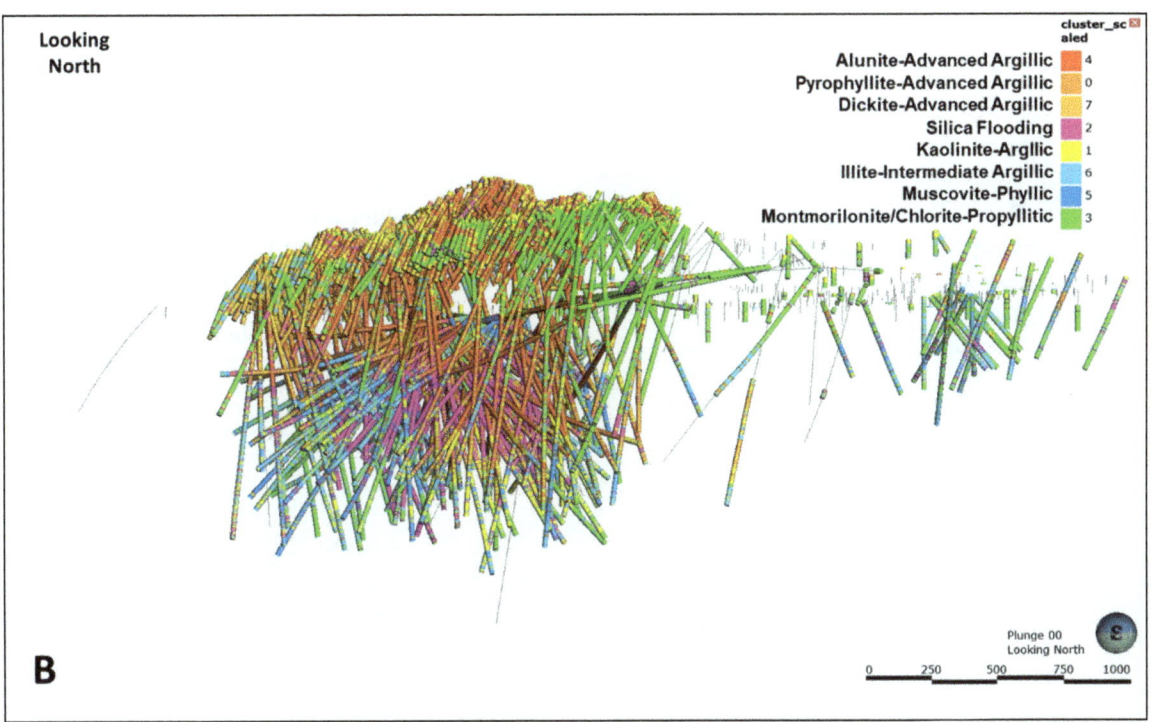

FIG 14 – K-means using SCI and Equotip data distribution (a). Merged data including K-means using SCI and Equotip, Neural Network using 34 assay and Equotip, Neural Network using 34 assay and AdaBoost using ten assay (b).

An examination of the modelling outcomes in both the plan view at RL5080 and the section view confirms a noteworthy enhancement in the model's definition, attributed to the availability of varied data support (Figure 15). Previously, limited SCI data at this elevation level imposed constraints on the implicit modelling performance, necessitating the simplification of the model. However, the current data set, which encompasses prediction results, boasts a significantly higher data density, leading to improved implicit modelling capabilities.

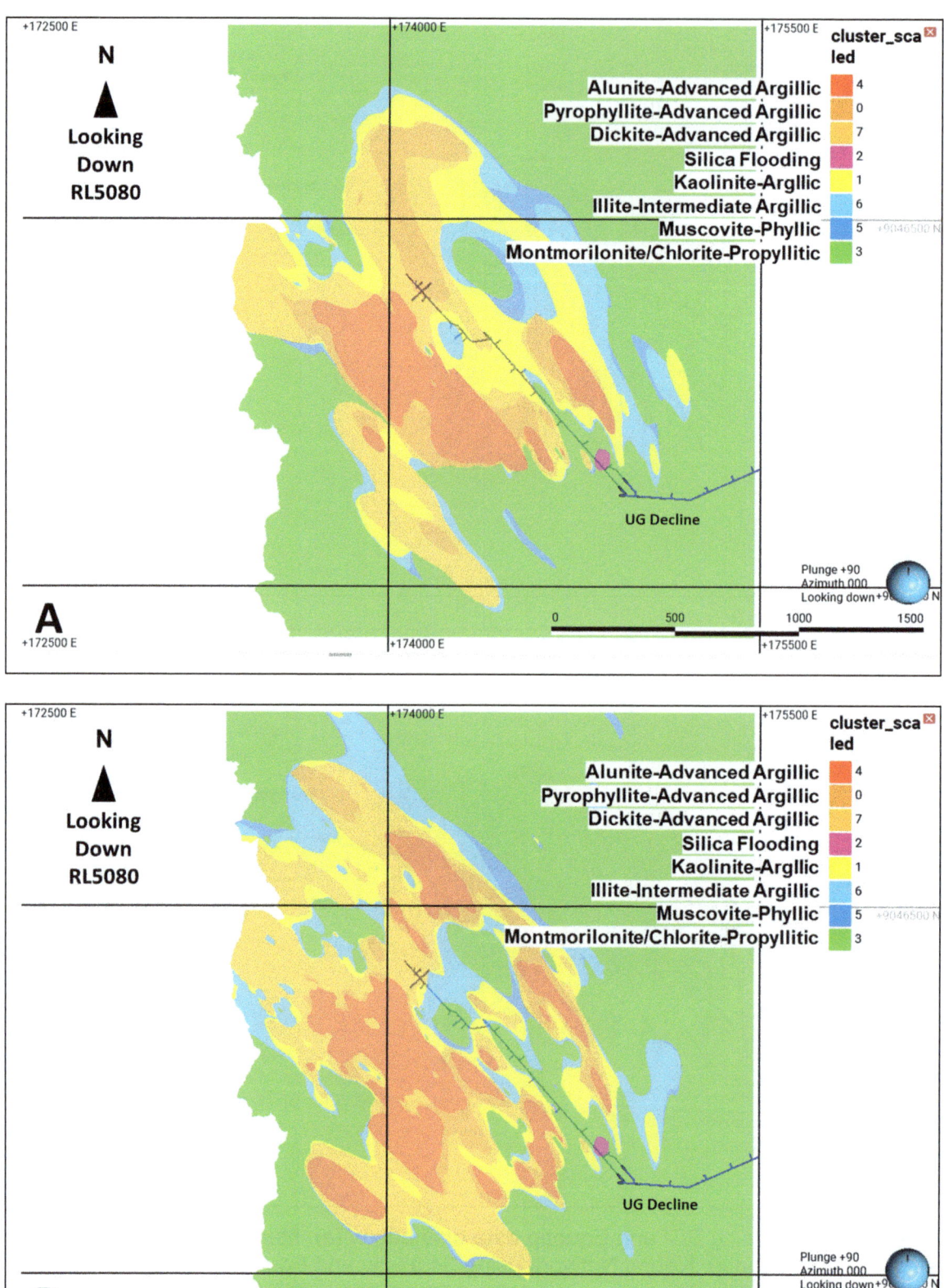

FIG 15 – Simplified alteration model due to limited data (a). Higher definition model based on additional artificial data from predictive machine learning model (b).

Similarly, the investigation of the model result in the section view also yielded similar findings (Figure 16). The model exhibited a remarkable improvement in capturing both the porphyry (deep) and high sulfidation (shallow) levels. This signifies the significant advancements achieved in accurately representing and characterising these distinct geological features within the model.

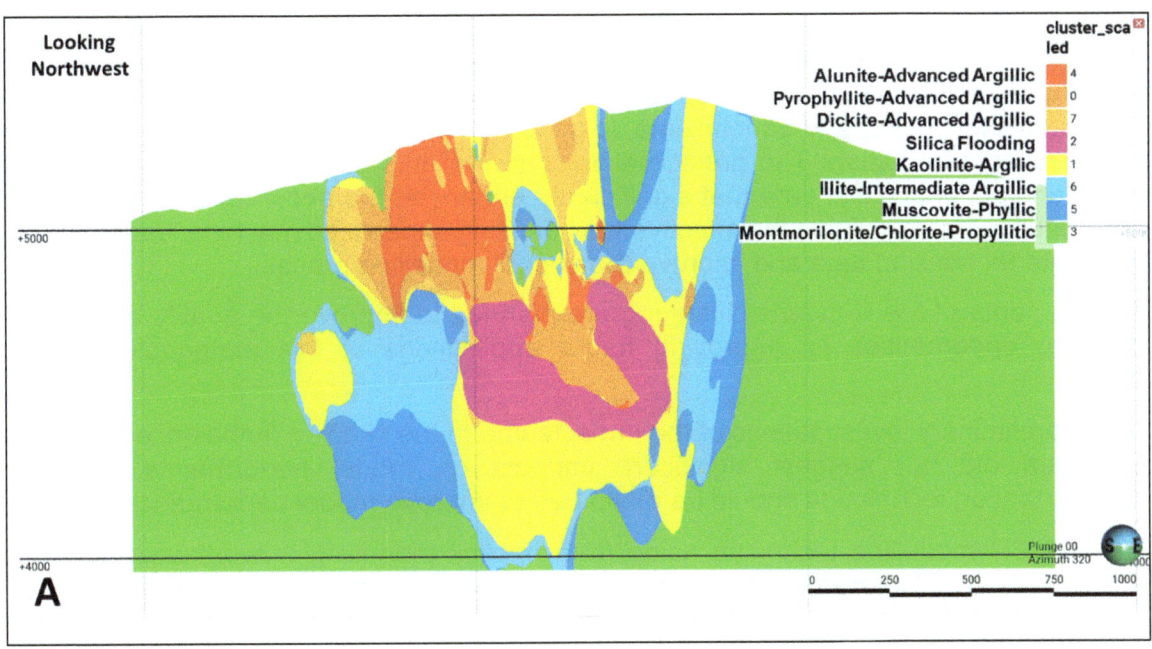

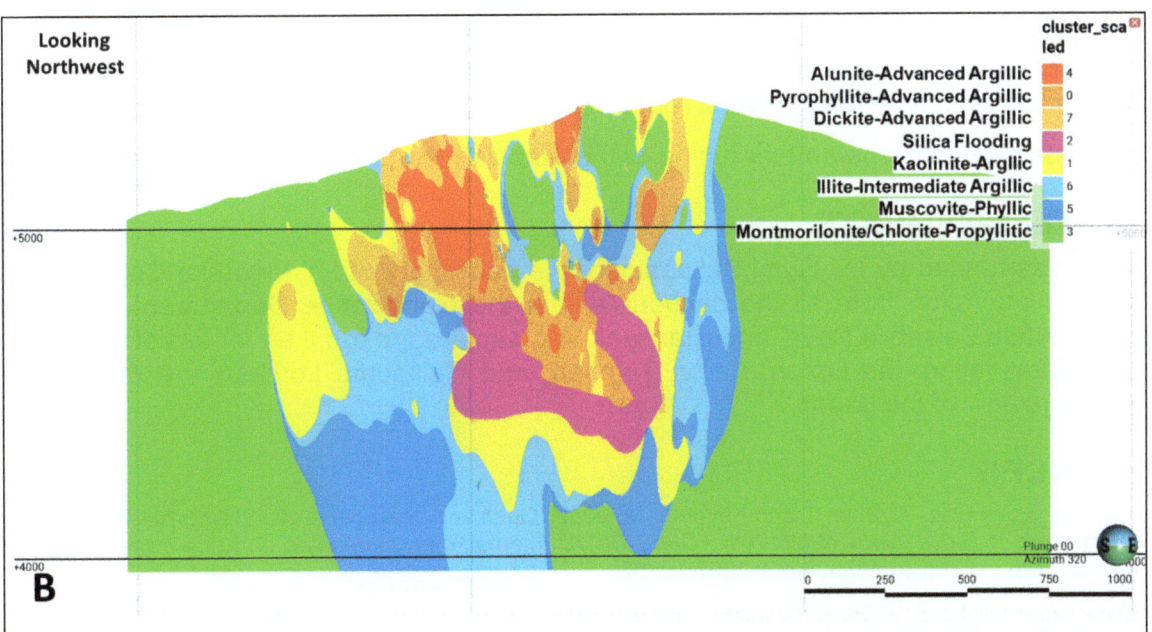

FIG 16 – Simplified alteration model due to limited data (a). Higher definition model based on additional artificial data from predictive machine learning model (b).

CONCLUSION

Using the K-means algorithm to classify samples into clusters based on spectrographic core imaging (Corescan) and dynamic rebound hardness (Equotip) is useful for improving alteration classification. Emphasising the quantitative approach, it significantly reduces the potential bias and uncertainties from qualitative visual logging by geologists.

From a practical point of view, based on high-quality data sets such as Corescan (SCI) and Equotip (DRH) data, machine learning-based domaining would be beneficial to provide accurate classification from data populations. It can eliminate significant potential human bias and uncertainty from conventional geological logging or interpretation. However, in very high quality data, K-means can generate a very high number of clusters which might be too complex for engineering purposes. Therefore, combining machine learning and technical understanding is critical to decide the most optimal number of clusters that will be useful during mining and metallurgical operation.

The K-means alteration (UML) has a limitation due to inadequate coverage of Corescan (SCI) data on historical drill holes. However, it was discovered that many drill holes lacking Corescan (SCI) data possessed assay or Equotip data. To address this data deficiency, supervised machine learning was employed, combining assay and Equotip data, to generate artificial K-means alteration data for areas without Corescan information. The predictive outcomes of this approach were highly successful, producing substantial artificial data that significantly enhanced the accuracy and quality of the 3D modelling results. This highlights the effectiveness of using supervised machine learning techniques to compensate for missing data and improve the overall modelling process.

Future refinements of the K-means clustering are planned to improve the quality of its domaining. The following opportunities for improving future applications of this methodology have been identified:

- While preliminary investigations of the impacts due to the skewed (imbalanced) distribution of the data did not suggest significant impacts on model performance, different data transformation techniques should be examined to improve model performance.

- Another 3D modelling technique aside from implicit modelling is worthwhile to explore for enhancing spatial interpolation such as neural network-based block modelling.

- The analysis has only been applied to records where Equotip_mean is not a missing value. The Equotip is not able to produce valid measurements below 200 Leeb. Below 200 Leeb, the material is very soft rock or soil. To apply the classification algorithm to all samples that have Corescan data, it will be necessary to select an average Leeb value less than or equal to 200 for the very soft samples. This will require some experimentation.

- More detailed hyperparameter tuning of the machine learning model could be conducted.

ACKNOWLEDGEMENTS

This study was funded by PT Bumi Suksesindo, the subsidiary of PT Merdeka Copper Gold. This work is a collaboration from all the sections within the Geoscience Department. Extivonus Kiki Fransiskus, Nada Salsabila Deva, and Riansyah Widisaputra are thanked for assisting with the core logging and validation. Ian Lipton, Aaron Tomsett, and Guillaume Lorilleux are thanked for their constructive review of the paper.

REFERENCES

Akiba, T, Sano, S, Yanase, T, Ohta, T and Koyama, M, 2019. Optuna: A Next-generation Hyperparameter Optimization Framework, in *Proceedings of the 25th ACM SIGKDD International Conference on Knowledge Discovery & Data Mining*, 10 p. doi.org/10.48550/arXiv.1907.10902.

Arthur, D and Vassilvitskii, S, 2006. *K-means++: The advantages of careful seeding* (Stanford).

Celebi, M E, Kingravi, H A and Vela, P A, 2013. A comparative study of efficient initialization methods for the K-means clustering algorithm, *Expert systems with applications*, 40(1):200–210.

Demsar, J, Curk, T, Erjavec, A, Gorup, C, Hocevar, T, Milutinovic, M, Mozina, M, Polajnar, M, Toplak, M, Staric, A, Stajdohar, M, Umek, L, Zagar, L, Zbontar, J, Zitnik, M and Zupan, B, 2013. Orange: Data Mining Toolbox in Python, *Journal of Machine Learning Research*, 14(Aug):2349–2353.

Dramsch, J S, 2020. 70 years of machine learning in geoscience in review, *Advances in Geophysics*, 61:1–55. doi:10.1016/bs.agph.2020.08.002.

Kotsiantis, S B, 2007. *Supervised Machine Learning: A Review of Classification Techniques, Emerging Artificial Intelligence Applications in Computer Engineering* (IOS Press).

Krose, B, 1996. An introduction to neural networks, Faculty of Mathematics & Computer Science, University of Amsterdam.

Schapire, R E, 2013. Explaining AdaBoost, in *Empirical Inference* (eds: B Schölkopf, Z Luo and V Vovk), (Springer: Heidelberg). doi:10.1007/978-3-642-41136-6_5.

Turing, A M, 1950. *Computing machinery and intelligence*, Mind, LIX (236), pp. 433–460. doi:10.1093/mind/LIX.236.433.

Van der Maaten, L and Hinton, G, 2008. Visualizing Data using t-SNE, *Journal of Machine Learning Research*, 9:2579–2605.

Verdhan, V, 2021. *Models and Algorithms for Unlabelled Data* (Manning Publications).

Implementation of the Chrysos PhotonAssay™ Method at Fosterville Gold Mine, Victoria

S P Hitchman[1], S Simbolon[2], D J Symons[3], J P Hoare[4] and J B Carpenter[5]

1. Principal Geologist, Agnico Eagle Mines, East Bendigo Vic 3550.
   Email: simon.hitchman@agnicoeagle.com
2. Superintendent Resource Geologist, Agnico Eagle Mines, East Bendigo Vic 3550.
   Email: saut.simbolon@agnicoeagle.com
3. Resource Geologist, Fosterville Gold Mine, Fosterville Vic 3557.
   Email: damon.symons@agnicoeagle.com
4. Senior Mine Geologist, Fosterville Gold Mine, Fosterville Vic 3557.
   Email: jason.hoare@agnicoeagle.com
5. Senior Consultant, SRK Consulting, Brisbane Qld 4000. Email: jcarpenter@srk.com.au

EXTENDED ABSTRACT

The Fosterville Gold deposit is situated 20 km east-north-east of Bendigo, Victoria, at latitude 36°42'S and longitude 144°30'E and is hosted by Ordovician Greenschist metasediments within the Bendigo Tectonic Zone of the Palaeozoic Lachlan Fold belt (Mernagh, 2001). The metasediments at Fosterville were deformed by regional east–west compression producing upright, north–south trending chevron folds and steeply west-dipping reverse faults and complex splay faults, which were the focus of multiple gold mineralisation events. This includes both sulfide-hosted micron sized gold within disseminated arsenopyrite and pyrite, and visible-gold (coarse gold) mineralisation contained in quartz shear veins often with the presence of stibnite and sulfosalt assemblages.

Gold production from the Fosterville area occurred in three periods: discovery in 1894 and ensuing 15 years of early mining; from 1988 to 2003 reprocessing and heap leach oxide gold open pit operations; and the current project, which from 2005 processed sulfide ores through a BIOX™ bacterial oxidation plant, with ore initially from open pits and from 2008 mostly from underground sources.

Up until 2015 the significant known gold mineralisation at Fosterville was exclusively micron-sized gold within sulfides (arsenopyrite and pyrite) disseminated in turbidite wall rock. However, very high-grade coarse gold in quartz-rich shear vein orebodies (eg Eagle and Swan) were discovered below 850 m depths. A gravity circuit was subsequently added to the Fosterville processing plant to improve the recovery of coarse gold.

The presence of nuggety gold caused a challenge in obtaining acceptably accurate assays due to well-known and documented issues in other nuggety gold terrains (eg Ingammells and Switzer, 1973; Gy, 2012; Pitard, 2019) and led the company to consider alternatives to fire assay (FA), such as the Chrysos PhotonAssay™ (PA) method, which uses a larger sample mass.

The PA method was trialled as a replacement of routine FA. Success would be achieved if the PA method:

- produced a total assay for gold (as opposed to a partial assay)
- was analytically accurate and precise
- reasonably free from interferences from matrix and gangue minerals
- have an equivalent or better assay turnaround time.

The above high standards and expectations were set, otherwise the PA method would be considered inferior to the FA method.

The PA technology, a non-destructive assay method, uses high energy X-Rays to irradiate jarred (nominal mass of 400 to 500 g) samples, and has a 0.03 to 35 000 ppm Au assay range (Tickner, Preston and Treasure, 2018). The method has operational and environmental benefits compared to FA; ie is less energy intensive (Frost and Sullivan, 2022) and does not require the use of lead (Pb) flux (Table 1). Any use of lead requires careful management for the health of laboratory personnel and the environment.

TABLE 1

Comparison of gold assay methods.

Assay method	Sample assay size (g)	Benefits	Shortfalls
Fire Assay (FA)	25 to 50 g aliquots	Trusted, well proven method, cheap cost	Small sample size unsuitable for coarse gold
Screen Fire Assay (SFA)	25 to 50 g aliquots	Large (1 kg) samples, proportionate pulp (fine) to mesh (nuggety) measure	High cost per sample
PhotonAssay™ (PA)	400 to 500 g jar	Large sample size, compared to FA and SFA, no lead (Pb) and less energy	New method, not widely used

Fosterville conducted comparison test work of the PA method against FA and screen fire assay (SFA) methods as early as 2017, and in 2019, using pulverised samples (-75 µm pulp). A paired data set of 1450 samples confirmed that the PA method correlated well with traditional FA. However, to gain an extensive Fosterville data set of the X-ray technology, a PA unit was installed in Bendigo, during the Covid-19 pandemic. Since commissioning of the unit in October 2020, benefits have included operational experience and allowing the mirroring QA/QC protocols in use for FA. In addition, all new Fosterville diamond drill core, and production (underground faces and stopes) samples were assayed by FA and then by PA.

Rigorous statistical evaluations of paired data, ie over 45 000 samples, confirmed the interchangeability of results obtained from FA and PA for all mineralisation styles, rock types and alteration mineral assemblages. However, the data shows a bias between the two methods at high-grades (Figure 1), and to determine which of the two is more reliable (true grade), a SFA comparison data set was assessed.

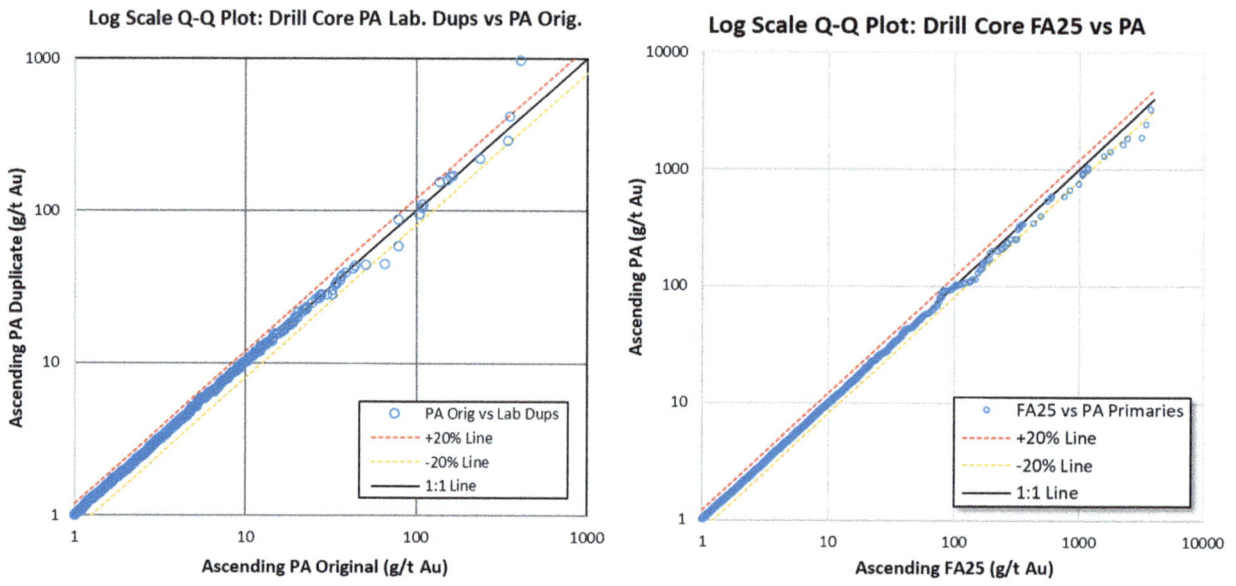

FIG 1 – Log Scale Q-Q plots showing PA gold results of Original versus laboratory duplicates (left) and FA versus PA (right). Note apparent bias of FA at high-grades.

Screen fire assay comparison confirmed that PA data is more similar to SFA than FA and that for high-grade samples FA biases high compared to SFA (Figure 2).

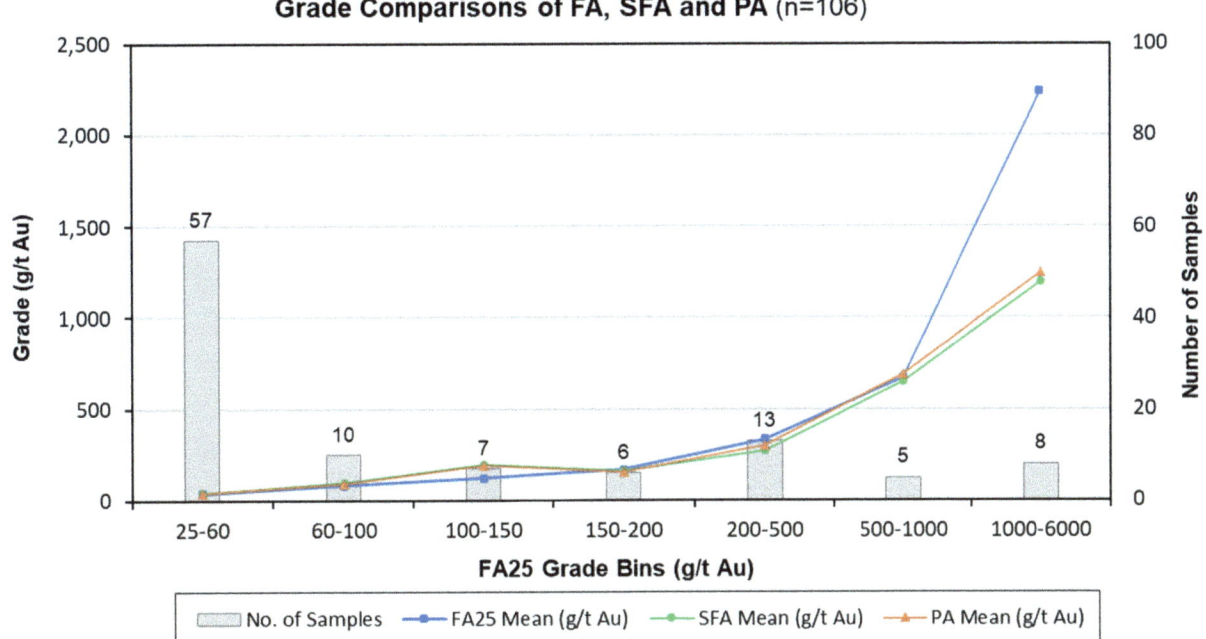

FIG 2 – Comparison chart of FA, SFA and PA sample gold results. Note that above the 200 g/t Au the FA method biases high compared to SFA and PA. There is strong agreement between SFA and PA across the entire grade range.

On gaining sufficient paired PA and FA data, mineral resource work was completed in November 2021 on eight mineralised zones (domains); categorised as low, medium and high-grade domains (Table 2). Statistical comparison of FA and PA 1 m composites for low, medium and high-grade domains returned lower grade and lower coefficient of variance in each case.

TABLE 2

Summary statistical comparison of 1 m composite data for PA and FA methods.

Domain Type (Grade)	Assay Type	No of Sample	Min Grade (g/t Au)	Max Grade (g/t Au)	Median Grade (g/t Au)	Mean Grade (g/t Au)	SD (g/t Au)	Coeff of variation
High	FA	173	0.01	1419.45	6.44	41.56	142.22	3.42
	PA	173	0.01	844.77	6.35	34.49	103.71	3.01
Medium	FA	849	0.01	1070.97	3.78	12.98	62.40	4.81
	PA	849	0.02	994.58	3.84	12.34	52.89	4.29
Low	FA	261	0.01	153.56	3.40	5.69	11.37	2.00
	PA	261	0.02	156.54	3.54	6.22	13.20	1.95

Variogram modelling for the three grade domains showed a similar underlying variogram structure, although, the variogram nugget was less for PA, compared to FA (Figure 3).

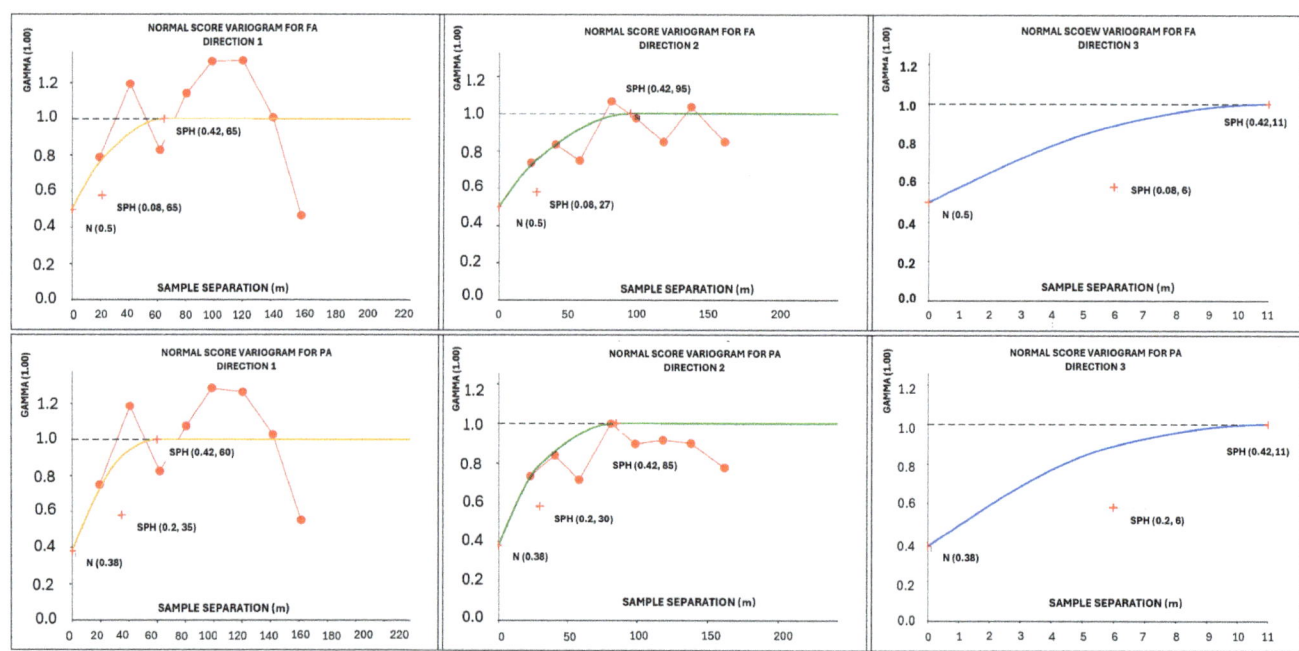

FIG 3 – Domain 61 (high-grade domain) omnidirectional normal scores variograms for FA (top) and PA (bottom). Note lower nugget for PA data compared to FA.

Mineral Resource estimation, using paired FA and PA data showed that comparisons of the mean grade of low, medium and high-grade domains are less than 3 per cent (Table 2). The estimation variance using PA data has shown improvement compared to FA data, as shown by the lower kriging variance using PA compared to FA (Figure 3). Global tonnes and grade plots show that PA returned similar values to FA across all grade ranges for each domain category (Figure 4).

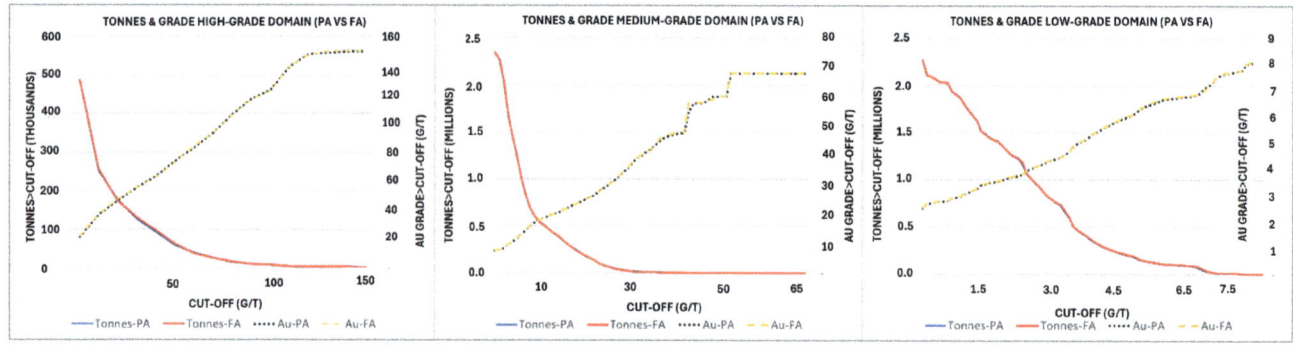

FIG 4 – Tonnage grade charts of high, medium and low-grade gold domains showing strong similarity of PA with FA.

An independent external review of the test work and resource modelling confirmed Agnico Eagle's findings.

The results of the PA study were considered sufficient to support a transition from FA to PA, and a change management process was initiated at Fosterville to further assess the PA technology. The process comprised four sections:

1. Reasons for change.
2. Risk assessments.
3. Expected outcomes.
4. Implementation.

Site approval was given to switch from FA to PA, and since June 2022, PA has been used for mineral resource estimation, and mine production decisions.

During and post the PA study, regular training and information sessions for on-site geologists ensured appropriate QA/QC protocols for PA were followed.

In November 2022, further optimisation test work occurred, with investigation of using a coarser grind (-2 mm crush) for potential reductions of sample turnaround time, energy, dust and cost. Data from five Fosterville mine production source areas indicate that -2 mm crush material could be used for production (stope and face) samples for sulfide ores, but not coarse gold ores (Figure 5). Of concern for coarse gold samples were occurrences of -2 mm crush samples returning average composite sample gold grades more than 10 per cent variance from pulp results.

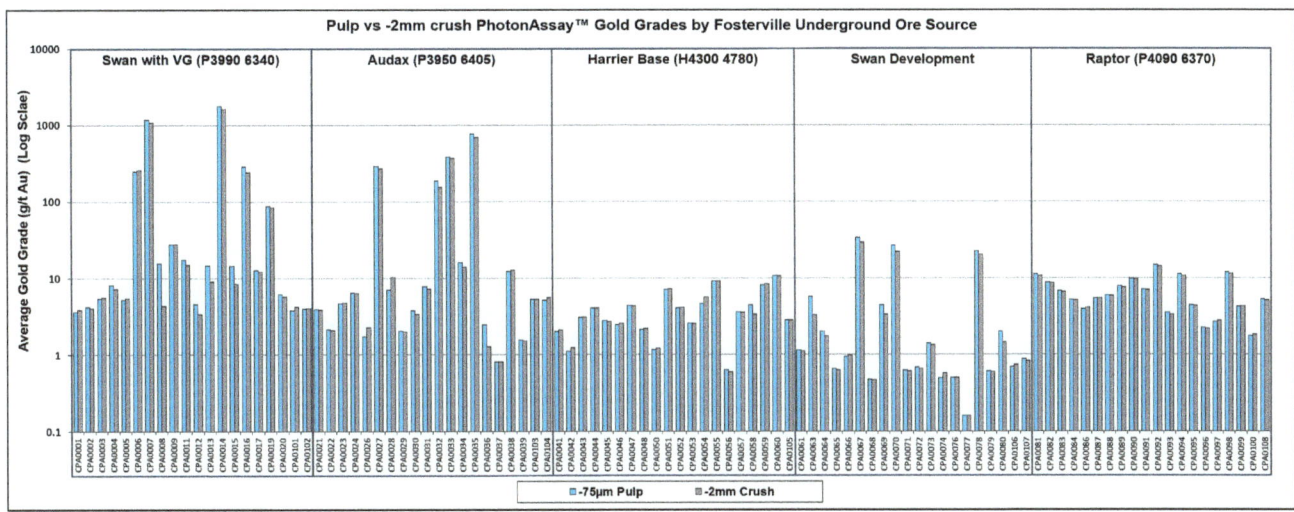

FIG 5 – PhotonAssay™ gold results comparison of -2 mm crushed material with pulp (-75 µm) for five underground Fosterville ore sources. Note higher variance for Swan and Audax visible-gold ore zones.

In conclusion, the key to the successful adoption of PA from FA for the Fosterville geological samples was through an extensive and rigorous trial, and a collaborative change management process with regular communication across all relevant stakeholders.

REFERENCES

Frost and Sullivan, 2022. Market Report, Gold Mining and Geochemistry Services, in *Chrysos Corporation Prospectus*, 14 April 2022, pp 197–212.

Gy, P, 2012. *Sampling of Particulate Materials Theory and Practice*, 450 p (Elsevier).

Ingammells, C O and Switzer, P, 1973. A proposed sampling constant for use in geochemical analysis, *Talanta*, 20(6:547–568.

Mernagh, T P, 2001. A fluid inclusion study of the Fosterville Mine; a turbidite-hosted gold field in the western Lachlan fold belt, Victoria, Australia, *Chemical geology*, 173:91–106.

Pitard, F F, 2019. *Theory of Sampling and Sampling Practice*, third edition, 726 p (Chapman and Hall/CRC).

Tickner, J, Preston, R and Treasure, D, 2018. Development and operation of PhotonAssay System for rapid analysis of gold in mineral ores, in *Proceedings of the ALTA 2018 Nickel-Cobalt-Copper-Uranium-REE-Lithium and Gold-PM Conference 2018*, pp 50–60.

Using representative elemental analysis on conveyed flows to complement grade control – a win-win for geologists and processors

H Kurth[1]

1. FAusIMM, Minerals Consultant and Chief Marketing Officer, Scantech International Pty Ltd, Adelaide SA 5038. Email: h.kurth@scantech.com.au

ABSTRACT

Mine geologists have limited control over ore quality dispatched to the mill. Mine designs may not align with spatial distribution of the ore. The mined volume often includes dilution by uneconomic internal waste or peripheral waste due to complex ore/waste boundaries. Blasting can cause ore and waste overbreak or underbreak. Overlying waste is expected to collapse and dilute the ore in caving operations as broken rock is drawn down during ore extraction. The effect is that the mine geologist (or anyone else) cannot predict with certainty the quality and variability of the ore sent to the mill.

Mines are successfully employing representative, continuous, real time high performance measurement technologies to determine multi-elemental content on primary crushed rock in conveyed flows. Bulk diversion systems separate small parcels (of a few tonnes) from the feed flow. Some parcels may be uneconomic to process, high in deleterious content, or be an ore type problematic to process. Different ores can be stockpiled separately and then blended into the process plant to reduce quality variability to optimise metal recoveries. High performance Prompt Gamma Neutron Activation Analysis (PGNAA) is used in many commodities (iron ore, manganese, copper, gold, zinc-lead, nickel, lithium, phosphate rock, bauxite, diamonds, chromium etc) to measure and control material quality in real time. Digitalisation of the conveyed flow provides useful data for ore reconciliation, metal accounting and feed forward control for metallurgists. Mines that have benefitted most through this technology have integrated geology, mining and processing teams to optimise quality control of process feed. Benefits include reduced processing of waste, higher and more consistent feed quality, increased metal recoveries, reduced fine tailings generation and lower carbon emissions per tonne of product. The paper details examples where the technology is complementing grade control in various commodities.

INTRODUCTION

Mined ore is often diluted because:

- mine designs may not align with spatial distribution of the ore due to practical mining limitations
- complex ore/waste boundaries or internal waste occur and selective mining is impractical
- blasting causes ore and waste overbreak or underbreak and mixing
- overlying waste rock collapses and dilutes ore in underground caving operations.

Reconciling ore and waste removed from each blast against a planned outcome is complicated by difficulties in quantifying the composition of the material removed.

The mine geologist (or grade controller) cannot predict with certainty the quality and variability of the material sent to the mill or the waste dump. Mine geologists therefore have limited control over ore quality produced. Grade control activities may include:

- pre-blast mapping and drill hole sampling to adjust a blast designs where possible
- defining ore and waste dig zones after blasting via observations of the rubble pile surface
- assessing ore and waste by shovel or truck inspections
- relying on multiple mining faces or sources to blend ore types to minimise variability of material quality sent to the mill. This strategy may not minimise dilution or optimise grade.

The tools available to the grade controller are limited and subjective decisions must be made which may result in sub-optimal ore and waste classification. Sampling blasted material (taking grab samples) is not representative and not ideal as a basis for ore/waste decisions. Ore quality becomes the responsibility of others after it leaves the mine.

The desire to increase confidence in ore and waste allocation has led to sensing technology adoption at some mines. Each technology has limitations and surface measurement systems are not representative as a large sampling error component exists which reduces effectiveness of ore/waste decisions. For technologies to be representative they should allow penetrative measurement and not bias results to different minerals, matrices, particle sizes, etc. Magnetic resonance has proven effective for a limited suite of minerals but each system can only be configured to measure one mineral. Prompt gamma neutron activation analysis (PGNAA) has proven representative and effective on primary crushed rock in conveyed flows. The paper details examples where this technology is complementing grade control in various commodities.

EXTENDING GRADE CONTROL

After the ore and waste leave the mine, waste is sent to a dump and ore is commonly sent to stockpiles or directly fed into a crusher to create a crushed ore stockpile (COS) which feeds a grinding mill. Ore effectively passes to the next silo at the site: the process plant. Different ores may be stockpiled separately and blended into mill feed. This is grade control performed by the process operators primarily for process benefits. Reducing variability of feed material improves process performance including reduced energy fluctuations, reduced variability in reagent demand, reduced wear and tear on equipment, fewer and less problematic process upsets, etc. Consistency in feed tonnage and quality results in optimised process performance and provides operators the best opportunity to increase metal recoveries, reduce plant downtime and minimise processing costs.

PGNAA

Representative, continuous, real time high performance measurement through PGNAA determines multi-elemental content on primary crushed rock in conveyed flows. High performance prompt gamma neutron activation analysis (PGNAA) has been used successfully in the minerals sector for over 20 years in many commodities (iron ore, manganese, copper, gold, zinc-lead, nickel, lithium, phosphate rock, bauxite, diamonds, chromium etc) to measure and control material quality in real time. The technique is suitable for primary crushed rock and unaffected by particle size, belt speed, mineralogy, dust, layering and moisture content.

High performance PGNAA utilises a source of neutrons beneath the conveyed flow. Elemental nuclei in the conveyed material capture neutrons and release gamma energy responses unique to each element. These spectral responses are accumulated over a measurement period of 30 seconds to two minutes using an array of high specification detectors enabling the proportion of each element in a conveyed parcel to be determined and reported (Figure 1). High performance PGNAA is the most representative analysis available to the minerals industry measuring many elements from carbon onwards in the periodic table at high precisions over short measurement times to low levels of detection. Recent developments include direct measurement of gold in conveyed ore using GEOSCAN-GOLD (Balzan *et al*, 2022).

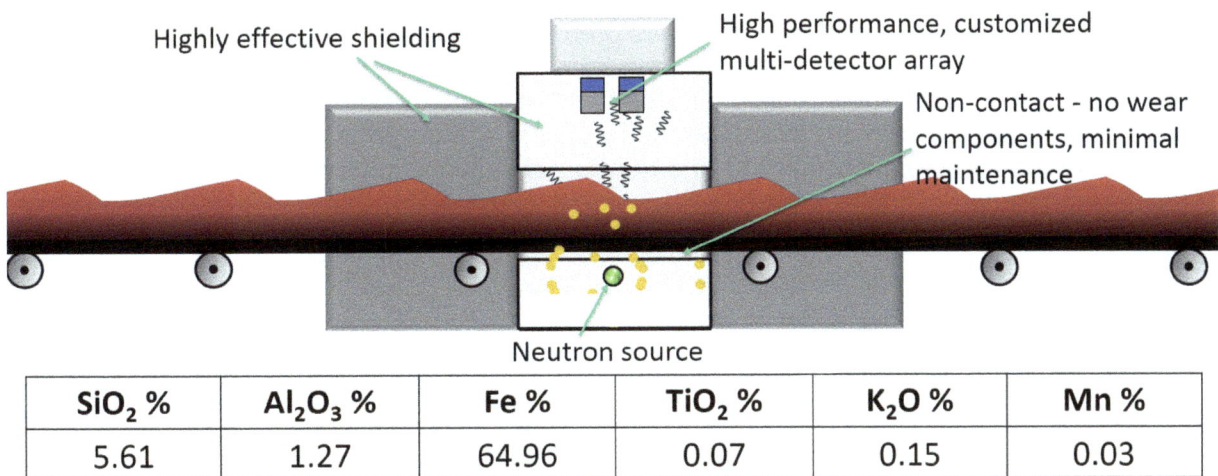

FIG 1 – Cross-section through GEOSCAN high specification PGNAA analyser showing main components and an example of elemental results for a two-minute measurement increment.

Digitalisation of the conveyed flow provides useful data for ore blending, bulk diversion, ore reconciliation, metal accounting and feed forward control for process operators. Mines that have benefitted most through this technology have integrated geology, mining and processing teams to optimise quality control of process feed. Benefits include reduced processing of waste, higher and more consistent feed quality, increased metal recoveries, reduced fine tailings generation and lower carbon emissions per tonne of product.

APPLICATIONS AND BENEFITS

High performance PGNAA has been successfully implemented in many applications and is customised to the data requirements for each. Measurement data from a single location is used for multiple concurrent uses. Some of these may involve instantaneous responses to the quality or in determining an average quality over a longer time frame, such as compositing multiple results, depending on the intended use of the data and the benefits available.

Bulk sensing for diversion (bulk sorting)

Conveyed material is measured at full production flows and unbiased measurement data provided to enable quality control. Shorter measurement times provide a higher level of selectivity, but it is proven that some technologies (not providing representative measurements) used for even shorter measurement times increase the amount of misallocated material. Shorter measurement times are only more beneficial if accompanied by high measurement precisions (Kurth, 2022; Scott *et al*, 2020). This highlights the need to customise the measurement solution to the application.

Iron Ore

At Assmang Khumani operations in South Africa iron ore quality on overland conveyors between two mines and a beneficiation plant is measured and parcels are diverted as needed (Matthews and Du Toit, 2011). Approximately one third of their average annual mine output is direct shipping quality and is diverted to bypass beneficiation saving US $5 M/annum in jig plant costs. Utilising analysers for the bypass application was considered in the original plant design so a smaller plant was built with the same planned output capacity at a significantly lower capital cost. The initial and ongoing savings have benefitted the operation for over 15 years. Jig plant feed consists of only the material that requires beneficiation.

Copper

Copper responds particularly well to high performance GEOSCAN and precisions of 0.02 per cent Cu have been achieved at multiple sites for 30 second measurement increments such as New Afton in British Columbia (Nadolski *et al*, 2018). Some ore types are considered relatively homogeneous (eg porphyry copper) and therefore potential for upgrading through bulk sorting has been expected by mine geologists as being quite low. The basis for these assumptions has been grade data from blocks of many hundreds or thousands of tonnes predicted using an orebody block model. Measured parcels are typically up to tens of tonnes and much higher variability at the measured parcel size than predicted from block models. Bulk ore sorting has been successfully applied to mines extracting porphyry copper ores. Copper ore grade variability occurs at different scales and this can be utilised for bulk ore sorting to increase ore grade and quality consistency in plant feed (Figures 2 and 3).

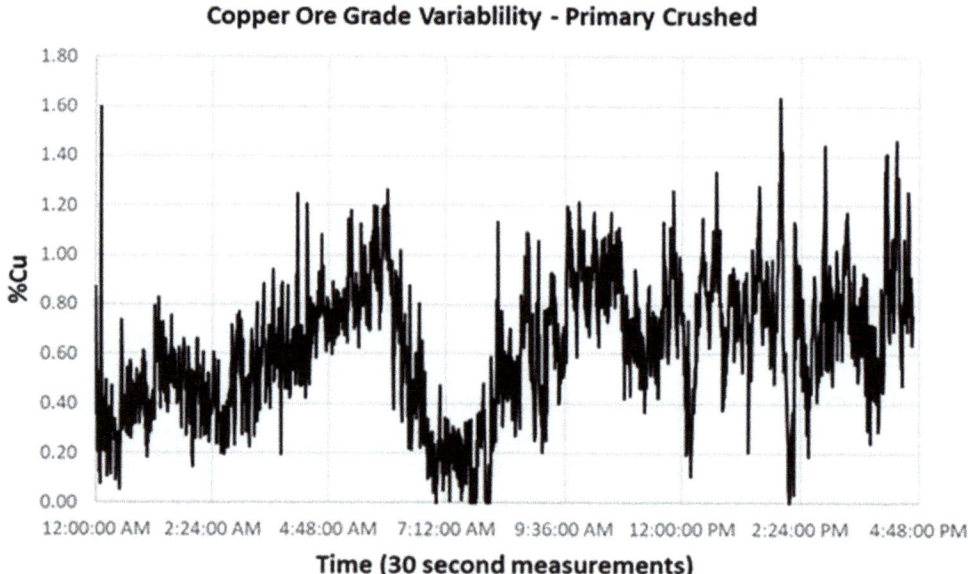

FIG 2 – Copper ore grade variability from each 30 seconds of conveyed flow from a block cave mine (Scantech).

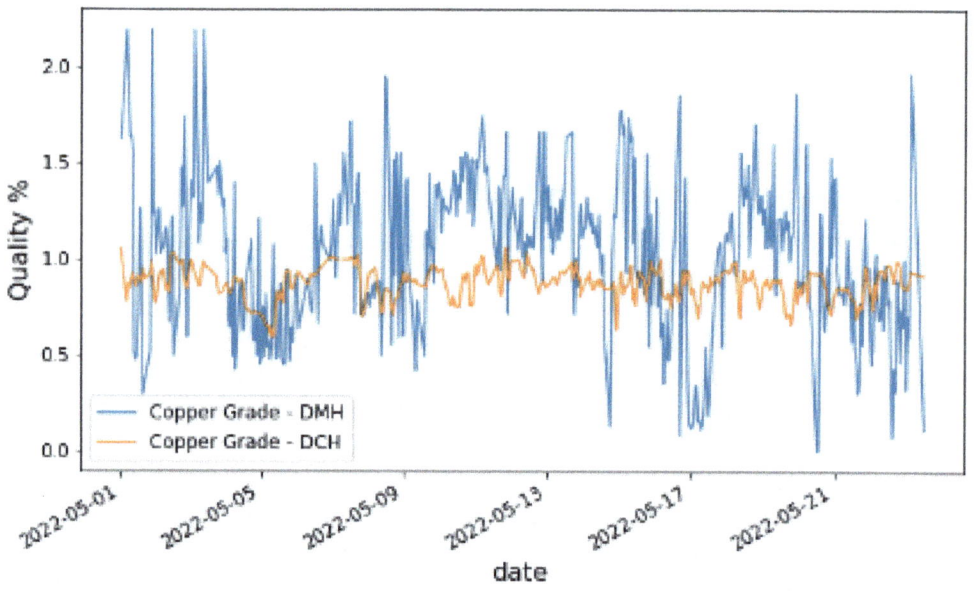

FIG 3 – Copper ore grade variability at Ministro Hales and Chuquicamata (Navarrete *et al*, 2022).

Bulk ore sorting at a copper mine in Chile resulted in increases of 20+ per cent in copper grade with 5–20 per cent of the mined ore being rejected as waste. Diverted coarse waste has saved up to 20 per cent of GHG emissions through avoidance of processing with only minor (up to 2.5 per cent) metal losses. Savings of over 40 000 t CO_2 e/annum are achieved at an ore processing rate of 1200 t/h where 14 per cent of the mined ore is discarded at waste. Figure 4 shows an example installation in a bulk sorting application.

FIG 4 – High specification GEOSCAN analyser generating representative, multi-elemental data used in a base metals bulk ore sorting plant for 30 second parcel diversion.

A diversion system (Figure 5) can utilise a screen to divert material to the waste stacker if grade by size analysis indicates that fines are predominantly above cut-off grade. The screen is raised or lowered to divert each 30 second parcel of flow as needed with fines always falling to the ore stacker side. The bulk sorting systems are designed to meet the requirements for each application and can be scaled according to required flow rates. Designs can cater for mobile plants as well as permanent installations.

FIG 5 – Example of a diversion system used in a copper ore bulk sorting plant for ore flows of 500 t/h (Hillyer, 2019).

Ore blending

Multi-elemental data used to improve ore blending in process feed allows:

- Control of copper metal in leach circuit feed to prevent overloading and copper losses to discard (Arena and McTiernan, 2011).

- Control of additives in copper leach feed, eg pyrite additive to copper carbonate ore in feed to a ferric leach process (Balzan *et al*, 2016).

- Improvements in metal recoveries of up to 15 per cent through a more consistent process feed quality (Goodall, 2021).

In ore blending applications the measurement need not be over very short parcels to be effective and longer measurement times of five minutes have been utilised to prevent more frequent responses to higher ore variations seen in shorter increments.

Metal accounting and ore reconciliation

Representative and precise multi-elemental measurement data provided for primary crushed conveyed flows is ideal for feedback to the mine to allow ore reconciliation to the mine schedule and block model. It represents the actual ore supplied from the mine to the process operation. It is the most effective way to measure mined material quality. Measurements of diverted parcels are tracked so that tonnage weighted averages of feed and reject material are captured.

This data is the most representative and precise to utilise for metal accounting for the process operations. When used for ore blending or bulk sorting the analyser data will be able to represent the material flow to the crushed ore stockpile or directly to the mill if measured after the crushed ore stockpile. Matthews and Du Toit (2011) discuss the application of analyser data for elemental balance at the Khumani iron ore operations as all relevant conveyed flows, including discard flows, utilise analysers and belt weighers to quantify tonnages and grades.

GEOSCAN GOLD is able to measure elements as proxies for gold in gold ores if they are present over each 30 seconds of flow and can measure gold directly over longer measurement times of five to ten minutes in successful installations to date (Balzan *et al*, 2022). The nature of the gold mineralisation determines the suitability of proxy elements. Gold measurement allows a previously unattainable level of metal accounting and ore reconciliation to be performed.

CONCLUSIONS

The mining industry's demand for increased resource efficiency and reduced environmental impact drives the need for innovative technologies. The use of high specification real-time, multi-elemental PGNAA has proven successful in optimising material diversion, improving ore blending, and therefore extending grade control beyond the mine and enhancing overall operational efficiency.

The findings in this paper underline the high specification GEOSCAN analyser's ability to increase ore grade by up to 20 per cent through bulk sorting, reduce GHG emissions, improve process recovery by 10 to 15 per cent through ore blending and provide valuable data for ore reconciliation and metal accounting.

High performance Prompt Gamma Neutron Activation Analysis (PGNAA) is proven in many mineral commodities (iron ore, manganese, copper, gold, zinc-lead, nickel, lithium, phosphate rock, bauxite, diamonds, chromium etc) to measure and control material quality in real time. Digitalisation of the conveyed flow provides useful data for ore blending, parcel diversion, ore reconciliation, metal accounting and feed forward control. Mining operations that have benefitted most through this technology have integrated geology, mining and processing teams to optimise quality control of process feed. Grade control has been extended beyond the mine to enable quality management on conveyed flows up to the mill feed. Benefits detailed in the paper include major operational savings in not processing waste or product quality material unnecessarily, blending ores to provide consistent process feed quality to improve many process performance parameters including metal recoveries and reducing GHG emissions. Extending grade control through representative measurement of conveyed flows has been a win-win for geologists and processors.

ACKNOWLEDGEMENTS

The authors wish to acknowledge permission of Scantech International Pty Ltd to publish this paper and acknowledge customers who have published their own papers on the successful implementation of the technology.

REFERENCES

Arena, T and McTiernan, J, 2011. On-belt analysis at Sepon copper operation, in *Proceedings MetPlant 2011*, pp 527–535 (The Australasian Institute of Mining and Metallurgy: Melbourne).

Balzan, L A, de Paor, A, Doorgapershad, A and Futcher, W, 2022. The end of the rainbow: real time direct gold analysis in run of mine ore at Newcrest's Telfer mine using GEOSCAN analysis, in *Proceedings of International Mineral Processing Conference – Asia Pacific 2022*, pp 1140–1149 (The Australasian Institute of Mining and Metallurgy: Melbourne).

Balzan, L, Jolly, T, Harris, A and Bauk, Z, 2016. Greater use of Geoscan on-belt analysis for process control at Sepon copper operation, in *Proceedings XXVIII International Mineral Processing Congress* (Canadian Institute of Mining, Metallurgy and Petroleum: Quebec).

Goodall, W, 2021. Understanding what is feeding your process: how ore variability costs money, in *Process Mineralogy Today blog*, March 10, 2021. Available from: <https://minassist.com.au/understanding-what-is-feeding-your-process-how-ore-variability-costs-money/> [Accessed: 26 May 2023].

Hillyer, L, 2019. Mining Automation – Mine Smarter, Load Quicker & Ore Sorting, presentation to Future of Mining Australia 2019, Aspermont Ltd. Available from: <https://www.youtube.com/watch?v=2NJ3HhphJc4/> [Accessed 26 May 2023].

Kurth, H, 2022. Ore quality measurement and control using Geoscan-M PGNAA real time elemental analysis, in *Proceedings International Mining Geology Conference 2022*, pp 338–345 (The Australasian Institute of Mining and Metallurgy: Melbourne).

Matthews, D and du Toit, T, 2011. Validation of material stockpiles and roll out for overall elemental balance as observed in the Khumani iron ore mine, South Africa, in *Proceedings Iron Ore Conference*, pp 297–305 (The Australasian Institute of Mining and Metallurgy: Melbourne).

Nadolski, S, Klein, B, Samuels, M, Hart, C J R and Elmo, D, 2018. Evaluation of cave-to-mill opportunities at the New Afton Mine, in *Proceedings of the 50th Annual Canadian Minerals Processors Operators Conference* (eds: B Danyliw, R Cameron and J Zinck), pp 270–281 (Canadian Institute of Mining, Metallurgy and Petroleum, Montreal).

Navarrete, M, del Rio, P, Aravena, P, Soto, A, Valido, J C and Allende, J, 2022. Stockpile Modeling and Ore Traceability with Advanced Analytics at DMH, paper presented to MineriaDigital 2022 – 9th International Congress on Automation, Robotics and Digitalization in Mining (Gecamin).

Scott, M, Rutter, J, du Plessis, J and Alexander, D, 2020. Operational deployment of sensor technologies for bulk ore sorting at Mogalakwena PGE Mine, in *Proceedings Preconcentration 2020*, pp 169–181 (The Australasian Institute of Mining and Metallurgy: Melbourne).

Directional change in underground drilling – a case study

V Mead[1], B Sharrock[2] and M Ayris[3]

1. Project Mine Geologist, Northern Star Resources, Kalgoorlie WA 6430.
   Email: vmead@nsrltd.com
2. Senior Mine Geologist, Northern Star Resources, Kalgoorlie WA 6430.
   Email: bsharrock@nsrltd.com
3. Senior Technical Advisor, IMDEX, Balcatta WA 6021. Email: mike.ayris@imdexlimited.com

ABSTRACT

Directional drilling has become common place for surface diamond drilling over the last two decades. This technique has allowed geologists and geotechnical engineers to not only intersect targets with precision and accuracy but significantly reduce costs when multiple targets are intersected from a single collar. However, it is rarely used underground as current methods require extremely high volumes of water and power. IMDEX subsidiaries, Devico and Downhole Surveys have addressed this challenge by developing a new directional drilling tool that requires similar water and power to standard underground drilling.

Northern Star Resources (NSR) and Australian Underground Drilling have partnered with IMDEX to test this tool at the Hampton-Boulder-Jubilee (HBJ) underground mine in Western Australia. At the HBJ underground mine, delineation of the down plunge extension to the northern high-grade lode has been difficult from existing mine development. This new tool may provide a viable solution in overcoming unfavourable intercept angles and significantly reduce the requirement for expensive underground drill platform development.

The aim of the study is to utilise this new directional drilling tool to intersect the orebody in multiple locations at improved angles of intersection without additional underground development. This study explores the tool's practical implications for diamond drillers, geologists and geotechnical engineers in an Australian underground setting, encompassing precise targeting, hole branching, flexibility and effective directional drilling.

Beyond HBJ, the tool's potential extends to sites like Kalgoorlie Consolidated Gold Mines (KCGM), where its use could be transformative. Navigating around historical underground voids with this tool could reduce large costs tied to ineffective holes and expedite critical data collection.

This innovative drilling tool has the potential to revolutionise underground directional drilling with significant benefits to orebody knowledge, resource growth and cost management. The ongoing collaboration between industry partners underscores the practical relevance of this DeviDrill technology.

INTRODUCTION

The Hampton-Boulder-Jubilee (HBJ) underground mine is located within Northern Star Resources (NSR) South Kalgoorlie Operations (SKO), 35 km south of Kalgoorlie, WA as seen in Figure 1. The mineralisation trend extends for over 4 km hosting multiple open pits and underground operations, with HBJ being the only mine currently operating today. Gold was first discovered near what is now HBJ in 1895, with multiple open pits running by 1919 (Nichols and Hagemann, 2014). Poor water availability and a weak gold price forced these mines shut in the mid-1920s (Copeland, 1998). Exploration of the area recommenced in the early 1960s by WMC Resources Limited with small scale mining recommencing in the late 1970s by Hampton Areas Australia. From 1984 the Jubilee and Hampton-Boulder pits and underground were mined as separate operations until the late 1990s where they were combined to form the Hampton-Boulder-Jubilee (HBJ) pit which now hosts the portal to the underground mine in this study. NSR purchased the tenement package and existing mines in 2018 and the site currently has resources and reserves of 1.09 Moz and 894 koz respectively (Northern Star Resources, 2023).

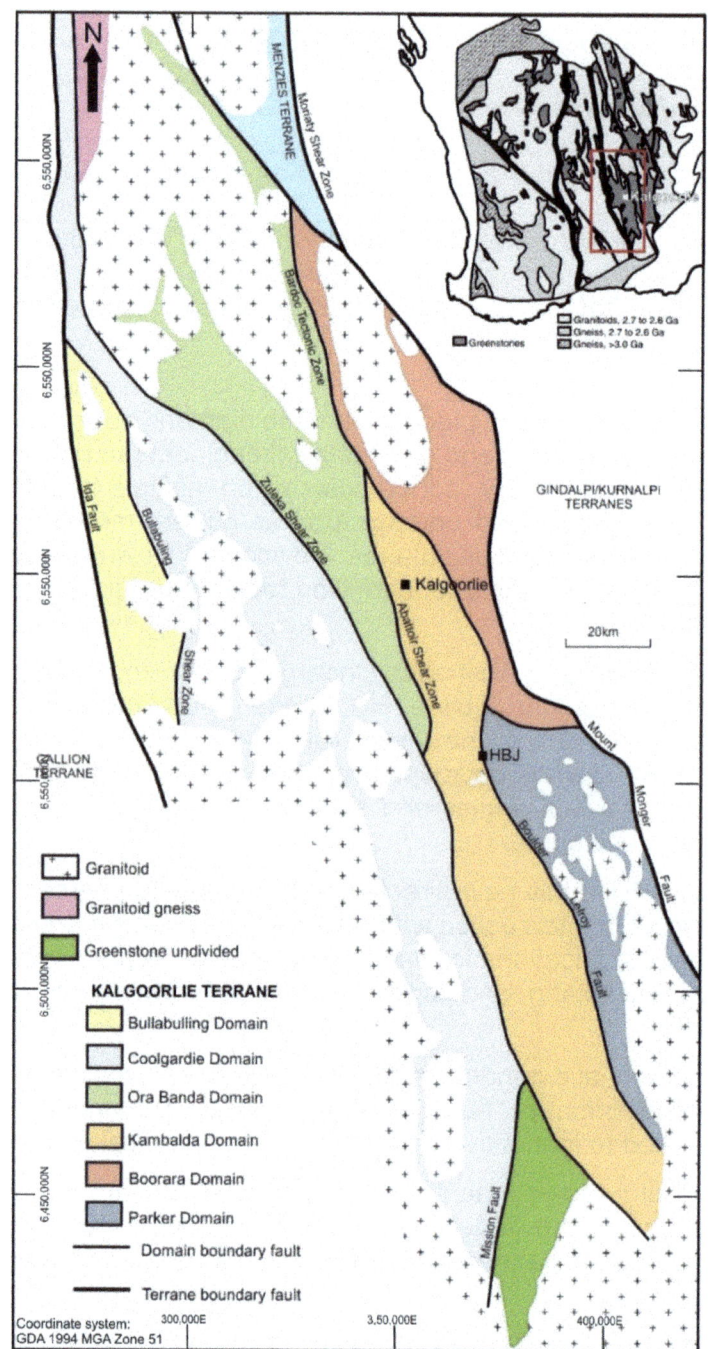

FIG 1 – Location of Hampton-Boulder-Jubilee (HBJ) underground mine in the Eastern Goldfields province, coloured by terrane (adapted from Tripp, 2013).

As the underground mine advances, targeting the northerly plunge of the orebody from current underground infrastructure becomes increasingly difficult. Developing drill drives to provide platforms to reach targets is expensive and removes machines that would otherwise have developed the mine for ore extraction. This problem is not unique to HBJ, mining companies like any business are driven to deliver the greatest value for shareholders which includes growing the life-of-mine and reducing the discovery cost per ounce. Underground directional drilling tools may provide the opportunity to underground miners to reach targets at acceptable angles without developing new drill drives. This paper will document the trial of Devico's directional drilling tool in the HBJ underground mine to test the northerly plunge of the orebody.

GEOLOGY

Regional geology

HBJ is located in the Eastern Goldfields within The Norseman-Wiluna Greenstone belt in the Yilgarn Craton, Western Australia. The deposit is hosted within the Kambalda sequence of the Kalgoorlie terrane, bound by the Ida fault to the west and the Mount Monger fault to the east, identified on Figure 1. The terrain comprises mafic volcanics and volcaniclastics that have been deformed and metamorphosed to greenschist facies (Nichols and Hagemann, 2014).

Local geology

The HBJ deposit is located along the Celebration Fault, a local splay of the Boulder Lefroy Shear Zone in an anastomosing section of the zone between the Celebration and Golden Hope pits, outlined in Figure 2 (Copeland, 1998). East of the fault is a highly strained ultramafic package (Kambalda Komatiite), younging to the east and intruded by an earlier syn-deformation, intermediate porphyry and later felsic quartz-feldspar porphyry (Hodge et al, 2004). West of the fault, stratigraphy youngs to the west. A thin package of sheared volcaniclastics is located directly adjacent to the Celebration Fault belonging to the lower Black Flag Group preceding a mafic intrusive dolerite (Condenser Dolerite) separating the upper and lower Black Flag Group sediments (Nichols and Hagemann, 2014; Dorling, 2017).

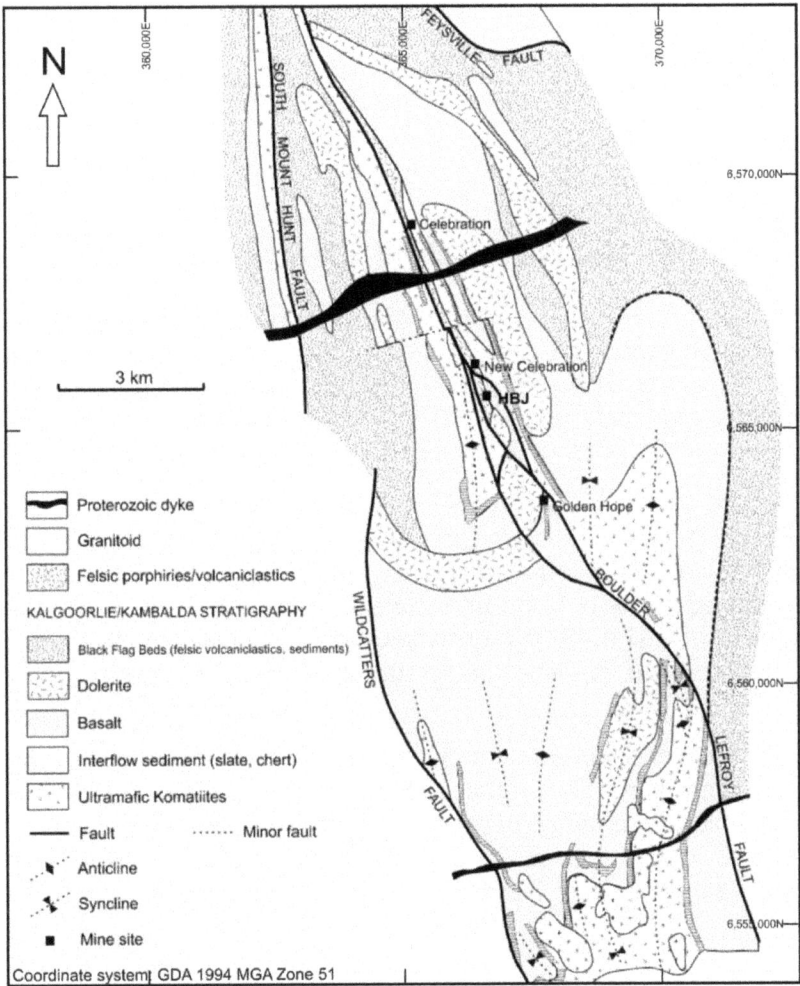

FIG 2 – Camp scale geology of the Kalgoorlie-Kambalda greenstone belt surrounding HBJ (adapted from Swager, 1989).

Mineralisation

There are two main types of mineralisation at HBJ, mylonitic and porphyry. A third late-stage vein set also hosts gold to a lesser degree.

Mylonitic type mineralisation is associated with proximity to the Celebration Fault. Gold is predominantly very fine grained and associated with pyrite and visible gold is rare. On the hanging wall side mineralisation is hosted in quartz-ankerite-biotite-sericite altered highly strained volcaniclastics with disseminated fine grained pyrite. On the footwall side, mylonite mineralisation is seen in the ultramafic unit directly adjacent to the fault but is generally a poor host for mineralisation, most mineralisation in the footwall is found in intermediate intrusives within the ultramafic host rock. Gold tenor in both hanging wall and footwall is higher closer to the fault (Nichols and Hagemann, 2014). Visually ore is difficult to identify, not sharing a strong or direct relationship to alteration or sulfide content. Mylonitic style mineralisation is actively mined in the northern ore zone (NOZ) and Mutooroo zones of the HBJ underground mine as seen in Figure 3.

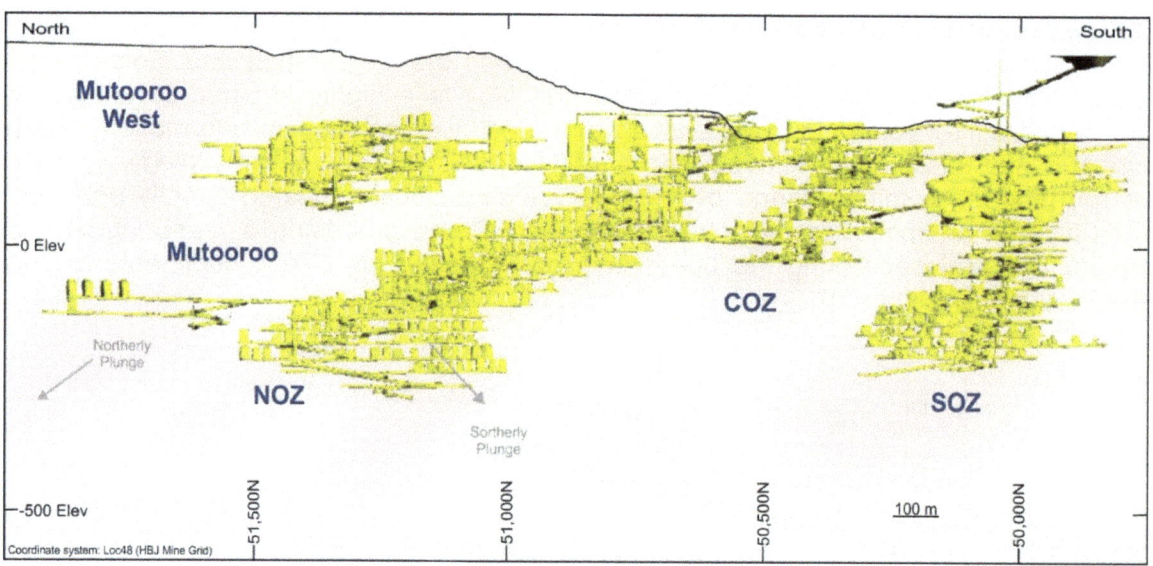

FIG 3 – Mining areas of the HBJ underground mine.

The plunge of the NOZ and Mutooroo orebodies are structurally controlled by the regional shallowly north dipping fold plunge and southerly plunge associated with perpendicular refolding. The orebody plunges are outlined in Figure 3. The width of the ore and high-grade pods are structurally controlled by the interaction of NNE and NNW striking faults with the Celebration fault (Ball, 2023). Locally this causes thickening of the ore up to 18 m wide and <1 m at its minimum.

Mineralisation also occurs within and along the contacts of porphyry dykes in the footwall ultramafics and hanging wall Black Flag Group sediments. Gold is very fine grained and associated with fine grained pyrite and quartz-carbonate veins (Nichols, Hagemann and Neumayr, 2003). This porphyry associated gold was mined in the HBJ pit and the southern and central ore zones (SOZ and COZ) underground mining areas, as shown on Figure 3. While no active mining of porphyry mineralisation is currently occurring at HBJ, it was mined in the footwall until late 2023 and continues to be a target for life-of-mine.

Later stage mineralisation is associated with clusters of brittle fracture fill quartz veining with quartz-sericite-pyrite±gold proximal alteration halos. Free gold is hosted in the quartz veins. These veins form clusters within and along the margins of felsic porphyry dykes and in the granophyric unit of the Condenser Dolerite (Ball, 2023). A small surface open pit, Mutooroo West, has previously focused on this late-stage mineralisation with the down dip extension now a new mining front at the HBJ underground.

CHALLENGES FACED

As the HBJ underground mine progresses, drilling the orebody from existing infrastructure becomes more challenging. Commonly, intercept angles are compromised to optimise the use of existing drill drives and platforms. This problem is not unique to the HBJ underground mine, many mining companies are motivated to extend the life-of-mine with limited capital expenditure in drill drives to reduce cost and improve margin. Underground directional drilling provides a promising alternative to developing new drill drives while maintaining favourable intercept angle on the lode.

Infrastructure limitations

As in many underground mines the effective extent of drill holes is limited by the existing infrastructure for drill positions and the angle of intercept considered appropriate for an orebody delineation. Once established, drill platforms are utilised to their maximum effective extent, as highlighted in Figure 4. Drilling oblique to the lode amplifies survey discrepancy, reduces the effectiveness of structural data and risks the hole getting dragged into the shear zone thus biasing the results and/or making the hole redundant. Often the only way to improve drill hole intercept angle on a lode is to develop new drill platforms, which requires significant capital expenditure. In the case of the HBJ mine, 185 m of further development would be required to achieve an acceptable intercept angle on the high-grade plunge of the northern lode from underground. Drilling from surface is a potential solution to improving the angle of intercept on the lode. An acceptable intercept is considered greater than 30° from the lode orientation, as it allows accurate measurements of veins and structural orientations as well as allowing for more accurate true width calculations. Furthermore, drill holes can often get dragged along the lode, particularly shear or vein hosted lodes further distorting the true width of the lode.

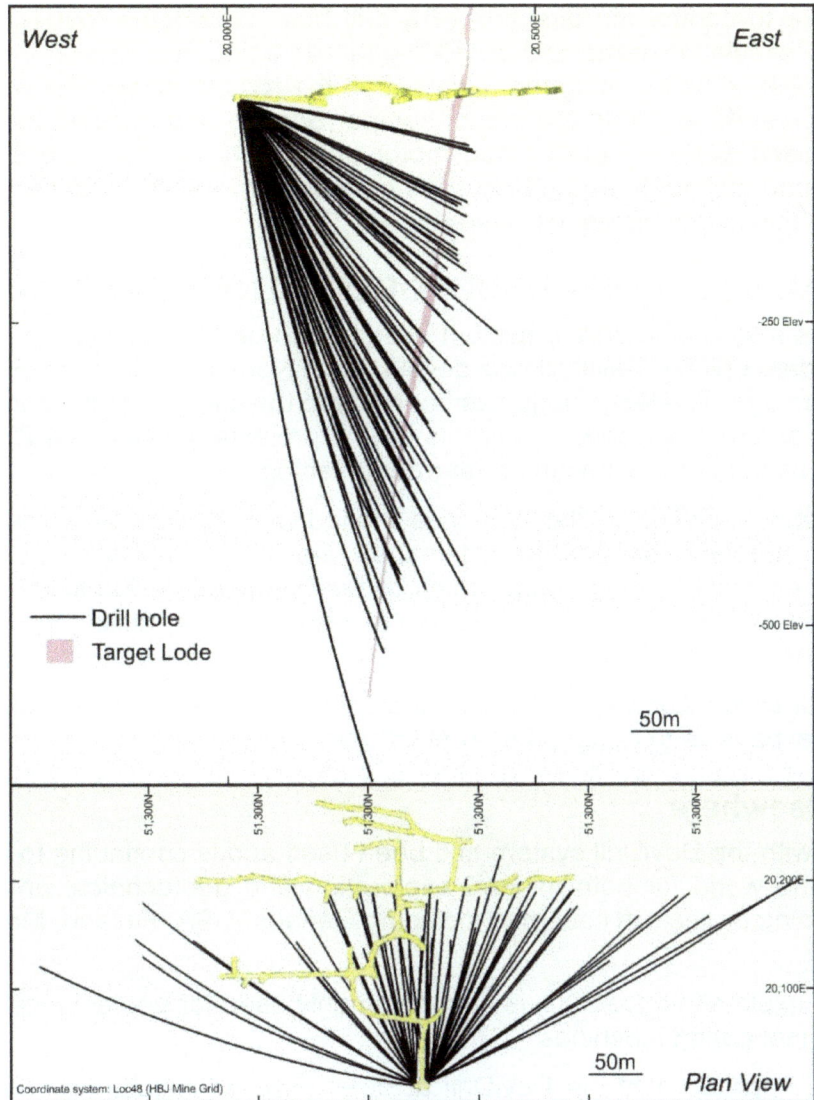

FIG 4 – An example of an exhausted drill platform in the HBJ underground mine.

Surface limitations

Surface drilling can be a viable solution to reaching near mine targets that cannot be effectively tested from underground. Often though there are geographical, infrastructure, environmental or financial limitations which control the application of surface exploration especially in deep underground mines.

Surface drilling comes with significantly more cost compared to underground drilling, approximately two to three times the cost per metre. Budgets are a finite resource providing motivation to drill more metres from underground for the same cost. The HBJ mine is now at depths over 600 m which requires holes drilled from surface to reach up to 1 km to test targets below the mine. At depths over 400 m, drillers' control on a hole is limited, leaving surface navigational drilling the only option to accurately reach targets. Surface navigational drilling is expensive and time consuming.

Conducting surface exploration in Australia is also becoming increasingly difficult. Increasing restrictions for environmental and heritage conditions and growing requirements for surveys and approvals prior to drilling can eliminate the application of surface drilling or create lengthy delays, increasing the appetite for underground exploration methods.

Mud motors in an underground environment

Directional drilling has become common place on surface drill rigs and is used at several operations to accurately reach targets and utilise branched holes.

Downhole mud-motors, more commonly referred to as navi tools, are non-coring bits operating under high water pressure to control the direction of a drill hole. Navi tools require an additional high pressure, high flow circulating pump and an additional rod string in a smaller diameter. High water pressure is accomplished by pumping more water into the rod string which is achievable in normal drilling conditions on surface due to the use of sumps (Scarlett and Pollock, 2013). Between 120–150 L of water is used every minute in mud motor drilling which is six to eight times more than standard underground drilling. In an underground environment water is not recycled so increasing the demand on the mine water limits the operation.

THE POTENTIAL SOLUTION – UNDERGROUND NAVIGATIONAL DRILLING

The DeviDrill tool is a steerable drilling system designed to be fitted to standard NQ drill rods and powered by a standard drill rig. DeviDrill was developed in Norway by Devico AS, part of the IMDEX group, over 20 years ago. The technology has been key to the success of directional drilling projects in various settings across the world. Despite its use in projects globally, the DeviDrill system has never before been used in an Australian underground setting.

Underground directional drilling in the past has trialled mud motors with limited success. High volumes of water and increased power demands are not compliant with the limitations of underground services. Unlike mud motors, which convert pumped drilling fluid energy into rotation, the DeviDrill is powered by the same method as standard diamond drilling where drill rod rotation provides rotation energy at the bit face.

The application of directional drilling in underground settings opens significant near mine exploration targets, difficult or expensive to reach off current infrastructure.

Applications elsewhere

Directional drilling with the DeviDrill system has been used and is continuing to be used in multiple countries around the world for both mineral exploration and geotechnical environments. Devico currently have ongoing projects in countries including Canada, USA, Finland, Mexico, New Zealand and Hong Kong.

Outside of mineral exploration, DeviDrill is also commonly used for geotechnical investigation prior to installation of tunnel boring machines.

An example of the application of the DeviDrill at other projects include its use in a geotechnical project at Jurong Island, Singapore for installation of an underground cable. Directional drilling was required to bend the 1000 m hole 90° and maintain a 10 m tolerance at an inclination of -25°, followed by horizontal through to end of hole. Directional coring was utilised for 35 per cent of total drilling on this hole.

Components

The DeviDrill system comprises two tools, the Directional Coring Drilling (DCD) and Rotary Steerable System (RSS). The DCD and RSS tools were both used in the underground trial at HBJ.

The DCD tool, which can be seen in Figure 5c, is fitted with a coring bit and contains a non-rotating innertube capable of collecting up to 3 m of AQ diameter core. More recently the RSS tool has been developed, seen in Figure 5b, which differs from the DCD tool, being fitted with a flat faced, non-coring bit which provides flexibility and robustness to the DeviDrill system.

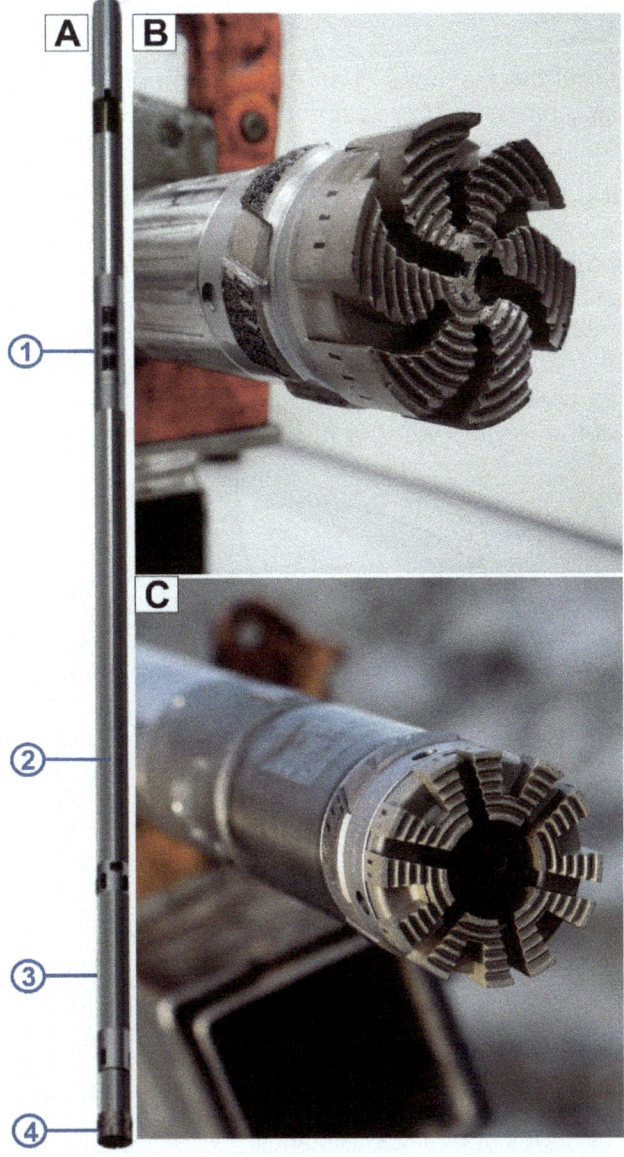

FIG 5 – (a) Components of the DCD tool, (1) locking tungsten knives, (2) internal rotating drive shaft, (3) stationary outer barrel, (4) rotating coring bit face. (b) RSS non-coring bit face. (c) DCD coring bit face.

Both tools operate in the same way, attaching to a standard NQ rod string through a drive shaft. The tool is set in place downhole by rotation of the drill rods until the desired directional heading is obtained. Once in place the outer housing of the DeviDrill is held in place by hydraulically engaged external tungsten knives, labelled in Figure 5a. These knives are expanded by an internal inflatable lifter system which is activated once required. Soft or incompetent ground conditions will not support the locking tungsten knives and the outer casing will not remain stationary. The drive shaft, powered by the rod string, runs internally within the stationary outer casing to power the drill bit which drills an NQ diameter hole.

Turn of the hole is determined by the bend of the drive shaft manually set by adjustable bearings while the tool is out of the hole. The maximum turn of a hole is limited by the drill rods.

NQ rods are designed to handle a maximum of between 10 to 12° of turn over 30 m, more turn is possible, however, it increases the risk of failure in the hole. Turn over this limit increases the wear on the drill rods against the external rock due to the rod torque created by the turn, which decreases the lifespan of the rods. Threads between drill rods are also put under strain at higher torque which increases the risk of rods separating downhole. Finding the correct setting for a project involves some trial and error to determine how the tool will respond. The recommended turn over 30 m is 9°.

OUTCOMES

The aim of this trial was to test the DeviDrill in an Australian underground environment for reaching distal near mine targets, using branched holes and to document the application to drillers, geologists and geotechnical engineers.

The targets of the trial at the HBJ underground mine were designed to test the northerly plunge of the Mutooroo orebody in the main mylonitic lode in open ground. Without further drill drive development conventional drilling from underground cannot effectively test the extents of this lode due to the poor intercept angle, highlighted in Figure 6. Drill holes ideally intercept a lode perpendicular to the strike and dip of the structure, however holes which intercept at greater than 30° to the strike of the lode are accepted. The limit of conventional drilling from the 1865 drill drive where the primary hole was collared is highlighted on Figure 7. The Devico directional drilling tool will allow effective testing of the targets without further drill drive extensions.

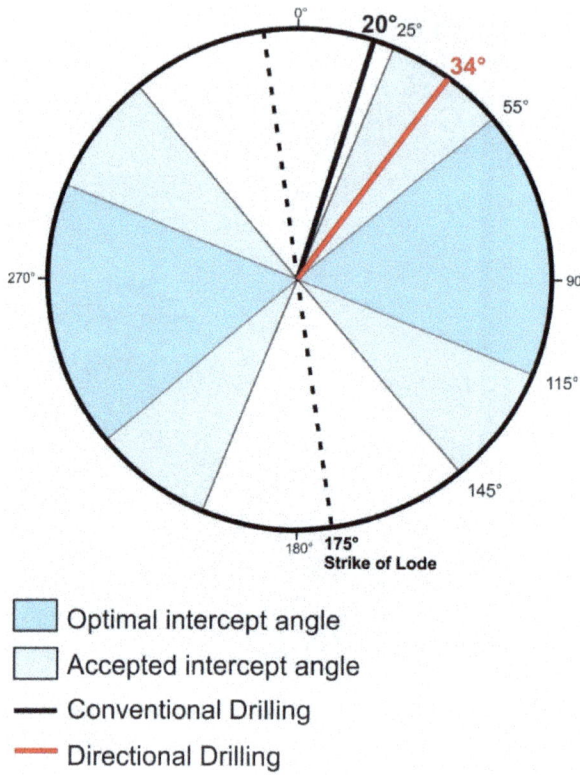

FIG 6 – Comparison between directional and conventional drilling angle of intercept for the primary target.

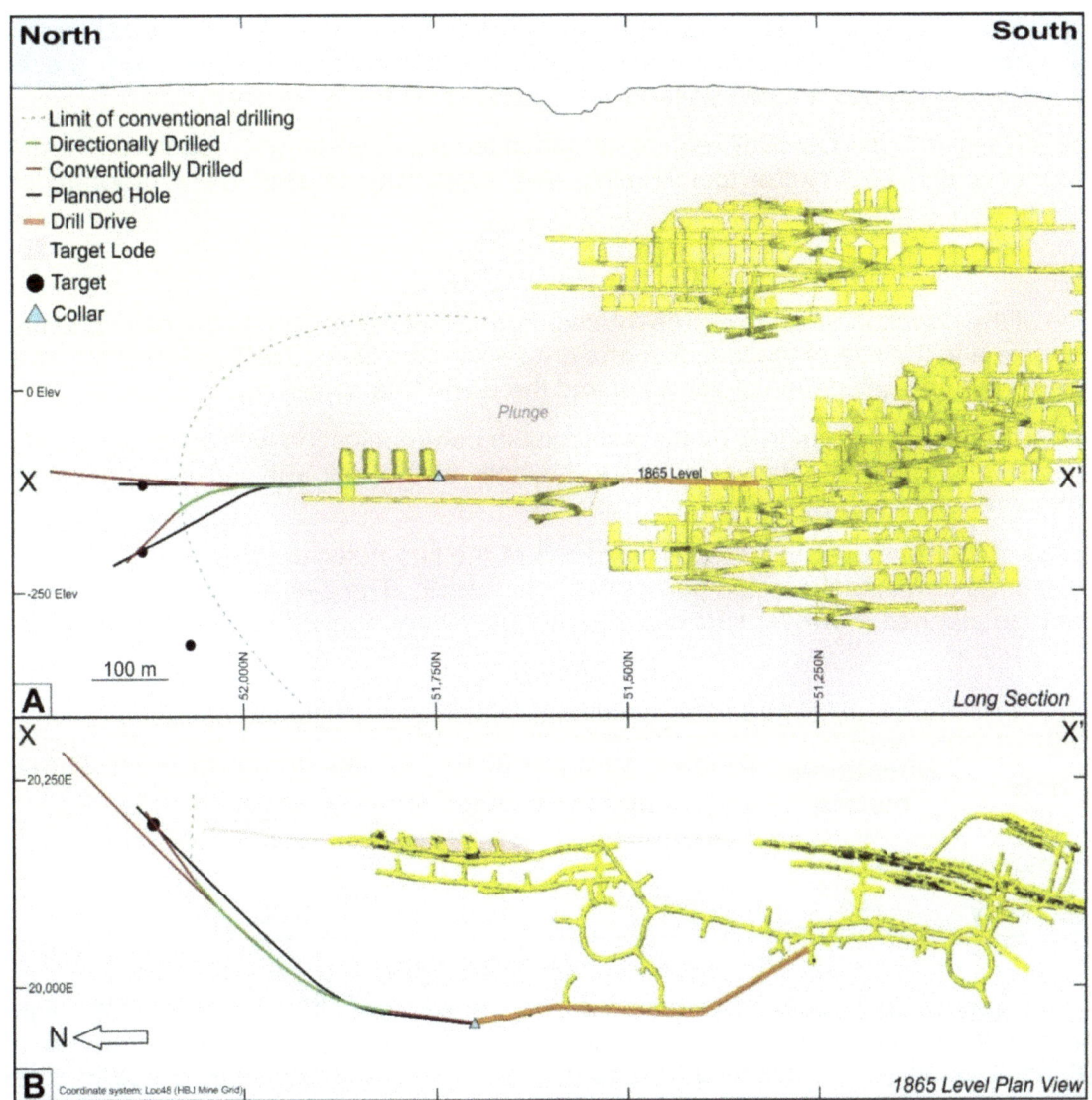

FIG 7 – (a) Long section of the HBJ mine; (b) plan section of the 1865 level. Both sections reference the current mine infrastructure including the 1865 dedicated drill drive where the directional holes were collared. Target points, in black, are down dip of the north plunge of the Mutooroo orebody. Dashed line highlights the limit of conventional drilling, the potential extension of the northern plunge is outside of this range.

Geologists and geotechnical engineers

The study conducted at the HBJ underground mine aimed to intersect the orebody in multiple locations at improved angles of intercept from existing mine infrastructure. Success of the trial was measured by the ability of the tool to meet the target, to reach the target in a reasonable time frame and to operate in normal underground conditions.

Directional drilling with the DeviDrill commenced on 24 November 2023. The primary and first secondary holes were drilled prior to the submission to this study summarised in Figure 7. The use of this tool is continuing at the HBJ mine.

Planning

Planning of the directional holes was completed by Devico, after on-site geologists provided Devico with a collar and target point. Planned drill strings were returned to the geologists with a breakdown of metres drilled conventionally and metres drilled with the directional tool. The plan was reviewed by the geology team and amendments requested until the plan achieved the required goal.

During the trial at HBJ, six holes were planned prior to drilling. The primary hole and first branched holes were completed from the original plan, while the remaining holes are currently underway.

Following the outcome of the first two holes, a revised plan was made for the remaining targets including a new primary hole. This revised plan used information collected from the first holes including the ground trends and the drill rod limitations.

Directional drilling with the DeviDrill requires a dedicated serviceman with specialised knowledge of the components and use of the tool. Timing and availability of staff for a program should be accounted for during planning.

Drilling

During use of the directional tool, a representative from Devico is on-site at the drill rig to monitor the turn of the hole and the use of the tool. Reports are sent by the Devico technician at the end of every shift including the time allocation of activities and the downhole survey.

Daily survey monitoring is required by the geologist in communication with Devico to adjust plans if needed. The Devico representative monitors the hole at the drill and makes adjustments to the equipment as required to reach target.

The average movement over the directed sections of the two holes drilled at HBJ is summarised in Table 1. Planned turn in the directed part of the hole was 9°. The average turn rate over both holes was lower than planned, however sections of both holes were above the 9° per 30 m planned rate.

TABLE 1

Summary of rate of turn in directional drilling conducted at HBJ.

Hole	Directional metres	Average turn per 30 m (°) Azimuth	Average turn per 30 m (°) Dip	Maximum turn per 30 m (°) Azimuth	Maximum turn per 30 m (°) Dip
Primary	138	6.8	0.01	10.7	0.23
Secondary	150	2.4	-6.0	0.86	-9.6

The turn was successful in both holes drilled during the study at HBJ. The intercept point at target depth on the primary hole was 26 m from target. Initially the hole was planned to be within 10 m of the target point, however the distance from target was considered acceptable due to issues faced with deviation earlier in the hole. The secondary hole was 10 m from target and within planned limits.

Core recovery

During the study at HBJ the primary hole was drilled using the coring tool (DCD) to collect the relevant geological information and the secondary holes were planned with the non-coring tool (RSS). The core retrieved during the primary hole and later part of the secondary hole was able to be logged and provide valuable geological information to the model. A section of this core is pictured in Figure 8.

FIG 8 – AQ diameter core retrieved during the turn on the primary hole. Light passing between the core and the flat surface highlights the bend that can be observed in the core.

Core recovery in both the directional part of the hole and at target can provide valuable information to geotechnical engineers. Rock-quality designation (RQD) measurements can be collected as well as making observations regarding the condition of the core. In areas of high strain, 'disking' may be observed within the recovered core, thus identifying locations where higher strain/stress are present during future development. AQ core from the directional cut can be sent for uniaxial compressive strength (UCS) testing if rock strength data is required.

In surface mud motor drilling, care must be taken to place directional turns in barren areas due to the non-coring nature of the tool, which can be difficult if multiple stacked lodes exist before the main target. Retrieving core from the DCD has significant applications in these mines including NSR's Fimiston operation at Kalgoorlie Consolidated Gold Mines (KCGM). It allows flexibility in directional cut placement to optimise the direction of the hole without compromising on logging and sampling zones of interest.

Orientation of the core retrieved during directional drilling is however not possible. Due to the tools configuration an orientation tool cannot be fitted while drilling.

Deviation

Rock fabric conditions were a significant problem during the trial at HBJ. The natural deviation in conventional drilling at the underground mine is strongly affected by the Celebration fault zone and accompanying foliation. Holes deviate both with and against rotation depending on the starting azimuth. The natural deviation encountered during the study, coupled with the set deviation of the tool combined causing turn rates higher than the rod string could withstand which contributed to several re-drills.

While deviation at HBJ is well understood, drilling in unfamiliar directions can lead to challenges. Once Devico understand and account for the ground conditions at HBJ, control on the holes should improve.

Branched holes

Branched holes are executed by drilling straight into a directed section of the hole to cut a new hole, visually represented in Figure 9. The site at which a new hole is cut is referred to as a cut lip. Once a secondary hole has been cut, the primary hole below that cut lip can no longer be accessed. The number of branches a primary hole can host is limited by the length of the directed section in a hole.

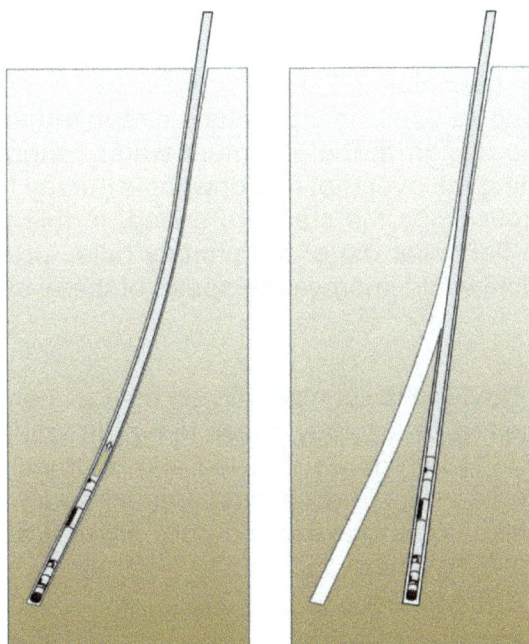

FIG 9 – Schematic of a branched hole (Devico, n.d.).

During the study period one branched hole was successfully cut from the primary hole. In addition to the planned branched hole, a total of five unplanned corrections were made to the primary and secondary holes to correct deviation. Each correction followed the same cut lip process as a planned branch.

Drilling specifications

The holes involved in the trial at HBJ underground were collared at -2° dip from the centre of the face in a 6 × 6 m drill drive. The difference in operational conditions while using the tool are summarised in Table 2.

TABLE 2

Comparison between operational conditions of conventional and directional drilling observed during the HBJ trial.

Operational condition	Standard core barrel	Devico DeviDrill	Post-turn standard core drilling
Water use (L/min)	20–25	30–40	20–25
Operating pressure (bar)	20	30	20
Penetration rate (m/shift)	60–110	3–27	30–50
Power		No notable difference	
Revolutions (rev/min)	1650	600–1000	1000–1300

Drilling of the directional hole was completed with a standard Australian Underground Drilling (AUD) rig with minor modifications to increase the electric pump and hydraulic feed ram.

Drilling at HBJ is typically within hard dolerite and silicified sediment lithologies. Where the rock is not affected by deformation, shearing or jointing is to be considered strong to very strong. UCS testing within the dolerite and silicified sediments return values of 103 MPa and 194 MPa respectively. These values can be used to construct a geotechnical setting of the mine to aid in further development and design.

Water pressure

The most notable difference in operation is the water pressure, which is increased by the tool itself downhole using the outer casing as consistent pressure is required to expand the packers and hold the tool in position. Due to the design of the tool more water cannot be pumped downhole when required, for example, for running the overshot or a downhole survey tool. Running these tools down the hole took approximately four times the standard speed. In this trial tools were required to be pumped downhole due to the flat collar dip of the primary hole. Gravity fed tools can be used on holes greater than 30° dip which would improve the speed of these operations.

Rod handling

The direction achieved by the DeviDrill is set manually on the tool before being run down the hole. If adjustments are required to the tool or a survey needs to be run to the end of the hole all rods must be pulled out. During the trial at HBJ rods were pulled and returned up to three times a shift. Rod pulls at the depths of the drilling during the study cost approximately one hour, a significant portion of the shift was allocated to this task. Increased rod pulls present an increased hazard of manual handling.

Broken ground

Broken ground was encountered while drilling with the DeviDrill at HBJ. Broken ground can cause blockages in the DCD coring tool and is recommended to be drilled with the RSS non-coring tool which eliminates the risk of blockage. In extremely broken ground the tool may not remain seated correctly and will fail to steer the hole, fortunately this did not happen during the trial. If water return is lost at any time including due to broken ground directing the hole will not be possible.

The DeviDrill tool does not have a back reamer like a conventional barrel, which can be risky in poor ground conditions. Following successfully drilling through a zone of broken ground on the secondary hole the tool became stuck in this section when pulling rods and could not be reamed out. Parts of the RSS tool were broken off and abandoned in the hole and other components damaged during the recovery process. Following this disruption, the DCD tool was used again to complete the turn for the second hole.

Time

Directional drilling at HBJ was on average 10–20 per cent of the normal metres drilled in a shift. Due to the slow nature of directional drilling, a second standalone rig was introduced to the mine so production drilling could continue uninterrupted.

The DeviDrill itself takes approximately 30 minutes to core 3 m, this is just under half the speed of normal drilling. The non-drilling activities associated with the directional drill contribute the most to the reduction in metres drilled in a shift. The RSS tool is slightly faster in the case of shallow holes because the overshot does not need to be pumped down to retrieve core every 3 m. Pulling rods more often and pumping down tools takes significantly more time than conventional drilling.

CONCLUSIONS

The trial of the Devico directional tool in the underground HBJ mine successfully tested the northerly plunge of the orebody as targeted. The angle of intercept on the target was within acceptable limits providing useful geological data to the model. Testing distal targets before committing to a new drill platform is an appealing prospect which will continue to be explored by the geology team at South Kalgoorlie Operations with the trial of the DeviDrill continuing at the time of writing.

Success of the tools' application was measured by the ability to reach a target at an acceptable intersection angle, operating under normal underground conditions and within a reasonable time frame. The initial target was reached although somewhat further than planned but the trial was considered successful. The tool operated in close to normal underground conditions and can be repeated at other operations. The speed at which the hole was drilled during directed drilling was however slower than anticipated. Significantly less targets were able to be drilled in the allocated time period than planned. This is a factor that is now understood and can be applied to future applications. Directional drilling to test distal near mine targets before committing to new drill platforms will have significant applications in underground mines. The speed of drilling however cannot compete with conventional drilling and could not replace new drill platforms all together.

Going forward directional drilling in an underground environment has applications for other NSR mines including the Fimiston operation where historic underground voids often terminate holes before meeting target. The use of directional drilling in this environment could eliminate the interaction with a known void, creating a primary hole in unmined ground to reach multiple targets on the other side of the void. Outside of the company, directional drilling has applications in most underground mines aspiring to extend the life-of-mine.

ACKNOWLEDGEMENTS

The authors would like to acknowledge Northern Star Resources for their permission to use the data from South Kalgoorlie Operations in this publication. The authors also would like to thank Geoff Muir, Managing Director of AUD, for taking on the trial. We acknowledge the contributions of Jamie Brown and Jed Smith for their assistance in explaining drilling information during the study, all AUD drillers working on the directional drilling rig who embraced the opportunity to learn a new technique, Megan Gillies for assisting in compiling and communicating information on the tool and to Emma Murray-Hayden for her guidance and review of the manuscript.

REFERENCES

Ball, A, 2023. South Kalgoorlie structural review, Internal technical report.

Copeland, I K, 1998. Jubilee gold deposit, Kambalda, *Geology of Australian and Papua New Guinea Mineral Deposits*, pp 219–224.

Devico, n.d. Devico AS – general presentation directional core drilling & borehole surveying solutions, internal report.

Dorling, S L, 2017. Structural setting of the central and southern ore zones (COZ & SOZ) in the Hampton-Boulder-Jubilee gold operations, CSA Global Report No. R105.2017.

Hodge, J, Nichols, S, Hagemann, S and Neumayr, P, 2004. The structural, hydrothermal alteration and fluid evolution of the New Celebration lode-gold deposits: multiple deformation events and two stages of gold mineralisation, in *Predictive Mineral Discovery CRC Conference*, pp 97–100.

Nichols, S J and Hagemann, S G, 2014. Structural and hydrothermal alteration evidence for two gold mineralisation events at the New Celebration gold deposits in Western Australia, *Australian Journal of Earth Sciences*, pp 113–141.

Nichols, S J, Hagemann, S G and Neumayr, P, 2003. Structural controls and hydrothermal alteration of multiple gold mineralising events at New Celebration, Kalgoorlie, Western Australia, Centre for Global Metallogeny, University of Western Australia.

Northern Star Resources (NSR), May 2023. Resource, reserves and exploration update, ASX. Available from: <https://www.nsrltd.com/investor-and-media/asx-announcements/2023/may/resources,-reserves-and-exploration-update>

Scarlett, M and Pollock, M, 2013. Directional drilling for multiple intersections, Subiaco: AIG Drilling Seminar.

Swager, C, 1989. Structure of Kalgoorlie greenstones – regional deformation history and implications for the structural setting of the Golden Mile gold deposits, Western Australia Geological Survey Report, 25:59–84.

Tripp, G I, 2013. Stratigraphy and structure in the Neoarchean of the Kalgoorlie district, Australia: critical controls on greenstone-hosted gold deposits, PhD thesis, James Cook University.

Open to anything

Exploring deep learning in resource estimation

N Battalgazy[1], R Valenta[2], P Gow[3], C Spier[4] and G Forbes[5]

1. PhD candidate, W H Bryan Mining and Geology Research Centre, Sustainable Minerals Institute, The University of Queensland, Brisbane Qld 4072. Email: n.battalgazy@uq.edu.au
2. Director, Sustainable Minerals Institute, The University of Queensland, Brisbane Qld 4072. Email: r.valenta@uq.edu.au
3. Adjunct Associate Professor, W H Bryan Mining and Geology Research Centre, Sustainable Minerals Institute, The University of Queensland, Brisbane Qld 4072. Email: p.gow@uq.edu.au
4. Associate Professor, School of the Environment, Faculty of Science, The University of Queensland, Brisbane Qld 4072. Email: c.spier@uq.edu.au
5. Senior Research Fellow, Julius Kruttschnitt Mineral Research Centre, Sustainable Minerals Institute, The University of Queensland, Brisbane Qld 4072. Email: g.forbes@uq.edu.au

ABSTRACT

3D spatial assessment of mineral resources is an important step that forms an integral part of any mining operation. Due to the complex nature of orebodies, the construction of a block model can often be challenging, and traditional interpolation methodologies may not be efficient with the increased variability expressed in the strong nugget effect and the inherent complexity in the data set. In this context, the integration of deep learning (DL) methods becomes an important and reasonable means. Leveraging the power of DL, skilfully trained on data sets with comparable characteristics, structures, or patterns to those the model intends to analyse and predict, provides a valuable alternative tool for 3D spatial prediction. The essence of DL lies in its ability to assimilate and understand the complex characteristics of data. Thus, the implementation of DL methodologies promises to solve the problems inherent in 3D spatial forecasting in the context of the mining industry. The research project presented in this study explores the potential of convolutional neural networks (CNNs) in the realm of resource estimation. The DL strategy has shown a remarkable ability to capture complex geological shapes and forms even with a limited number of samples. This strength becomes particularly evident when comparing the overall discrepancies between the actual and predicted block models, particularly in the context of grade versus tonnage relationships. The research findings highlight the potential of employing a CNN with the capability to integrate additional input for delivering more accurate and reliable tonnage predictions. While the initial findings of employing CNN in 3D spatial predictions demonstrate promising capabilities, it also reveals limitations, particularly when dealing with non-gridded samples in 3D space.

INTRODUCTION

In mineral resource estimation, both machine learning (ML) techniques and, more specifically, deep learning (DL) approaches, a subset of ML, are being increasingly utilised. This adoption is evident in research initiatives and, to a limited extent, within industry practices. Although DL methods may exhibit high accuracy under favourable conditions (regular grids without missing data and robust ground truth) in mineral resource estimation, their application to real-world scenarios with poor data quality and limited data poses a major challenge. A key question arises: Can DL method be effectively adapted in situations where data quality and quantity are suboptimal? This necessitates the development of strategies and considerations for implementing DL methodologies under less-than-ideal circumstances, aiming to bridge the gap between theoretical promise and real-world feasibility.

METHODOLOGY

Constructing a convolutional neural networks (CNNs) for 3D prediction tasks involves developing a custom architecture. In this study, the U-Net encoder-decoder style is employed as presented in Figure 1, with symmetrical structure for high-level feature extraction. CNN structure is composed of specific convolutional blocks with two convolutional layers for extracting features through kernels or filter, average pooling to preserve more information and batch normalisation for stability (Figure 1) (Battalgazy *et al*, 2023). The size of the architecture depends on the size of the 3D input and output data. In this study, the input data is 64 × 64 × 64, which is drill hole data in 64 m³ volume. The output is an 8 × 8 × 8 block model, representing the sub-blocks of the block model, each 8 × 8 × 8 metres in size.

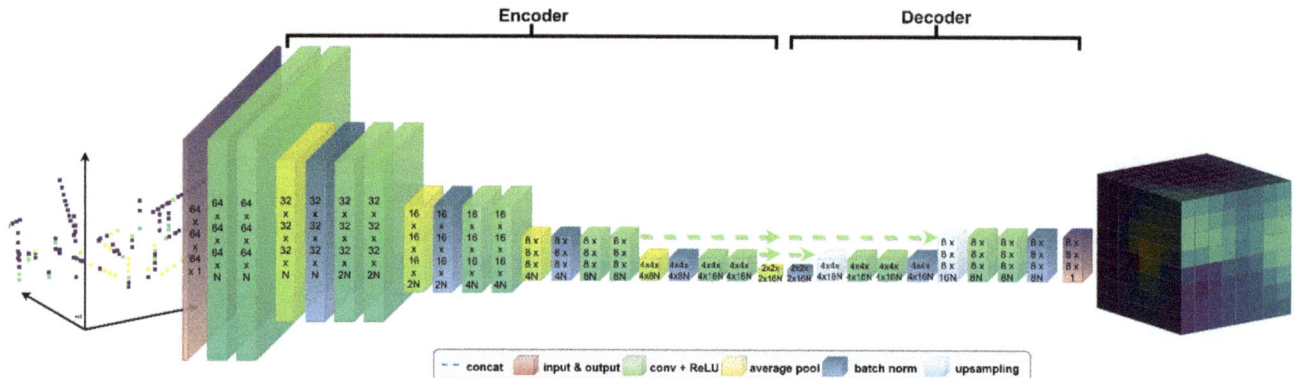

FIG 1 – CNN U-net architecture with input (64 × 64 × 64) and output with 8 × 8 × 8 (8 × 8 × 8 m subblock size) voxel size.

Adapting CNNs for resource estimation requires specific steps illustrated in Figure 2. The workflow involves transforming drill hole samples and block models into a workable format, partitioning the data, training the model and applying the trained model for predictions on new data.

Step 1 – Converting into a workable format

Unlike geostatistics or other DL techniques, CNNs require data in tensor or multi-dimensional array format. Drill hole samples and block models are converted into 3D tensors. This step ensures the data's compatibility with CNN processing.

Step 2 – Training CNNs

Training the CNN involves several steps: data preprocessing, partitioning, data split and holdout data. The CNN architecture in U-Net style is recommended for resource estimation (Pan *et al*, 2020). Training metrics include Mean Squared Error (MSE) loss, R2 score and histogram intersection score.

Step 3 – Prediction on new data set

Once trained, the CNN model can predict block models on new data sets. Predictions are converted from TIFF format back to geospatial dataframe format for analysis and comparison with geostatistical interpolation techniques.

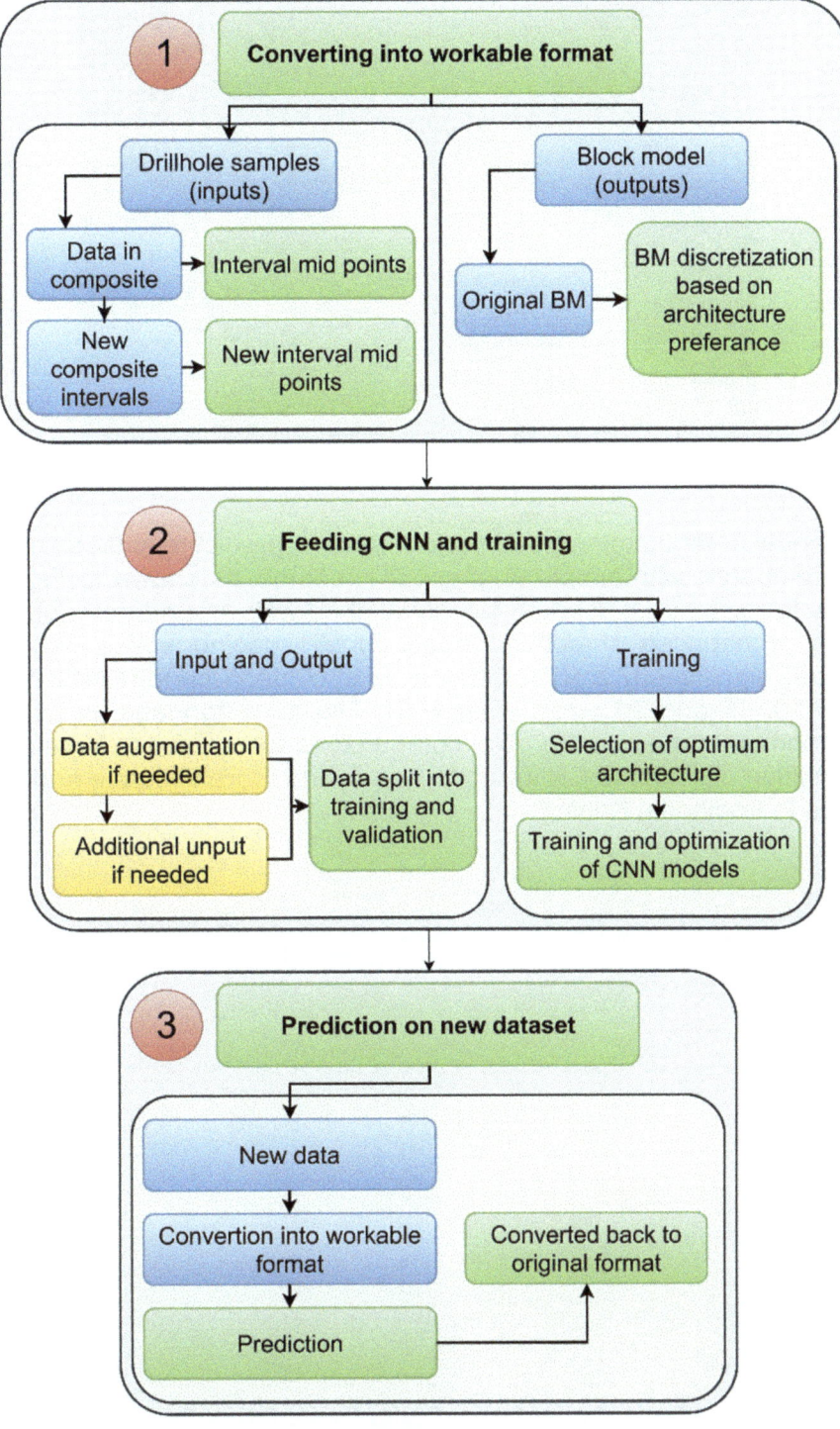

FIG 2 – Workflow for convolutional neural network (CNN) application.

RESULTS AND COMPARISONS

CNN results in controlled environment

In the research phase involving the Estaillades carbonate data (Spurin *et al*, 2021), 3D micro-CT image of carbonate sample shown in Figure 3 were strategically manipulated to simulate drill hole samples by selectively removing pixels, creating scenarios with 10-, 15- and 20-pixel differences. This approach facilitated the evaluation of the CNN methodology for predicting 3D values within the context of 3D images. These manipulated images served as a controlled environment to assess the model's performance, offering insights into its efficacy under various conditions. The micro-CT images, capturing geological characteristics and incorporating fault structures, provided a robust foundation for testing and refining the CNN methodology.

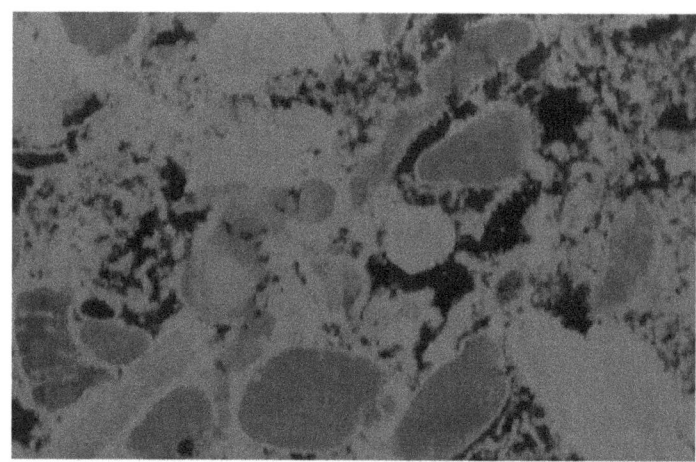

FIG 3 – Micro-CT image of porous carbonate rock (Estaillades) (Spurin *et al*, 2021).

The trained CNN model demonstrated good performance in predicting unseen data in all scenarios, which then were compared with simple kriging (SK) method as shown in Figure 4, where CNN exhibited the ability to reconstruct 3D data closely aligned with ground truth, outperforming Kriging which displayed a pronounced smoothing effect in all spacing scenarios. Notably, the CNN consistently produced significantly smaller differences in content-tonnage relationships in scenarios with differences of 10, 15 and 20 pixels (Figure 5). The term 'tonnage' in this context reflects a simulation of real tonnage curves, since pixel comparisons were used to simulate tonnage values. However, this simulation highlighted CNN's ability to provide more accurate and consistent tonnage forecasts compared to traditional Kriging.

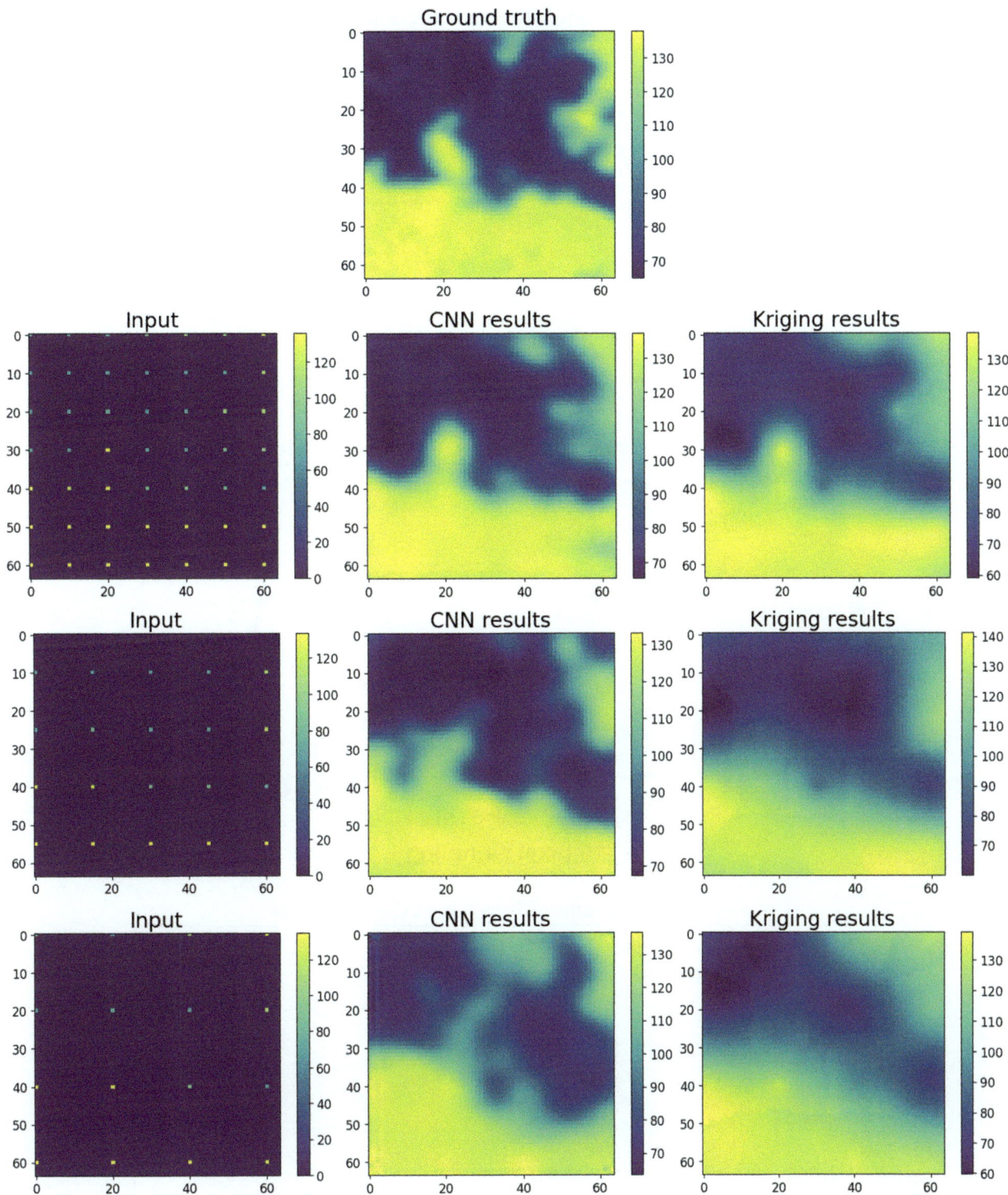

FIG 4 – Results of CNN and Simple Kriging in conditions with regular grid and 10, 15, 20 pixel sampling.

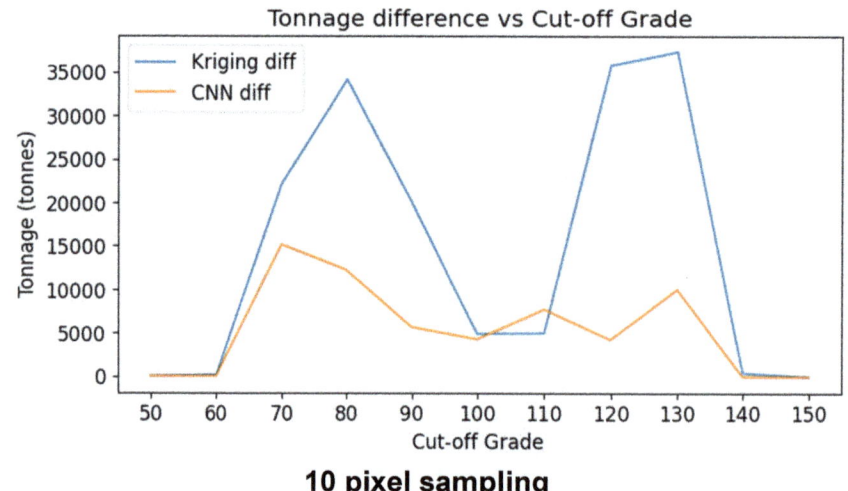
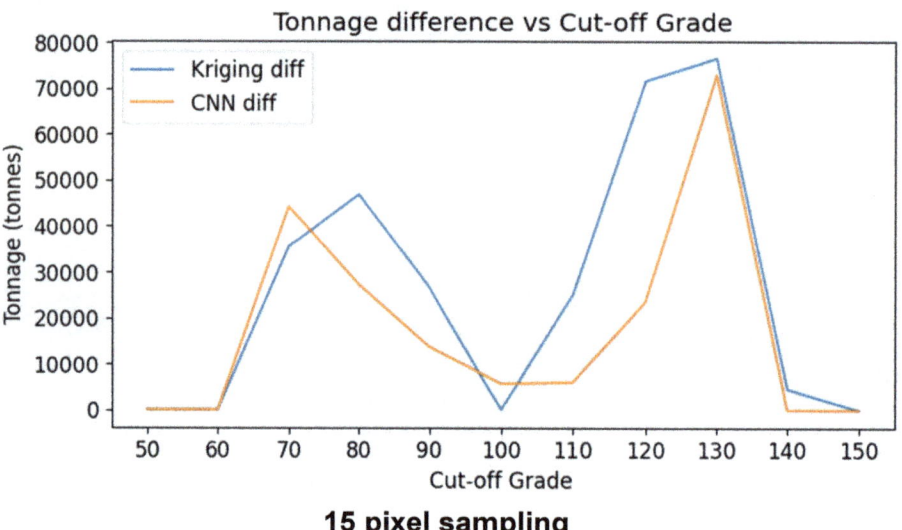
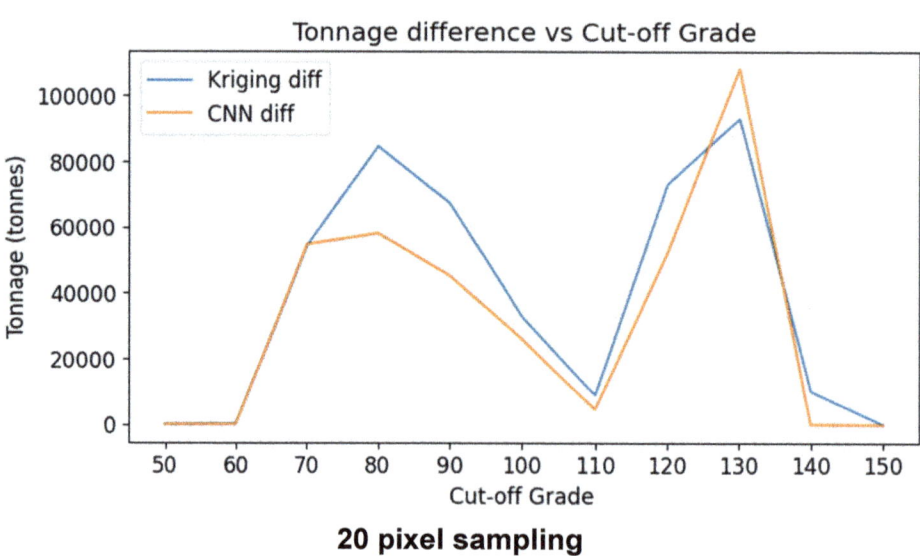

FIG 5 – The tonnage differences at each cut-off grade between ground truth and predicted by Kriging and convolutional neural network (CNN) methods.

CNN results in real world

The application of CNN in the Lisheen Mine, located in Ireland, involved training on a block model generated using ordinary kriging (OK). In contrast to the evidence-based research approach with available ground truth, the problem here is the uneven spatial distribution of drill hole samples. Creating appropriate data pieces that included both input (drill hole samples) and output (OK block model) information proved challenging due to the irregular and sparse nature of drill hole samples in different directions as shown in Figure 6. This lack of consistency posed a significant challenge and results obtained by CNN in this case were not satisfactory as it can be seen in one of the results in Figure 7.

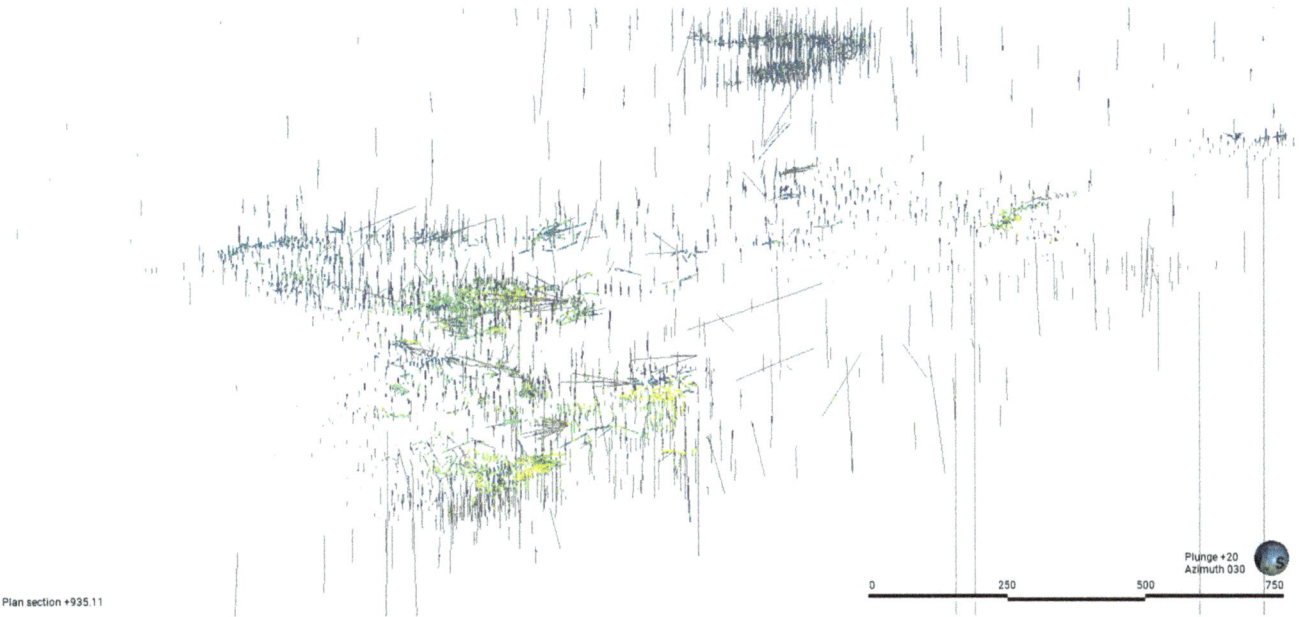

FIG 6 – Lisheen Mine data.

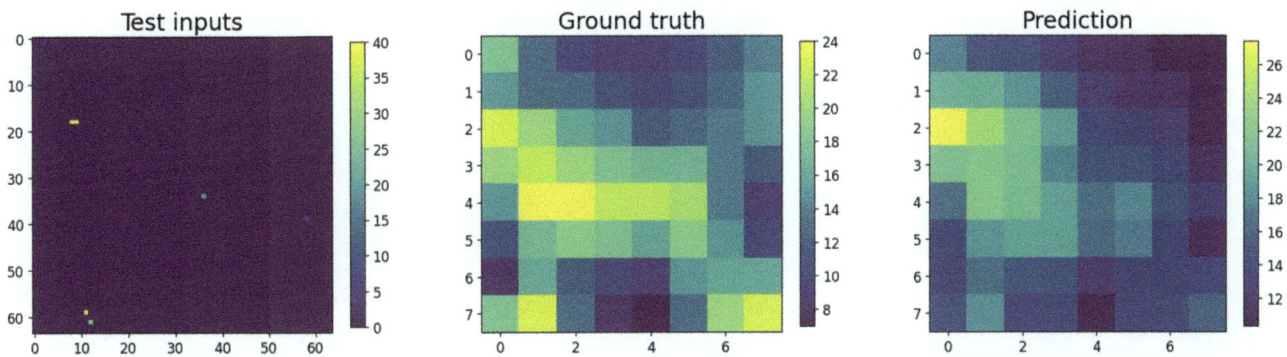

FIG 7 – Input drill hole sample (left), ground truth (OK block model) (middle) and CNN block model (right).

DISCUSSION AND CONCLUSION

The reliability of the CNN approach is highlighted in scenarios with accessible ground truth and well-defined input conditions. The outstanding performance in minimising differences, particularly in small spacings with more samples, showcases the potential for accurate tonnage predictions. However, it's crucial to acknowledge that this research-focused success might not directly translate to real-world applications with irregular and sparse data, as demonstrated in the Lisheen Mine case study. The challenges encountered in irregular data distribution and sparse drill hole samples underscore the importance of considering data characteristics in CNN applications.

Future work could involve exploring ways to enhance CNN adaptability to irregular data sets, potentially through data augmentation techniques or hybrid approaches that integrate CNN strengths with geostatistical methods. Additionally, it may be worthwhile to investigate applications of CNN in open pit mines with regularly gridded data, where its strengths in capturing spatial relationships could shine. Exploring the application of CNNs in open pit mines with regular grid RC data could be a promising avenue. The uniform distribution of RC samples in open pit mines provides a structured data set akin to regular grids, aligning well with CNN strengths observed in research scenarios.

ACKNOWLEDGEMENTS

We express our gratitude to the Dan Alexander Memorial Prize for their generous support, contributing to the advancement of our research endeavours at The University of Queensland's Sustainable Minerals Institute.

REFERENCES

Battalgazy, N, Valenta, R, Gow, P, Spier, C and Forbes, G, 2023. Addressing Geological Challenges in Mineral Resource Estimation: A Comparative Study of Deep Learning and Traditional Techniques, *Minerals*, 13(7):982.

Pan, Z, Xu, J, Guo, Y, Hu, Y and Wang, G, 2020. Deep learning segmentation and classification for urban village using a worldview satellite image based on U-Net, *Remote Sensing*, 12(10):1574.

Spurin, C, Krevor, S, Blunt, M and Bultreys, T, 2021. Decane and brine injected into Estaillades carbonate – steady-state experiments, Digital Rocks Portal (February 2021). doi:10.17612/cd7a-y955.

AUTHOR INDEX

Afonseca, B C	17	Gulay, R	201
Armstrong, B J	131	Hargreaves, R	3
Ayris, M	355	Harvison, A	289
Ball, J	93	Heriyanto, M	319
Bananga, H	103	Hertrijana, J	263
Battalgazy, N	371	Hidayat, M I	263
Beckett, D	39	Hitchman, S P	341
Biven, M	155	Hoare, J P	341
Boyce, G	115	Ireland, J	187
Bright, S	301	Jackson, A	39
Brown, A	301	Khariyah, N	263
Carpenter, J B	341	Kurth, H	347
Carter, J	131	Lucien, M	221
Carvalho, D	17	Lumaad Paras, R	201
Chapman, A	143	Machado, W	3
Chiquini, A	17	Magautu, D	221
Chitambala, J	103, 181	Maguire, S	67
Claflin, A	155	Maxted, J	289
Clarke, D	143	Mead, V	355
Cox, B	39	Megebry, M	301
Cox, N	39	Merello, G	85
Curtis-Morar, C	279	Minniakhmetov, I	85
Dale, R	155	Mogilny, D	47
Darvall, M	173	Mondo, M	221
Dimond, A	39	Morley, C	3, 209
Edmond, S	309	Mudinzwa, E	221
Finch, R	85	O'Brien, S	67, 309
First, D M	47	O'Rielly, D	231
Forbes, G	371	Owers, M	243
Gagau, L	221	Puerto, M	155
Gilchrist, G	103, 181	Rahadi, M H	319
Goldman, M	253	Rajcoomar, Y H	47
Gonzalez, J	155	Ravella, M	309
Gonzalez, O	39	Reid, R	221
Gordon, R	201	Rolley, P	231
Gow, P	371	Rowett, C	231

Saala, N	263	Townsend, J	243
Schouten, D	243	Triyunita, A	263
Sharrock, B	355	Tunnadine, A	243
Sidabutar, J	253	Turner, G	301
Simbolon, S	341	Valenta, R	371
Snell, R	71	Van Ryt, M	173
Souza, J	3	Vedrik, S	47
Spier, C	371	Vermaak, M C	279
Sternadt, G	173	Vigor-Brown, W	289
Strang, J	201	Western, E	243
Symons, D J	341	Whaanga, A	289
Taube, R	319	Wiguna, A	319
Taylor, J	243, 253	Williams, C	85
Tod, A	309	Yeates, D	309

Milton Keynes UK
Ingram Content Group UK Ltd.
UKHW051312250524
442964UK00013B/18